Intelligent Computing Techniques and Applications

Edited by

Tusharkanta Samal

Ambarish Panda

Manas Ranjan Kabat

Ali Ismail Awad

Suvendra Kumar Jayasingh

Deepak K Tosh

Intelligent Computing Techniques and Applications

A proceeding of ICETICT – 2024

Edited by

Tusharkanta Samal
Ambarish Panda
Manas Ranjan Kabat
Ali Ismail Awad
Suvendra Kumar Jayasingh
Deepak K Tosh

CRC Press
Taylor & Francis Group
Boca Raton London New York

CRC Press is an imprint of the
Taylor & Francis Group, an **informa** business

First edition published 2026
by CRC Press
44 Park Square, Milton Park, Abingdon, Oxon, OX14 4RN

and by CRC Press
2385 NW Executive Center Drive, Suite 320, Boca Raton FL 33431

British Library Cataloguing-in-Publication Data
A catalogue record for this book is available from the British Library

ISBN: 9781041110859 (pbk)
ISBN: 9781041110835 (hbk)
ISBN: 9781003658221 (ebk)

DOI: 10.1201/9781003658221

Typeset in Sabon LT Std
by HBK Digital

Contents

List of Figures

List of Tables

Preface

This Taylor & Francis, CRC Press volume contains the papers presented at the International Conference on Emerging Trends in Intelligent Computing Techniques (ICETICT – 2024) held during 27th and 28th December 2024 organized by DRIEMS University, Tangi, Cuttack, Odisha, India. A lot of challenges at us and no words of appreciation is enough for the organizing committee who could still pull it off successfully.

The conference draws the excellent technical keynote talk and many papers. The keynote talks by Prof. Sanjeevikumar Padmanaban, University of South-Eastern Norway and Prof. Bidyadhar Subudhi, Director, NIT, Warangal are worth mentioning. We are grateful to all the speakers for accepting our invitation and sparing their time to deliver the talks.

We received 238 full paper submissions and we accepted only 66 papers with an acceptance rate of 28%, which is considered very good in any standard. The contributing authors are from different parts of the globe that includes UAE, Philippines, Uganda, Jordan, Malaysia, and India. The conference also received papers from distinguished authors from the length and breadth of the country including 20 states and many premier institutes. All the papers are reviewed by at least three independent reviewers and in some cases by as many as five reviewers. All the papers are also checked for plagiarism and similarity score. It was really a tough job for us to select the best papers out of so many good papers for presentation in the conference. We had to do this unpleasant task, keeping the Taylor & Francis guidelines and approval conditions in view. We take this opportunity to thank all the authors for their excellent work and contributions and also the reviewers who have done an excellent job.

On behalf of the technical committee, we are thankful to Prof. Durga Prasad Mohapatra, NIT, Rourkela, to accept the invitation as General Chair of the Conference, for his timely and valuable advice. We also thankful to Prof. Srinibash Sethy for his constant support and guidance. We cannot imagine the conference without his active support at all the crossroads of decision-making process.

The university management of the host institute, the honorable Chairman Dr. Pramod Chandra Rath, Patron, Vice Chairman Durga Prasad Rath and Vice Chanceller, Prof. Prakash Kumar Hota, Director Admin Sj Balaram Kar, Director Coordination Mrs Chinmayee Rath, Co-Patron of the conference and Dean Prof. Nayan Ranjan Samal, Associate Dean Prof. Alok Ranjan Biswal, Organising Chair, Dean Research Prof. Biswaranjan Mohanty have extended all possible support for the smooth conduct of the Conference. Our sincere thanks to all of them.

We would also like to place on record our thanks to all the keynote speakers, reviewers, session chairs, authors, technical program committee members, various Chairs to handle finance, accommodation and publicity and above all to several volunteers. Our sincere thanks to all press, print and electronic media for their excellent coverage of this conference.

We are also thankful to Taylor & Francis, CRC Press publication house for agreeing to publish the accepted and presented papers.

Best wishes.
December, 2024

Tusharkanta Samal
Ambarish Panda
Manas Ranjan Kabat
Ali Ismail Awad
Suvendra Kumar Jayasingh
Deepak K Tosh

Editors Biography

Dr. Tusharkanta Samal
Associate Professor
DRIEMS University, Cuttack

Dr. Tusharkanta Samal holds a Bachelor of Technology in In received MTech and Ph.D. degree in Computer Science and engineering from Veer Surendra Sai University of Technology, Burla, Odisha, India. Currently he is working as an associate Professor in Department of Computer Science and engineering, DRIEMS University, Odisha, India. He has more than 12 years of teaching and research experience in different Engineering colleges and Universities. Dr. Samal has authored over 30 scientific papers published in SCI, Scopus Indexed Journals and Conferences.

Dr. Ambarish Panda
Assistant Professor
SUIIT, Burla

Dr. Ambarish Panda holds a Bachelor of Engineering in Electrical Engineering and Master of Technology in Power System Engineering from Sambalpur University and V.S.S University of Technology. He completed his Ph.D. in Electrical Engineering from V.S.S University of Technology since April, 2016. He has more than 14 years of teaching and research experience in different Engineering colleges and Universities. To his credit, he has published multiple research articles in SCI/SCIE indexed journals with high repute. Besides his research activities he serves as reviewer and Editor in SCI/SCIE Indexed Journal.

Prof. (Dr.) Manas Ranjan Kabat
Professor and Principal
IMIT, Cuttack

Manas Ranjan Kabat, is currently working as the principal of IMIT, a constituent college of BPUT. He has received his M.E. degree in Information Technology and Computer Engineering from Bengal Engineering College, India, and the Ph.D. degree in Computer Science and Engineering from Sambalpur University He has more than two decade of teaching experience both at undergraduate and postgraduate level. He has published more than 75 research paper in various referred international journals and conferences. He has guided more than 20 M.Tech and 9 Ph.D. students. Chaired and organized many international and national conferences. Contributed as editor and reviewer of many peer-reviewed international journals His research interests include QoS in internet, Wireless Sensor Network, Body Area Network, Cloud Computing etc.

Dr. Ali Ismail Awad
Associate Professor, College of Information Technology, United Arab
Emirates University (UAEU), Al Ain, United Arab Emirates

Ali Ismail Awad (Ph.D., SMIEEE, MACM) is an Associate Professor at the College of Information Technology, United Arab Emirates University (UAEU), Al Ain, United Arab Emirates, where he has been coordinating the master's program in Information Security since 2022. He is also an Honorary Associate Professor at the University of Nottingham, Nottingham, U.K. From 2018 to 2023, he served as a Visiting Researcher at the University of Plymouth, Plymouth, U.K. Dr. Awad joined the Department of Computer Science, Electrical and Space Engineering at Luleå University of Technology (LTU), Luleå, Sweden, in 2013. In 2017, he was promoted to Associate Professor (Docent) and served as the coordinator of the master's program in Information Security from 2017 to 2020. In recognition of his teaching merits and pedagogical achievements, he was promoted to the rank of Recognized University Teacher at LTU in 2021. His research interests include cybersecurity, network security, Internet of Things (IoT) security, and image analysis with biometrics and medical imaging applications. He has edited or co-edited several books and authored or co-authored numerous journal articles and conference papers in these areas of interest. From 2021 to 2024, he was recognized among the top 2% of influential scientists worldwide. Dr. Awad serves on the Editorial Boards of Future Generation Computer Systems, Computers & Security, Internet of Things; Engineering Cyber-Physical Human Systems, Health Information Science and Systems, and Security, Privacy and Authentication (Frontiers). Dr. Awad is an IEEE Senior Member and an ACM Professional Member.

Dr. Suvendra Kumar Jayasingh
Associate Professor and HOD in the Department of Computer Science and Engineering, Institute of Management and Information Technology (IMIT), Cuttack (A Constituent College of BPUT, Govt. of Odisha)

Dr. Suvendra Kumar Jayasingh is working as Associate Professor and HOD in the Department of Computer Science & Engineering, Institute of Management and Information Technology (IMIT), Cuttack (A Constituent College of BPUT, Govt. of Odisha) after being selected in OPSC (Orissa Public Service Commission) in 2005. He has obtained his Bachelor of Engineering the year 2003 from University College of Engineering (UCE), Burla (Now VSSUT). He got his M. Tech. in Computer Science & Engineering in 2007 from RVU, Udaipur and Ph. D. in Computer Science & Engineering in 2020 from North Orissa University, Baripada (Now Maharaja Sriram Chandra Bhanja Deo University). He is having 20 years of teaching experience in Computer Science & Engineering and MCA. He has published several articles, book chapters in reputed National and International journals and periodicals including Springer and Taylor & Francis and has presented research papers in National and International Seminars and Conferences. He has also participated in many National and International Workshops, FDPs, Industrial Training Programs organized by IITs, NITs and NITTTRs. His research interests include Artificial Intelligence, Data Mining, Soft Computing, Machine Learning, Computational Intelligence, Database Management System and Algorithm Analysis and Design. He has published a book on "Introduction to Machine Learning" and a UK patent on "Smart Home Air Quality Monitoring Device".He is a life member of Indian Society for Technical Education (ISTE).

Dr. Deepak K Tosh
Assistant Professor, Department of Computer Science, University of Texas at El Paso, 500 W. University Avenue, El Paso, TX 79968-0518

Dr. Deepak K Tosh is an assistant professor in Computer Science at the University of Texas at El Paso. Before that he was a Cybersecurity Researcher at the DoD Sponsored Center of Excellence in Cybersecurity, Norfolk State University (NSU), Norfolk, Virginia. During that time, he has closely collaborated with researchers from Air Force Research Lab and Army Research Lab to establish data provenance mechanisms in cloud computing in addition to addressing research challenges in the arena of distributed system security, Blockchain, cyber-threat information sharing, cyber-insurance, and Internet of Battlefield Things (IoBT). He was appointed shortly as a postdoctoral researcher at the Tennessee State Univerity, where he worked with Dr. Sachin Shetty in the

CyberViz Laboratory. He has been collaborating with Dr. Shetty since then focusing on research topics such as: distributed consensus models in Blockchain technology, cyber-resiliency in battlefield environment, and various practical issues in cloud computing security. He has received my Ph.D. in Computer Science and Engineering from University of Nevada, Reno, under the supervision of Dr. Shamik Sengupta. His dissertation was focused on designing market based models to enable cybersecurity information sharing among organizations. He received my masters in Computer Science from University of Hyderabad, India in 2012. His master thesis addressed the issue of cognitive radio transmission parameter adaptation problem using multi-objective optimization techniques.

Reviewers

BIRESH KUMAR DAKUA
DR. SOHAN KUMAR PANDE
DR. CHINMAYEE PADHY
DR ASHUTOSH RATH
DR. RAM CHANDRA BARIK
MANAS RANJAN MISHRA
DR. PRASANT KUMAR DASH
SAKAMBARI MISHRA
DR. DEBASHREET DAS
SASMITA TRIPATHY
SAROJA KUMAR ROUT
DEBABRATA SINGH
SIGMA RAY
ASHISH SINGH SALUJA
DEEPAK KUMAR ROUT
BIJAYINI MOHANTY
SUSHREE BIBHUPRADA B
    PRIYADARSHINI
DR- SAMEER KUMAR DAS
DR. BISWAJIT JENA
PRASANTINI SAMAL
DEBASISH SWAPNESH KUMAR
    NAYAK
SUSHANTA MEHER
DR. ARABINDA PRADHAN
NILAMBAR SETHI
DR BHRAMARA BAR BISWAL
SACHIKANTA DASH
DR. PAVAN KUMAR A V S
BIDUSH KUMAR SAHOO
SOUMYA RANJAN MISHRA
APARNA BABOO
CHANDRAKANTA MAHANTY
DR. JAGANNATH PATRA
PRADYUT KUMAR BISWAL
SHANTILATA PALEI
MITALI SINHA
SUBHASHREE PRIYADARSHINI
NITEN KUMAR PANDA
SUJATA SWAIN
DIBYARANJAN DAS
SHUVAM DAS
RITU ROUMYA SAMAL

BHABANI PATNAIK
ABDUL KHADAR SHAIK
ANUP KUMAR NANDA
ANUPAM DAS
SIDHARTHA SANKAR DORA
SRITOSH KUMAR SAHOO
JYOTI RANJAN NAYAK
SUBASISH MOHAPATRA
NUZHAT PROVA
RASHMI RANJAN RANJAN
    SAHOO
RASHMI R MAHARANA
ADITI BHATEJA
SOUMYA SNIGDHA
    MOHAPATRA
SUJIT BEBORTTA
AMIYA KUMAR SAHU
ALPESH KUMAR DAUDA
DEEPAK KUMAR PATEL
MONALISA MOHANTY
CHIDURALA SAIPRAKASH
DR. BUDDHADEVA SAHOO
RAHUL RAY
SWARNABALA. UPADHYAYA
MANOJ KUMAR KAR
JOGESWARA SABAT
BISWARANJAN NAYAK
BHABENDU KUMAR
    MOHANTA
AMARDEEP DAS
SOUMYA RANJAN MAHANTA
RASMITA MOHANTY
KISHORE KUMAR SAHU
SAGARIKA PRADHAN
DR. SANTOSH KUMAR MAJHI
DEEPAK KUMAR LAL
DILLIP KHAMARI
SANJIB KUMAR NAYAK
SONALIKA MISHRA
ALINA DASH
LADU KISHORE SAHOO
DEVISMITA SAHOO
SAMITA BALA ROUTRAY
ANITA MOHANTY

SACHIDANANDA SAMAL
SHEKHARESH BARIK
SUBHENDU BEHERA
DR. TUSHAR KANTA SAMAL
DR. DEBAYANI MISHRA
DR SATYABRAT KAR
MANAS MAHAPATRA
SAILENDRA PATTNAIK
SATYABRAT SAHOO
SUSHANTA KUMAR SETHY
DEEPAK RANJAN BISWAL
NIVA TRIPATHY
DR. BAJRA PANJAR MISHRA
BIDYADHAR SWAIN
JAYASHREE BHUYAN
MOUSUMI ACHARYA
SUBHASIS MOHAPATRA
DR ANUREKHA NAYAK
RAGHUNATH ROUT
ROJALIN DASH
SUBHASISH MOHAPATRA
SURAJIT MOHANTY
UTTAM KUMAR TARAI
DR.BIPLAB BEHERA
SANDIP PADHIARI
DR. PRIYADARSHI KANUNGO
DR PRATAP CH. PRADHAN
DR. PRATEEVA MAHALI
DR. JIBANANANDA MEHENA
SUKENDER REDDY
    MALLREDDY

Committee Members

Patron
Dr. Pramod Chandra Rath, Founder Chairman, DRIEMS University

Co-Patron
Er. Durga Prasad Rath, Pro-Chairman, DRIEMS University
Prof. (Dr.) Prakash Kumar Hota, Vice-Chancellor, DRIEMS University
Sj. Balaram Kar, Director Admin, DRIEMS University
Mrs. Chinmayee Rath, Director Co-Ordination, DRIEMS University

General Chair
Prof. (Dr.) Durga Prasad Mohapatra, NIT, Rourkela
Prof. (Dr.) Manas Ranjan Kabat, IMIT, Cuttack
Prof. (Dr.) Himansu Sekhar Behera, VSSUT, Burla

Organising Chair
Prof. (Dr.) Priyadarshi Kanungo, DRIEMS University
Prof. (Dr.) Nayan Ranjan Samal, DRIEMS University
Prof. (Dr.) Biswa Ranjan Mohanty, DRIEMS University
Prof. (Dr.) Alok Ranjan Biswal, DRIEMS University

Program Chairs
Prof. (Dr.) Sidhartha Panda, VSSUT, Burla
Dr. Suvendra Kumar Jayasingh, IMIT, Cuttack
Prof. Deepak Tosh, University of Texas, USA
Prof. (Dr.) Ali Ismail Awad, United Arab Emirates University, UAE
Prof. (Dr.) Tarlochan Sidhu, University of Ontario Institute of Technology, Canada

Convenor
Dr. Tusharkanta Samal, DRIEMS University

Co-Convenor
Dr. Pratap Chandra Pradhan, DRIEMS University

Editorial Board
Dr. Tusharkanta Samal, DRIEMS University
Dr. Ambarish Panda, SUIIT, Burla
Prof. (Dr.) Manas Ranjan Kabat, IMIT, Cuttack
Dr. Ali Ismail Awad, United Arab Emirates University, UAE
Dr. Suvendra Kumar Jayasingh, IMIT, Cuttack
Dr. Deepak K Tosh, University of Texas, USA

Advisory Board
Dr. Bhakta Charan Pradhan, Registrar, DRIEMS University
Prof. (Dr.) Chittaranjan Tripathy, Ex-VC, BPUT & SU
Prof. (Dr.) Amiya Kumar Rath, VC, BPUT
Prof. (Dr.) Susanta Kumar Das, DRIEMS University
Prof. (Dr.) Kathleen B. Aviso, De La Salle University, Manila, Philippines
Prof. (Dr.) Laxmidhar Behera, Director, IIT, Mandi
Prof. (Dr.) Bidyadhar Subudhi, Director, NIT Warangal

Prof. (Dr.) Kshiti Bhusan Das, VC, Central University, Jharkhand
Prof,(Dr.) Pramod Kumar Meher, CGU, Bhubaneswar
Prof. Mrutyunjay Bhuyan, University of Malaya
Prof. (Dr.) Ajit Kumar Panda, Vice President, Engineering RF VVDN Technologies Pvt. Ltd.
Prof. (Dr.) Subhranshu Samantaray, IIT, Bhubaneswar
Prof. (Dr.) Srinibash Shethy, IGIT, Sarang
Prof. (Dr.) Ajit Kumar Barisal, OUTR, BBSR

Local Organising Chairs
Prof. Surajit Mohanty, DRIEMS University
Dr. Dipak Ranjan Biswal, DRIEMS University
Dr. Biplab Kumar Behera, DRIEMS University
Dr. Bidyadhar Swain, DRIEMS University
Dr. Ranjan Kumar Jati, DRIEMS University
Prof. Raghunath Rout, DRIEMS University
Prof. Rajeev Agarwal, DRIEMS University
Prof. Sekharesh Barik, DRIEMS University
Prof. Rashmi Ranjan Biswal, DRIEMS University

Publicity Chair
Dr. Suraj Sharma, Guru Ghasidas Central University, Bilaspur
Dr. Ashok Kumar Bhoi, GCEK, Kalahandi
Prof. Manas Ranjan Mishra, CGU, Bhubaneswar
Prof. Bivasa Ranjan Parida, Silicon University, Bhubaneswar
Dr. Deepak Ranjan Nayak, MNIT, Jaipur
Dr. Rahul Dixit, SVNIT, Surat
Dr. Sipra Swain, SOA University, Bhubaneswar
Dr. Rajashree Nayak, BGU, Bhubaneswar
Dr. Sujata Swain, KIIT University, Bhubaneswar
Dr. Debashreet Dash, BPUT, Rourkela
Dr Sibarama Panigrahi, NIT Rourkela
Dr. Preeti Ranjan Sahu, NIST University

Publishing Chairs
Dr. Ambarish Panda, SUIIT, Burla
Dr. Satya Prakash Sahoo, VSSUT, Burla
Prof. (Dr.) Brojo Kishore Mishra, NIST University
Dr. Rakesh Ranjan Swain, SOA University
Dr. Rakesh Kumar Lenka, Central University, Koraput
Dr. Bunil Balabantaray, NIT, Meghalaya
Dr. Bhabandu Kumar Mohanta, United Arab Emirates University
Dr. Manoj Kumar Kar, Tolani Maritime Institute, Pune
Prof. Satyabrat Sahoo, DRIEMS University
Dr. Bajra Panjar Mishra, DRIEMS University
Prof. Niva Tripathy, DRIEMS University
Dr. Anurekha Nayak, DRIEMS University
Dr. Debayani Mishra, DRIEMS University
Prof. Jayashree Bhuyan, DRIEMS University
Dr. Prateeva Mahali, DRIEMS University
Dr. Uttam Kumar Tarai, DRIEMS University

Finance Chairs
Prof.(Dr.) Jibanananda Mehena, DRIEMS University
Prof. Surajit Mohanty, DRIEMS University

Prof. Sudhanshu Bhusan Kar, DRIEMS University
Mr. Mrutyunjay Parida, DRIEMS University

Hospitality & Logistics Chair
Prof. Sachidananda Samal, DRIEMS University
Prof. Rojalin Das, DRIEMS University
Prof. Mousumi Acharya, DRIEMS University
Prof. Kulshan Pattnaik, DRIEMS University
Dr. Pradeep Kumar Sahoo, DRIEMS University
Prof. Sushanta Kumar Sethy, DRIEMS University
Prof. Karamjeet Patra, DRIEMS University
Prof. Maheswar Mishra, DRIEMS University
Prof. Satyabrat Sahoo, DRIEMS University
Prof. Subbhasis Mohapatra, DRIEMS University
Prof. Anita Mohanty, DRIEMS University

1 An SRR metamaterial for focusing enhancement of W-band lens antenna for microwave imaging

Debapriya Nilakantha Padhy[1,a], Bajra Panjar Mishra[2,b], Sudhakar Sahu[3,c], Sakambari Mishra[4,d], Subhranshu Kumar Singh[2,e] and Sudhanshu Bhusan Kar[2,f]

[1]Department of Electronics and Communication, GIET, Bhubaneswar, Odisha, India

[2]School of Engineering and Technology, DRIEMS University, Cuttack, Odisha, India

[3]School of Electronics Engineering, KIIT University, Bhubaneswar, Odisha, India

[4]Department of Mathematics, College of Basic Science and Humanities, OUAT, Bhubaneswar, Odisha, India

Abstract

This paper introduces a hybrid split ring resonator (SRR)-based periodic metamaterial structure developed for a W-band lens antenna, designed to detect electromagnetic radiation emitted by plasma or any cancerous cell. This detection enables localized analysis of the electron temperature from the radiation source. Here the metamaterial is tailored to operate within the W-band spectrum (70–110 GHz). The unit cell in the metamaterial structure is engineered for low loss, minimal dispersion, broad bandwidth, and high gain, optimizing lens performance. The structure's design, simulation, and optimization were conducted using ANSYS HFSS software. A composite metamaterial (CMM) was selected for the unit cell, achieving a near-zero refractive index to construct a low-loss, broadband, high-gain NZMTM (near-zero refractive index metamaterial) lens antenna that significantly enhances focusing. The lens aperture dimensions are 1100 μm × 550 μm × 250 μm.

Keywords: CMM, NZMTM, SRR, W-band

Introduction

Metamaterials have made significant strides in antenna systems, lens antenna, microwave detectors and filters that envisages for modern biomedical experiments for cancerous tissue imaging. Recently, lens antennas effectively focus electromagnetic energy into precise directions, achieving a narrow beam profile that maximizes signal strength in targeted areas as proposed by Abdul Mahmood [1]. Traditionally, dielectric materials are used as lens but the demerits of these materials as lens are their heaviness and bulkiness property, and their machining process is also expensive. Resonant and nonresonant sub wavelength periodic structures and artificial materials called metamaterial depicted by Bray et al. [2] are extensively used as lens for collimating the spherical wave into a focusing beam effectively for breast cancer detection. By choosing the specific sub wavelength metamaterial structure, we can control the effective material parameter. Thus, by manipulating the propagation of the EM wave, one can design different types of lens, e. g., negative index metamaterial lens (NIM), zero index metamaterial lens (ZIM), and gradient refractive index metamaterial lens (GRIN) for bringing high resolution imaging in cancerous cell diagnosis proposed by Dixit and Pinto [3]. Metamaterials in microstrip patches are some specialized geometrical structures that do not exist naturally but are being designed artificially that depicts the negative or near zero value of permittivity, permeability and refractive index that enhances the tomographic microwave imaging quality as depicted by Grzegorczyk et.al, Ibrahim et.al, Islami et.al and Islam et.al [4-7]. Hence, these are also called left-handed materials (LHM) to enhance microwave imaging in biomedical instruments used particularly for cancer cell treatment proposed by Maamoun et al. [8]. This study introduces a multi-layered metamaterial design made up of unit cell components to develop a low-loss, high-efficiency, and broadband thin lens antenna. To achieve enhanced focusing and high-gain broadband performance, careful design and optimization of both the illuminator and unit cell are essential and suitable for biomedical imaging. These processes are conducted using ANSYS HFSS software to maximize performance.

[a]hodece@gietbbsr.edu, [b]bajrapanjar@gmail.com, [c]ssahufet@kiit.ac.in, [d]sakambari@ouat.ac.in, [e]subhranshu@driems.ac.in, [f]director@driems.ac.in

DOI: 10.1201/9781003658221-1

Unit Cell Design and Parameter Extraction

The proposed metamaterial unit cell, featuring split ring resonators (SRRs) positioned on one side of a cost-effective substrate, while a metallic strip is placed on the opposite side, directly aligned with the SRRs. The substrate, Rogers RT/Duroid 5880, has a relative permittivity of 2.0 and a loss tangent of $\tan \delta = 0.0009$. The SRR design incorporates two concentric rings, each with a strip width, separation, and split width measuring 55 μm. The outer ring has a diameter of 275 μm while the inner ring has a diameter of 110 μm. The substrate dimensions extend 550 μm along the x-axis and 1100 μm along the y-axis, with a thickness of 250 μm. The metallic strip on the unit cell measures 550 μm along the x-axis and 275 μm along the y-axis, with a copper layer thickness of 9 μm applied to both the SRRs and metallic strip. Simulations are conducted using a linearly polarized wave propagating in the Y-direction, with boundary conditions set for the E-field (PEC) in the X-direction and the H-field (PMC) in the Z-direction, as illustrated in Figure 1.1.

The simulation utilizes ANSYS HFSS software, which operates on the principles of the finite element method (FEM) to analyze and optimize the design as considered by Afridi et.al.[9]. The effective medium parameters from the scattering parameter data are being extracted by the Nicolson–Ross–Weir (NRW) approach is shown in Figure 1.2. It is clear from the Figure 1.2(a) that the permittivity and permeability are near zero from 70-110 GHz with a resonance at 94 GHz that shows low electric and magnetic loss. In Figure 1.2(b), the dispersion diagram shows a phase (β) change of 1800 each from 87-92 GHz (LH band) and from 92-97 GHz (RH band) reveals near zero metamaterial behavior. In these ranges, the refractive index is almost near zero.

From the simulation of the unit cell structure, the transmission (S21) and reflection (S11), coefficients are shown in Figure 1.3(a). It shows that, in the band of 70-89 GHz, the S21 parameter shows lossy transmission and after that, S21 is apparent. The phase in angle degree for S11 and S21 is shown in Figure 1.3(b).

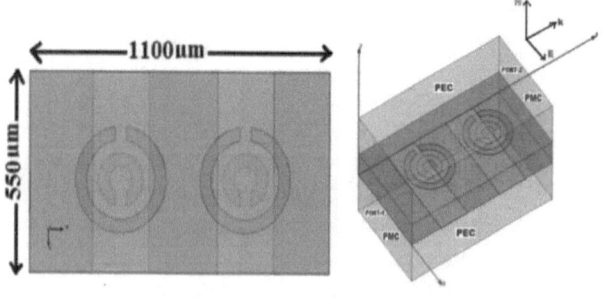

Figure 1.1 The proposed Unit Cell SRR structure and its simulation model with boundary conditions
Source: Author

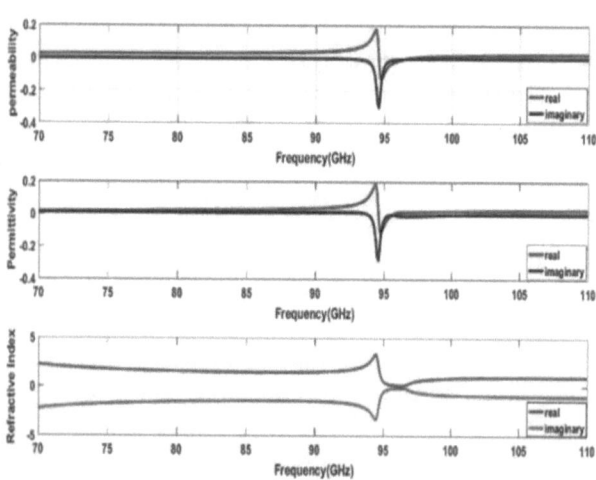

Figure 1.2 (a) Extracted medium parameters of the unit cell
Source: Author

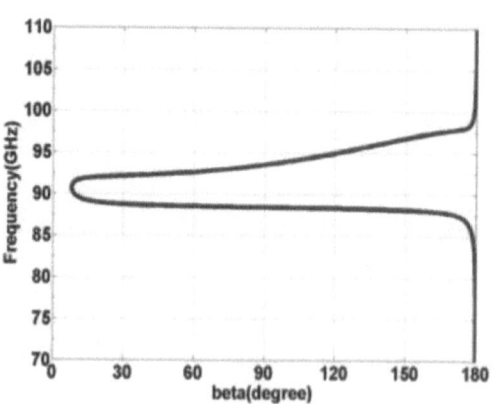

Figure 1.2 (b) Dispersion diagram for the proposed structure
Source: Author

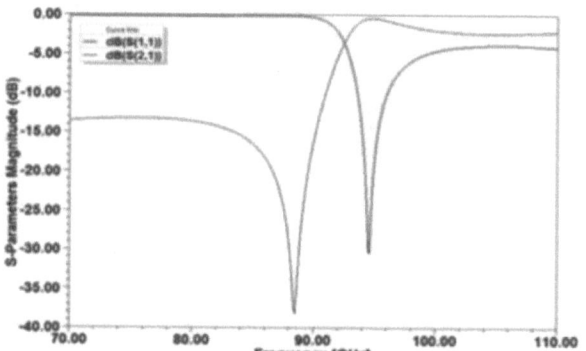

Figure 1.3 (a) Simulation results of magnitude of S11 and S21
Source: Author

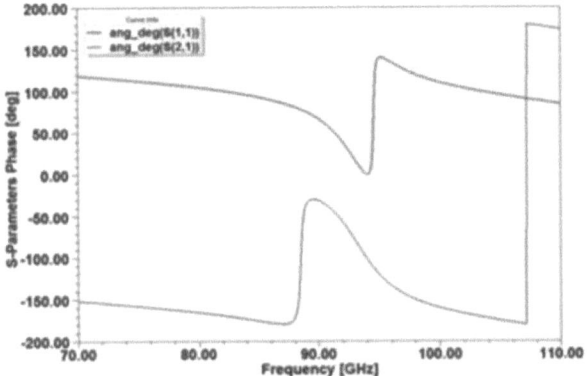

Figure 1.3 (b) Simulation result of phase of S11 and S21
Source: Author

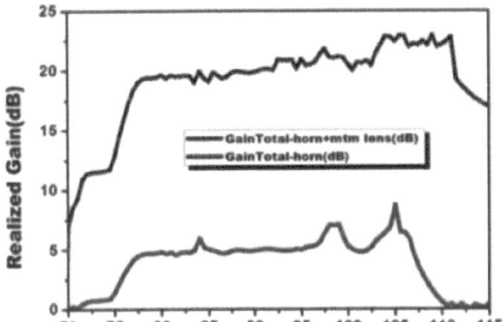

Figure 1.4 Maximum gain versus frequency plot
Source: Author

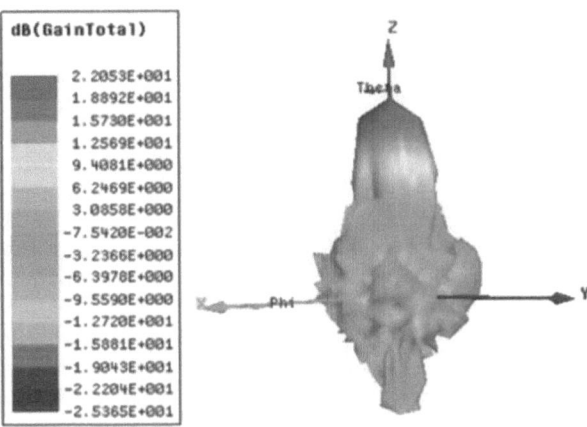

Figure 1.5 The 3D polar plot of gain at 104.5 GHz
Source: Author

Lens Design and Simulation Result and Discussion

Metamaterial structure comprising of 10x10 unit cells in 10 layers is considered here to achieve high gain and directivity for a pyramidal horn in this W-band. The SRR lens-based horn is simulated for the desired frequency band and resulted gain versus frequency graph is depicted in Figure 1.4. It shows a consistency in gain enhancement of 10-12 dB in horn over a wide band of 77.5-102.5 GHz after the introduction of the parasitic SRRs.

In Figure 1.5, the 3D polar plot of gain of the lens antenna shows a gain of 22.053 dB in the propagation direction at 104.5 GHz. The lens antenna efficiently focuses electromagnetic waves in that direction, achieving a high concentration of signal strength compared to other directions. This would be beneficial in applications that require directed, high-intensity signal transmission, or reception. The gain is measured at a specific frequency, 104.5 GHz, which lies within the W-band (75–110 GHz). This implies that the antenna is designed or optimized to perform particularly well at this frequency, making it suitable for applications in that band, such as radar, satellite communications, and high-frequency imaging systems.

Conclusion

This paper presents a periodic metamaterial structure based on a hybrid split ring resonator (SRR) designed for a W-band lens antenna aimed at absorbing and detecting electromagnetic radiation from plasma sources. The hybrid design incorporates double-negative (DNG) material properties, enabling it to act as an effective focusing lens with an innovative approach within this frequency range. Optimized for the W-band (70–110 GHz), this metamaterial lens holds potential applications in the development of wideband, high-gain antennas suitable for fusion plasma diagnostics, temperature measurement, radar, and imaging sensor systems.

Acknowledgment

The authors extend their sincere gratitude to the Department of Atomic Energy-Board of Research in Nuclear Sciences (DAE-BRNS), Government of India, for the financial support provided for this research project, referenced as 39/32/2015-BRNS/39012.

References

[1] Abdul Mahmood, S. M. (2015). Assessment of plasma lipid profile and prolactin level in sudanese female with breast cancer. Doctoral dissertation thesis, Alza-eim Alazhari University.

[2] Bray, F., Ferlay, J., Soerjomataram, I., Siegel, R. L., Torre, L. A., & Jemal, A. (2018). Global cancer statistics 2018: GLOBOCAN estimates of incidence and

mortality worldwide for 36 cancers in 185 countries. *CA: A Cancer Journal for Clinicians*, 68, 394–424.

[3] Dixit, A., & Pinto, M. S. (2019). Simulation of microstrip patch antenna for detection of abnormal tissues in thyroid gland. *International Journal of Innovations in Engineering and Technology*, 13(3), 50–55.

[4] Grzegorczyk, T. M., Meaney, P. M., Kaufman, P. A., & Paulsen, K. D. (2012). Fast 3-D tomographic microwave imaging for breast cancer detection. *IEEE Transactions on Medical Imaging*, 31, 1584–1592.

[5] Ibrahim, A. S. (2014). Cancer incidence in Egypt: results of the national population-based cancer registry program. *Journal of Cancer Epidemiology*, 2014(437971), 18. [Online]. Available at: https://doi.org/10.1155/2014/437971.

[6] Islami, F., Torre, L. A., Drope, J. M., Ward, E. M., & Jemal, A. (2017). Global cancer in women: cancer control priorities. *Cancer Epidemiology, Biomarkers and Prevention*, 26(4), 458–470. cebp.0871, 2016.

[7] Islam, M. T., Mahmud, M. Z., Misran, N., Takada, J.-I., & Cho, M. J. (2017). A microwave breast phantom measurement system with compact side slotted directional antenna. *IEEE Access*, 5, 5321–5330.

[8] Maamoun, W., Badawi, M. I., Aly, A. A., & Khedr, Y. (2021). Nanoparticles in enhancing microwave imaging and microwave Hyperthermia effect for liver cancer treatment. *Reviews on Advanced Materials Science*, 60(1), 223–236.

[9] Afridi, M. A. (2015). Microstrip patch antenna – designing at 2.4 GHz frequency. *Biological and Chemical Research*, 2015, 128–132.

2 Development of a visitor tracking system for tourist areas to monitor crowd density, peak hours and visitor demographics

Anshuman Sahoo[a], Ajay Kumar Yadav[b], Rana Sujeet Kumar[c], Jyoti Ranjan Swain[d], Adyasha Panda[e] and Saswat Gaudu[f]

Department of Computer Science and Engineering, C.V. Raman Global University, Bhubaneswar, Odisha, India

Abstract

This study presents an integrated visitor tracking system for tourism, utilizing Internet of Things technologies, real-time analytics, and smartphone applications. The system monitors crowds, identifies peak times, and provides accurate visitor data to help authorities improve staffing, enhance visitor experiences, and ensure safety. Tested in popular tourist destinations, it demonstrates its potential to revolutionize tourism management and aid future planning.

Keywords: Crowd density, real-time data analytics, tourism management, visitor demographics, visitor tracking

Introduction

The rapidly growing tourism industry requires effective visitor management to enhance experiences and ensure development. This article explores a visitor tracking system for tourist destinations, utilizing Internet of Things (IoT) sensors, real-time data processing, and smartphone apps [1]. The system detects busy periods, tracks visitor movement, and provides personalized updates. Tourism authorities benefit from better crowd management, efficient resource allocation, and enhanced security, while visitors enjoy reduced waiting times and improved experiences. The collected data also aids in marketing and process optimization to meet tourists' needs.

Literature Review

The proposed system architecture enables cooperative monitoring of crowded areas, utilizing computer vision with SVM classifiers to estimate occupancy and user feedback to train crowd-detection models [2]. It integrates environmental data such as UV strength, wind, and temperature, supported by field tests and Markovian agent models to predict system behavior and crowd dynamic [7].

Innovative methods like wearable biosensors, facial recognition, and real-time feedback apps are combined with VR/AR simulations to assess visitor responses to congestion, aiding urban planning. The system uses FIESTA-IoT [9], an optimized IoT platform, for interoperable crowd control applications tested in regions like Spain, Gold Coast, and Santander, offering population mobility insights.

Designed for tourism, the integrated visitor tracking system employs mobile apps, real-time analytics, and IoT sensors to monitor crowds, identify peak times, and enhance safety and visitor experience [5]. Field tests highlight its potential to revolutionize tourism management [10], providing valuable data for planning and resource optimization.

Proposed Methodology

We propose a visitor control system using sensors and microcontrollers to monitor crowds, rush hours, and visitors in tourist areas, providing real-time data for better management. The system's process includes design, sensor integration, data collection, processing, user interface development, testing, and deployment.

Key Components

Arduino Uno R3: Central processor managing data collection and control.

LCD Display: Displays visitor numbers, peak hours, and feedback in real-time.

PIR Sensors: Detect visitors at entry and exit points using infrared radiation.

Potentiometer and resistor: Adjust LCD contrast and prevent circuit damage.

Software development: Includes algorithms for data analysis and user-friendly dashboards for data visualization.

[a]sahooo.anshuman999@gmail.com, [b]azaykumaryadav1@gmail.com, [c]ranasujeet905@gmail.com, [d]jyotiranjanswain3181@gmail.com, [e]adyaa0301@gmail.com, [f]gaudusaswat@gmail.com

DOI: 10.1201/9781003658221-2

System architecture

The system architecture integrates hardware and software to gather, analyze, and present visitor data. Its modular, scalable design allows for future expansions and adaptability. Focused on efficient data processing, real-time monitoring, and reporting, it provides a robust, flexible solution for dynamic environments.

Data collection and processing

Data collection: PIR sensors detect movement and count people entering and exiting the area. The data is sent to Arduino Uno for processing.

Data processing: Arduino processes data to calculate crowd and peak times. Algorithms analyze patterns to determine access time and target audience.

Data storage: Processed data is stored locally on an SD card or transmitted to a central server for long-term analysis.

Real-time display: The LCD displays the current visitor count and peak hours. This real-time information is crucial for on-site decision-making.

Testing and calibration

Testing involves simulating different visitor scenarios to ensure the accuracy of the sensors and processing algorithms. Measures are taken to adjust sensor sensitivity, data processing to ensure performance in various conditions.

Deployment and maintenance

After testing, the system is deployed and regularly monitored to ensure proper operation. It provides valuable insights for managing passenger traffic [8], improving resource allocation, and enhancing the visitor experience.

Schematic view

The system monitors data, compares it with thresholds, and triggers alerts for overcrowding. A schematic view shows component integration for seamless

functionality. This design enhances decision-making and crowd management.

Mechanism

To find people entering and exiting

Crowd control relies on monitoring people entering and exiting an area using various IoT devices [6]. When the input sensor detects a person, the count increases.

4.1 To find people entering and exiting

Tracking people through the counting process helps control the crowd. A counter increase and decreases based on responses from PIR sensors at the entrance and exit.

Threshold checking

Monitoring crowds and flows for safety and comfort. When the population density of a given area surpasses a pre-established threshold prompt action must be implemented to ensure successful management of the situation [10]. distributed the alert. Threshold

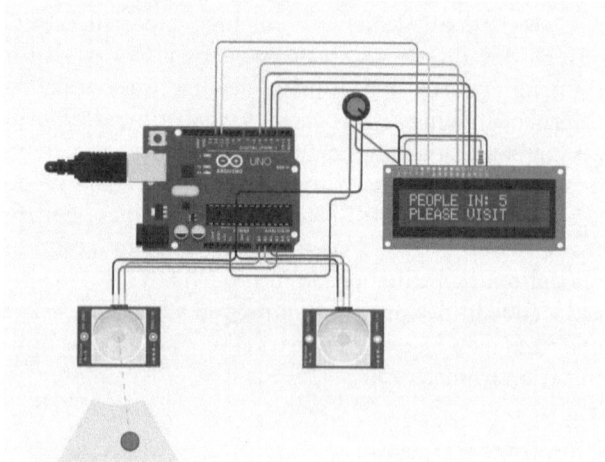

Figure 2.2 System architecture of the visitor tracking system
Source: Author

Name	Quantity	Component
U1	1	Arduino Uno R3
U2	1	LCD 16 x 2
PIR1	1	-37.75118794326235 , -122.10235714752889 , -212.21164007092233 PIR Sensor
PIR2	1	-4.33805471124609 , -142.7387157718025 , -212.67243287740678 PIR Sensor
Rpot1	1	250 kΩ Potentiometer
R1	1	220 Ω Resistor

Figure 2.1 Components list used for implementation
Source: Author

detection is important for monitoring crowds and flows for safety and comfort.

Figure 2.1 illustrates the list of components used in the system, including the Arduino Uno R3, LCD 16x2, PIR sensors, a potentiometer, and a resistor, along with their respective quantities and Figure 2.5 represents the data flow diagram of the visitor detection system, illustrating the process from detecting incoming and outgoing persons to updating the visitor count

and displaying entry restrictions when the maximum limit is reached.

Results and Discussion

System performance

Accuracy of visitor counting: The sensor system achieved over 98% accuracy in detecting and counting visitors, with PIR sensors effectively tracking entries and exits in real-time.

Real-time data processing: The microcontroller processed data instantly, enabling real-time visitor counts for efficient crowd management and fast decision-making.

User interface and display: The LCD display showed visitor numbers and peak hours clearly, with adjustable contrast for readability [3]. Test users found it

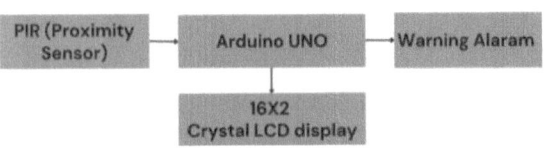

Figure 2.3 Data process path
Source: Author

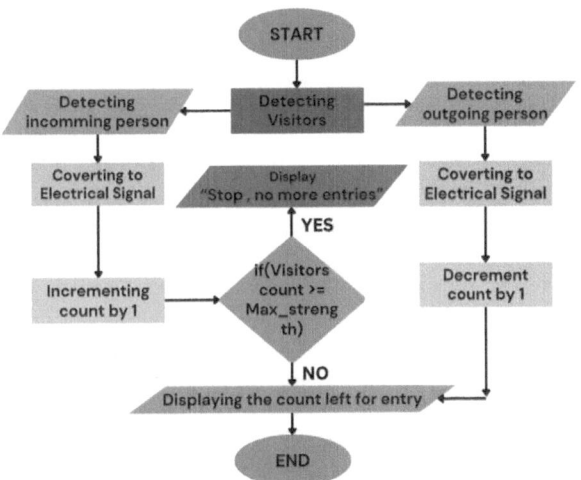

Figure 2.4 Schematic view
Source: Author

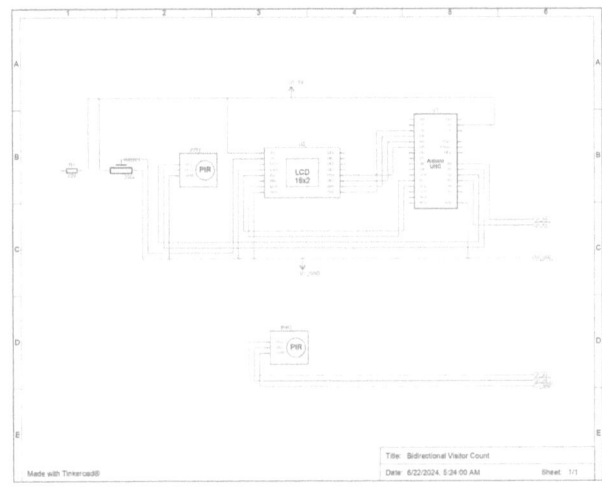

Figure 2.5 Data flow diagram
Source: Author

Table 2.1 Performance metrics.

Metric	Description	Result
Accuracy of visitor counting	Comparison of sensor data with manual counts	Error margin = 2%
Real-time data processing	Ability of Arduino Uno R3 to process and display data without delay	Real-time processing achieved
Usability of user interface	Feedback on the ease of understanding and operating the system	High usability reported by test users
Peak visiting hours	Identification of times with highest visitor traffic	Specific peak hours identified
Visitor flow patterns	Analysis of common entry and exit times	Clear patterns observed
System reliability	Continuous operation without failures	No failures over extended testing period
Environment all sensitivity	Instances of PIR sensors triggered by non-human movement	Some false positives detected
Scalability	Performance in larger areas with multiple entry/exit points	Needs enhancement for larger scalability

Source: Author

highly usable. Results of the visitor control system, tested through a reliable simulation, are discussed.

Table 2.1 presents key performance metrics of the proposed system, including accuracy, real-time processing, usability, peak hours, visitor flow, reliability, environmental sensitivity, and scalability. The system achieved real-time data processing with Arduino Uno R3 and high usability based on user feedback. Peak hours and visitor flow patterns were identified, with no failures observed in system reliability. However, false positives in environmental sensitivity and scalability limitations were noted.

Data analysis and insights
The gathered information was examined to find trends in the behavior of the visitors. Important realizations comprised of:

Peak VISITING HOURS: The data made it possible for authorities to more efficiently allocate resources by identifying the specific times of day with the greatest visitor traffic.

Visitor flow patterns: Common entry and exit times were identified by the analysis, which made it easier to comprehend visitor movement and improve entry and exit locations to ease traffic [4].

System reliability
The system was evaluated over an extended period, showing no failures. The sensors, microprocessor, and display functioned flawlessly, ensuring long-term reliability for real-world deployment.

Limitations on board
Not withstanding the general success, a few drawbacks were found:

Environmental sensitivity
From time to time, movements of non-human objects, like animals, might trigger the PIR sensors. To lower false positives, more filtering techniques might be used.

Scalability
Even though the system worked effectively in a controlled setting, more sensors and processing power could be needed to scale it to bigger regions with many entry/exit points.

Conclusion

The proposed visitor management system demonstrates high accuracy and reliability in tracking and managing visitors, making it an invaluable tool for tourism authorities. Its real-time data processing and intuitive interface enhance crowd control, resource allocation, and visitor experiences. Future improvements include

advanced data analytics for forecasting trends [12], integration with smart city systems for safety and urban planning, scalability for larger areas, real-time mobile applications for visitor convenience, and environmental adaptability to reduce false positives. These advancements position the system as a key element in sustainable and efficient tourism management.

References

[1] Chianese, A., & Piccialli, F. (2014). Designing a smart museum: when cultural heritage joins IoT. In 2014 Eighth International Conference on Next Generation Mobile Apps, Services and Technologies, (pp. 300–306).

[2] Kalisch, D. (2012). Relevance of crowding effects in a coastal national park in Germany: results from a case study on Hamburger Hallig. *Journal of Coastal Conservation*, 16, 531–541.

[3] Barbierato, E., Gribaudo, M., & Iacono, M. (2016). Modeling and evaluating the effects of big data storage resource allocation in global scale cloud architectures. *International Journal of Data Warehousing and Mining (IJDWM)*, 12, (2), 1–20.

[4] Jacques, Jr, , Musse, , & Jung, (2010). Crowd analysis using computer vision techniques. *IEEE Signal Processing Magazine*, 27(5), 66–77.

[5] Böhlen, M., Clark, B., Dalton, J., Atkinson, J., Blersch, D., & Yang, L. (2013). Another day at the beach: combining sensor data with human perception and intuition for the monitoring and care of public recreational water resources. In 2013 9th International Conference on Intelligent Environments, (pp. 37–44).

[6] Gribaudo, M., Iacono, M., & Levis, (2017). An IoT-based monitoring approach for cultural heritage sites: the matera case. *Concurrency and Computation: Practice and Experience*, 29(11), e4153.

[7] Yannuzzi, M., van Lingen, F., Jain, A., Lluch Parellada, O., Mendoza Flores, M., Carrera, D., et al. (2017). A new era for cities with fog computing. *IEEE Internet Computing*, 21(2), 54–67.

[8] Girau, R., Ferrara, E., Pintor, M., Sole, M., & Giusto, D. (2018). Be Right Beach: A Social IoT System for Sustainable Tourism Based on Beach Overcrowding Avoidance. In Proceedings of the 2018 IEEE International Conference on Internet of Things (iThings), Green Computing and Communications (GreenCom), Cyber, Physical and Social Computing (CPSCom), and Smart Data (SmartData) (pp. 9-14) IEEE. DOI: 10.1109/Cybermatics_2018.2018.00036. Hyperlink: https://ieeexplore.ieee.org/document/8726793

[9] Girau, R., Martis, S., & Atzori, L. (2016). Lysis: a platform for IoT distributed applications over socially connected objects. *IEEE Internet of Things Journal*, 4(1), 40–51.

[10] Kawamoto, Y., Yamada, N., Nishiyama, H., Kato, N., Shimizu, Y., & Zheng, Y. (2017). A feedback control-based crowd dynamics management in IoT systems. *IEEE Internet of Things Journal*, 4(5), 1466–1476.

3 A study on extending IoT applications in multiprocessing agricultural systems

G. Sophia Reena[1,a], R. Jeevitha[2,b], S. Hemalatha[3,c] and O. Sujitha[3,d]

[1]Associate Professor and Head, Department of IT, PSGR Krishnammal College for Women, Coimbatore, Tamilnadu, India

[2]Assistant Professor, Department of IT, PSGR Krishnammal College for Women, Coimbatore, Tamilnadu, India

[3]Student, Department of IT, PSGR Krishnammal College for Women, Coimbatore, Tamilnadu, India

Abstract

Internet of Things (IoT) applications in multiprocessing agricultural systems will not only increase agricultural productivity, but it will also significantly improve the quality of the products, reduce labor costs, and increase farmers' income, which could actually lead to the modernization and intelligence of agriculture. Farmers are also in a very disadvantaged position economically, unable to purchase tractors and other expensive machinery, and are forced to farm in traditional ways. Therefore, we believe that some form of modern mechanization can replace human and animal labor from an economic and labor point of view. Developing a multipurpose agricultural machine (with FMB mapping) to overcome the fragmentation prevalent in farmland is the objective of this research work. This vehicle will be efficient for small farmers, as it can perform a variety of tasks related to agricultural activities. The mechanism uses Bluetooth connected applications as a management tool for manual management and to support navigation of the mechanism department. Robotics is the best solution to overcome all the shortcomings, i.e. automation with machines performing all the tasks to significantly improve the yield. This paper systematically summarizes the status of IoT in multi-processing agricultural machinery.

Keywords: Arduino, FMB, humidity sensor, lithium battery, multi-processing

Introduction

Internet of Things (IoT) is a network of everything where devices, sensors, machines, software, and people are integrated to communicate, exchange information, and interact through the Internet environment to provide a holistic solution between the real and virtual world. IoT has been recently used in various sectors like smart agriculture, autonomous vehicles, smart homes, smart cities, smart energy, campus management, healthcare, logistics, etc. Kevin Ashton believes that in the future, IoT will be established and installed in everyday objects with RFID and sensors. The aim is to help farmers address the imbalance between demand and supply by integrating IoT solutions for agriculture. This can be achieved by ensuring high yields, increasing revenues and protecting the environment. To sustain farmers and resources, sustainable agriculture promotes agricultural methods and practices that are economically rational, increase rural biodiversity, conserve water resources, reduce soil erosion and ensure a natural and healthy environment. India's largest economic sector is agriculture. It generates 18% of the country's GDP and around 57% of the population is employed in rural areas.

Though India's total agricultural production has increased over time, the share of farmers has declined from 71.9% in 1951 to 45.1% in 2011. Agreeing with the economic study of 2018, as it were 25.7% of the workforce is anticipated to be utilized in agriculture in 2050 [2].

When India gained independence in 1947, the production and productivity of groundnuts, corn, and other crops was very low. India's population could not have been fed by current production. To meet the needs of its people from many countries, the country imported them in large quantities. Lack of machinery in cultivated land contributed to the low production and productivity. Compared to other countries, most of the agricultural work in India is done manually. At that time, there were no facilities for sowing peanuts and corn, so only human labor was required. More money was spent on labor, but the process was relatively slow. Users of the Blynk app can communicate with robots that most people are familiar with. The advantages of these robots include hands-free, smartphone-controlled operations and quick data entry. In the field of agricultural robotic vehicles, a concept was developed to test whether a number of small robotic machines can outperform traditional large tractors and human labor [2].

[a]sophiareena@psgrkcw.ac.in, [b]jeevitha@psgrkcw.ac.in, [c]22bit034@psgrkcw.ac.in, [d]22bit114@psgrkcw.ac.in

DOI: 10.1201/9781003658221-3

Literature Review

Field mapping (FMB Mapping)

Using the IoT, field mapping entails producing intricate maps of agricultural areas to maximize crop management and resource allocation. Drones and GPS-capable sensors are examples of IoT devices that gather data on crop health, environmental factors, and soil characteristics. Precision agriculture maximizes productivity and reduces waste by customizing agricultural operations to certain locations within a field. It is made possible by IoT-based field mapping [14]. High-resolution photographs are taken by drones with multispectral sensors, and these images are processed to provide comprehensive field maps. According to Matese et al. [7], these maps offer information on soil variability and crop health.

Sprinkler systems

By automating irrigation, IoT-enabled sprinkler systems ensure efficient water utilization and lower labor expenses. These systems modify watering schedules based on information from weather forecasts and soil moisture sensors. According to Pereira, Cordery, and Iacovides [9], IoT-based smart irrigation systems precisely control water distribution by utilizing real-time data. This minimizes water waste and increases agricultural yields. Farmers can make timely modifications to sprinkler systems based on field conditions by using computers or cell phones for remote monitoring and control [3].

Remote control functions

In agricultural, remote-control operations improve operational efficiency and safety by enabling farmers to operate machinery and equipment from a distance.

According to Sørensen et al. [11], the IoT has made it possible to construct autonomous tractors and other agricultural machinery that can be remotely monitored and controlled to carry out operations like planting, harvesting, and spraying. Farmers can identify problems early and avoid failures with the assistance of IoT devices, which offer real-time data and alerts on machinery performance [13].

Soil moisture sensor

Sensors for Soil Moisture In order to maximize water usage during precision irrigation, soil moisture sensors are essential since they provide real-time data on soil moisture levels. By placing a network of soil moisture sensors throughout a field, extensive data on the state of the soil can be obtained, allowing for accurate irrigation scheduling [6]. Making better decisions and managing resources is made possible by integrating

Table 3.1 Comparative analysis between traditional manual systems and IoT based systems [2][3][4][5][6].

Aspect	Traditional farming	IoT Based multi-processing agricultural system
Efficiency	Relies on manual labor and standalone machinery, leading to slower operations	Combines sowing, ploughing, and irrigation in one system, increasing efficiency.
Cost	Requires multiple machines and accessories, increasing the overall expenditure.	Minimizes accessories and costs through an integrated multi-machine system.
Power source	Operates using fossil fuels or manual effort, leading to high energy consumption and emissions.	Powered by lithium batteries, offering a sustainable and cost-effective energy solution.
Automation	Minimal automation; most tasks require human intervention.	Integrates IoT for automated control and monitoring, reducing the need for manual involvement.
Flexibility	Limited to specific tasks, requiring multiple tools for different farming activities.	A single machine performs multiple tasks, enhancing adaptability and reducing dependency.
Monitoring and control	Requires farmers to be physically present to supervise activities.	IoT applications enable remote monitoring and control through connected devices.
Scalability	Difficult to scale due to high costs and manual dependency.	Scalable with IoT, as the system can be expanded or modified with additional features.
Environmental impact	High carbon footprint due to traditional machinery and fossil fuel use.	Reduced environmental impact through renewable energy sources and optimized operations.
Accuracy	Prone to human errors during sowing, ploughing, and irrigation.	IoT ensures precise and consistent operations, improving yield and reducing resource wastage.
Maintenance	Maintenance is frequent and often costly due to the use of multiple machines.	A single integrated system reduces maintenance frequency and costs.

Source: Author

Figure 3.1 Multi-processing agriculture machine
Source: Author

soil moisture data with other IoT's devices, like irrigation controllers and weather stations [2].

Methodology

The aim of this research is to develop a multi-machine that can be used for sowing, ploughing and irrigation while requiring as few accessories as possible and at the lowest possible cost. The modules used are Bluetooth module-HC05, Arduino UNO- r3, Humidity sensor-DHT11, Servo motor-SG90, Single relay module-NRP18-CT2DH, Lithium Battery-ICR-18650.

Liquid Crystal Display-JHD162ALithium batteries power the entire system of the machine. The basic frame of the machine consists of three wheels that are connected and operated. The two front wheels are driven by direct current and the rear wheel can rotate 360 degrees. Arduino - UNO R3 can be integrated into many electronic projects and is used to create standalone interactive objects. A dielectric is placed between the two electrodes on the moisture storage board of the humidity sensor, DHT 11 module, which is the sensor capacitor. The Bluetooth module, HC-05, is identical to the HC-04 and HC-06 but is a separate module. Rechargeable lithium batteries also charge quickly and are very safe. These batteries are 50-60% lighter than lead acid batteries and have a capacity of 12V and 3V respectively, and are mounted in the center of the wooden plate on the left. The Single Relay Board is an electrochemical device that works by switching power between contacts. The direction and speed of the DC motor is controlled by the driver board. This board is located in the middle of the wooden board. The output of the motor driver is digital, so the speed of the motor is adjusted using PWM (Pulse Width Modulation). The motor driver basically acts as a current amplifier with an input signal to drive the motor.

Working Principle

This research is a timely and essential exploration into the role of IoT in addressing critical challenges in agriculture, such as resource wastage, low productivity, and the economic limitations faced by small-scale farmers. By proposing an IoT-based multi-processor agricultural machine, the paper effectively highlights the following strengths: The multi-purpose machine designed to handle sowing, ploughing, and irrigation provides a cost-effective alternative to traditional farming methods. This is particularly beneficial for economically disadvantaged farmers, making modern mechanization more accessible. The use of components like Arduino boards, Bluetooth modules, and soil moisture sensors showcases the integration of affordable yet effective technologies. Such systems empower farmers to monitor and control processes remotely, reducing labor dependency.

This paper emphasizes sustainable agricultural practices, such as precision irrigation and smart farming. IoT-enabled sensors and sprinklers help conserve water and other resources, addressing environmental concerns while increasing productivity. The research introduces Field Measurement Book (FMB) mapping and autonomous machinery controlled through IoT, promoting accuracy in land usage and operational efficiency. The discussion on next-generation IoT advancements, including AI, machine learning, and 5G, indicates the transformative potential of these technologies in agriculture. This forward-looking approach positions the research as a bridge to future innovations [14].

FMB mapping

Every surveyor dealing with land or real estate must use FMB (Field Measurement Book Sketch), which can be used to map locations and record the dimensions of a particular area. FMB drawings are collections of cartographic information kept in bulk by the government in every Tahsildar's office. Individual sketches with survey numbers are stored in the FMB sketches at a scale of 1:1,000 or 1:2,000. Each subdivision in a tehsil has a different survey number, and each subdivision number is allotted to a landowner. In other words, FMB gives the exact dimensions of the land or property [4].

Components of FMB

A sketch consists of several elements such as G-lines which represent the perimeter of the survey area and F-lines which represent the land boundaries. Plot lines divide the surveyed area into more manageable sections. Ladders help in determining the elevation of points within the area surveyed. The points on the sketch are connected by construction lines [10].

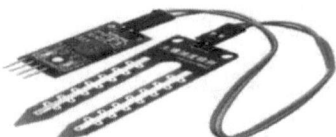

Figure 3.2 Soil moisture sensor
Source: https://www.researchgate.net/publication/315755117_Flood_recovery_system_in_agriculture_using_internet_of_things

Remote control

A remote control (also called a remote or clicker) is an electrical device that can be used to wirelessly control another device remotely. It can be used to control equipment that is difficult to operate directly with a controller. Remote controls are typically made from modern consumer infrared devices that transmit digitally encoded infrared pulses. These devices often come with a portable, compact wireless remote control with a variety of buttons. The LED lights on the remote-control flash very quickly to send a message to the TV. The TV acts as the transmitter and the remote control as the receiver. Single-channel remote controls (single-function, one-button remote controls) can be used to activate functions when a carrier signal is present. For multi-channel remote controls (usually multi-function remote controls), more complex processes are required, one of which is to modulate the carrier wave with signals of different frequencies. The receiver first demodulates the incoming signal before applying appropriate frequency filters to separate the different signals [12].

Soil moisture sensor

Soil moisture sensors work in a simple way. The resistance of the forked probe changes depending on the moisture content of the soil, acting as a variable resistor (like a potentiometer). The pressure between the soil particles and the water particles is measured by a soil dampness sensor - a tensiometer. Soil dampness sensors utilize capacitance to degree the dielectric consistent of the encompassing medium. The dielectric constant of the soil depends on the amount of water present. The sensor produces a voltage proportional to the moisture content and dielectric constant of the soil [5]. Depending on the technology used, soil moisture sensors can be divided into two groups:

1. volumetric water content sensors
2. sensors that measure soil tension when embedded in the soil profile.

Sprinkler

According to Newton's third law of motion, an object experiences an equal and opposite force when it

Figure 3.3 Sprinkler
Source: https://biovoicenews.com/ipl-biologicals-and-spains-afepasa-partner-for-global-registration-of-bio-pesticides/

applies a force to another object. Sprinklers experience an equal and opposite force when water leaves the device. Sprinkler/spray irrigation is a technology that provides precise water delivery similar to rain. Among the parts of the water distribution network are sprinklers, pipes, pumps, and valves. Irrigation sprinklers can be used in residential, commercial, and agricultural environments [5].

Next-Generation IoT-based Agriculture

Future developments in IoT technology are expected to have a significant impact on agriculture. Farmers can use precision agriculture to increase crop yields and reduce resource waste by making data-driven decisions. IoT sensors can be used to monitor plant health, weather, and soil conditions in real time. Equipped with IoT sensors, drones and self-driving cars can complete tasks like planting, harvesting and pest control more efficiently and with less human effort. IoT devices can track the location and health of animals, allowing for earlier diagnosis of disease and improved overall herd management [8]. By using sensors to analyze soil moisture levels and adjust irrigation schedules accordingly, IoT can help farmers optimize water usage.

IoT-powered tracking and monitoring improves produce traceability, improving food safety and reducing waste. Machine learning and AI enable analytics that predict agricultural diseases, weather patterns and market trends, enabling proactive decision-making [1]. Using IoT data to identify and minimize risks could potentially reduce insurance premiums for farmers. IoT technology has the potential to completely transform the agricultural sector by improving productivity, sustainability, and efficiency. With the advent of next-generation IoT-based agriculture, the potential and impact of IoT in agriculture is expected to grow even more [1].

Real-time monitoring and control of IoT devices in remote agricultural areas will be possible with the advent of 5G networks, which will allow for faster and more dependable data transmission. Edge Computing diminishes idleness and speeds decision-making by preparing information locally on gadgets set closer to the information source, such as Field sensors. IoT systems progressively integrate AI and machine learning algorithms to generate more accurate predictions, use resources more effectively, and automate advanced tasks such as pest detection and diagnosing plant diseases. Robotics and automation: Smart robots and drones are becoming increasingly common in the agricultural sector, as IoT sensors control their behavior in operations such as planting, harvesting, and weed control [1]. Blockchain technology allows the entire agricultural supply chain, from planting to distribution, to be transparently and immutably recorded, ensuring food safety and traceability. IoT sensors are being developed to enable precise and individual control of plants by monitoring a wide range of factors such as pH, plant DNA, and nutrient levels. IoT technology can help farms maximize energy consumption by monitoring and controlling equipment and renewable energy sources. To drive innovation, problem-solving, and recommend customizable solutions, farmers are increasingly collaborating with researchers, agronomists, and other stakeholders to share IoT-generated data [8].

IoT enables "recipe farming," where each field and crop receives customized treatment based on its individual needs. With IoT data confirming the products' quality and place of origin, farmers can sell their goods directly to businesses and consumers through online marketplaces. Farmers are less likely to face fines and other penalties when they use IoT technology to adhere to environmental regulations and sustainable agricultural practices. IoT-powered agriculture could lead to more sustainable, profitable, and productive farming in the future. Increasing sustainability, productivity and profitability for farmers, as well as contributing to environmental protection and global food security, are some of the expectations of the next generation of IoT-based agriculture [8].

Conclusion

IoT-based multi-processor agricultural tools represent a significant advancement in modern agricultural technology, combining real-time data, automation and sophisticated algorithms to make the most of agricultural cultivation and resource management. They have the potential to transform agriculture by increasing productivity, reducing labor costs and promoting sustainable agricultural practices. This ultimately leads to increased yields and improved food security, while reducing environmental impact. As this technology advances, it will enable farmers to meet growing global food demand more efficiently and sustainably, holding great potential for the future of agriculture. The extension of IoT applications in multi-processing agricultural systems holds significant potential for improving agricultural productivity and sustainability. Field mapping, sprinkler systems, remote control operations, and soil moisture sensors are key areas where IoT technologies are making a substantial impact. Continued research and development in these areas will further enhance the capabilities and benefits of IoT in agriculture.

References

[1] Biradar, H. B., & Shabadi, L. (2019). Review on IOT based multidisciplinary models for smart farming. In *Proceedings of the 2019 2nd IEEE International Conference on Recent Trends in Electronics, Information & Communication Technology (RTEICT)*, (pp.1923–1926).

[2] Evett, S. R., Tolk, J. A., & Howell, T. A. (2012). A depth control stands for improved soil water content measurement with time domain reflectometry. *Vadose Zone Journal*, 1(1), 11–18.

[3] Goldstein, A. (2016). The impact of remote monitoring and control in agricultural water management. *Agricultural Water Management*, 178, 32–39.

[4] Swati, J. S., & Shinde, T. A. (2017). Design and development of automatic seed sowing machine. *International Journal of Electronics and Communication Engineering (IJECE)*, Special Issue, 40–44. ISSN 2348-8549.

[5] Sahu, J., Sahu, S. K., & Kumar, J. (2012). Microcontroller based Dc motor control. *International Journal of Engineering Research and Technology (IJERT)*, 1(3), 1–4.

[6] Jones, H. G. (2004). Irrigation scheduling: advantages and pitfalls of plant-based methods. *Journal of Experimental Botany*, 55(407), 2427–2436.

[7] Matese, A., Toscano, P., Di Gennaro, S. F., Genesio, L., Vaccari, F. P., Primicerio, J., et al. (2015). Intercomparison of UAV, aircraft and satellite remote sensing platforms for precision viticulture. *Remote Sensing*, 7(3), 2971–2990.

[8] Butet, P. V., Deshmukh, S., Rai, G., Patil, C., & Deshmukh, V. (2018). Design and fabrication of multipurpose agro system. *International Research Journal of Engineering and Technology (IRJET)*, 5(1), 865–868.

[9] Pereira, L. S., Cordery, I., & Iacovides, I. (2012). Coping with Water Scarcity: Addressing the Challenges. Springer Science and Business Media.

[10] Saharawat, Y. S., Singh, B., Malik, R. K., Ladha, J. K., Gathala, M., et al. (2010). Evaluation of alternative

tillage and crop establishment methods in a rice, wheat rotation in north western IGP. *Field Crops Research*, 116, 26.

[11] Sørensen, C. G., Fountas, S., Nash, E., Pesonen, L., Bochtis, D., Pedersen, S. M., et al. (2010). Conceptual model of a future farm management information system. *Computers and Electronics in Agriculture*, 72(1), 37–47.

[12] Balaji, T., Rajappan, R., & Senthil, S. (2015). Mechanical design and development of agricultural robot. *International Journal of Advancements in Mechanical and Aeronautical Engineering*, 2(1), 94–97.

[13] Walter, A., Finger, R., Huber, R., & Buchmann, N. (2017). Opinion: smart farming is key to developing sustainable agriculture. *Proceedings of the National Academy of Sciences*, 114(24), 6148–6150.

[14] Zhang, N., Wang, M., & Wang, N. (2002). Precision agriculture—a worldwide overview. *Computers and Electronics in Agriculture*, 36(2-3), 113–132.

4 A novel approach for mask face detection using convolutional neural network

Ram Gopal Sharma[1,a], Hitendra Garg[2,b] and Law Kumar Singh[3,c]

[1]Research Scholar, GLA University, Mathura, UP, India

[2]Professor, GLA University, Mathura, UP, India

[3]Associate Professor, Amity University Punjab, Mohali, India

Abstract

One way to reduce the spread of virus infections from an infected person is to wear a mask. Wearing masks is required in many locations to stop the spread of viruses. Therefore, in order to check whether someone is donning a mask, we must require an automatic method. The goal of this study is to create a real-time, extremely accurate method for identifying faces in public that are not wearing masks and subsequently requiring mask wear. A novel CNN architecture for detecting face masks is introduced in this paper. The studies were carried out using the Kaggle face mask detection dataset as the benchmark dataset. Our work's predictability and robustness are demonstrated by the outcomes for the suggested model.

Keywords: Convolutional neural network, deep learning, machine learning, masked face detection

Introduction

The method of identifying whether or not someone is wearing a mask is known as face mask detection. Researchers were able to extract specific features that most accurately represented the problem with great assistance from deep learning algorithms. Numerous applications, including speech recognition, facial identification, cancer diagnosis, etc., have seen success with the application of neural networks. Deep learning demonstrated its effectiveness in identifying several object classes for each of these applications. Convolutional neural network (CNN)-based face identification techniques have been extensively created in the last few years to enhance detection performance. In order to efficiently recognize face masks, we suggested a unique CNN approach.

The following sections make up the remaining portion of this work. Present literature on face mask detection is covered in Section 2. The recommended methodology is presented in Section 3. The effectiveness of the suggested technique is assessed over a range of parameters in Section 4. Section 5 provides a final summary of the work.

Literature Review

The most accurate approach now in use is believed to be deep neural network-based categorization algorithms, which outperform other traditional approaches. Two novel deep learning models are presented by Abdulmunem et al. [2]. The first is based on MobileNetv2 model, which provides 99.05% accuracy and the second is new CNN architecture, which provides 97.59% accuracy. Abbas et al. [1] addresses the problem of face occlusion by creating a Convolutional mixer approach for face identification. They use DenseNet-16, ResNet-50 and Inception-v3 models to extract features.

Agarwal et al. [3] suggested a deep learning model with the ability to distinguish between people wearing masks and those who are not. The mask detector in the suggested work is designed using a CNN and support vector machine, which yields 99.11% accuracy. Allam and Jones [4] discuss how smart city networks are handling the COVID-19 pandemic, highlighting the significance of AI and common data sharing standards.

In order to detect face masks, Chavda et al. [5] created a multi-stage CNN architecture. Their model is made to accurately determine whether individuals are wearing masks in various situations. Chong et al. [6] suggested the approach that is intended to address the issue of underfitting while minimizing computing demands using large-scale dataset training.

Gao et al. [7] offer a method for re-identifying obscured people using feature fusion and sparse reconstruction. Their method improves recognition accuracy by successfully addressing occlusions, indicating its suitability for security needs. Ge et al. [8] present a method that enhances the convolution operator by utilizing self-attention. In particular, this is accomplished by joining a self-attention feature map.

[a]cs.ramgopal@gmail.com, [b]hitendra.garg@gmail.com, [c]lawkumarcs@gmail.com

DOI: 10.1201/9781003658221-4

In comparison to existing methods, this technique enhances mask face recognition performance.

Hussain et al. [9] recognize face masks using MobileNetV2-based transfer learning and deep CNN. Their approach attains a high degree of efficiency and accuracy. A face mask identification model was presented by Joshi et al. [10] in video data. Their system will aid in public health efforts during the pandemic by monitoring in real time.

Ketab et al. [11] combines best practices for feature extraction to handle the low-resolution face recognition task. This study makes use of the characteristics that are infused into the training process using customizable and learnable filters in order to better align them with the functioning of the human brain. Khan et al. [12] used deep transfer learning and noise-based data augmentation to create a convolutional architecture for recognition of faces. Their approach is useful for security applications with increasing face recognition accuracy.

Kumar [13] suggested face mask detector device that may be combined with security cameras and utilized as a tool in self-sufficient monitoring situations to help locate and apprehend lawbreakers. Kumar et al. [14] provide an extended framework that integrates the RDGF descriptor and BoCNN. To improve efficiency even further, a dynamic classifier is used during runtime to decide between hand-crafted features and the CNN framework. A partial occlusion thermal face dataset is also proposed.

Mamieva et al. [15] suggested the one-stage face detector based on Retina net as a solution to the challenging face detection problem. They enhanced the network, which increased the accuracy and speed of detection. Mosayyebi et al. [16] looks at the increasing demand for smart devices to identify user data, namely gender, in order to improve user comfort and engagement. Naser et al. [17] attempted to accurately identify faces that are hidden under masks and eyeglasses. They put up a solution to the problem of faces that are partially concealed in facial recognition.

Shukla et al. [18] addresses the issue of identifying face mask using transfer learning utilizing MobileNet V2 as a base model. The recognition accuracy of this suggested model is 99.82%. Yu et al. [19] offers recognition of face mask images and proposed algorithm based on the enhanced YOLO-v4 to address issues of low accuracy, poor robustness, and others issues imposed by the complicated environment. Zhao et al. [20] presents a straightforward but effective architecture that maximizes loss through contrastively learning the distorted characteristics against individual labels to reduce the need for manual labeling.

Proposed Methodology

To improve the accuracy of detection of face masks, a unique CNN is developed. The following steps are included in the suggested method.

3.1 Data input and preprocessing: To detect face masks, we employed the standard dataset available on Kaggle. There are two classes in this dataset. One wears a mask, whereas the other does not. There is a total of 4095 images in this dataset, with 2165 images belonging to the mask class and 1930 without. RGB images with 224 × 224 pixels are input into our model.

3.2 Data augmentation: To improve the model's robustness, apply data augmentation techniques like rotation, flipping, zooming and scaling.

3.3 Prepare CNN model: There are seven layers in the suggested model, wherein convolutional layers make up the first four layers, max pooling layers come next, and fully connected layers make up the final three.. The model has three different kinds of layers. An overview of the recommended architecture is given in Figure 4.1.

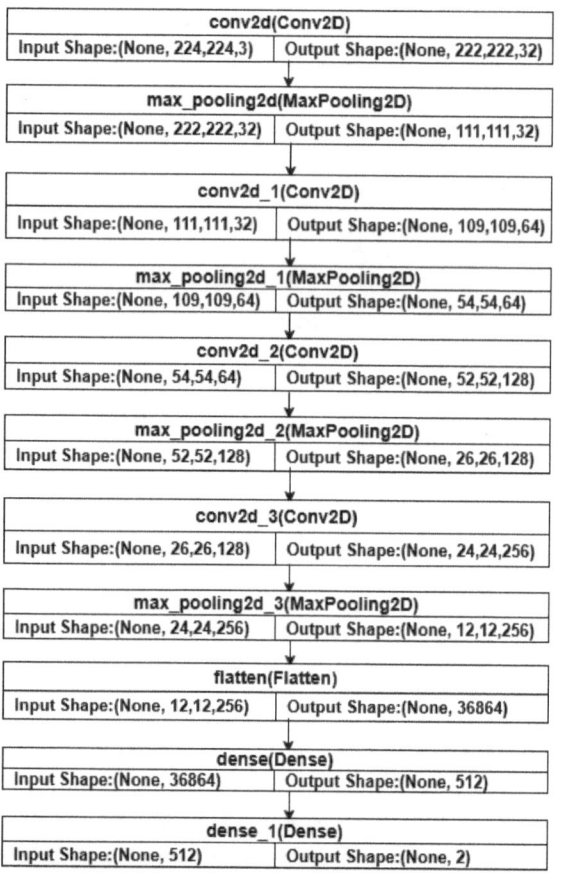

Figure 4.1 Proposed model architecture
Source: Author

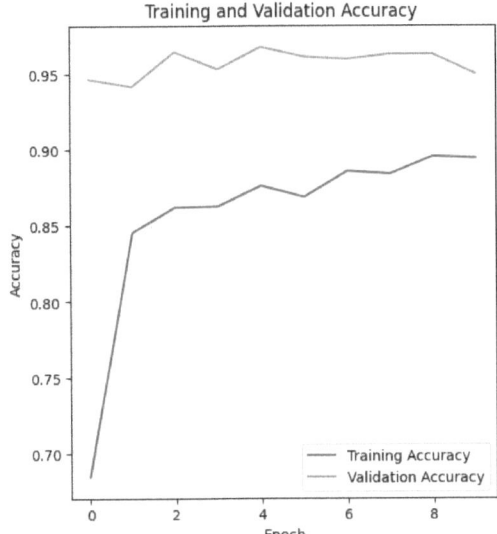

Figure 4.2 Validation and training accuracy
Source: Author

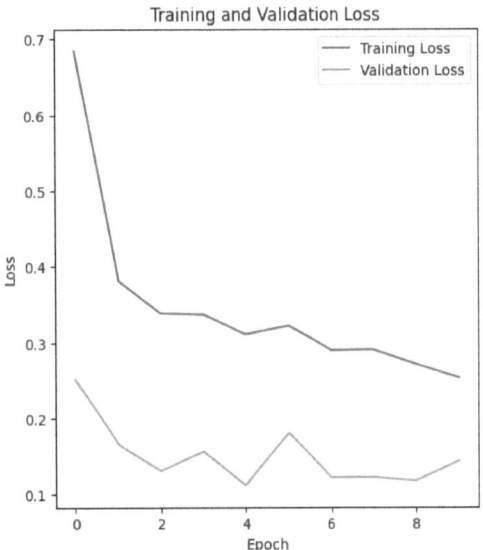

Figure 4.3 Validation and training loss
Source: Author

3.3.1 Convolutional layer: The convolutional layer makes use of tiny (3x3) filters. A linear activation function with RELU follows each convolutional layer.

3.3.2 Pooling layer: In this architecture, Max Pooling was used after every convolutional layer with 2x2 strides. These layers are designed to down-sample the output from earlier.

3.3.3 Fully connected layer: Two fully-convolutional layers that are employed in this architecture.

3.4 Model compile and train: We utilize the Adam optimizer to compile the model. We take batch size=100 and Epochs=10 for training. We divided the dataset into a subset of training and testing data (85:15).

3.5 Model assessment: Precision, accuracy, recall and ROC curve were among the evaluation measures used to check the effectiveness of the suggested model.

Dataset

In our work, we utilized Kaggle's standard dataset. The dataset can be downloaded from https://github.com/chandrikadeb7/Face-Mask-Detection/tree/master/dataset

Results and Discussion

The proposed new CNN model is evaluated with different standard matrices like recall, precision, accuracy etc. Anefficient method for comprehending the behavior of the suggested models during their learning on a

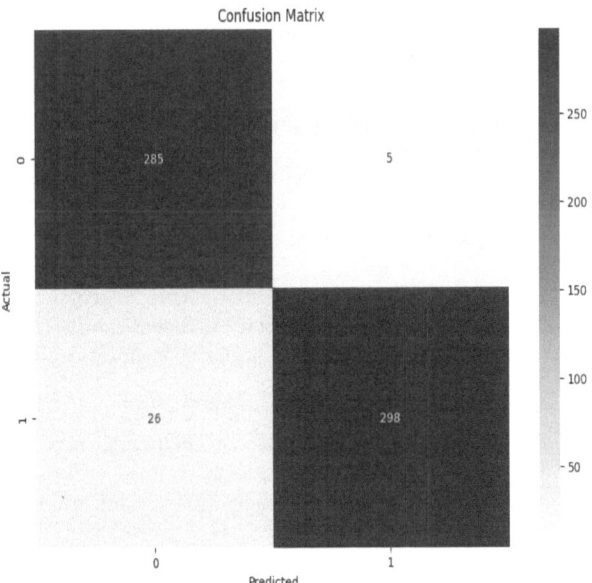

Figure 4.4 Confusion matrix
Source: Author

Table 4.1 The suggested CNN model's performance.

Precision	Recall	Specificity	F1-Score	Accuracy
0.9835	0.9198	0.9828	0.9506	0.9495

Source: Author

particular dataset isto assess the training and validation datasets at each epoch and graph the outcomes.

The accuracy of the suggested CNN model's validation and training is displayed in Figure 4.2. The new CNN model'svalidation and training losses are displayed in Figure 4.3. Model's predictions for the actual outcomes are shown in Figure 4.4.

The values of the various evaluation matrices of the new CNN model are displayed in Table 4.1.

Conclusion

In contrast to other traditional techniques, the deep neural network-based classification system can be thought of as the most accurate method. The proposed model achieves good results in identifying the presence of a face mask in masked or non-masked images. Salient features are extracted, and weak features are ignored using the CNN-integrated activation function called ReLU in the suggested method, which leads to the resolution of sample noise. The suggested model identifies the face images with accuracy rates of 94.95% over 10 epochs.

References

[1] Abbas, Q., Albalawi, T. S., Perumal, G., & Celebi, M. E. (2023). Automatic face recognition system using deep convolutional mixer architecture and AdaBoost classifier. *Applied Sciences*, 13(17), 9880.

[2] Abdulmunem, A. A., Al-Shakarchy, N. D., & Safoq, M. S. (2023). Deep learning based masked face recognition in the era of the COVID-19 pandemic. *International Journal of Electrical and Computer Engineering*, 13(2), 1550.

[3] Agarwal, C., Kaur, I., & Yadav, S. (2022). Hybrid CNN-SVM model for face mask detector to protect from COVID-19. In Artificial Intelligence on Medical Data: Proceedings of International Symposium, ISC-MM 2021. Springer Nature, Singapore.

[4] Allam, Z., & Jones, D. S. (2020). On the coronavirus (COVID-19) outbreak and the smart city network: universal data sharing standards coupled with artificial intelligence (AI) to benefit urban health monitoring and management. *Healthcare*, 8(1), 46.

[5] Chavda, A., Dsouza, J., Badgujar, S., & Damani, A. (2021). Multi-stage CNN architecture for face mask detection. In 2021 6th International Conference for Convergence in Technology (i2ct), (pp. 1–8). IEEE.

[6] Chong, W. J. L., Chong, S. C., & Ong, T. S. (2023). Masked face recognition using histogram-based recurrent neural network. *Journal of Imaging*, 9(2), 38.

[7] Gao, F., Jin, Y., Ge, Y., Lu, S., & Zhang, Y. (2024). Occluded person re-identification based on feature fusion and sparse reconstruction. *Multimedia Tools and Applications*, 83(5), 15061-15078.

[8] Ge, Y., Liu, H., Du, J., Li, Z., & Wei, Y. (2023). Masked face recognition with convolutional visual self-attention network. *Neurocomputing*, 518, 496–506.

[9] Hussain, S., Yu, Y., Ayoub, M., Khan, A., Rehman, R., Wahid, J. A., et al. (2021). IoT and deep learning-based approach for rapid screening and face mask detection for infection spread control of COVID-19. *Applied Sciences*, 11(8), 3495.

[10] Joshi, A. S., Joshi, S. S., Kanahasabai, G., Kapil, R., & Gupta, S. (2020). Deep learning framework to detect face masks from video footage. In 2020 12th International Conference on Computational Intelligence and Communication Networks (CICN). (pp. 435–440). IEEE.

[11] Ketab, F., Russel, N. S., Selvaraj, A., & Buhari, S. M. (2023). Parallel deep learning architecture with customized and learnable filters for low-resolution face recognition. *The Visual Computer*, 39(12), 6699–6710.

[12] Khan, M. J., Khan, M. J., Siddiqui, A. M., & Khurshid, K. (2022). An automated and efficient convolutional architecture for disguise-invariant face recognition using noise-based data augmentation and deep transfer learning. *The Visual Computer*, 38, 509–523.

[13] Kumar, A. (2023). A cascaded deep-learning-based model for face mask detection. *Data Technologies and Applications*, 57(1), 84–107.

[14] Kumar, S., Singh, S. K., & Peer, P. (2023). Occluded thermal face recognition using BoCNN and radial derivative Gaussian feature descriptor. *Image and Vision Computing*, 132, 104646.

[15] Mamieva, D., Abdusalomov, A. B., Mukhiddinov, M., & Whangbo, T. K. (2023). Improved face detection method via learning small faces on hard images based on a deep learning approach. *Sensors*, 23(1), 502.

[16] Mosayyebi, F., Seyedarabi, H., & Afrouzian, R. (2024). Gender recognition in masked facial images using EfficientNet and transfer learning approach. *International Journal of Information Technology*, 16(4), 2693–2703.

[17] Naser, O., Ahmad, S., Samsudin, K., Shafie, S., & Zamri, N. (2023). Facial recognition for partially occluded faces. *Indonesian Journal of Electrical Engineering and Computer Science*, 30(3), 1846–1855.

[18] Shukla, R. K., & Tiwari, A. K. (2023). Masked face recognition using mobilenet v2 with transfer learning. *Computer Systems Science and Engineering*, 45(1).

[19] Yu, J., & Zhang, W. (2021). Face mask wearing detection algorithm based on improved YOLO-v4. *Sensors*, 21(9), 3263.

[20] Zhao, C., Qin, Y., & Zhang, B. (2023). Adversarially learning occlusions by backpropagation for face recognition. *Sensors*, 23(20), 8559.

5 Enhancing 5G network management through association rule mining: Uncovering network traffic patterns

Sushanta Meher[1,a], Bijayini Mohanty[2,b], Bharat Jyoti Ranjan Sahu[3,c], Santilata Champati[4,d], Swadhin Kumar Barisal[3,e] and Shatarupa Dash[3,f]

[1]Department of CSE, CUTM, SoET, Bhubaneswar, Odisha, India

[2]Department of CDS, ITER, SOA Deemed to be University Bhubaneswar, Odisha, India

[3]Department of CSE, ITER, SOA Deemed to be University Bhubaneswar, Odisha, India

[4]Department of Mathematics, ITER, FET, SOA Deemed to be University Bhubaneswar, Odisha, India

Abstract

The 5G wireless networks has brought about a significant increase in data complexity, necessitating advanced techniques for effective network management and optimization. This research focuses on applying association rule mining (ARM) to uncover hidden patterns and relationships within the vast amounts of network traffic data generated in 5G systems. By enhancing ARM methodologies, the study aims to address critical challenges in 5G operations, including efficient resource allocation, quality of service (QoS) management, and anomaly detection. The proposed research is to develop scalable and computationally efficient of processing high- dimensional, real-time data in dynamic wireless environments. These advanced ARM techniques are expected to provide actionable insights that enhance network performance, reduce latency, and support intelligent, data-driven decision-making in 5G and future wireless networks.

Keywords: 5G wireless networks, anomaly detection, association rule mining, quality of service (QoS) management

Introduction

The rapid rollout of 5G wireless networks has revolutionized the telecommunications landscape, enabling unprecedented levels of connectivity and data transfer. This evolution has significantly increased with the complexity of network traffic data, resulting in challenges related to efficient management and optimization. As the volume of data generated by 5G systems, the traditional network management techniques struggle to handle the large scale data with dynamic information [1]. In this manner, advanced analytical approaches are essential to uncover hidden patterns and relationships within this vast dataset. Association rule mining (ARM) has emerged as a promising technique for analyzing large volumes of transactional data, allowing for the discovery of meaningful patterns that can inform network management decisions. By leveraging ARM methodologies, this research aims to tackle critical operational challenges, including efficient resource allocation and quality of service (QoS) management. ARM will facilitate the real-time processing of high-dimensional data, enabling network operators to make informed decisions that improve overall network performance and user experience. Recent technical reports (TRs) have highlighted the importance of leveraging data analytics to optimize network performance [2]. This research not only address current challenges but also lay the groundwork for future advancements in wireless network management. By integrating ARM into the operational framework of 5G networks, this study generates actionable insights that enhance network performance, and support intelligent, data-driven information for decision- making. These outcomes are expected to contribute significantly to the 5G wireless technologies and enhance their operational efficiencies, innovations that can be leveraged in upcoming generations of wireless networks.

Related Work

The 5G networks has introduced unprecedented complexity in network management, leading researchers to explore advanced techniques such as machine learning and data mining for optimization. Periyathambi and Ravi [4] took a significant step forward by applying machine learning to enhance spectral efficiency and underscoring the critical need for intelligent resource allocation strategies in managing high data traffic [1]. Similarly, Gao and Chen [1]

[a]sushant.meher9@gmail.com, [b]bijayinimohanty7@gmail.com, [c]bharatjyotisahu@soa.ac.in, [d]santilatachamapti@soa.ac.in, [e]swadhinbarisal@soa.ac.in, [f]shatarupadash@soa.ac.in

DOI: 10.1201/9781003658221-5

demonstrated how ARM could be effectively utilized to uncover hidden patterns within 5G network data, offering deeper insights for improving network decision-making processes [2]. Rahman et al. [5] focused on anomaly detection in 5G networks, using mathematical models to identify traffic deviations and mitigate potential threats [3]. Meanwhile, Yuan et al. [7] emphasized the role of AI in optimizing QoS by empowering network association processes with data-driven decision-making [4]. The scalability challenges inherent in processing vast 5G datasets were tackled by Zhao et al. [8], who introduced an improved ARM capable of handling large data volumes [5]. Huang et al. [2] added to this by proposing a service-oriented architecture for 5G-enabled IoT, aiming to seamlessly integrate diverse services and devices into the network framework [6]. Wang et al. [6] contributed by applying frequent itemset mining in the IoT era, focusing on efficient data processing [7]. Finally, Liu et al. [3] provided a foundational outlook on user association in 5G networks, highlighting the pivotal role of intelligent algorithms in navigating the complexities of [8]. These studies collectively emphasize the growing importance of ARM techniques to overcome the challenges in 5G network management.

Research objectives

The primary objectives of this research are:

- To enhance ARM methodologies for the analysis of network traffic data in 5G systems.
- To provide actionable insights that support resource allocation, QoS, and anomaly detection.

Proposed Model

Let D represent the dataset containing all network traffic records. Each record $t \in D$ is a transaction that includes various attributes such as signal strength, bandwidth usage, latency, etc. Let, $I = \{I_1, I_2, I_3, I_4, \ldots \ldots I_n\}$ & $D = \{t_1, t_2, t_3, \ldots \ldots t_m\}$ be a set of binary items representing different network conditions or states (e.g., low signal strength, high latency, high bandwidth usage). To ensure the ARM can handle large-scale and high-dimensional data, techniques such as: Incremental ARM for real-time updates: By leveraging ARM for both anomaly detection and resource optimization, the proposed model ensures efficient network management by adapting dynamically to changing conditions. Figure 5.1 illustrates the complete work process.

Methodology

Optimizing 5G network resource allocation and anomaly detection using ARM begins with the collection of

network traffic data. In the data set each transaction represents network attributes such as signal strength, bandwidth usage, and latency. These attributes are converted into binary items, allowing the identification of frequent item sets and generation of association rules that capture relationships between different network conditions. The strength of these rules is evaluated using support, confidence, and lift metrics, with a lift value greater than 1 indicating strong positive associations between conditions. To ensure the scalability, ARM is employed by allowing. The rule set to be dynamically updated with new traffic data in real time Anomalies are detected by identifying transactions that deviate from normal behavior based on established rules. The ARM results are then integrated into an optimization model that minimizes network latency while maintaining quality of service (QoS) and resource utilization constraints. Parameters such as bandwidth allocation and signal strength thresholds are adjusted accordingly.

Finally, the model continuously adapts to changing conditions, ensuring efficient, real-time management of 5G network resources.

Result and Discussion

The proposed ARM for both resource optimization and anomaly detection in 5G networks. It demonstrates significant improvements in several key performance metrics.

The following sections discuss the outcomes of various experiments, supported by graphical representations. Figure 5.2 shows support values for the different items represented. The support values were calculated based on the frequency of network conditions observed in the dataset. This step was critical in identifying frequently occurring patterns that help higher support values indicate more commonly

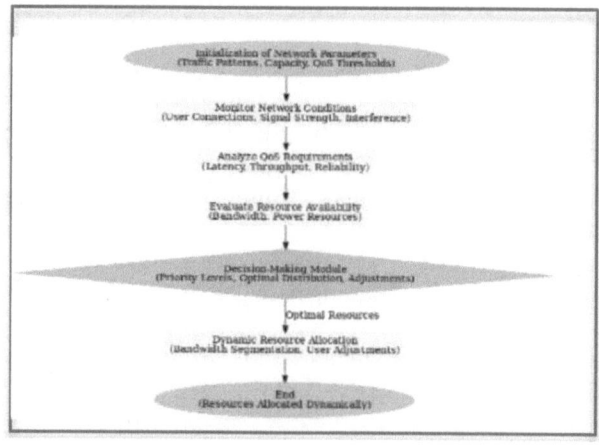

Figure 5.1 Workflow of the proposed wok
Source: Author

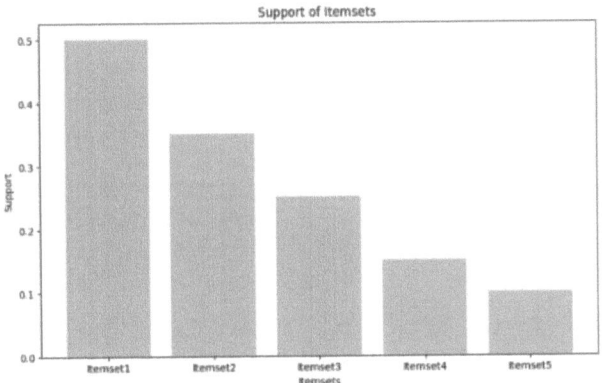

Figure 5.2 Representation supports item sets
Source: Author

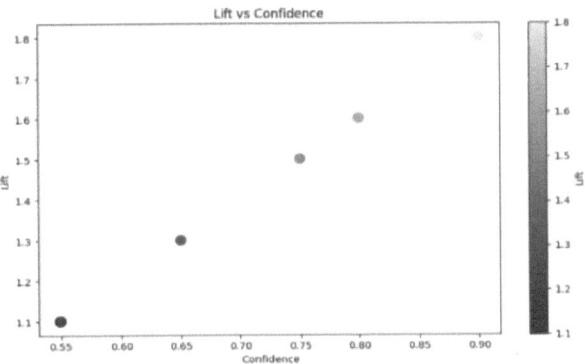

Figure 5.3 Representation lift vs. confidence
Source: Author

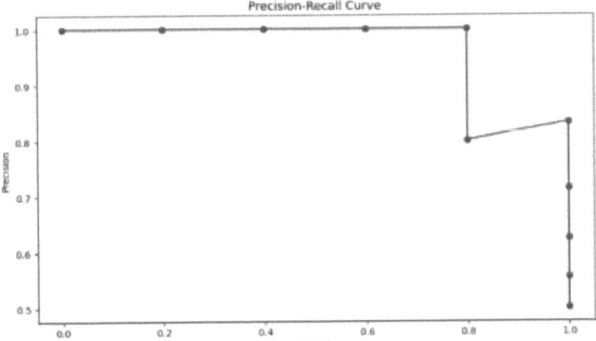

Figure 5.4 Precision-recall curve for anomaly detection
Source: Author

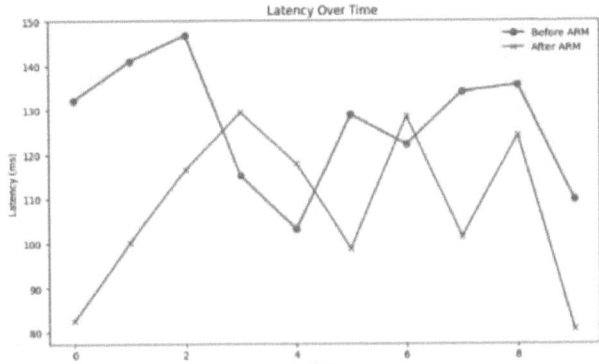

Figure 5.5 Representation of latency over the time
Source: Author

encountered network conditions, which were subsequently used for rule generation.

Figure 5.3 shows the relationship between lift and confidence for the association generated rules. As expected, rules with higher confidence tend to have higher lift values. This indicates that the strong associations discovered are not merely coincidental but have a meaningful correlation between the network conditions. Lift values greater than 1 confirm the positive association, providing insights into which conditions influence each other in the 5G network. Figure 5.3 depicts the reduction in latency over time after applying ARM- based optimizations. Prior to optimization, the latency fluctuated significantly due to suboptimal resource allocation. After implementing the ARM-driven model, latency became more stable and consistently lower, demonstrating the system's ability to dynamically adapt to changing conditions and improve response times. For evaluating the anomaly detection module, the ROC curve is plotted in Figure 5.4, illustrating the trade-off between true positive rate (TPR) and false positive rate (FPR).

A higher area under the curve (AUC) indicates that the model is proficient in distinguishing

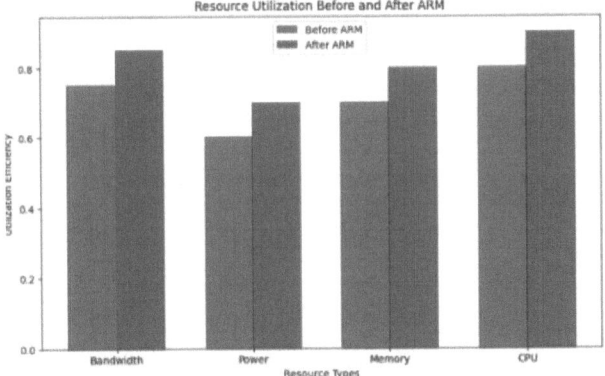

Figure 5.6 Resource utilization before and after ARM
Source: Author

between normal and anomalous traffic patterns. Additionally, Figure 5.5 shows the precision-recall curve highlighting the model's ability to maintain high precision while improving recall, which is crucial in minimizing false alarms while catching true anomalies.

The impact of ARM-based optimizations on resource utilization is displayed in Figure 5.6. The figure clearly shows that after applying the ARM algorithm, resource utilization becomes more efficient.

Conclusion

The efficacy of applying association rule mining (ARM) to enhance network management and optimization in 5G wireless systems. By leveraging advanced ARM techniques, the study successfully identifies hidden patterns and relationships within complex network traffic data, addressing critical challenges such as resource allocation, quality of service (QoS) management, and anomaly detection. The findings underscore the potential of ARM methodologies to transform network operations into 5G and future wireless environments, ultimately leading to more resilient and efficient communication systems. Future research could focus on integrating machine learning techniques with ARM to enhance predictive capabilities and adaptability.

References

[1]   Gao, J., & Chen, Z. (2023). Application of association rule mining algorithm based on 5G technology in information management system. *Scalable Computing: Practice and Experience*, 24(2), 139–149.

[2]   Huang, M., Liu, A., Xiong, N. N., Wang, T., & Vasilakos, A. V. (2020). An effective service-oriented networking management architecture for 5G-enabled internet of things. *Computer Networks*, 173, 107208.

[3]   Liu, D., Wang, L., Chen, Y., Elkashlan, M., Wong, K. K., Schober, R., et al. (2016). User association in 5G networks: a survey and an outlook. *IEEE Communications Surveys and Tutorials*, 18(2), 1018–1044.

[4]   Periyathambi, P., & Ravi, G. (2024). Optimizing resource allocation in 5G wireless networks for enhanced spectral efficiency and energy conservation using machine learning methods. *Signal, Image and Video Processing*, 18(6), 4961–4977.

[5]   Rahman, A. U., Mahmud, M., Iqbal, T., Saraireh, L., Kholidy, H., Gollapalli, M., et al. (2022). Network anomaly detection in 5G networks. *Mathematical Modelling of Engineering Problems*, 9(2).

[6]   Wang, Z., Liang, W., Zhang, Y., Wang, J., Tao, J., Chen, C., et al. (2019). Data mining in IoT era: a method based on improved frequent items mining algorithm. In 2019 5th International Conference on Big Data and Information Analytics (BigDIA), (pp. 120–125). IEEE.

[7]   Yuan, X., Yao, H., Wang, J., Mai, T., & Guizani, M. (2021). Artificial intelligence empowered QoS-oriented network association for next-generation mobile networks. *IEEE Transactions on Cognitive Communications and Networking*, 7(3), 856–870.

[8]   Zhao, Z., Jian, Z., Gaba, G. S., Alroobaea, R., Masud, M., & Rubaiee, S. (2021). An improved association rule mining algorithm for large data. *Journal of Intelligent Systems*, 30(1), 750–762.

6 A binarization method for Odia palm leaf manuscripts by using the Odia language accelerator model

Aloka Natha[1,a], Bichitrananda Behera[1,b] and Nishanta Ranjan Nanda[2,c]

[1]Department of Computer Science and Engineering, C. V. Raman Global University, Bhubaneswar, Odisha, India

[2]Tata Consultancy Services, Bhubaneswar, Odisha, India

Abstract

A binarization technique is suggested for improving the images of palm leaf manuscripts. It can efficiently eliminate unnecessary lines from the photos of palm leaf manuscripts and perform image binarization. Using machine learning, an enhanced background elimination technique is first introduced to address the issue of various backdrop colors in a palm leaf. Following that, the image is binarized using the AI-based large language model (LLM) technique. However, the result photographs produced by the above include a lot of noise, therefore noise is eliminated using a noise reduction technique. Lastly, a connected domain-based OCR technique is explained for removing unnecessary lines with impressions and Odia letters on palm leaves. Experimental findings from Python programming demonstrate how successful the suggested approach is.

Keywords: AI, large language model, OCR, OLA

Introduction

Since ancient times, India has maintained a magnificent heritage of transmitting knowledge orally and in writing. Also, it is thought to be one of the sources of conventional wisdom worldwide. Knowledge was communicated using a wide range of writing surfaces, from copper plates to cave walls and from different types of leaves to tree bark. Because palm trees are found all over the Indian subcontinent, palm leaves were the most widely used writing material of all the leafy writing materials. Transforming handwritten materials into electronically readable data is a possible case for the digital world. Numerous discoveries made in the present day can be traced back to the work of our ancestors' scholars. Knowledge from ancient times was originally transmitted orally, and writing systems weren't created until much later.

As we know, these manuscripts made from palm leaves retain our old cultural heritage and are considered to be important sources of knowledge for reconstructing a country's history and culture. However, the preservation of these manuscripts is fraught with difficulty because, as organic materials, palm leaves [17] are more prone to deterioration than other materials. We refer to these texts as the written messengers of our history. Employing strains, yellowing, low-intensity changes, random sounds, fading, and degradation, antique documents like palm leaf manuscripts and handwritten history records become an enormous task.

Manuscript alludes [13] to an old, manually written document. Originating from the Latin phrase "manu scriptum," which means "written by hand," comes the word "manuscript." Our explanation of what a manuscript is appears to be related to the Antiquities and Art Treasures Act of 1972 (India), which attempts to control the trade in antiquities and art treasures as well as the preservation and protection of items with artistic, historical, and scientific significance. This Act does, in fact, define a manuscript as a document of substantial scientific, historical, literary, or artistic value that is at least 75 years old in order to be considered a "antiquity." This period guarantees the legal protection of objects of significant cultural significance. The Act covers a variety of antiquities that constitute a part of India's legacy, such as sculptures, coins, paintings, and manuscripts. Its goal is to stop these cultural objects from being smuggled or traded illegally.

A manuscript may be a written composition on paper, cloth, metal, bark of a tree, palm leaf, or other material. In Bharat manuscripts area units written in varied languages and scripts. The foremost manuscripts area unit found in the Sanskrit language. The Sanskrit language assortment of Sarasvati Mahal is that the largest manuscripts of India that contain the foremost works of Sanskrit literature starting with the Vedas. These manuscripts were written on palm leaf and paper.

In the image below, many aspects of the manuscript recognition process are illustrated.

---

[a]aloknath23@gmail.com, [b]bichitrananda.b@cgu-odisha.ac.in, [c]nishantaranjan.nanda@tcs.com

DOI: 10.1201/9781003658221-6

By taking high-quality photos, the digitization process [2, 15–17] would convert the manuscripts from their current physical palm leaf form to a digital image. A palm leaf's entire text will be accessible as photos, but it won't be feasible to perform a straightforward search on these images. For instance, if you are reading a palm leaf manuscript on a horoscope and would like to look up your horoscope prediction, you will need to manually browse through each leaf image, which will be a taxing mental workout. As a result, the manuscripts must be accessible in textual form in addition to visual form.

As far as we are aware, there aren't many studies in the literature that use machine learning techniques for Odia palm leaf manuscript binarization. In this paper, we have employed an important machine learning methods and a model for Odia palm leaf manuscript binarization [13–15], and we have examined their accuracy. Therefore, the following is how our research paper's contribution is presented.

1. To scan manuscripts on palm leaves written in Odia.
2. To evaluate how accurate various models and algorithms are in digitizing Odia palm leaf manuscripts.
3. Identify a research gap and suggest a model to increase precision.

Literature Review

An overview of earlier study techniques and findings pertaining to the binarization of Odia palm leaf manuscripts is given in this section [4, 12].

My current research focusses on a number of well-known low-resource languages, including Odia. Validating a natural language processing model's capacity to extrapolate results from data and languages other than those used for training and testing is preferable, nonetheless. Models that may be made to function quite well across languages are described by the concept of language independence, which is comparable to this one.

Odia palm leaf manuscript binarization covers the areas of preprocessing, noise removal, segmentation, digitization, etc.

Devanathan et al. [3] emphasize that the earliest scripts were written on difficult-to-preserve materials like Talapatra and Bhurjapatra. Many texts in the vast corpus of Ayurvedic literature are either unavailable or only partially available. Later manuscripts only mention a few phrases or references from numerous similar passages. Sadly, a great deal of Ayurvedic literature that has not yet been fully studied is probably

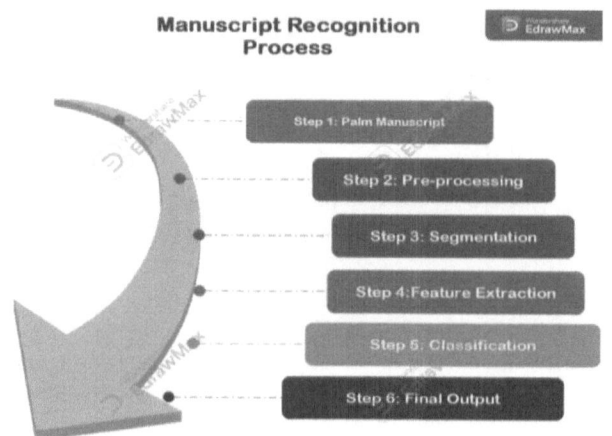

Figure 6.1 The steps of manuscript recognition
Source: Author

contained in palm-leaf manuscripts [9, 11] that are deteriorating or destroyed. As a result, a great deal of important and distinctive information found in these manuscripts is being lost. Within this study, the author also addresses the preservation and protection of manuscripts in their natural state.

Nicholas et al. [1] explain a different method for handling manuscripts made of palm leaves, which are vulnerable to deterioration from chemicals, physical factors, and biological processes [15, 17]. Because of its high cellulose content and suitability as filler material, the mulberry plant's inner white bark offers better material permanence. According to the tests, using white bark instead of traditional building materials has produced better outcomes, resulting in an infill that is both aesthetically pleasing and structurally sound.

Majumdar et al. [10] offer information about India's documented knowledge, literary and cultural traditions, and history. According to him, the identical vocabulary employed by various Indian languages gives the cultural depictions in Indian literature a special significance. Ancient manuscripts have been unearthed on a variety of materials, including copper plates, bamboo, wood, cotton, silk, and palm leaves. It is admirable that the Indian Government established the National Mission for Manuscripts in order to preserve these culturally significant works [11]. The mission's ultimate goal, according to the author, is to find such a rich legacy, register it wherever it is possible, conserve it, and supply surrogates for global distribution.

Razdan et al. [15] analyzed the various 3D methods for digital library analysis of handwritten manuscripts. Concluded that these technologies offer strong and novel analytical skills for more thorough comprehension and examination of intricate and highly changeable handwritten materials. These technologies not only help with character and letter deciphering but also enhance the comparison and analysis of different manuscripts and help identify writers across documents.

Proposed Methodology

Finding the unprocessed Odia palm leaf manuscripts and convert the same into the binarized format is the aim of this work. For this suggested process, we must create a collection of Odia palm leaf manuscripts using a digital camera. A number of gathered images of unprocessed palm leaf manuscripts are then utilized for additional analysis and processing, which includes noise reduction, character detection, and conversion of the entire palm leaf manuscript into a neatly arranged binarized format [8].

The above framework for the OLA model in which manual annotation and digital camera collection create the PLM images. Following that, it was split up into the test, validation, and training sets. Next, different loss functions and pre-processing techniques are contrasted to determine the most effective way to construct damage-free outputs. Lastly, the prediction outcomes in the third stage are optimized using test-time enhancement techniques and various post-processing techniques.

The following problems are going to be solved by using the OLA Model for Odia palm leaf manuscripts binarization.

1. Our approach will solve the problem of reading and analyzing old palm leaves and rock inscriptions with technology.
2. Our system will automatically detect the letter and words and provide analytics and frequency of each letter analyzing the entire palm leaf with letters and pictures.
3. No human interference is needed; the system can track and detect the letter based on its AI and machine learning-based judgment models.
4. Mutilated, wet, water spill, termite attack, and wasted leaf are tracked and auto-corrected to provide letters as per human readability.

This is a real-life Smart AI-based PLM system here uses an AI&ML-based model and image processing; it acts like a correction system without any real human being to read and understand the Odia letter. It can do much more than image processing, our system intelligently trains itself to adjust to mutilated, water spills, and termite destruction areas carefully and collects data from images, applies AI based model for processing, applies its large language models, lets the user understand, generate meaningful Odia words and sentences later can be translated to English also.

The key aspects of the OLA model are defined as follows:

1. Noise removal
2. Feature extraction
3. Segmentation technique
4. Digitization technique
5. Language translator or language conversion

The architecture of OLA Model mainly focuses on:
Noise removal focuses on a) Mutilated images b) wet spots c) damage due to water spills d) termite attack e) wasted palm leaf etc.

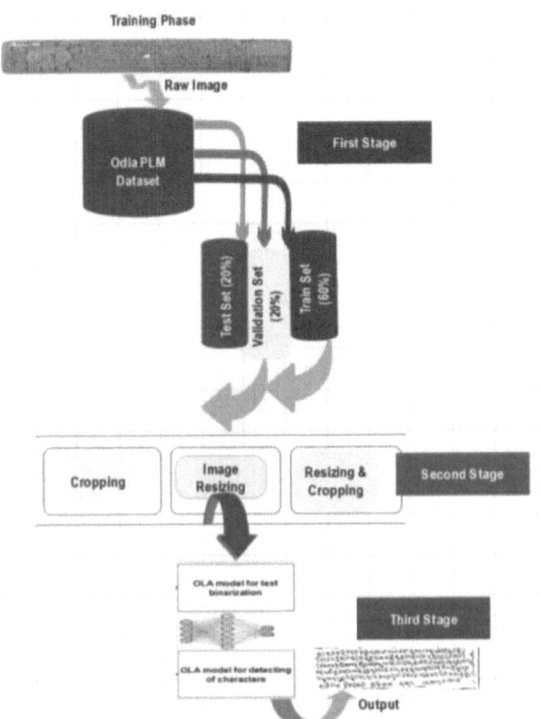

Figure 6.2 The pictorial description of different stages
Source: Author

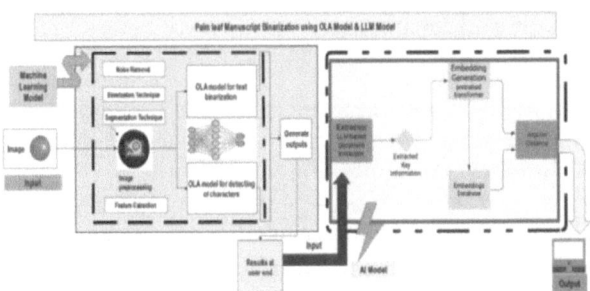

Figure 6.3 The architecture of the OLA model
Source: Author

Figure 6.4 Odia palm leaf as input
Source: Author

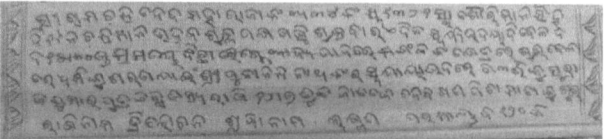

Figure 6.6 Odia palm leaf as input
Source: Author

Figure 6.5 Odia palm leaf with meaningful words as output
Source: Author

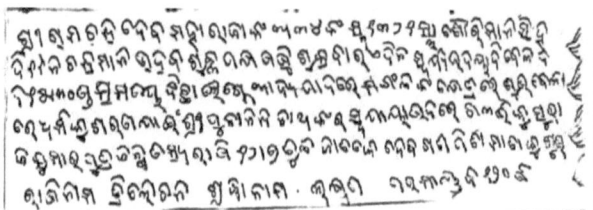

Figure 6.7 Odia palm leaf with meaningful words as output
Source: Author

Table 6.1 The rate of prediction for different machine learning techniques and the associated forecast time.

Method of classification	Rate of prediction (%)	Rate of incorrect predictions (%)	Time prediction (minutes)
SVM	84.46	15.54	2.06
k-NN	72.21	27.29	3.04
CNN	95.21	4.79	1.64
OLA	96.23	3.77	1.45

Source: Author

It also solves the problem of reading and analyzing old palm-leaf manuscripts.

The AI mechanism extracts the rock inscriptions with technology implemented by our ML mechanisms.

It detects the letters and words and provides analytics and frequency of each letter analyzing the entire palm leaf with letters and pictures.

The entire model is also controlled by a robotics mechanism so that its entire mechanism is automated.

It can detect the letters in Odia and convert the same into meaningful words in different languages.

Result and Discussion

Experimental setup

We implemented machine learning methods using a variety of Python modules. Machine learning techniques are implemented using the Scikit-learn Python module. Google Integration handles end-to-end testing. The images below demonstrate that we utilized unprocessed Odia palm leaf manuscripts as an input image, and then our OLA Model applied machine learning algorithms, the converted digitalized Odia palm leaf manuscripts as an output image.

Machine learning computer techniques are implemented using Python's Scikit-Learn [19] packages. All of the testing is done on the Google Collaboratory cloud service, which features a GPU with an integrated Tesla K20, 2496 CUDA cores, 16GB of RAM, and a 512GB SSD. The studies described in this paper were conducted on a laptop featuring an Intel Core i5-11320H 11th generation processor running Windows 11, 64-bit operating system and x64-based processor. This CPU's clock speed is 3.19 GHz.

The performance of various machine learning methods, including CNN, k-NN, and support vector machine (SVM), has been compared with the proposed OLA architecture in terms of palm-leaf manuscript character recognition. The rate of prediction, method of classification, rate of incorrect predictions and time prediction for identifying the character found in the Odia palm leaf manuscripts by our proposed OLA model using different machine learning techniques are displayed in Table 6.1. The results indicate that the created OLA architecture has a 96.23% rate of prediction, a 3.77% rate of incorrect predictions, and a 1.45 minutes single-character prediction time. When compared to other machine learning techniques now in use, the results achieved are far better.

Conclusion

In this work, we have put forth an algorithm for the identification and recognition of ancient Odia Palm Leaf Manuscripts from several dynasties. This effort creates a palm-leaf manuscript binarization. Additionally, an OLA architecture is designed for this research, and its performance was assessed using the binarization method that was created. The same created raw palm leaf images have also been applied to other machine learning techniques to compare the prediction rate.

In printed form, the cursive Odia script is easily readable. But because of their writing technique, handwritten Odia characters appear alike. The development of a recognition system is so made more difficult. With a 96.23% accuracy rate, this paper presents one method for identifying handwritten Odia Palm Leaf Manuscripts. Expanding the sample size, implementing novel feature extraction and classification strategies, and so on. Also, it allows the work to be further developed. Additionally, it may be argued that creating a standard data set for the Odia script is a significant first step toward further research and work.

References

[1] Azimi, H., & Nazi, A. (2021). The status of cataloging manuscripts in large libraries in Iran. *Library Collections, Acquisitions, and Technical Services*, 35(4), 106–117.

[2] Challa, N. P., & Mehta, R. V. K. (2017). Applications of image processing techniques on palm-leaf manuscripts—a survey. *Helix: The Scientific Explorer*, 7(5), 2013–2017.

[3] Chamnongsri, N., Manmart, L., & Wuwongse, V. (2020). Implementation and evaluation of palm leaf manuscript metadata schema (PLMM). In Proceedings of the 9th ACM/IEEE-CS Joint Conference on Digital Libraries, (pp. 367–368).

[4] Comment, J. (2021). Archiving cultural heritage and history through digitization: case studies from russia (comintern archives) and Albania (national archives). In Proceeding of the Conference on Knowledge Creation, Preservation, Access, and Management, (pp. 387–391).

[5] Das, M., Panda, M., & Sahoo, S. (2024). Enhancing the Odia handwritten character and numeral recognition system's performance with an ensemble of deep neural networks. *International Journal of Advanced Computer Science & Applications*, 15(2).

[6] Das, M., Panda, M., & Dash, S. (2022). Enhancing the power of CNN using data augmentation techniques for Odia handwritten character recognition. *Advances in Multimedia*, 2022(1), 6180701.

[7] Devanathan, R. (2019). Conservation of manuscripts–the natural way. *International Journal of Current Pharmaceutical Review and Research*, 3(4), 99–104.

[8] El Makhfi, N., El Bannay, O., Benslimane, R., & Rais, N. (2021). System of indexing, annotation, and search in the old Arabic manuscripts. In 2021 Colloquium in Information Science and Technology, (pp. 9–9). IEEE.

[9] Harinarayana, N. S., & Gangdharesha, S. (2021). Metadata standards available for cataloguing indian manuscripts: comparative study. In Proceedings of the Conference on Recent Advances in Information Technology (READIT 2021), (pp. 259–270).

[10] Majumdar, S. (2019). Preservation and conservation of literary heritage: a case study of India. *The International Information and Library Review*, 37(3), 179–187.

[11] Nichols, K. (2021). An alternative approach to loss compensation in palm leaf manuscripts. *The Paper Conservator*, 28(1), 105–109.

[12] Rani, N. S. (2024). A modified deep semantic binarization network for degradation removal in palm leaf manuscripts. *Multimedia Tools and Applications*, 83:62937–62969, 1–33.

[13] Sasikala, C. (2020). Digitizing the historical and valuable collections: a case study of Dr.V. S. K. Library, Andhra University. In Proceeding of the Conference on the Knowledge Creation, Preservation, Access and Management, (Vol. 2, pp. 1058–1060).

[14] Shafi, S. M. (2020). Digital archiving of medieval manuscripts in india: problem and perspectives. In Proceeding of the Conference on the Knowledge Creation, Preservation, Access and Management, (Vol. 2, pp. 1061–1062).

[15] Sharma, U., Bansal, R., Sarangi, P. K., Gupta, D., Rani, S., Khan, F., et al. (2023). Handwritten Odia digit recognition using learning systems: a comparison of neural networks and support vector machine models. *ACM Transactions on Asian and Low-Resource Language Information Processing*. 2375-4699/2023/1.

[16] Takagi, N., Chudo, Y., Maeda, R., & Saito, Y. (2020). Preservation cooperation in Nepal: from training to conservation and digitization of rolled palm leaf manuscripts. In Preservation and Conservation in Asia in Pre-Conference of WLIC 2020, (pp. 1–9).

[17] Tensmeyer, C., & Martinez, T. (2020). Historical document image binarization: a review. *SN Computer Science*, 1(3), 173.

[18] Torre, M. (2019). Cataloging and classification of illuminated manuscripts. *Library Student Journal*, 2, 007355.

[19] Tripathy, S. S., & Behera, B. (2023). Performance evaluation of machine learning algorithms for intrusion detection system. arXiv preprint arXiv:2310.00594.

7 FRMM-LSTM multimodal fusion based caption recurrent network for vision analysis

Harapriya Kar[1,a], P. Viswanathan[1,b], Bajra Panjar Mishra[2,c] and K. S. Kuppusamy[3,d]

[1]Computer Science and Engineering, Vellore Institute of Technology, Tamilnadu, India

[2]Electronics and Telecommunication Engineering, DRIEMS University, Odisha, India

[3]Computer Science and Engineering, Pondicherry University, Puducherry, India

Abstract

People with visual impairments worldwide often face challenges in accessing information about different environments. Captions based on scene content, along with image descriptions, have been found to help address these difficulties. Combining natural language processing with computer vision plays a vital role in improving scene understanding. However, without proper captions, image descriptions can become inconsistent. To address this issue, this paper introduces a multiregional fusion-based FRMM-LSTM neural system that generates descriptive scene captions. The fully connected recurrent multimodal long short-term memory (LSTM) gates learn from each modality present in the image. The model operates in two distinct phases, one for image input and the other for language, followed by a shared fusion phase to learn the relationship between the two multimodal features. A fully connected LSTM layer processes the language component, while the fusion stages in FRMM are used for training. The language features are then mapped onto a shared vector space, pre-trained using FRMM and LSTM, to produce coherent captions. Additionally, Caption Crawler technology uses reverse image search to find captions from the Web and global repositories, making them accessible to users with screen readers. The interface for the fused model is implemented using the flask framework, a Python-based web development tool. This work aims primarily to assist visually impaired individuals by offering a region-specific approach to image captioning.

Keywords: CNN, FRMM, image description, long short-term memory, RNN, visual impairment

Introduction

Advancements in factual dialect modelling Wu et al. [8] and image recognition, as noted by Amirian et al. [6], have made this area a complex and crucial field of research. The task of generating captions from images, as discussed by Ma and Wang [3] and Wang et al. [10], offers numerous applications, ranging from providing support to visually impaired individuals to automating the labelling of millions of images uploaded to the Internet daily in a more cost-efficient manner. Natural language processing and computer vision, two core pillars of artificial Intelligence. The two fundamental approaches to picture captioning that are bottom-up and top-down. State-of-the-art models designed and incorporated in the model. Our method is based on the success of top-down image captioning models, as outlined by Wang et al. [10]. This approach leverages a deep convolutional neural network to create a vectorized representation of an image, which is processed through a long short-term memory (LSTM) network to generated descriptive captions. Image Captioning is a challenging task which faces overfitting while generating information. as it were having 160000 labelled illustrations of the data, from which any top-down Yu et al. [4] approach must learn (a) strong picture representation (b) a vigorous hidden-state LSTM based representation to capture picture semantics (c) dialect modelling Wu et al. [8] for grammatical factor plan of the reason of image description. An unrolled LSTM arrange for our FRMM-LSTM Xian and Tian [11]. All LSTMs share the same parameters where vectorized picture representation Han and Choi [7] is nourished into the organize, after uncommon begin of sentence token. The covered-up state created is at that point utilized by the LSTM, predict the caption for the given image. The training memorizes inputs and utilized the comparative sounding captions Hsieh et al. [2]. The illustration enhances, a picture of a man on a skateboard on a slope may get the same caption has a picture of a man on a skateboard on a table.

Addressing the issue requires a cross-modal embedding of the image, caption, and subject matter. The proposed model has achieved by applying state of the art methods in both image description retrieval and caption generation tasks, tested on the MSCOCO and Flickr8K datasets, as shown by Jamil et al. [1]. we introduce a domain-specific image caption tool,

[a]harapriya1999sa@gmail.com, [b]pviswanathan@vit.ac.in, [c]bajra@driems.ac.in, [d]kskuppuusamy.csc@pondiuni.edu.in

DOI: 10.1201/9781003658221-7

which represents a novel approach to captioning. By incorporating both visual and semantic features into account, model replaces standard terms in captions Hsieh et al. [2] with domain specific, resulting in a caption.

The original video caption recognition proposed by Yang [5] was achieved using a similarity degree equation based on line and column projection analysis. The test results indicate that this method can effectively segment video descriptions and improve caption generation efficiency as demonstrated by Kar et al. [9].

Other part of the paper as follows with Section 2 reviews the extant literature. Section 3 describes the sample and variables. Section 4 explains the research methodology. Section 5 discusses the empirical findings. Section-6 summarises the paper.

Proposed Model

The detection of text segment alignments with corresponding representative areas in the image shown in Figure 7.1. Its objectives and problem definition of the work show that, in addition to translation, various scoring metrics can be applied to address a wide range of language generation tasks using deep learning techniques. including tasks such as text generation, image captioning, summarization, speech recognition, and regional language captioning. The models distinct impact makes visual impaired community empowered which is as follows,

- The people with outwardly impeded frequently confront troubles in accessing pictures virtually locales while surfing.
- Text portrayal of a image varied from each other, creates contradiction for captioning.
- The expanding web-based data assets has moved from a basic level to dynamic level.

CNN is mostly developed with the premise that the input will be photos. CNN is made up of three separate layers, convolution, pooling, and fully connected layers. When these layers are stacked, they form the CNN architecture. CNNs consist of multiple layers of artificial neurons, which process weighted inputs to generate an output or activity value. The convolution

operation itself involves multiple pixel weights by a set factor, typically two, and summing the results. This process is key to CNN embedding, leading to higher-level feature recognition, such as objects in the image using Eq. 1.

$$\theta = arg_{\theta max} Y \sum \log(Y|i, \theta) \tag{1}$$

Where i denoted as input of the image, Y = Y1. . . . . .. Yn is the actual output sentence consisting of n words. θ shows the parameter of the model. p (Y |i, θ) denotes the probability ofgenerating sentence n from the input image i, and the parameter used as θ. The log likelihoodis used to evaluate the predicted rate of the caption of the image to understand which word ismapped well for the caption analysis using CNN model. For simplification of the sentence θ is removed using Eq. 2.

$$\log p(Y|i) = \sum_{t=1}^{T} log p(Y_t|Y_1, Y_2 \dots, Y_{t-1}, i) \tag{2}$$

The top-layer features F of the input image are extracted by the encoder as Eq. 3

$$F = CNNLSTM(i) \tag{3}$$

The probability of the model is processed using Eq. 4 with the FRMM-LSTM model to process image caption, which can be resembled as a unique model analysis for visual impaired person.

$$log p (Y_t | Y_1, Y_2 \dots, Y_{t-1}, i) \cdot F \tag{4}$$

The architecture of FRMM and LSTM is shown in Figure 7.2. and flow diagram shown in Figure 7.3.

The model leverages four key advantages: enhanced feature extraction, improved classification accuracy, end-to-end training capability, and effective generalization. This architecture processes images through a pipeline that includes caption generation with defined start/stop boundaries, text embedding, and speech

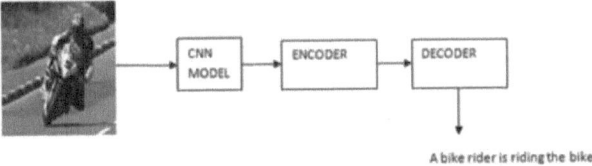

Figure 7.1 Architectural captioning model
Source: Author

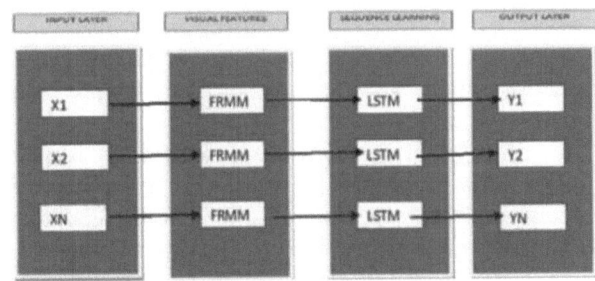

Figure 7.2 FRMM-LSTM architectural diagram
Source: Author

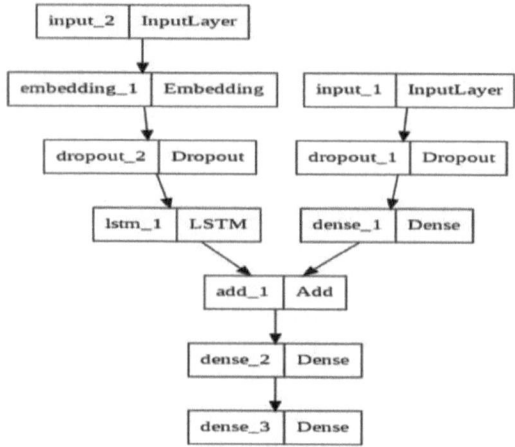

Figure 7.3 Flow diagram of FRMM model
Source: Author

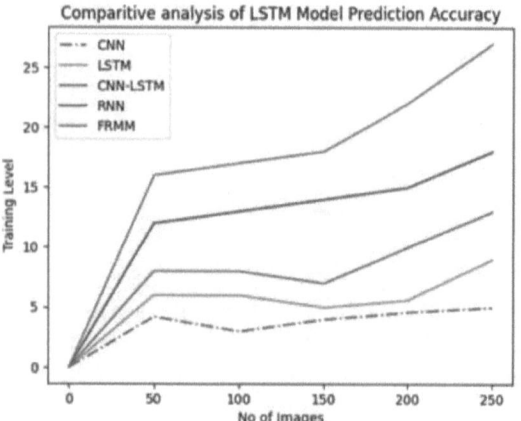

Figure 7.4 Comparative FRMM accuracy
Source: Author

framework uses the tensor stream library as a backend to design and prepare advanced neural systems.

$$N = \frac{\% \; ofngrams}{totalnumberofngrams} \qquad (5)$$

Conclusion

The FRMM model demonstrates superior accuracy in image caption generation, effectively addressing visual accessibility needs. Future enhancements will integrate RCNNs with CNNLSTM architectures, expanding into video descriptions and multi-regional language support. The model shows promise for scaling with larger datasets, aiming for human-like prediction accuracy. Key developments will focus on multimedia accessibility features, including language translation and audio conversion capabilities. By addressing deep learning challenges in image analysis and caption generation, this research provides a foundation for future advancements in assistive technology. Specific performance matrices will ensure continued improvement in accuracy and efficiency across diverse applications.

References

[1] Jamil, A., Saif-Ur-Rehman, Mahmood, K., Villar, M. G., Prola, T., Diez, , et al. (2024). Deep learning approaches for image captioning: opportunities, challenges and future potential. *IEEE Access*, 4, 1–27.

[2] Hsieh, HY., Huang, SA. & Leu, JS (2021). Implementing a real-time image captioning service for scene identification using embedded system. Multimed Tools Appl, 80, 12525–12537.

[3] Ma, M., & Wang, B. (2017). A grey relational analysis-based evaluation metric for image captioning and video captioning. In 2017 International Conference on Grey Systems and Intelligent Services (GSIS), (pp. 76–81). IEEE.

[4] Yu, N., Hu, X., Song, B., Yang, J., & Zhang, J. (2019). Topic-oriented image captioning based on order embedding. *IEEE Transactions on Image Processing*, 28(6), 2743–2754.

[5] Yang, Q. (2011). An improved algorithm of news video caption detection and recognition. In Proceedings of 2011 International Conference on Computer Science and Network Technology. 3, 1549–1552. IEEE.

[6] Amirian, S., Rasheed, K., Taha, , & Arabnia, (2019). Image captioning with generative adversarial network. In 2019 International Conference on Computational Science and Computational Intelligence (CSCI), (pp. 272–275). IEEE.

[7] P, V., Kar, H., Gautam, S., Mekala, M., Rahimi, M., & Gandomi, A.H. (2023). Sign Language Translator for Dumb and Deaf. 2023 10th International Conference on Soft Computing & Machine Intelligence (ISCMI), 198–202, IEEE.

synthesis. It enables users to navigate websitesthrough comprehensible audio feedback generated from visual content.

Result Analysis and Discussion

The image description of people with visual impaired models result is resembled to the realpers on word of the flickr 8k dataset. It includes five distinct generation with 8,000 snaps common-place situations, actions, and objects. The results are compared using the work's outcomes, and it can evaluate the picture's human-generated descriptions Figure 7.4.

Here the precision metrics which are used to calculate the works result, instead of dividing by the n-gram count(n), The model n-gram count calculated by the following formulae using Eq. 5 The extension was created with Jupiter environment and jar system in Python, picture processing and contained the vgg net. The model was implemented using Keras 2.0 and

[8] Wu, S., Zhang, X., Wang, X., Li, C., & Jiao, L. (2020). Scene attention mechanism for remote sensing image caption generation. In 2020 International Joint Conference on Neural Networks (IJCNN), (pp. 1–7). IEEE.

[9] Kar, V. P. H., Gautam, S., Mekala, M., Rahimi, M., & Gandomi, (2023). Sign language translator for dumb and deaf. (pp. 198–202).

[10] Wang, Y., Shen, Y., Xiong, H., & Lin, W. (2019). Adaptive hard example mining for image captioning. In 2019 IEEE International Conference on Image Processing (ICIP), (pp. 3342–3346). IEEE.

[11] Xian, Y., & Tian, Y. (2019). Self-guiding multimodal lstm when we do not have a perfect training dataset for image captioning. *IEEE Transactions on Image Processing*, 28(11), 5241–5252.

8 Convolutional neural network-based classification of potato leaf diseases

Prateek Mahapatra[a] and Madhumita Panda[b]

School of Computer Science, Gangadhar Meher University, Odisha, India

Abstract

Early and precise detection of potato leaf diseases is critical for crop health and increased agricultural productivity. Despite several research efforts in this area, reaching high levels of precision in disease classification remains challenging. Utilizing the PlantVillage dataset, we propose a convolutional neural network (CNN) model to classify potato leaf images. Effective disease diagnosis is crucial because it allows for early intervention, which helps to reduce crop losses and increase output. The proposed CNN model was specifically designed to capture the unique characteristics of each disease class, resulting in a 96.28% classification accuracy during training and validation. These findings highlight the model's utility in precision agriculture, where accurate disease identification is essential for optimal crop management and performance.

Keywords: Agricultural productivity, classification, convolutional neural network (CNN), PlantVillage dataset, precise detection, precision agriculture

Introduction

Potatoes contribute significantly to global food security and agriculture. Diseases in potato leaves can drastically damage potato production if not treated in time [9]. Timely and precise identification of diseases is essential for restraining their spread and reducing production losses.

Standard procedures for identifying crop diseases rely mainly on agricultural specialists' manual checks, which are time-consuming and error-prone [5]. As farming operations expand, there is an increased demand for more efficient, automated methods. The advancement in the deep learning introduced as a powerful tool in the field of disease analyzing and identifying. Deep learning uses CNN which can be learned the features automatically and identifying diseases through image analysis [6].

The study is based on deep learning architecture using the variant of CNN which is used to classify the potato leaves of PlantVillage dataset. Our proposed model aims to provide a robust and reliable disease detection solution by using CNNs' ability to recognize detailed patterns inside images. Implementing such a system could assist farmers and agricultural professionals in better managing crop health, ensuring timely intervention, and boosting overall agricultural productivity.

This study is organized into six parts. Some existing studies are briefly given in part 2. The information about the dataset is given in part 3. Our proposed CNN model is described in part 4 while part 5 presents the result. Finally, the present study is concluded in part 6 followed by a short discussion on the future scope.

Literature Review

The rapid growth of deep learning motivated researchers to explore the agriculture field. Many research have been conducted and a number of algorithms have been developed for the classification of the agricultural data. A CNN model was suggested by Kusumawati et al. [10] for potato leave classifications with an achieved accuracy of 91%.

By fine-tuning the Xception model, Ambhaikar et al. [1] achieved 95.37% accuracy in classifying the leaf images of potatoes. It revealed that advanced CNN architectures improve disease detection performance.

Bhowmik et al. [2] detected potato diseases having 95% success rate employing image segmentation and multiclass SVM on the PlantVillage dataset, confirming the method's applicability in agricultural applications.

In the study of Meena et al. [11], a support vector machine (SVM) was utilized for potato blight disease detection, achieving 91.4% accuracy.

Chug et al. [3] employed CapsNet to categorize potato diseases and examined its efficiency to popular pre-trained models such as ResNet18, VGG16, and GoogleNet. CapsNet achieved a 91.83% accuracy rate on the PlantVillage dataset, indicating its competitive performance in disease identification.

In another work, a faster R-CNN with a SVM have been employed by Feng et al. [4] to identify the potato

[a]prateekmahapatra75@gmail.com, [b]mpanda.gmu@gmail.com

DOI: 10.1201/9781003658221-8

leaves disease. The model performs a superior accuracy compared to the traditional model.

Gupta et al. [7] employed machine learning to identify potato leaf diseases. They tested the VGG16 model and achieved an accuracy of 95%. The dataset included leaf images of potatoes from the PlantVillage dataset. This study demonstrated the efficacy of CNNs for disease management in precision agriculture.

Several studies have been conducted regarding potato disease classification, but the desired level of accuracy has not been achieved yet. So, here we tried to propose a CNN model for classifying potato leaves utilizing the PlantVillage dataset.

Dataset Description

This study utilized the PlantVillage dataset [8] for classifying potato leaves. A sample of the tested dataset in this study is presented in Figure 8.1.

Figure 8.1 Sample of the tested dataset in this study
Source: Author

Data augmentation

The ImageDataGenerator was used to enrich data by resizing pixel values and making other changes. These included up to 20-degree rotations, 20% width and height changes, 20% shear and zoom adjustments, and horizontal flips.

Methodology

Proposed CNN architecture

The pixel size of the supplied images is 150×150. The proposed model incorporates a single convolutional layer with a 3×3 filter, and a 2×2 max pooling layer, respectively. To achieve non-linearity, the ReLU activation function is implemented after each convolutional operation.

64 filters with 3 × 3 kernel and the ReLU activation algorithm is used for the second convolutional layer. To further downsample the feature maps, an additional 2 × 2 max-pooling layer was incorporated. The

third convolutional layer used the ReLU activation function in conjunction with 128 filters and a 3 × 3 kernel. The extracted features were compressed using a 2 × 2 max-pooling, that reduced the spatial dimensions while preserving the most significant features.

The fully connected layer comprised a dense layer with 128 units and employed a ReLU activation function, enabling the model to capture nonlinear relationships within the extracted features. The final output layer utilized the softmax activation. Figure 8.2 shows a detailed structure of the present model.

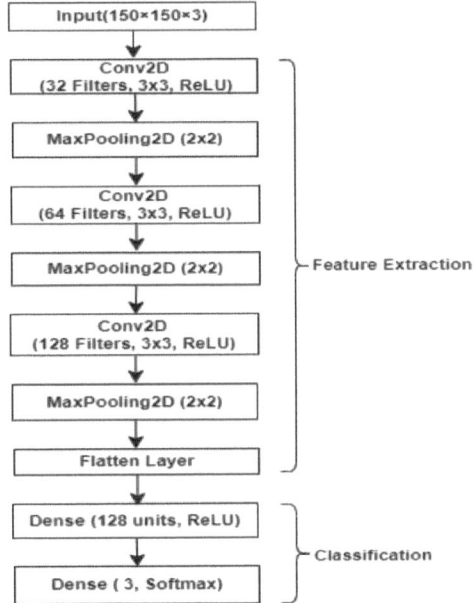

Figure 8.2 Structure of the proposed model
Source: Author

Hyperparameters

To ensure uniform input dimensions, all images were adjusted to 150 × 150 pixels. 70%, 15%, and 15% of images were chosen for training, validation, and testing respectively. A batch size of 32 throughout 100 epochs was utilized during the training. The categorical cross-entropy loss function was utilized. With an initial learning rate of 0.001, Adam optimizer was employed.

Result and Analysis

The effectiveness of the proposed model is described in this section. The confusion matrix and different performance metrics are presented in Figure 8.3 and Table 8.1, respectively.

Confusion matrix

Figure 8.3 shows 100% accuracy for healthy and early blight classes, while late blight achieved 92%

accuracy, with 4% misclassified as healthy and early blight.

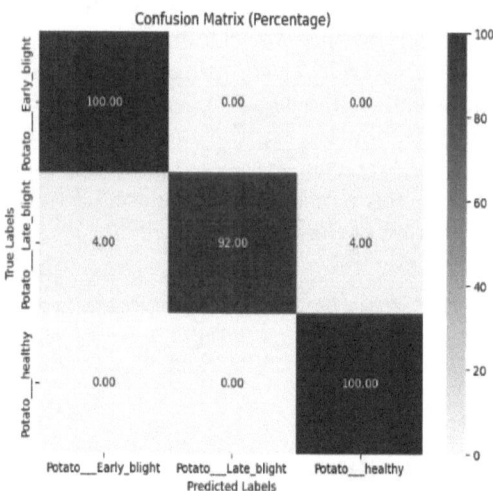

Figure 8.3 The achieved confusion matrix
Source: Author

Classification report

Table 8.1 Classification report of the proposed work.

Class	Precision	Recall	F1-Score
Healthy	0.96	1.00	0.98
Late Blight	1.00	0.92	0.96
Early Blight	0.79	1.00	0.88

Source: Author

Conclusion and Future Work

The importance of classifying potato leaf diseases and the benefits of using deep learning models over conventional techniques were the first points of emphasis in this study. Then all the related works were presented followed by the information of the dataset used i.e. PlantVillage, after that, the proposed model architecture was presented in detail followed by all the results achieved during this study. The model achieved 96.28% accuracy for potato disease classification outperforming some previous studies. In the future, we will explore various plant leaf datasets to classify other plant leaves and apply our model to real-world environmental situations.

References

[1] Ambhaikar, A., & Shinde, N. (2023). Fine-tuned Xception model for potato leaf disease classification. In International Conference on Cognitive Computing and Cyber Physical Systems, (pp. 663–676). Singapore: Springer Nature Singapore.

[2] Bhowmik, P., Dinh, A., Islam, M., & Wahid, K. (2017). Detection of potato diseases using image segmentation and multiclass support vector machine. In 2017 IEEE 30th Canadian Conference on Electrical and Computer Engineering (CCECE), (pp. 1–4). IEEE.

[3] Chug, A., Singh, A. P., & Verma, S. (2020). Exploring capsule networks for disease classification in plants. *Journal of Statistics and Management Systems*, 23(2), 307–315.

[4] Feng, Q., Sun, W., Wang, G., Yang, S., & Zhang, J. (2020). Identification method for potato disease based on deep learning and composite dictionary. *Transactions of the Chinese Society for Agricultural Machinery*, 51(7), 22–29.

[5] Ferentinos, K. P. (2018). Deep learning models for plant disease detection and diagnosis. *Computers and Electronics in Agriculture*, 145, 311–318.

[6] Gülmez, B. (2024). A comprehensive review of convolutional neural networks based disease detection strategies in potato agriculture. *Potato Research*, 1–35.

[7] Gupta, U., Jodan, J. S., Kumar, S., Roy, N. R., & Vijh, S. (2023). Potato leaf disease detection using machine learning techniques for precision agriculture. In 2023 Seventh International Conference on Image Information Processing (ICIIP), (pp. 913–918). IEEE.

[8] Hughes, D., & Salathé, M. (2015). An open access repository of images on plant health to enable the development of mobile disease diagnostics. arXiv preprint arXiv:1511.08060.

[9] Hughes, D. P., Mohanty, S. P., & Salathé, M. (2016). Using deep learning for image-based plant disease detection. *Frontiers in Plant Science*, 7, 1419.

[10] Kusumawati, E., Risnumawan, A., Sholihati, R. A., & Sulistijono, I. A. (2020). Potato leaf disease classification using deep learning approach. In 2020 International Electronics Symposium (IES), (pp. 392–397). IEEE.

[11] Meena, S. M., Patil, P., & Yaligar, N. (2017). Comparison of performance of classifiers—SVM, RF, and ANN in potato blight disease detection using leaf images. In 2017 IEEE International Conference on Computational Intelligence and Computing Research (ICCIC), (pp. 1–5). IEEE.

9 Content-based deep neural image retrieval system for remote sensing images

Sai Bhargav Kasetty[1,a], B. Gouthami[2,b], Sravani Nalluri[3,c],
Rajesh Kumar Ojha[4,d], Issac Neha Margret[1,e] and K. Rajakumar[1,f]

[1]SCOPE, V.I.T, Vellore, Tamilnadu, India

[2]Department of CSE, Gokaraju Rangaraju Institute of Engineering and Technology, Hyderabad, Telangana, India

[3]Department of CSE, Siddhartha Academy of Higher Education, Vijayawada, Andhra Pradesh, India

[4]Department of CSE, Silicon University, Bhubaneswar, Odisha, India

Abstract

Rapid innovations in remote sensing expertise have escalated the availability of a wide range of imaging modalities. Qualitative analysis of these images is essential for various applications such as disaster management, urban planning, and environmental monitoring. This article presents a content based deep neural image retrieval system for multispectral remote-sensing images. This paper evaluates the performance of five deep learning models, including a neural deep network specifically designed for such tasks. The deep neural network model outperformed other models by achieving 98–99% accuracy in all tests. The results show that deep neural networks can improve the performance of the CBIR system in detecting multi-dimensional images from remote locations.

Keywords: Content-based image retrieval, deep neural networks, generative adversarial nets, long short-term memory, multi-spectral remote sensing. recurrent neural networks

Introduction

With the development of remote sensing technology, monitoring, and observation of the earth's surface has been greatly improved, providing various information for use such as environmental monitoring, urban planning, and disaster management [6]. Midst the diverse types of remote sensing (RS) data, multispectral images stand out with their ability to capture data in multiple spectral bands. The combination of these bands will allow different features of the image to be extracted for detailed analysis and retrieval; This unique approach is ideal for remote sensing applications where the data volume is too large to handle manual annotation. The emergence of deep learning has led to substantial advances in the content-based image retrieval (CBIR) field, especially in remote sensing. In processing multispectral images, deep learning models need to address the additional complexity of multiple spectral bands [13]. Effective and retrievable multispectral images require models that utilize the spectral diversity and precision of the retrieved images. This paper presents a CBIR system using deep neural networks for multiple satellite images.

Literature Review

Deep learning (DL) has revolutionized CBIR, especially in the field of RS. Keisham and Neelima [6] developed a deep search method to achieve efficient and accurate results, while Alshehri [2] used neural networks for correlation estimation.

Ahmad and Ahmad [1] combined CNN with geometric augmentation to achieve good results. Sukhia et al. [9] employed retrieval using multi-scale local ternary models. Sambre (2021) discusses DL models that can be applied to big data.

Hameed et al. [5] combined feature fusion and support vector machines for accuracy. Semenov et al. [8] used DL for SAR retrieval in GPS-limited scenarios, whereas Xiong et al. [11] overcame cross-modality problems with deep hashing. Byju et al. [3] improved CBIR efficiency in compressed archives.

Kumar et al. [7] discussed CBIR techniques, and Devulapalli et al. [4] applied hybrid features for accuracy.

The DNN-based system in this study for multispectral images obtains the accuracy of 98–99% and outperforms five other DL models.

[a]kasettisai.bhargav@vit.ac.in, [b]gouthami1852@grietcollege.com, [c]drsravani@vrsiddhartha.ac.in, [d]rajesh.ojha@silicon.ac.in, [e]issacneha.margret@vit.ac.in, [f]rajakumar.krishnan@vit.ac.in

DOI: 10.1201/9781003658221-9

Methodology

This study underscores the potential of deep neural networks (DNN) as shown in the below Figure 9.1 in boosting the effectiveness and efficiency of CBIR systems for multispectral remote sensing images, offering valuable insights for future research and applications in this domain.

Dataset

The dataset is made up of multispectral satellite images over visible (RGB) and NIR bands, supporting applications like land use classification, vegetation monitoring, and environmental analysis. It contains N images with dimensions W×H×B, where B is the total no. of spectral bands and W, H are image height and width [12]. A 21-class land use dataset comprises 100 images per class, such as agriculture, forests, rivers, buildings, and airplanes, each sized 256 × 256 pixels.

Preprocessing

Preprocessing the data is a vital stage to ensure the consistency and quality of the images or data before feeding it to a deep neural network. The following steps describe the pre-process: data standardization or normalizing, resizing, and enhancement or augmentation of the data.

Deep neural network

The CBIR method is proposed to use DNN to obtain features from multiple images and evaluate their similarity. This model is built on a neural network (CNN) that can capture the spatial hierarchy in image data.

Feature extraction: DNN uses multiple convolutional layers along with filters to create feature maps, and capture textures and patterns. This scheme is achieved without any competition from powerful algorithms like ReLU, represented as:

$$F_l = ReLU\ (W_l * F_{l-1} + b_l)$$

where F_l is the feature map at layer l, W_l, b_l are the weights and biases of the filters at layer l, and * denotes the convolution operation.

Similarity measurement: effective in high-dimensional spaces, is used to compare query and database image features. It is calculated as:

$$Cosine\_similarity(f_q, f_i) = \frac{f_q \cdot f_i}{\|f_q\| \, \|f_i\|}$$

where (.) represents the dot product and ∥ . ∥ denotes the Euclidean norm of the vectors.

Retrieval process: The query image passes through the DNN to generate feature vectors. Cosine similarity scores are computed against database images, ranking and retrieving the top k most similar results.

Results and Discussions

LSTM networks, CNNs, RNNs, autoencoders, and GANs were benchmarked against the proposed DNN model for CBIR. The evaluation used metrics like recall (r), f1-score (f), Mean average precision (mAP) and precision (p). As illustrated in Table 9.1, the DNN outperformed other models, achieving 99.2%

Table 9.1 Performance measure.

Model	p (%)	mAP (%)	f (%)	r (%)
RNN	85.7	82.9	84.7	83.8
GAN	90.1	87.9	89.4	88.7
CNN	92.4	89.7	90.5	91.4
Autoencoder	88.2	85.6	87.3	86.4
LSTM	87.6	84.8	86.7	85.9
Proposed DNN	99.2	98.4	98.8	98.5

Source: Author

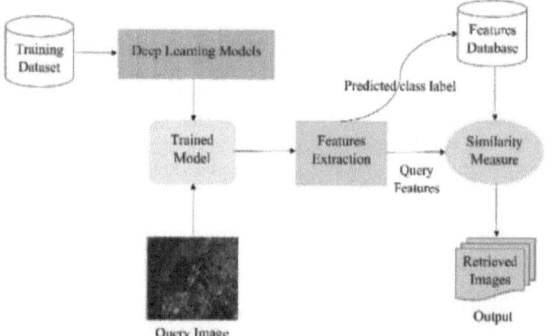

Figure 9.1 Proposed methodology for the CBIR system
Source: Author

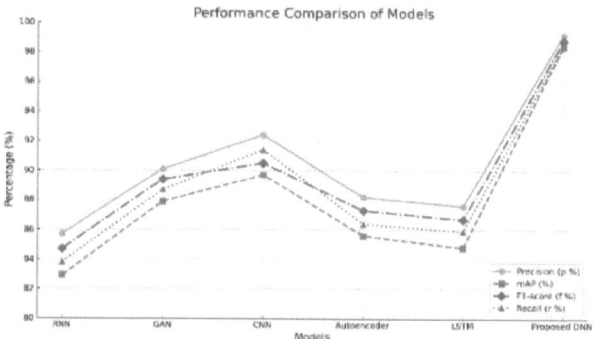

Figure 9.2 Performance comparison of different models
Source: Author

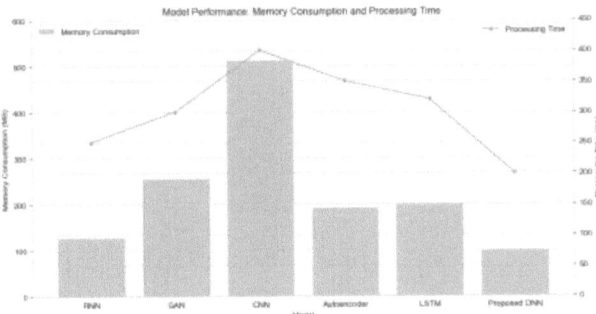

Figure 9.3 Performance comparison of efficiency measures
Source: Author

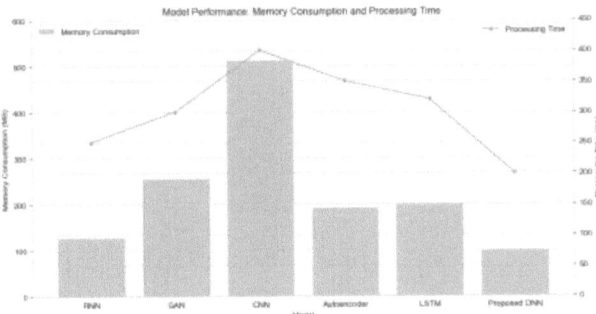

Figure 9.4 Results of the CBIR system for remote sensing images
Source: Author

precision, 98.5% recall, 98.8% F1-score, and 98.4% mAP. These results highlight DNN's effectiveness in accurately retrieving relevant images.

Figure 9.2 shows how the different models performed across the different metrics in comparison.

Figure 9.3 depicts the memory consumption in m and processing time in milliseconds for all the models, where the proposed DNN is the most efficient model, requiring only 100 MB of memory and completing tasks in 200 milliseconds. Its optimized structure and design make it ideal for applications with limited resources.

Using the DNN, Figure 9.4 shows the result of retrieving the topmost similar images to a given query image. In practical applications of CBIR, the retrieved images display high relevance and similarity to the query image, validated by the model's effectiveness.

Conclusion

Concerning content-based image retrieval for multispectral remote sensing images, the thorough evaluation and comparison demonstrate the suggested deep neural network model's great advantage. The strong recall, accuracy, F1-score, mean average precision, and precision all highlight its potential for real-world applications where relevance and accuracy are critical.

Subsequent efforts will concentrate on refining the model and investigating its use in additional remote sensing and image retrieval areas.

Future work would be beneficially applied to extend improve the model's working and adaptability for application in other remote sensing domains, as well as tasks such as image retrieval. This generalizes the work to establish a robust and scalable system on diverse datasets and practical challenges.

References

[1] Ahmad, F., & Ahmad, T. (2021). Content based image retrieval system based on deep convolution neural network model by integrating three-fold geometric augmentation. *Optical Memory and Neural Networks*, 30, 236–249.

[2] Alshehri, M. (2020). A content-based image retrieval method using neural network-based prediction technique. *Arabian Journal for Science and Engineering*, 45(4), 2957–2973.

[3] Byju, A. P., Demir, B., & Bruzzone, L. (2020). A progressive content-based image retrieval in JPEG 2000 compressed remote sensing archives. *IEEE Transactions on Geoscience and Remote Sensing*, 58(8), 5739–5751.

[4] Devulapalli, S., Potti, A., Krishnan, R., & Khan, M. S. (2023). Experimental evaluation of unsupervised image retrieval application using hybrid feature extraction by integrating deep learning and handcrafted techniques. *Materials Today: Proceedings*, 81, 983–988.

[5] Hameed, I. M., Abdulhussain, S. H., Mahmmod, B. M., & Hussain, A. (2021). Content based image retrieval based on feature fusion and support vector machine. In 2021 14th International Conference on Developments in Esystems Engineering (DeSE), (pp. 552–558). IEEE.

[6] Keisham, N., & Neelima, A. (2022). Efficient content-based image retrieval using deep search and rescue algorithm. *Soft Computing*, 26(4), 1597–1616.

[7] Kumar, M. S., Rajeshwari, J., & Rajasekhar, N. (2022). Exploration on content-based image retrieval methods. In Pervasive Computing and Social Networking: Proceedings of ICPCSN 2021, (pp. 51–62). Singapore: Springer.

[8] Semenov, A., Rysz, M., & Demeyer, G. (2024). Deep learning approach for SAR image retrieval for reliable positioning in GPS-challenged environments. *IEEE Transactions on Geoscience and Remote Sensing*. 62, 1–11.

[9] Sukhia, K. N., Riaz, M. M., Ghafoor, A., & Ali, S. S. (2020). Content-based remote sensing image retrieval using multi-scale local ternary pattern. *Digital Signal Processing*, 104, 102765.

[10] Sumbul, G., Kang, J., & Demir, B. (2021). Deep learning for image search and retrieval in large remote sensing archives. Deep learning for the earth sciences: a

comprehensive approach to remote sensing, climate science, and geosciences, 150–160.

[11] Xiong, W., Xiong, Z., Zhang, Y., Cui, Y., & Gu, X. (2020). A deep cross-modality hashing network for SAR and optical remote sensing images retrieval. *IEEE Journal of Selected Topics in Applied Earth Observations and Remote Sensing*, 13, 5284–5296.

[12] Yang, Y., & Newsam, S. (2010). Bag-of-visual-words and spatial extensions for land-use classification. In Proceedings of the 18th SIGSPATIAL International Conference on Advances in Geographic Information Systems, (pp. 270–279).

[13] Payal, Payal, et al. (2024). Medicare: A telemedicine healthcare website. 7th International Conference on Circuit Power and Computing Technologies (IC-CPCT). Vol. 1. IEEE, 2024.

10 Modelling non-linear time series data using deep learning techniques

Masrath Parveen[1,a], Mohammed Ali Shaik[2,b], Nanditha Boddu[1,c], Akheel Mohammad[3,d], Udaya Kiran Mandhugula[1,e] and Vupulluri Sudha Rani[2,f]

[1]Department of IT, Vidya Jyothi Institute of Technology, Hyderabad, Telangana, India

[2]Department of CS&AI, SR University, Warangal, Telangana, India

[3]Department of CSE, Dr. Vrk Women's College of Engineering and Technology, Telangana, India

Abstract

This paper introduces a novel approach for employing deep learning techniques to apply a mathematical model for performing forecasting on non linear time series (NLTS). The algorithm's objective is to remove any unwanted interference and progressively transform the area into a more advanced state. By performing calculations on the transformations, the iterative set values can be utilized to transition from a declining trend to an ascending trend or from an ascending trend to a declining trend. And the proposed model must efficiently process voluminous multi dimensional data that demonstrates the dynamic character as it is essential to associate the outcomes that are derived from the suggested model with the multifaceted ARIMA model through petroleum dataset of India using a mathematical model is proposed.

Keywords: Classification, forecasting, machine learning, time series

Introduction

In the contemporary age of computers, which encompasses an extensive quantity of historical data time series (TS) is a term utilized to designate data that is gathered and arranged according to the element of time. Automated classification is employed in petroleum data [1] analysis to evaluate the production and distribution of petroleum products through market analysis. The primary emphasis lies in comparing ARIMA model [2] by employing various classification techniques to perform decisions cantered on various observations derived from fractional samples. As the TS classification is utilized in the domains of health or vigor diagnostics to implement classification monitoring systems [3] and predictive maintenance [4].

An artificial neural network (ANN) [5] is composed of interconnected processors called neurons. Every neuron has a role in creating a series of real-life situations that stimulate the input neurons [4].

Literature Review

As per the reference [2] authors proposed the aspect of various time series observations in their work which is represented by the observation that denotes xt, where x symbolizes a detailed observation and machine time is denoted by t [3]. Thes two observations are implemented through continuous and discrete methodologies over time series data standardized through a scale that ranges from 0 to 1 [4–8].

As per the reference [9] authors have intended a better technique to assess the efficiency of the Nearest Neighbour algorithm to perform binary classification, the process of it merely relies on the labelled data for categorizing the unlabelled data in an iterative manner [10]. The methodology persists up to a predetermined termination condition is met and more over instances are collected on the basis of the obtained outcomes while performing classification [9].

The researchers has developed a 1NN classification method which is specifically developed for implementing and handling univariate time series data accurately [5]. This method utilizes the Euclidean distance calculation and the Frobenius distance estimate is used to calculate distances for time series data [6]. The authors present a new method to improve the k-nearest neighbours (KNN) search algorithm for dealing with multivariate time series (MTS) data and also they propose the use of a unique sliding windows terminology is used to enhance the efficiency of the KNN algorithm [1]. By creating correlation matrices as the objective is to provide a singular value that precisely captures the characteristics of the time series with multiple variables across a defined time interval [2].

[a]masrath19@gmail.com, [b]niharali@gmail.com, [c]nanditha.boddu@gmail.com, [d]alikhancrm@gmail.com, [e]udayait@vjit.ac.in, [f]vupullurisudharani1703@gmail.com

DOI: 10.1201/9781003658221-10

In correlated research, investigators present a deep neural network (DNN) [5] which uses convolutional neural networks (CNNs) [6] to tackle diverse facets of visual processing in computers along with the distinct attributes, such as sparse connectivity, wherein particular areas in the input image (known to be responsive fields) belong to a distinct layers [7]. The study utilizes deep belief networks (DBNs) [8].

Proposed Model

Proposed ARIMA model

One of the robust frameworks is proposed autoregressive integrated moving average (ARIMA) [12] structure is a robust framework used for investigating and predicting time series data as the model will integrate and accelerating average elements to effectively represent the fundamental trends and interconnections present in the data. ARIMA is a forecasting method that uses former values of a TS to predict imminent values as the process entails finding and classifying the necessary modifications required.

The integrated differencing component (IDC) [9] is a statistical method which is a part of ARIMA model that requires to measure the vast amount of variation in time series data by using passively differentiation, auto regressive (AR) [10]: This element provides the correlation among present values calculated according to past data. The moving average (MA) is a statistical calculation that has been utilized to analyze data by performing smoothing out fluctuations within a particular amount of time [12].

Proposed mathematical model

This work presents a model being developed through NLTSD and employs stationarity modifications to make accurate predictions. TS data frequently requires transformation into stationary information before it can be directly utilized in applications. The NLTSD is subsequently transformed to attain stationarity aspect which is indicated by the equation

$$P = Ae^{(bq)} \qquad (1)$$

To calculate the SD at an explicit occurrence (T+1), the undistinguished data at the T+1 value needs to be resolved. The purpose of the SD is to simplify the quadratic equation:

$$ax^2 + bx + c = 0 \qquad (2)$$

with $a \leftarrow n(n+1)$, $b \leftarrow -2s(n+1)$, $c \leftarrow -(ns^2) -s^2 + m(n+1)^2$ $(nd^2) (n+1)^2 (4)$

The value of (n+1) corresponds to the cumulative sum of all element values in the time series that are calculated by adding up the data value of m is calculated by summing the squares of the data, whereas d indicates the SD of the (n+1) attribute values with the component which has been deliberately incorporated into the time.

Proposed algorithm

Algorithm proposal: Development of a predictive mathematical model for nonlinear time series data

Step 1: Begin

Step 2: Compute the initial amounts of ch (latest high value) and cl (latest low value) for each of the single instance by utilizing the NLTD

Step 3: Compute the mean values over the time series dataset using ch and cl

Step 4: Compute the proportion of evidence inside an appropriate time range denoted by t.

Step 5: Compute the SD ratio over the above equation 2

Step 6: Evaluate the gradient

Step 7: perform regression analysis

Step 8: Evaluate the total amount

Step 9: Generate sum of squares of the obtained value in step 8.

Step 10: Compute the SD

In Step 11, equations 3 and 4 are employed to analyze the TS data at time T+1 and attain the desired results

Step 12: Determine the value of 'a' by multiplying 'n' by ('n+1').

Step 13: Determine the value of b by multiplying -2 by s and then multiplying the resulting product by the sum of n and 1.

Step 14: Determine the value of c by subtracting the product of n, s, n and s from the square of s and then adding the product of m, the square of the sum of n and 1 & the negative square of nd times the square of the sum of n and 1.

Step 15: Compute the sum of the time series (a, b, c) and assign it to the variable s.

Step 16: Compute the total of the squares of the values in s and apply the outcome to the variable m.

Step 17: Calculate the standard deviation by accessing the element at location (m, n+1).

Step 18: Determine the value of x by computing the median of the maximum and minimal cost at time t-1.

Step 19: End.

Experimental Results

The performance assessment was conducted using pre-established metrics, specifically the Mean Square

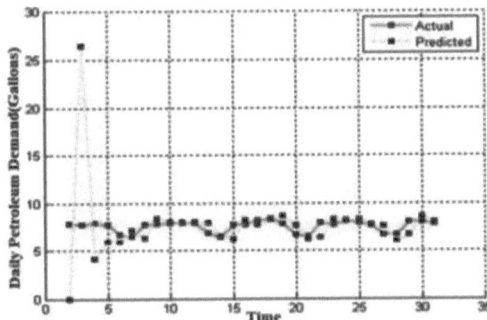

Figure 10.1 Daily tangible and projected petroleum usage in gallons
Source: Author

Table 10.1 Performance comparison of the proposed and ARIMA mathematical model.

Parameter model	ARIMA	Proposed
MSE	1.66	1.38
RMSE	1.44	1.16
MAPE	1.65	1.11
Accuracy predicted	6.95%	4.92%
speedup	2.8	12.4

Source: Author

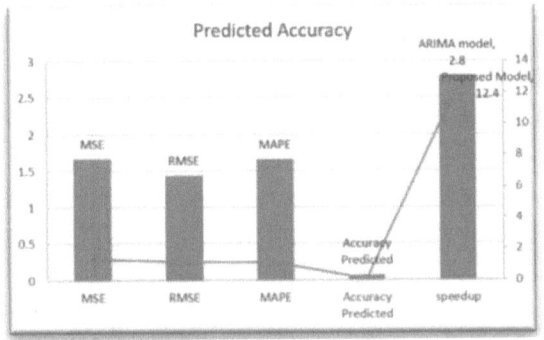

Figure 10.2 Prediction of accuracies for petroleum demand
Source: Author

Error (MSE) [11]. The Root Mean Square Error (RMSE) [11] and the Mean Absolute Percentage Error (MAPE) [11] are the metrics employed for assessing the precision of a model or prediction. The accuracy of predictions (PA) and the inclusive time required to finish the forecasting tasks [13].

Table 10.1 presents an association of the performance of the alluded mathematical model with the ARIMA model by using deep learning technology using filters such as MSE, RMSE and MAPE.

The forecast generated by the recommended model is improved by increasing the order of ARIMA. The improvement results in a 7.6% and 4.5% increase in prediction accuracy when assessed to the ARIMA model. It is noteworthy that the proposed model requires 13 times the amount of time compared to the ARIMA model.

Conclusion

This study introduces a systematic strategy for applying deep learning techniques to create a mathematical model for forecasting non-linear time series. The algorithm's objective is to remove unwanted interference and improve the data transformation by repeatedly exchanging the downward trend with the upward trend, or vice versa, according to specified values. By employing these techniques, the projected algorithm is capable of forecasting future prices without the necessity of professional assistance.

References

[1] Kaur, A., Kukreja, V., Upadhyay, D., Aeri, M., & Sharma, R. (2024). A deep learning-based long short-term memory technique for Google stock price prediction. In 2024 Asia Pacific Conference on Innovation in Technology (APCIT), Mysore, India, (pp. 1–5). doi: 10.1109/APCIT62007.2024.10673419.

[2] Binkowski, M., Marti, G., & Donnat, P. (2018). Autoregressive convolutional neural networks for asynchronous time series. In Proceedings of the International Conference on Machine Learning (ICML), (pp. 580–589).

[3] Liu, H., Qian, Y., Yang, G., & Jiang, H. (2022). Super-resolution reconstruction model of spatiotemporal fusion remote sensing image based on double branch texture transformers and feedback mechanism. *Electronics (Switzerland)*, 11(16), 247.

[4] Ibarra-Fiallo, J., & Lara, (2024). Contextual deep learning approaches for time series reconstruction. In 2024 IEEE International Conference on Omni-layer Intelligent Systems (COINS), London, United Kingdom, (pp. 1–4). doi: 10.1109/COINS61597.2024.10622120.

[5] Fawzy, R., Eltrass, , & Elkamchouchi, (2024). A new deep learning hybrid model for accurate web traffic time series forecasting. In 2024 6th Novel Intelligent and Leading Emerging Sciences Conference (NILES), Giza, Egypt, (pp. 403–406). doi: 10.1109/NILES63360.2024.10753194.

[6] Bandyopadhyay, S., Sarma, M., & Samanta, D. (2024). Prediction of neurodevelopmental disorder by combining data-driven ROI extraction and frequency specific brain functional connectivity approach. In 2024 15th International Conference on Computing Communication and Networking Technologies (ICCCNT), Kamand, India, (pp. 1–7). doi: 10.1109/ICCCNT61001.2024.10723846.

[7] Nagpal, V. Y., Jindal, V., Kukreja, V., Vats, S., & Sharma, R. (2023). Deep learning multiclassification model: recognizing monuments. In International Conference on Augmented Intelligence and Sustainable Systems, (pp. 315–319).

[8] Arangi, V., Krishna, S. J. S., Gongada, , Basha, M., Santosh, K., & Muthuperumal, S. (2024). Hybrid deep learning and time series analysis for stock market prediction: a multilayer perceptron and ARIMA modelling approach. In 2024 10th International Conference on Advanced Computing and Communication Systems (ICACCS), Coimbatore, India, (pp. 421–427). doi: 10.1109/ICACCS60874.2024.10716884.

[9] Vimal, V. (2023). Integrating IoT-based environmental monitoring and data analytics for crop-specific smart agriculture management: a multivariate analysis. In Proceedings - International Conference on Technological Advancements in Computational Sciences ICTACS 2023, (pp. 368–373).

[10] Wang, Z., Yan, W., & Oates, T. (2017). Time series classification from scratch with deep neural networks: a strong baseline. In Neural Networks (IJCNN) International Joint Conference USA, (pp. 1578–1585). IEEE.

[11] Li, X., Metsis, V., Wang, H., & Ngu, (2022). Tts-gan: a transformer-based time-series generative adversarial network. In International Conference on Artificial Intelligence in Medicine (pp. 133–143). Cham: Springer International Publishing.

[12] Pang, X., Zhou, Y., Wang, P., Lin, W., & Chang, V. (2020). An innovative neural network approach for stock market prediction. *The Journal of Supercomputing*, 76(3), 2098–2118.

[13] Zhao, B., Huanzhang, L., Chen, S., Liu, J., & Dongya, W. (2017). Convolutional neural networks for time series classification. *Journal of Systems Engineering and Electronics,* 28(1), 162–169.

11 Intelligent strategies for the techno-economic evaluation of large-scale energy storage integrated hybrid power systems

Ashribad Pattnaik[1,a], Alpesh Kumar Dauda[1,b], Ambarish Panda[1,c] and Kathleen B. Aviso[2,d]

[1]Department of EEE, Sambalpur University Institute of Information Technology, Burla, Odisha, India

[2]Department of Chemical Engineering, De La Salle University, Manila, Philippines

Abstract

Most of the present power demand is obtained from fossil fuels like coal, oil, and natural gas which have volatile prices. To address this volatility while reducing the climate change, hybrid power systems with cleaner energy sources become very meaningful. In this context, solar and wind power have gained attention as viable options to meet the power demands. However, the intermittency of renewable energy sources is a challenge for power generation scheduling. Therefore, integration of large-scale storage with HPS may address the operational issues. In an attempt to model the thermal-wind generation with the inclusion of largescale storage, this work focuses on minimizing generation cost and voltage deviation within an optimal power flow framework. The optimal operational paradigm of HPS is obtained using the brown bear optimization algorithm and the flower pollination algorithm. IEEE30 bus system has been considered as system for validation of this work.

Keywords: Brown bear optimization algorithm, flower pollination algorithm, hybrid power system, optimal power flow, pumped hydro storage

Introduction

The absence of solar power during the evening and night makes wind power one of the preferred choices as acrucial source for Hybrid Power Systems (HPS). In this context, Wind and Thermal (WT) systems can address bothoperational and environmental challenges [2][4][7]. The sporadic characteristics of renewable energy (RE)underscore the need for integrating Electrical Energy Storage (EES) facilities with HPS [15]. Utilizing high storagecapacity, Battery Energy Storage (BES) with grid-connected (GC) HPS involves substantial initial costs [5]. In [6], athree-hybrid con-fifiguration (WST, WHT, and WHST) is analyzed, demonstrating voltage deviation minimization andoperating cost reduction using the Jaya algorithm. This has drawn the attention of researchers to explore possiblealternative solutions.

Large-capacity storage systems, such as pumped hydro storage (PHS) (Nyeche and Diemuodeke, 2020; Javed et al., 2020; Panda et al., 2021) [3] are considered as efficient options for reducing operational cost (OC) associated with HPS while maintaining the generation load balance (Pattnaik et al., 2023). In Panda and Tripathy [7], the optimization of a WT system is solved in OPF framework to minimize the OC and voltage deviation (VD) while implementing the modified bacteria foraging algorithm (MBFA). Fuel cost and emission minimization for WT operation were successfully accomplished in Shilaja and Arunprasath [2] with "gravitational search algorithm" and "moth swarm algorithm" (MSA-GSA). Emission minimization of ST system is investigated in Dauda et al. [4] using MBFA. Moreover, assessment of the impact of BES on OC and VD of a wind-photovoltaic-thermal (WPT) system, is depicted by Dauda and Panda, [5]. The study in Das et al., [3] investigated the impact of PHS in a PV-Diesel-PHS based SA configuration while minimizing the OC. Panda et al. (2021) have shown that optimal scheduling of a wind-PV-hydro-thermal system with PHS and compressed air energy storage (CAES) facilities minimizes VD, power loss, and OC. Attention is focused on minimizing the VD and operating costs using MBFA in a WPT based hybrid system in (Pattnaik et al., 2023). The identified optimal configuration was WPT+BES+PHS. An extensive analysis of the use of the flower pollination algorithm (FPA) for tackling various challenges in electrical power systems has been conducted by Lalljith et al. (2022). In Prakash et al. [8], a novel brown bear optimization algorithm (BBOA) is presented for solving the economic dispatch problem [1]. Additionally, Ojha and Maddela (2023) demonstrated the application of BBOA to address the load frequency control of a WPT system.

The literature survey reveals that a substantial amount of research is done on SA systems with and without considering energy storage facilities.

[a]ashripattnk@gmail.com, [b]alpesh.d123@gmail.com, [c]ambarish101@gmail.com, [d]kathleen.aviso@dlsu.edu.ph

DOI: 10.1201/9781003658221-11

However, very limited work is validated on standard IEEE benchmark systems. The uniqueness of the proposed work is to validate the results in IEEE 30 bus power system configuration and present the comparative OPF solution among WT and WT+PHS configuration while employing multiple intelligent techniques. In this framework, a comparison of the results from BBOA and FPA has been outlined. Figure 11.1 illustrates the basic layout of the HPS.

Problem Formulation

To provide a real-time analysis, the intermittence of wind power has been modelled for underestimation (UE) and overestimation (OE) scenarios. This is done by including the penalty and reserve cost functions. Mathematical modelling for the two HPS configurations considered is represented by (1) and (2). Eq (1) represents the cost function for the WT system (C1), while (2) represents the cost function for the WT+PHS system (C2).

$$TC_1^{op} = \sum_{i=1}^{N_i} C_i^{The} + \sum_{j=1}^{N_j} C_j^{Wi} + pf_e \tag{1}$$

$$TC_2^{op} = \sum_{i=1}^{N_i} C_i^{The} + \sum_{j=1}^{N_j} C_j^{Wi} + \sum_{k=1}^{N_k} C_k^{PHS} + pf_e \tag{2}$$

The minimization of voltage deviation of HPS is described in Equation (3).

$$F^{vd} = \sum_{b=1}^{N_b} V_b^{ac} - V_{nom} \tag{3}$$

The objective functions indicated in (1) and (2) are subject to the following constraints.

$$\sum_i^{N_i} P_i + \sum_j^{N_j} P_j + \sum_k^{N_k} P_k^{es} = P_{loss} + P_{load} \tag{4}$$

$$\sum_i^{N_i} Q_i + \sum_j^{N_j} Q_j = Q_{loss} + Q_{load} \tag{5}$$

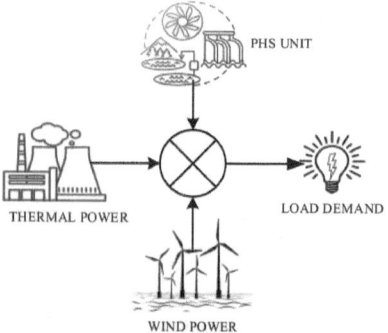

Figure 11.1 Schematic representation of proposed HPS
Source: Author

$$P_i^{min} \le P_i \le P_i^{max} \tag{6}$$

$$Q_i^{min} \le Q_i \le Q_i^{max} \tag{7}$$

$$V_b^{min} \le V_b \le V_b^{max} \tag{8}$$

For wind units

$$P_j \le P_j^{max} \tag{9}$$

$$Q_j^{min} \le Q_j \le Q_j^{max} \tag{10}$$

For pumped hydro storage

$$f_{m,t}^p = \frac{\eta_p \, P_{ph,p}}{\rho g H_t} \tag{11}$$

$$\ddot{Y}_{ur}(t+1) = \ddot{Y}_{ur}(t)(1-\gamma) + f_{m,t}^p - f_{m,t}^g \tag{12}$$

$$0 \le f_{m,t}^p \le x_{m,t}^p \cdot f_{m,p}^{max} \tag{13}$$

$$0 \le f_{m,t}^g \le x_{m,t}^g f_{m,g}^{max} \tag{14}$$

$$\sum_{m=1}^{N_m} x_{m,t}^p \, x_{m,t}^g = 0 \tag{15}$$

$$x_{m,t}^p + x_{m,t}^g \le 1 \tag{16}$$

$$P_{H\,m,t}^g = \eta_t f_{m,t}^g \, \rho \, g H_t \tag{17}$$

The details of the cost function of thermal, wind and PHS unit are presented in (Pattnaik et al., 2023).

Application of Optimization Techniques

Flower pollination algorithm
The vital stages in FPA are: universal pollination and local pollination. The various phases of FPA are as follows:

Step 1: Cross-pollination or biotic pollination or global pollination involves pollinators transporting pollen while exhibiting movement patterns consistent with Lévy flight.
Step 2: Local pollination is carried out through abiotic means self-pollination.
Step 3: Pollinators contribute to flower stability, influencing reproductive success depending on the similarity of the flowers.
Step 4: The balance between local and global pollination is controlled by a switching probability "p," which varies from 0 to 1.

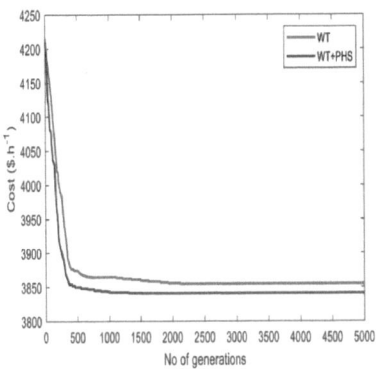

Figure 11.2 CCC of both the HPS with FPA
Source: Author

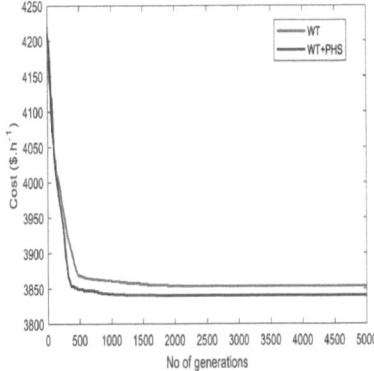

Figure 11.3 CCC of both the HPS with BBOA
Source: Author

Brown bear optimization algorithm

A nature-inspired optimization method known as BBOA, detailed in Prakash et al. [8], is implemented in this work. The steps for BBOA are as follows:

Step 1: During the initial one-third of the total iterations, the paw scent marks, mimicking the bear's walking gait, are continuously refreshed.

Step 2: From one-third to two-thirds of the iterations, the paw imprints are carefully refreshed.

Step 3: During the final one-third of the iterations, the paw imprints are refreshed with twisting foot movements.

Step 4: The sniffing behavior enables bears to communicate by sniffing the paw-scented imprints.

Findings and Analysis

This section provides an in-depth performance evaluation of two HPS, namely C1 and C2, following their optimization using BBOA and FPA.

Table 11.1 Comparison of operating cost ($/hr.) of various HPS with different optimization techniques.

Technique	WT	WT+PHS
FPA	3854.78	3841.21
BBOA	3853.62	3840.53

Source: Author

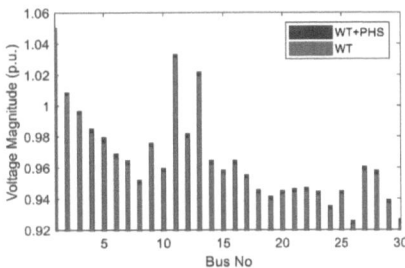

Figure 11.4 A comparative depiction of the bus voltage profile across various HPS
Source: Author

Comparative evaluation of economic operation

By optimizing equations (1) and (2), the objective related to the techno-economic operation of the HPS has been achieved.

The cost convergence characteristics (CCC) of C1 and C2, optimized using FPA and BBOA, are illustrated in Figures 11.2 and 11.3, respectively. Table 11.1 shows that with FPA schedule, C1 has a higher operating cost compared to C2. When optimized with BBOA, C1 depicts a cost of 3853.62$/h, and C2's operating cost of 3840.53$/h demonstrates its superiority over C1. With both optimization techniques, C2 showcased its dominance over C1. Moreover an annual cost saving (hourly cost saving × 8760 h) of 5956.8$ has been achieved with C2 optimized with BBOA when compared to FPA. This shows the supremacy of BBOA over FPA.

Comparative evaluation of voltage secure operation

The comparative voltage for both hybrid configurations is assessed using the BBOA, as depicted in Figure 11.4. The analysis reveals that, for configuration C1, the average bus voltage is 0.9620 p.u., while the minimum bus voltage is 0.9218 p.u.

In comparison, the mean and minimum bus voltages for configuration C2 are 0.9658 p.u. and 0.9249 p.u., respectively.

Conclusion

To achieve the techno-economic and voltage secure HPS operation, FPA and BBOA techniques have been implemented to evaluate the optimal generation schedule.

It was observed that, with BBOA optimized schedule, the operating cost of C2 was less than C1 compared to that of the FPA schedule. A relatively improved bus voltage profile was observed with C2 compared to that of C1. The implementation of both optimization techniques and the validation of results are done on IEEE 30 bus benchmark configuration. It may be summarized that, with better convergence characteristics BBOA has shown its superiority over FPA in terms of fulfilling different operational objectives of large-scale storage integrated systems.

References

[1] Abdelaziz, A. Y., Ali, E. S., & Abd Elazim, S. M. (2016a). Implementation of flower pollination algorithm for solving economic load dispatch and combined economic emission dispatch problems in power systems. *Energy*, 101(April), 506–518.

[2] Shilaja, C., & Arunprasath, T. (2019). Optimal power flow using moth swarm algorithm with gravitational search algorithm considering wind power. *Future Generation Computer Systems*, 98(September), 708–715.

[3] Das, B. K., Hasan, M., & Rashid, F. (2021). Optimal sizing of a grid-independent PV/diesel/pump-hydro hybrid system: a case study in Bangladesh. *Sustainable Energy Technologies and Assessments*, 44(April), 100997.

[4] Dauda, A. K., Panda, A., & Mishra, U. (2023). Synergistic effect of complementary cleaner energy sources on controllable emission from hybrid power systems in optimal power flow framework. *Journal of Cleaner Production*, 419(September), 138290.

[5] Dauda, A. K., & Panda, A. (2022). Impact of small scale storage and intelligent scheduling strategy on cost effective and voltage secure operation of Wind+pv+thermal hybrid system. *Advanced Theory and Simulations*, 6(1), 2200427.

[6] Dauda, A. K., & Panda, A. (2023). Techno-economic and voltage secure operation of hybrid power systems in optimal power flow framework. In International Conference on Advanced Engineering Optimization Through Intelligent Techniques, (pp. 31–40). Singapore: Springer Nature Singapore.

[7] Panda, A., & Tripathy, M. (2014). Optimal power flow solution of wind integrated power system using modified bacteria foraging algorithm. *International Journal of Electrical Power and Energy Systems*, 54(January), 306–314.

[8] Prakash, T., Singh, P. P., Singh, V. P., & Singh, S. N. (2023). A novel brown-bear optimization algorithm for solving economic dispatch problem. In Advanced Control and Optimization Paradigms for Energy System Operation and Management, (pp. 137–64), River Publishers.

[9] Evangeline, S. I., & Rathika, P. (2022). Wind farm incorporated optimal power flow solutions through multi-objective horse herd optimization with a novel constraint handling technique. Expert Systems with Applications, 194, 116544.

[10] Javed, M. S., Zhong, D., Ma, T., Song, A., & Ahmed, S. (2020). Hybrid pumped hydro and battery storage for renewable energy based power supply system. Applied Energy, 257, 114026.

[11] Lalljith, S., Fleming, I., Pillay, U., Naicker, K., Naidoo, Z. J., & Saha, A. K. (2021). Applications of flower pollination algorithm in electrical power systems: a review. IEEE Access, 10, 8924–8947.

[12] Nyeche, E. N., & Diemuodeke, E. O. (2020). Modelling and optimisation of a hybrid PV-wind turbine-pumped hydro storage energy system for mini-grid application in coastline communities. Journal of cleaner production, 250, 119578.

[13] Ojha, S. K., & Maddela, C. O. (2024). Load frequency control of a two-area power system with renewable energy sources using brown bear optimization technique. Electrical Engineering, 106(3), 3589–3613.

[14] Panda, A., Aviso, K. B., Mishra, U., & Nanda, I. (2021). Impact of optimal power generation scheduling for operating cleaner hybrid power systems with energy storage. International Journal of Energy Research, 45(10), 14493–14517.

[15] Panda, A., Dauda, A. K., Chua, H., Tan, R. R., & Aviso, K. B. (2023). Recent advances in the integration of renewable energy sources and storage facilities with hybrid power systems. Cleaner Engineering and Technology, 12, 100598.

[16] Pattnaik, A., Dauda, A. K., Padhee, S., Panda, S., & Panda, A. (2023). Security constrained optimal power flow solution of hybrid storage integrated cleaner power systems. Applied Thermal Engineering, 232, 121058.

12 Revolutionizing cancer diagnosis using machine learning techniques

Amiya Kumar Sahoo[1,a], Sucharita Prusty[1,b], Ashish Kumar Swain[1,c] and Suvendra Kumar Jayasingh[4,d]

[1]Department of CSE, Aryan Institute of Engineering and Technology, Odisha, India

[2]Department of CSE, IMIT, Cuttack, BPUT, Odisha, India

Abstract

Cancer continues to be a significant global health issue, highlighting the urgent need for new approaches in early detection and risk evaluation. In this research, we introduce a thorough methodology for forecasting lung cancer, employing a meticulously selected dataset of 1000 participants sourced from Kaggle. We leverage cutting-edge machine learning models including support vector machines (SVM), meta bagging, PART, and random forest (RF), to enhance the accuracy of our predictions. The dataset we've compiled includes a wide array of patient characteristics, encompassing demographics, lifestyle factors, medical history, and health data gathered through Internet of Things (IoT) devices. By harnessing the capabilities of IoT technology, we enable real-time and continuous health monitoring, facilitating a dynamic assessment of lung cancer risk. Our findings reveal that the PART model achieves an impressive accuracy of 95%, surpassing other models like NBM (67%), RF (93%), SVM (92%), and bagging (86%). This pioneering method holds potential for primary diagnosis of lung cancer and the delivery of tailored risk assessments. These advancements could contribute to better patient consequences and alleviate the burden on health care systems.

Keywords: Cancer prediction, support vector machines, machine learning, Naive Bayes multinomial, meta- bagging, PART, random forests

Introduction

Cancer presents an enduring and formidable global public health challenge, demanding a continuous search for innovative strategies focused on early detection and precise risk assessment. Its pervasive nature and severe consequences underscore the critical need for fresh and advanced approaches. In response, our research endeavors to introduce a pioneering solution that merges. The capabilities of data science and Internet of Things (IoT) technology to transform lung cancer prediction and evaluation. Despite notable advancements in medical science and unwavering efforts in public health, lung cancer remains a significant in death due to cancer on a global scale. Its insidious nature, often remaining asymptomatic until reaching advanced, and frequently untreatable stages, emphasizes the immediate urgency of early detection. In this context, our study establishes a novel paradigm, with a primary focus on utilizing a meticulously curated dataset encompassing 1000 individuals. This dataset serves as the foundation of our research, including a diverse range of patient attributes such as demographics, lifestyle factors, medical history, and data collected from IoT devices. These multifaceted data sources provide the basis for constructing our predictive models, equipping them with the necessary information for making more informed and precise assessments.

Central to our groundbreaking methodology are advanced machine learning models such as support vector machines (SVM), meta bagging, Partial decision tree (PART), and random forest (RF). The integration of these advanced models substantially enhances the accuracy of our predictions, providing healthcare professionals and individuals with a potent tool for making critical decisions pertaining to lung cancer detection and risk assessment. A distinctive feature of our research is the seamless integration of IoT technology for real-time and continuous health monitoring. This dynamic approach transcends the limitations of intermittent snapshots, ensuring an ongoing and adaptive evaluation of lung cancer risk. By synergizing the strengths of robust data analysis with the capabilities of the IoT, our research aims not only to identify lung cancer in its earliest, most treatable stages but also to offer personalized risk assessments tailored to everyone's unique health profile. Our study highlights the remarkable efficacy of the PART model, which achieved the accuracy rate of 95% which beats the performance of other contemporary models, including NBM (67%), RF (93%), SVM

[a]dramiya79@gmail.com, [b]bandana.sucharita@gmail.com, [c]ashishswain001@gmail.com, [d]sjayasingh@gmail.com

DOI: 10.1201/9781003658221-12

(92%), and bagging (86%). These results underscore the effectiveness of our approach and its potential to revolutionize the landscape of lung cancer detection and risk assessment.

In summary, our research offers a groundbreaking opportunity to advance the field of early lung cancer detection, empowering individuals to take proactive measures for their health and contributing to improved patient outcomes. By mitigating the healthcare burdens associated with advanced-stage lung cancer, our comprehensive approach holds the potential to transform the status quo and make lung cancer a more manageable and ultimately preventable condition on a global scale. In light of this urgent global health challenge, our research aims to leave a meaningful and lasting impact, revolutionizing the way we approach the diagnosis and risk assessment of lung cancer.

Literature Review

Jayasingh et al. [1] presented a groundbreaking method for data classification, utilizing neural networks and investigating diverse soft computing models. Their research demonstrates the strategic implementation of neural networks in an innovative way, offering valuable insights into effective data classification techniques. Illness prognosis, focusing on diabetes and heart diseases in the elderly [2]. It focuses on the potential of IoT and AI to aid in diagnosis and prevention. Additionally, it highlights the disconnect between healthcare and technology and suggests that new technologies, including soft computing models and artificial neural networks, can significantly benefit heart disease treatment. The same team focused on smart weather prediction for Delhi [3]. They utilized an array of machine learning techniques such as support vector, decision tree, neural network, machine random forest, Naive Bayes, Adaboost, gradient boosting, Xgboost and logistic regression. Their results demonstrated that RF surpassed the other models, improving the accuracy of weather forecasting. Furthermore, exploration of weather prediction using hybrid soft computing models revealed that these hybrid approaches are better than the traditional soft computing models [4]. Prusty et al. [5] delved into the domain of fraud detection in SMS using various machine learning methods. Their study showed the superiority of RF in achieving the highest accuracy among the methods considered, contributing to more effective fraud detection. In a comparative analysis of different soft computing models for weather forecasting, Jayasingh et al. [6] assessed J48 DT etc. for classification. Their

conclusions highlighted that SVM exhibited superior performance, outshining the other soft computing models in the context of weather forecasting accuracy. Mantri et al. [7] has shown the application of soft computing algorithms for predicting weather changes in Delhi. Their research emphasized that the SVM model yielded the highest accuracy and minimal error values, making it the optimal choice for enhancing the accuracy of weather change predictions in this specific setting. These studies collectively contribute to the broader field of machine learning and soft computing applications, showcasing the effectiveness of specific algorithms in diverse domains, from data classification to weather prediction and fraud detection [8]. Swain. et al. [2] utilized deep learning techniques for prediction of the effectiveness of lung cancer treatment by analyzing serial medical imaging data. Avanzo et al. [9] demonstrated the application of radiomics and deep learning in the context of lung cancer. Hyun et al. [10] elucidated one approach using machine learning that utilizes PET-based radiomics for making accurate prediction of historical lung cancer subtypes. Radhika et al. [11] conducted an analytical study regarding detection of lung cancer with the utilization of machine learning algorithms.

Research Design

How ML differs from traditional programming
Traditional programming relies on explicit rules written by developers, whereas ML is driven by data, learning patterns and making predictions. ML is particularly suited for complex, data-rich tasks like image recognition. In contrast, traditional programming excels in well-defined roles, such as database management, where explicit rules suffice. Moreover, in ML, the model training phase includes assigning new datasets to train the algorithm for predictions, showcasing the adaptability of ML systems. The choice between the two approaches depends on problem complexity and data availability as shown kin Figure 12.1 and 12.2.

Voting model
The main goal of employing voting models is to enhance the accuracy and strength of machine learning systems. After combining several individual model predictions, voting models significantly reduce the risks of overfitting and bolster the system's resilience against noise and outliers in the data. In the context of our research, we have incorporated five distinct machine learning models with the aim of predicting lung cancer.

Navie Bayes multinomial

The Bayes theorem is a mathematical formula that lets us figure out how likely it is for an event to happen based on the knowledge of an earlier event.

Formula can be used to calculate:

$$P(B|A)/P(B) * P(A) = P(A|B)$$

Support vector machine

For problems involving regression and classification, SVM is one of the best techniques under supervised category of machine learning. In order to maximize the margin between data points, it finds a hyperplane that best divides them into different classes as shown in Figure 12.3.

SVM divided into two types, based on the training sets,

- Linear SVM – A linear line makes it simple to segregate data points.
- Non-linear SVM – It is difficult to divide data points using a straight line.

Partial decision tree

PART is also tailored for classification tasks. It employs a decision tree approach to construct concise and intelligible trees by emphasizing the selection and refinement of attributes and values that offer maximum classification insight. Beginning with a complete decision tree, PART trims it by eliminating branches with limited impact on classification precision.

Random forest

This is the prevailing ensemble learning technique that is widely working in machine learning. It functions by

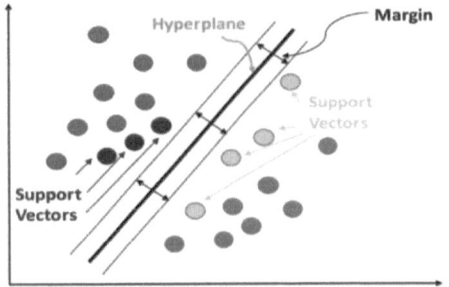

Figure 12.3 Working principle of SVM
Source: Author

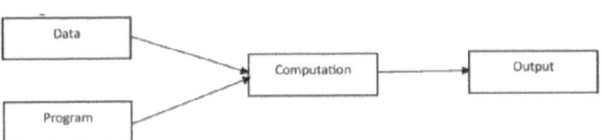

Figure 12.1 Traditional program workflow
Source: Author

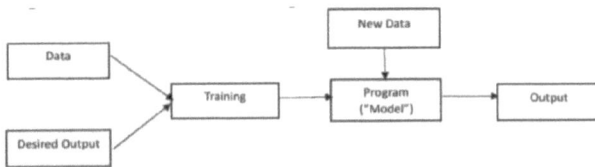

Figure 12.2 Workflow of machine learning processing
Source: Author

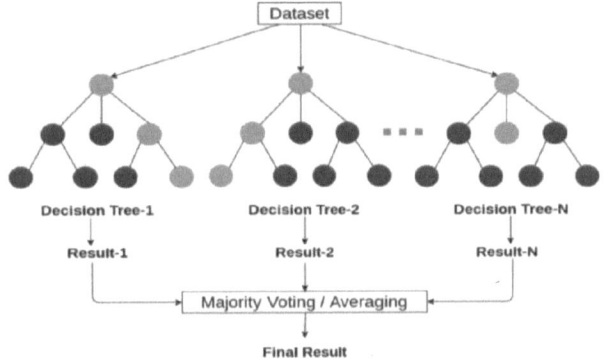

Figure 12.4 Graphical representation of RF
Source: Author

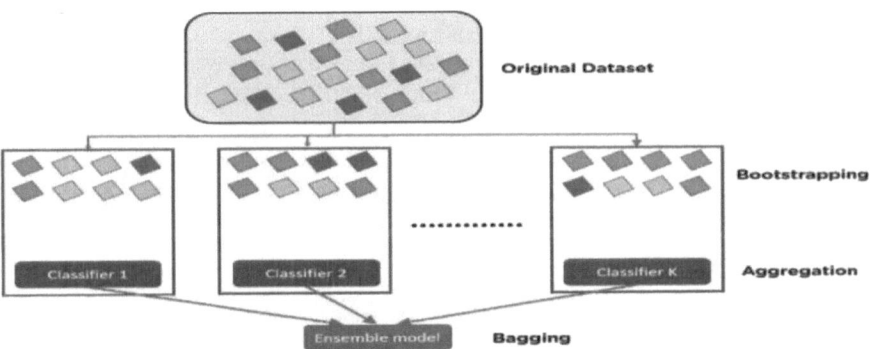

Figure 12.5 Working principle of bagging model
Source: Author

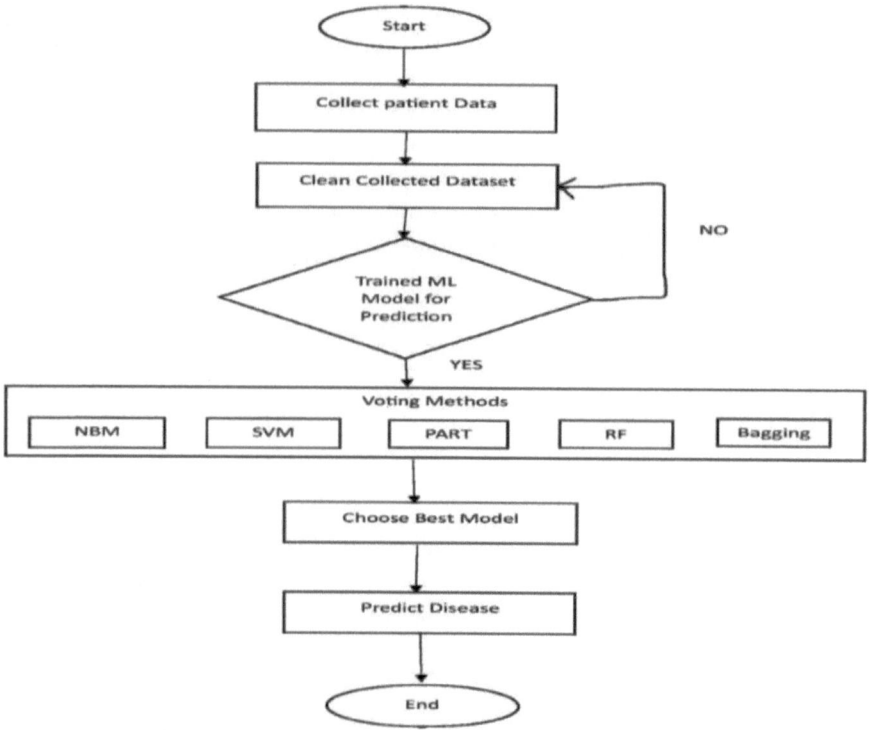

Figure 12.6 The flow chart of our proposed work
Source: Author

building many numbers of decision trees during training and amalgamating their forecasts for enhanced accuracy and robustness as shown in Figure 12.4. We can use it in encompassing image classification, bioinformatics, medical and financial analysis as shown in Figure 12.5.

Bagging

It is an ensemble learning method that helps decision trees and other machine learning models become more accurate and stable. It operates by bootstrapping (random sampling with replacement) to create several subsets of the training data, then training a base model on each subset. Bagging reduces overfitting, enhances model robustness, and is particularly effective when applied to unstable or high- variance algorithms as shown in Figure 12.6.

Research Methodology

Flow of work
Our proposed work is shown below in the flow chart in Figure 12.6.

Data collection
We gathered a dataset comprising 1000 individuals diagnosed with lung cancer, sourced from a Kaggle dataset. This dataset includes various parameters such

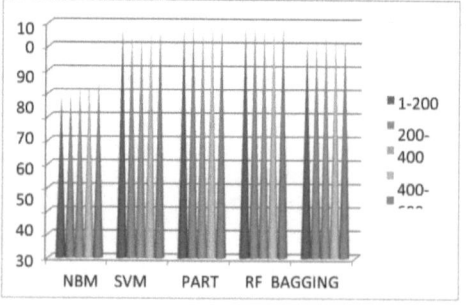

Figure 12.7 Graphical representation of NRM SVM PART
Source: Author

as age, lifestyle factors, gender and medical history, which encompass attributes like air pollution, smoking, chronic lung disease, and more. Our main objective is to leverage this data for predicting the stages of lung cancer. We strive to create a model that will aid to detect early and finding the staging of lung cancer, ultimately enhancing treatment effectiveness and patient care.

Clean dataset
By using the mean, mode, and median we clean the collected dataset, which helps ensure the dataset is more complete and ready for analysis by minimizing the impact of missing or outlier values.

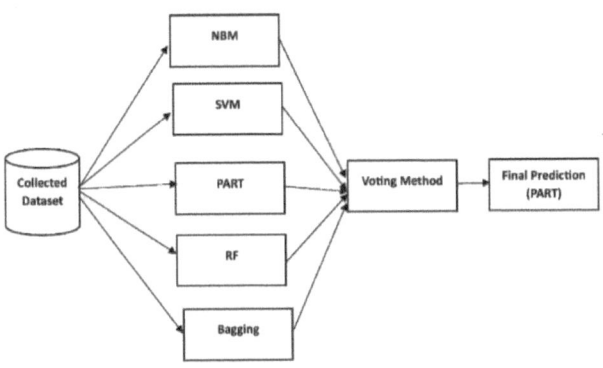

Figure 12.8 Workflow of voting model
Source: Author

Table 12.1 Comparison of accuracy.

Sl.	Name of model	Accuracy
1	Navie Bayes multinomial (NBM)	67.3%
2	Support vector machine (SVM)	92.96%
3	PART	95.23%
4	Random forest (RF)	93.17%
5	Meta bagging:	86.23%

Source: Author

Table 12.2 Accuracy table of voting model of NBM, SVM, PART, RF and bagging over five equal divided datasets.

Dataset	NBM	SVM	PART	RF	Bagging
1-200	68.34	97.98	97.48	98.22	91.95
200-400	69.69	93.93	98.98	98.98	90.4
400-600	72.73	93.93	95.95	97.97	94.44
600-800	73.73	95.45	98.48	97.97	93.43
800-1000	74.39	97.58	98.55	99.03	93.71

Source: Author

Mean: It is the average of all values in a dataset. Example: For the dataset {1,2,3,4,5}\{1, 2, 3, 4, 5\}, the mean is (1+2+3+4+5)/5=3(1 + 2 + 3 + 4 + 5) / 5 = 3.

Mode: Most frequently appearing value in a dataset. Example: In the dataset {1,2,3,4,5}\{1, 2, 3, 4, 5\}, the mode is 3 because it appears the most times.

Median: In one ascendingly ordered dataset, the value appearing in middle is called median. Example: For the dataset {1,2,3,4,5}\{1, 2, 3, 4, 5\}, the median is 3, as it is the middle value when the data is sorted.

Comparison Analysis

We compare used machine learning models in our research based on various evaluation metrics, which is a common practice in data analysis and model selection. Here are some key evaluation metrics like accuracy, RMSE, MAE, RRSE and RAE used for model comparison as shown in Figure 12.8.

It achieved individual accuracy scores for all the constituent models. Here are the accuracy scores for all the models in Table 12.1.

Here we found PART gives best accuracy than other four models.

Here we split our collected dataset into five equal set and apply models individually and store the error value in Tables 12.1–12.2. The pictorial representations of comparisons of different models are shown in Figure 12.8.

Figure 12.7 is the pictorial representation of comparison of NBM, SVM, PART, RF and bagging on the basis of accuracy.

Conclusion

In the field of lung cancer prediction, our comprehensive research has led to valuable insights and encouraging results. The PART model stood out as the top performer, achieving an outstanding accuracy of 95.23%. This underscores its potential as a powerful tool for early detection and risk assessment in lung cancer. However, it's important to recognize the significant contributions of other models such as support vector machine (SVM) and random forest (RF), which also demonstrated strong performances with accuracies of 92.96% and 93.17%, respectively.

Looking ahead, our research opens up several exciting avenues for further exploration. Firstly, incorporating larger and more diverse datasets promises to enhance the robustness of our predictive models. Additionally, integrating advanced deep learning techniques could further enhance precision and accurately prediction of lung cancer prediction. The exploration of real-time IoT data streaming for continuous monitoring and immediate risk assessment paves the way for dynamic and proactive healthcare. Lastly, emphasizing the interpretability and explainability of our models is crucial to ensure that both the medical community and patients can trust and benefit from these predictive tools. Our ongoing journey in the fight against lung cancer will undoubtedly be propelled forward by these initiatives, advancing this crucial field of research.

References

[1] Jayasingh, S. K., Gountia, D., Samal, N., & Chinara, P. K. (2021). A novel approach for data classification us-

ing neural networks. *IETE Journal of Research*, 69(9), 6022–6028.

[2]   Swain, S., Behera, N., Swain, A. K., Jayasingh, S. K., Patra, K. J., Pattanayak, B. K., Mohanty, M. N., Naik, K. D., & Rout, S. S. (2023). Application of IoT framework for prediction of heart disease using machine learning. *International Journal on Recent and Innovation Trends in Computing and Communication*, 11(10s), 168–176.

[3]   Avanzo, M., Stancanello, J., Pirrone, G., & Sartor, G. (2020). Radiomics and deep learning in lung cancer. *Strahlentherapie und Onkologie*, 196, 879–887.

[4]   Senapati, C., Patra, K. J., Swain, S., Jayasingh, S. K., Senapati, S., &Ray, N. K. (2024). A comprehensive approach to groundwater quality prediction and management in the state of Telangana, India, using stacking ensemble approach. *2024 International Conference on Emerging Systems and Intelligent Computing (ESIC)*, Bhubaneswar, India, 2024, pp. 118-123,

[5]   Jayasingh, S. K., Mantri, J. K., & Gahan, P. (2016). Comparison between J48 Decision Tree, SVM and MLP in weather forecasting. *International Journal of Computer Science and Engineering*, 3(11), 42–47.

[6]   Mantri, J. K. & Jayasingh, S. K. (2019). Soft computing techniques for weather change predictions In Delhi. *International Journal of Recent Technology and Engineering*, 8(4), 793–800.

[7]   Pradhan, K. & Chawla, P. (2020). Medical Internet of things using machine learning algorithms for lung cancer detection. *Journal of Management Analytics*, 7(4), 591–623.

[8]   R. P.R., R. A. S. Nair and V. G., "A Comparative Study of Lung Cancer Detection using Machine Learning Algorithms," *2019 IEEE International Conference on Electrical, Computer and Communication Technologies (ICECCT)*, Coimbatore, India, 2019, pp. 1–4,

[9]   Xu, Y., et al. (2019). Deep learning predicts lung cancer treatment response from serial medical imaging. *Clinical Cancer Research*, 25(11), 3266–3275.

[10] Hyun, S. H., Ahn, M. S., Koh, Y. W., & Lee, S. J. (2019). A machine-learning approach using PET-based radiomics to predict the histological subtypes of lung cancer. Clinical Nuclear Medicine, 44(12), 956–960.

[11] Radhika, P. R., Nair, R. A., & Veena, G. (2019, February). A comparative study of lung cancer detection using machine learning algorithms. In 2019 IEEE international conference on electrical, computer and communication technologies (ICECCT) (pp. 1–4). IEEE.

13 An ensemble approach for heart disease prediction

Avishek Mishra[1,a], Gopal Behera[1,b], Basanta Ku. Swain[1,c], Binayak Panda[2,d] and Monalisha Sahu[2,e]

[1]Department of CSE, GCEK, Bhawanipatna, Odisha, India
[2]Department of CSE, BPUT, Rourkela, Odisha, India

Abstract

Heart disease remains the leading cause of death globally. Early detection and diagnosis are vital for effective treatment and prevention. This research focuses on predicting the likelihood of heart disease using a voting ensemble method that combines multiple machines learning models. The dataset used for this study is sourced from publicly available medical records, containing various clinical and demographic attributes such as age, cholesterol, and sugar levels. The model was trained and evaluated using performance measures such as precision, accuracy, recall, and F1-score. The voting ensemble method demonstrated improved prediction performance by exploiting the strengths of specific models. This system also suggests additional medical tests for patients at high risk, contributing to more personalized healthcare recommendations. The proposed technique enhances reliability and precision compared to the baselines and thereby bolstering medical decision-making processes.

Keywords: Healthcare analytics, heart diseases, machine learning, medical diagnosis, voting ensemble

Introduction

According to WHO cardiovascular diseases (CVDs) are the foremost source of death globally, with an estimate of 17.9 million fatalities per year. The economic impact is equally intense, with countries lavish billions of dollars every year on treatments, hospitalizations, and etc. [14]. Given the multifactorial nature of heart disease, early detection is critical to managing the condition, improving patient outcomes, and reducing mortality rates. However, predicting heart disease is not straightforward, as it involves understanding a wide array of clinical factors, such as cholesterol, sugar, medical history, age, and lifestyle choices like diet, and physical activity. Traditional methods of diagnosing heart disease, though effective, often involve invasive procedures, high costs, and time-consuming processes, which may delay intervention [8]. With large datasets available from healthcare organizations and research studies, machine learning models have the potential to offer faster, more accurate predictions than traditional methods [3]. These models can continuously improve with more data, enabling them to adapt to the evolving landscape of healthcare [10]. While machine learning holds significant promise for heart disease prediction, its application is accompanied by several challenges [12,13]. Further, logistic regression may effectively model linear relationships between features and outcomes, while these models hold promise, they often struggle to effectively capture non-linear interactions [18].

To address this limitation, ensemble learning techniques, particularly voting systems, have emerged as a compelling solution [9]. The key purpose of this article is to enhance the accuracy of predicting disease using an ensemble voting technique that incorporates the diverse strengths of multiple algorithms.

Literature Review

Early machine learning models, especially logistic regression, have been widely used in this domain. Nichols et al. [14], split data into several branches using decision tress (DT) technique based on feature values, providing an intuitive representation of complex interactions between clinical variables.

Detrano et al. [8] demonstrated that random forests improved predictive accuracy for heart disease by averaging the biases of individual trees, thus creating a more balanced and generalizable model. Additionally, ensemble methods have shown promise in combining various models to enhance predictive capabilities. In a study by Weng et al. [18], ensemble methods leveraging voting systems demonstrated improved predictive accuracy in heart disease detection by aggregating the outputs of classifiers like decision trees, SVMs, and logistic regression. Each model contributes unique strengths: logistic regression is effective for capturing linear relationships, decision trees excel in handling non-linear patterns, random forests provide

[a]avishekmishra@gmail.com, [b]gbehera@gcekbpatna.ac.in, [c]bkswain@gcekbpatna.ac.in, [d]binayakpanda.cse@gmail.com, [e]rkmonalisha@gmail.com

DOI: 10.1201/9781003658221-13

robustness against overfitting, and SVMs are adept at processing high-dimensional data.

Proposed Model: Majority Voting Ensemble Model

In this section, we explored the proposed majority voting ensemble techniques for enhancing the prediction accuracy of heart disease. The proposed architecture is shown in Figure 13.1 and illustrates the complete data flow for the heart disease prediction system, starting from the collection of the training dataset and ending with the predicted output.

Once the data has been cleaned and transformed, it moves into the Model Training phase, where various machine learning techniques are trained. These techniques, are KNN [1], random forest, logistic regression, SVM [7], XGBoost [4–6], LightGBM, etc [15].

The system does not rely on a single model for its final decision. Instead, it uses a Majority Voting system, where the predictions from all models are combined to make the final prediction. evaluation, enhances the system accuracy and robustness, ensuring a reliable prediction for patients. Finally, the predicted output is generated, providing the user with the final diagnosis and any potential missing data points that were assumed during preprocessing. This system's design ensures that the decision-making process is both comprehensive and data-driven. The dataset utilized in this study was obtained from the Kaggle Heart Disease Dataset [11].

Majority voting

After all the individual models have made their predictions, a majority voting mechanism is adopted to estimate the final outcome as shown in Figure 13.2. In this ensemble approach, the final result is not determined by any single model but is based on the consensus of all models. If the majority of models (more than 50%) predict heart disease (1), the final output will be "heart disease detected."

The diagram showcases a majority voting ensemble approach for enhancement of the accuracy (ACC) of heart disease prediction by an ensemble approach that is majority voting technique. The individual outcomes are then aggregated to make a final decision based on the majority vote.

In this specific case, the output from each model is [1, 0, 1, 1, 1, 1, 0, 0, 1], indicating that five models predict heart disease (1), while three models predict no heart disease (0). The majority rule determines the final prediction: since the majority (5 out of 8) of the models favor the existence of heart disease, the ensemble predicts that the patient is likely to have heart disease. The key advantage of this ensemble approach is that it leverages the diversity and strengths of multiple algorithms. While some models may make errors or misclassify.

Result Analysis

We have verified the significance of our model on real world data and found that the majority voting mechanism improves the accuracy as compared to the baseline models as shown in Table 13.1. Raza [16] reported an accuracy (ACC) of 82.79%, a precision (76.70%), a high recall (91.83%), and an F1-score (83.20%) indicating moderate precision but strong recall. Amjed Almousa achieved 90% accuracy but did not provide specific values for precision, recall, or F1-score, limiting a full comparison. Atallah & Al-Mousa [2] demonstrated an accuracy of 90.25%, with a precision of 88.74%, recall of 91.15%, and an F1-score of 90.20, suggesting balanced performance across the metrics.

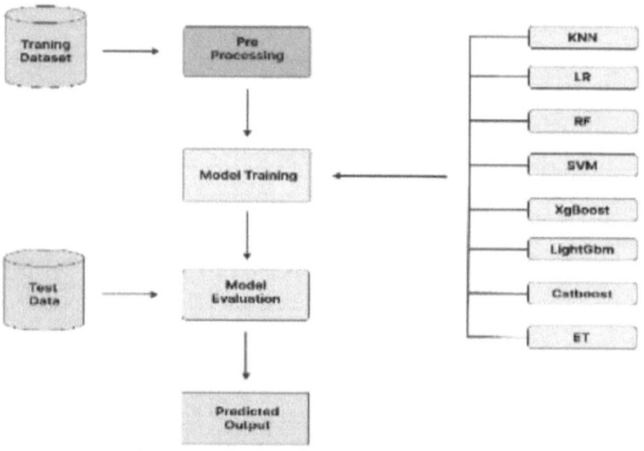

Figure 13.1 Model architecture of proposed ensemble voting method
Source: Author

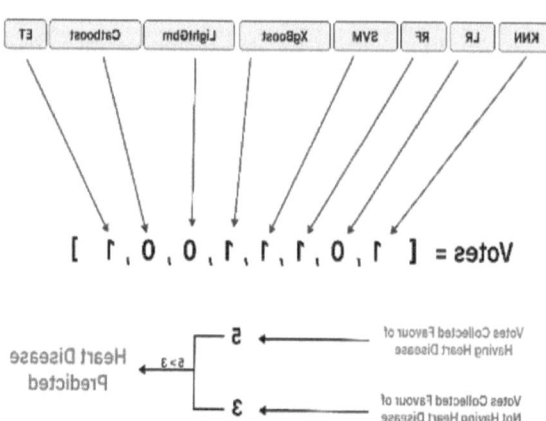

Figure 13.2 Working principle of majority voting model
Source: Author

Table 13.1 Performance comparison of MVE with eight algorithms with baseline models.

Authors	Techniques/algorithms	ACC	Precision	Recall	F1-score
Raza, 2019 [16]	Ensemble method (RFEC, Pruned J48, ELM, HGF, LR)	82.79	76.70	91.83	83.20
Atallah & Al-Mousa, 2019 [2]	Ensemble method (SGD, RF, KNN, LG)	90.25	88.74	91.15	90.2
Rafsun, et al., 2022 [17]	Ensemble method (XGB, LR, RF, K-NN)	96.75	97.24	95.91	96.77
Proposed	MVE (KNN, LR, RF, SVM, XGB, LightGBM, CatBoost, Extra Trees)	99.70	100	99.42	99.71

Source: Author

Rafsus et al. (2022) performed remarkably well, achieving 96.75% of accuracy, precision (97.24%), recall (95.91%), and F1-score of 96.77, reflecting strong predictive capability. In comparison the proposed majority voting with eight algorithms significantly outperformed the others, achieving precision of 100%, recall of 99.42%, accuracy of 99.70%, and F1-score of 99.71.

Conclusion

In this research problem, we have examined various machine learning (ML) approaches for the prediction of heart disease, leveraging advanced techniques such as ensemble approach to enhance prediction accuracy. The findings suggest that machine learning will play a significant factor in early diagnosis. Future work can focus on incorporating larger and more diverse datasets to improve generalizability, and incorporate advance machine learning techniques, may enhance performance by capturing more intricate features in patient data.

References

[1] Altman, N. S. (1992). An introduction to kernel and nearest-neighbor nonparametric regression. *The American Statistician*, 46(3), 175–185. KNN Overview.

[2] Atallah, R., & Al-Mousa, A. (2019). Heart disease detection using machine learning majority voting ensemble method. In Amman, Jordan, 2nd International Conference on new Trends in Computing Sciences (ICTCS), (pp. 1–6). IEEE.

[3] Attia, Z. I., Noseworthy, P. A., Lopez-Jimenez, F., Asirvatham, S. J., Deshmukh, A. J., Gersh, B. J., et al. (2019). An artificial intelligence-enabled ECG algorithm for the identification of patients with atrial fibrillation during sinus rhythm: a retrospective analysis of outcome prediction. *The Lancet*, 394(10201), 861–867.

[4] Behera, G., Panda, S. K., Hsieh, M. Y., & Li, K. C. (2024). Hybrid collaborative filtering using matrix factorization and XGBoost for movie recommendation. *Computer Standards and Interfaces*, 90, 103847.

[5] Behera, G., & Nain, N. (2019). Grid search optimization (GSO) based future sales prediction for big mart. In 2019 15th International Conference on Signal-Image Technology & Internet-Based Systems (SITIS). IEEE.

[6] Chen, T., & Guestrin, C. (2016). XGBoost: a scalable tree boosting system. In *Proceedings of the 22nd ACM SIGKDD International Conference on Knowledge Discovery and Data Mining*, (pp. 785–794).

[7] Cortes, C., & Vapnik, V. (1995). Support-vector networks. *Machine Learning*, 20(3), 273–297.

[8] Detrano, R., Janosi, A., Steinbrunn, W., Pfisterer, M., Schmid, J. J., Sandhu, S., et al. (1989). International application of a new probability algorithm for the diagnosis of coronary artery disease. *The American Journal of Cardiology*, 64(5), 304–310.

[9] Dietterich, T. G. (2000). Ensemble methods in machine learning. In International Workshop on Multiple Classifier Systems, (pp. 1–15). Berlin, Heidelberg: Springer.

[10] Gulshan, V., Peng, L., Coram, M., Stumpe, M. C., Wu, D., Narayanaswamy, A., et al. (2016). Development and validation of a deep learning algorithm for detection of diabetic retinopathy in retinal fundus photographs. *The Journal of the American Medical Association (JAMA)*, 316(22), 2402–2410.

[11] Kaggle (2019). Heart Disease Dataset. [Online] Available at: https://www.kaggle.com/datasets/johnsmith88/heart-disease-dataset [Accessed 02-10-2024].

[12] Malik, S., Patro, S. G. K., Mahanty, C., Lasisi, A., & Al-Sareji, O. J. (2024). Hybrid raven roosting intelligence framework for enhancing efficiency in data clustering. *Scientific Reports*, 14(1), 20163.

[13] Marateb, H. R., Mansourian, M., & Rezaei Tavirani, M. (2014). A hybrid intelligent system for diagnosing coronary artery disease based on rules and decision trees. *Journal of Medical Systems*, 38(8), 1–12.

[14] Nichols, M., Townsend, N., Scarborough, P., & Rayner, M. (2014). Cardiovascular disease in Europe: epidemiological update. *European Heart Journal*, 35(42), 2950–2959.

[15] Prokhorenkova, L., Gusev, G., Vorobev, A., Dorogush, A. V., & Gulin, A. (2018). CatBoost: unbiased boosting with categorical features. *Advances in Neural Information Processing Systems*, 31, 6638–6648.

[16] Raza, K. (2019). Improving the prediction accuracy of heart disease with enseble learning and majority vot-

ing rule. In U-Healthcare Monitoring Systems, (pp. 179–196), s.l.: Elsevier Inc.

[17] Jani, R., Shanto, M. S. I., Kabir, M. M., Rahman, M. S., & Mridha, M. F. (2022). Heart disease prediction and analysis using ensemble architecture. In 2022 International Conference on Decision Aid Sciences and Applications (DASA), (pp. 1386–1390). Chiangrai, Thailand, IEEE.

[18] Weng, S. F., Reps, J., Kai, J., Garibaldi, J. M., & Qureshi, N. (2017). Can machine-learning improve cardiovascular risk prediction using routine clinical data? *Plos One*, 12(4), e0174944.

14 Hierarchically vectorized algorithmic paradigm for probabilistic node localization and pervasive threat detection in synthetically optimized WSN-assisted IoT architectures

Sudheer Nidamanuri[1,a], Balingan Sangameshwar[1,b], U. Ganesh Naidu[2,c], Balu Srinivasulu[3,d], Syed Muqthadar Ali[4,e] and Mummalaneni Rajasekar[1,f]

[1]Department of CSE(CyS,DS) and AI&DS,VNR Vignana Jyothi Institute of Engineering and Technology, Hyderabad, India

[2]Department of CSBS, B V RAJU Institute of Technology, Narsapur, Telangana, India

[3]Department of Computer Science and Engineering, PACE Institute of Technology and Sciences, Ongole, India

[4]Department of Computer Science and Engineering, CVR College of Engineering, Telanagana, Hyderabad, India

Abstract

Wireless sensor networks (WSNs) play a key role in IoT systems, where node localization and security are essential for network functionality and protection against intrusions. Existing approaches like WOGRU-IDS and DEEC-G face challenges such as high false positive rates and reduced efficiency in large networks. To address these limitations, the paper introduces the Hy-CapSW framework, which integrates the hydrodynamic crayfish optimization algorithm for probabilistic node localization and a dual threat detection approach combining a vectorized adaptive capsule neural network with strategic warfare algorithmic optimization. Trained on the WSN-DS dataset, Hy-CapSW improves node localization accuracy to 97.5%, attack detection to 96.7%, and prediction accuracy to 98.7%, outperforming current methods in both efficiency and adaptability. This innovative system provides a robust solution for WSN-assisted IoT architectures.

Keywords: Attack detection, hydrodynamic crayfish optimization, node localization, strategic warfare algorithmic optimization, vectorized adaptive capsule neural network, wireless sensor networks, WSN-DS

Introduction

Wireless sensor networks (WSNs) have been applied in various fields including health and manufacturing due to their efficiency in data acquisition and communication [3]. However, as these networks based on wireless communication, they are vulnerable to a vast host of security threats – including malicious ones [2]. A key activity in WSNs is node localization, to ensure that the network continues to be core [1]. Effective attack detection mechanisms are crucial to improving WSN security and reliability [6]. WSN-DS provides network traffic and attack data for threat detection and prevention in WSNs [4, 5]. WSN-DS enables ML and DL to detect attacks and identify malicious nodes [3] these methods increase the network's ability availability and integrity.

Related Works

In 2022, Ramana et al. [7], introduced WOGRU-IDS, an intelligent intrusion detection system for IoT-assisted WSN.

In 2022, Jingjing et al. [8], introduced an intrusion detection model for WSNs using Multiple Convolutions and GRU (MC-GRU).

In 2023, Aroba et al. [9], developed a hyper-heuristic DEEC-Gaussian gradient distance algorithm to enhance node localization accuracy in WSNs.

In 2024, Niranjan et al. [10], introduced PSODESADV-Hop for sensor node localization, enhancing hop size estimation in three phases.

Proposed Methodology

WSNs in IoT leverage hydrodynamic crayfish optimization for efficient localization and V-AdCapNet

---

[a]nidamanuri.sudheer@gmail.com, [b]sagameshwar_b@vnrvjiet.in, [c]ganesh613@gmail.com, [d]balusrini@gmail.com, [e]sma.cvrce@gmail.com, [f]rajasekar_m@vnrvjiet.in

DOI: 10.1201/9781003658221-14

with SWAO for precise attack detection. Tests on the WSN-DS dataset confirm effective threat identification and positioning.

Data gathering
Our study uses the WSN-DS dataset, which captures sensor node activity in (WSNs), positions, signal intensities, and traffic flows.

Node localization
The collected data is then used to cluster the nodes for localization of the node through (HyCO) [11].

Initializing node population
In WSN, the position of each node L_n is defined as (1),

$$L_n = \{L_{n,1}, L_{n,2}, ..., L_{n,p}\} \tag{1}$$

where the position $L_{n,p}$ of node n in dimension p is calculated as (2),

$$L_{n,p} = lb_p + (ub_p - lb_p) \times z \tag{2}$$

Defining node fitness
The fitness F_{node} measures how close a node is to the best position, which is defined as (3),

$$F_{node} = A_1 \times \frac{1}{\sqrt{2\pi\sigma}} \times \exp\left(-\frac{(\Theta - \mu)^2}{2\sigma^2}\right) \tag{3}$$

Exploration phase
The new position L_{shade} is calculated as the average of the best-known global position L_G and best-known local position L_L as given in (4),

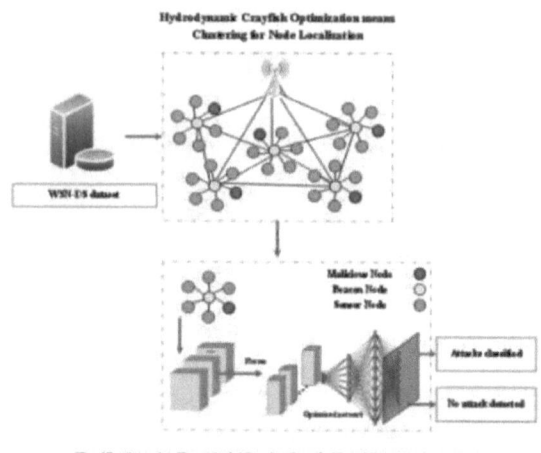

Figure 14.1 The architecture of the proposed Hy-CapSW

Source: Author

$$L_{shade} = \frac{L_G + L_L}{2} \tag{4}$$

Then the nodes move towards the new position using (5),

$$L_{n,p}^{(u+1)} = L_{n,p}^{(u)} + A_2 \times z \times (L_{shade} - L_{n,p}^{(u)}) \tag{5}$$

Exploitation phase
The node adjusts its position based on the randomly chosen node L_y as in (6),

$$L_{n,p}^{(u+1)} = L_{n,p}^{(u)} - L_y + L_{shade} \tag{6}$$

Where y is determined as (7),

$$y = round(z \times (M - 1)) + 1 \tag{7}$$

error size ξ is calculated as (8),

$$\xi = A_3 \times z \times \left(\frac{F_{node}}{F_{signal}}\right) \tag{8}$$

node adjusts its position using (9),

$$\begin{aligned} L_{n,p}^{(u+1)} &= L_{n,p}^{(u)} + L_{signal} \times \lambda \times (\cos(2\pi z) - \sin(2\pi z)) \; if \; \xi \; is \; large \\ L_{n,p}^{(u+1)} &= \left(L_{n,p}^{(u)} - L_{signal}\right) \times \lambda + \lambda \times z \times L_{n,p}^{(u)} \qquad if \; \xi \; is \; managable \end{aligned} \tag{9}$$

Attack detection
After node localization, attack detection is performed using V-AdCapNet and SWAO to ensure network security.

Vectorized adaptive capsule neural network
In attack detection for WSNs, the (V-AdCapNet) uses capsules to encode node coordinates or attack information.

a.   Transformation to Vectorized Capsule
V-AdCapNet, convolutional features capturing perception details are stacked to generate final capsule vectors for attack detection. (10),

$$H_{p,q} = T_{p,q} \times node_p \tag{10}$$

The vector is then summed with a weight r_q as (11),

$$r_q = \sum_p C_{p,q} H_{p,q} \tag{11}$$

Additionally, an agreement score $B_{p,q}$ is introduced to represent the consistency between the two capsules, updated as (12),

$$C_{p,q} = \frac{\exp(B_{p,q})}{\sum_K \exp(B_{p,K})} \tag{12}$$

The value of $B_{p,q}$ is updated on the match between the low-layer and high-layer capsules as (13),

$$B_{p,q} = B_{p,q} + H_{p,q} \cdot node_q \tag{13}$$

b. Learning Process in V-AdCapNet

Firstly, we initialize the learnable transformation matrix $T \in \Re^{p \times q \times r}$, The projection z on all the subspaces can be obtained as (14),

$$Z = \arg \min_{Z \in span(T)} \|H - Z\| \tag{14}$$

T^+ can be defined as (15),

$$T^+ = \left(T^H T\right)^{-1} T^H \tag{15}$$

class of each node in feature map $p \times q$ is given as (16),

$$\|Z\|^2 = \sqrt{Z^T Z} = \sqrt{H^T T (T^H T)^{-1} T^H H} \tag{16}$$

where $\|Z\|^2 \in \Re^{p \times q \times r}$.

SoftMax function defined as (17),

$$P(class - i|Z) = \frac{e^{\|z_i\|^2}}{\sum_j e^{\|z_j\|^2}} \tag{17}$$

actual attack class labels for the nodes, which is defined as (18),

$$loss = -\sum_i y_i \log P(class = i|Z) \tag{18}$$

Strategic warfare algorithmic optimization

The SWAO optimizes the loss function of V-AdCapNet.

Fitness function: The fitness function plays scalar role in optimization process by assessing quality solution defined in (19),

$$Fitness = \lambda_1 A(\theta) - \lambda_2 loss(\theta) \tag{19}$$

a. Attack strategy for rank and weight update mechanism

At the start of a network, all nodes within the network are equal and possess the same weight and rank. The position update for a sensor node $Z_k(u+1)$ at iteration $u+1$ is given by (20),

$$Z_k(u+1) = Z_k(u) + 2 \times \gamma \times (M - Q) + \varepsilon(X_k \times M - Z_k(u)) \tag{20}$$

The rank R_k and weight X_k of the sensor node are updated as (21) and (22),

$$R_k = \begin{cases} (R_k + 1), & V_{new} \geq V_{prev} \\ R_k & V_{new} < V_{prev} \end{cases} \tag{21}$$

$$X_k = X_k \times \left(1 - \frac{R_k}{Max\_iter}\right)^\beta \tag{22}$$

b. Defense strategy for exploration and replacement of weak nodes

The second strategy updates the sensor nodes' positions using the given equation (23),

$$Z_k(u+1) = Z_k(u) + 2 \times \gamma \times (M - Z_{rand}(u)) + \varepsilon \times X_k \times (Q - Z_k(u)) \tag{23}$$

A simple replacement strategy involves replacing weak node with a randomly selected node as (24),

$$Z_{weak}(u+1) = L_b + \varepsilon(U_b - L_b) \tag{24}$$

Results and Discussion

This section presents experimental findings and Hy-CapSW simulation results in Table 14.1.

Table 14.1 shows that our research is performed in sensing area of 120×120 m^2, in which WSN is set up.

Dataset description

The WSN-DS dataset benchmarks IDS in WSNs, simulating various attack types. It contains 18 features and five classes.

Performance analysis

Our study evaluates Hy-CapSW approach based on localization accuracy, security, and computational complexity.

Figure 14.2 Hy-CapSW outperforms others, lowest localization error (0.12%) 100% detection accuracy in Normal, 98–99% in others.

Table 14.1 Parameters in Hy-CapSW simulation [7–10].

Parameters	Values
Sensing area	120×120 m^2
No. of sensor nodes	200–1000
Communication radius	500 m
Software	MATLAB
No. of clusters	15
Data size	5000 KB

Source: Author

Table 14.2 Performance analysis of Hy-CapSW [7-10].

Methods	Clustering Accuracy (%)	Energy Efficiency (%)	Node Localiz- ation Accuracy (%)	Prediction Accuracy (%)
WOGRU-IDS [7]	91.5	86	96.2	95.7
MC-GRU [8]	89.7	83	94.5	94.0
DEEC-G [9]	87.9	80	93.1	92.5
PSODESA [10]	84.2	77	90.7	89.2
Hy-CapSW (proposed)	95	91	97.5	98.7

Source: Author

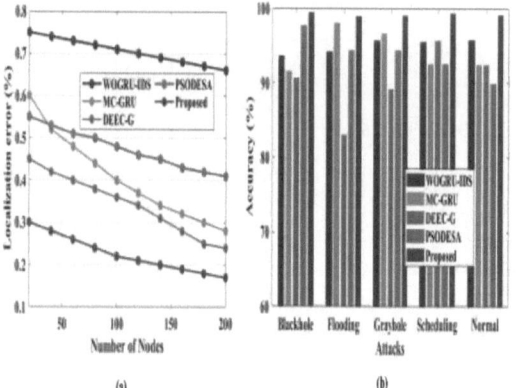

Figure 14.2 Performance analysis of Hy-CapSW
Source: Author

Conclusion

Hy-CapSW framework enhances WSN security in IoT, achieving 97.5% localization accuracy and 98.7% prediction accuracy. Simulations on WSN-DS dataset validate its scalability, future work focusing real-time advanced ML techniques.

References

[1] Gebremariam, G. G., Panda, J. & Indu, S. (2023). Localization detection multiple attacks in wireless sensor networks ANN. *Wireless Communications and Mobile Computing*, 2023(1), 2744706.

[2] Gebremariam, G. G., Panda, J. & Indu, S. (2023). Secure localization techniques in wireless sensor networks against routing attacks based on hybrid machine learning models. *Alexandria Engineering Journal*, 82, 82–100.

[3] Zheng, L. (2024). An improved localization approach based on Sybil attack for WSN. *Physical Communication*, 63, 102283.

[4] Vellela, S. S., & Balamanigandan, R. (2024). Efficient attack detection and prevention approach for secure WSN mobile cloud environment. *Soft Computing*, 28(19), 11279–11293.

[5] Jayadhas, S. A., Roslin, S. E., & Florin, W. (2024). Emerging network communication for malicious node detection in wireless multimedia sensor networks. *Optical and Quantum Electronics*, 56(1), 59.

[6] Moundounga, A. R. A., & Satori, H. (2024). Stochastic machine learning based attacks detection system in wireless sensor networks. *Journal of Network and Systems Management*, 32(1), 17.

[7] Ramana, K., Revathi, A., Gayathri, A., Jhaveri, R. H., Narayana, C. L., & Kumar, B. N. (2022). WOGRU-IDS—an intelligent intrusion detection system for IoT assisted WSN. *Computer Communications*, 196, 195–206.

[8] Jingjing, Z., Tongyu, Y., Jilin, Z., Guohao, Z., Xuefeng, L., & Xiang, P. (2022). Intrusion detection model for wireless sensor networks based on MC-GRU. *Wireless Communications and Mobile Computing*, 2022(1), 2448010.

[9] Aroba, O. J., Naicker, N., & Adeliyi, T. T. (2023). Node localization in wireless sensor networks using a hyperheuristic DEEC-Gaussian gradient distance algorithm. *Scientific African*, 19, e01560.

[10] Niranjan, M., Gupta, S., & Singh, B. (2023). Sensor node localization with improved hop-size using PSODE-SA optimization. *Wireless Networks*, 29(4), 1911–1934.

[11] Jia, H., Rao, H., Wen, C. &Mirjalili, S., 2023. Crayfish optimization algorithm. *Artificial Intelligence Review*, 56(Suppl 2), 1919–1979.

15 A six element antenna array for 5G application

Sarmistha Satrusallya[1,a], Shaktijeet Mahapatra[2,b], Sharmila, K. P.[3,c] and Mihir Narayan Mohanty[1,d]

[1]Applied Signal Processing Lab, ITER, Siksha 'O' Anusandhan (deemed to be University), Bhubaneswar, Odisha, India

[2]Center for Internet of Things, CSE, ITER, Siksha 'O' Anusandhan (deemed to be University), Bhubaneswar, Odisha, India

[3]Department of Electronics and Communication Engineering, CMR Institute of Technology, Bengaluru, Karnataka, India

Abstract

For fulfilling the need to increase the transmission distance, lower propagation and widening cellular coverage, sub-6G has been implemented. In this paper, we propose a six-element series antenna array. The antenna has a dimension of 20 mm × 25 mm × 0.8 mm. It resonates at 30 GHz and 40 GHz with a gain of 6 dBi and 6.5 dBi. It exhibits a bandwidth of 1.75 GHz and 2.24 GHz around respective resonant frequencies. The antenna has radiating efficiency of 69.7% and 70.8% at resonating frequencies. The antenna was found to be a suitable candidate for applications in 5th generation communication.

Keywords: 6G, array antenna, gain, series feed

Introduction

The requirement of antenna in any form of communication is essential for transmission and reception of signal. The concept of 5G in wireless communication has become a new trend in improving communication. 5G wireless communication is capable for using higher frequencies as compared to 4G communication. The 5G frequency range includes the frequencies used in current cellular networks. The higher frequency operation requires a miniature antenna with wide bandwidth.

The miniature micro strip antennas are considered as one of the significant components for high frequency transmission. The choice of antenna depends on the device's characteristics and ensures connectivity. That's why the design of antenna is a major research issue among researchers.

A variety of antenna for different application with different design are proposed by researchers to enhance the communication [6]. The micro strip antennas design contains patches and ground plane of various shapes, material of the substrate and different feeding technique to match with the device. A compact micro strip antenna for multi band operation was analyzed for a frequency range of 2.39GHz –3.15GHz. The antenna patch is of rectangular shape along with two slotted patch om RT Duroid 5880. The antenna gain is of 5.01dB [4]. A compact monopole antenna was designed with a size of 9.45x18.5 mm². The antenna had a rectangular radiator with a ground plane. It also has a four-sided slit. The L shaped stub was used to enhance the performance in Bekasiewicz and Koziel, [2].

Higher gain and wide bandwidth are the requirement of all devices to be used in any communication band. Antenna array is the one solution to full fill the demand for enhanced gain and bandwidth in Kumar et al. [5], Nagendra and Swarnalatha [10], Palanisamy and Subramani [12], Raad, [14]. It can provide both space diversity and time diversity. The MIMO antenna in Raad, [14] had four number of elements. A sickle-shaped radiator with a L-slit was considered as the antenna elements and a CSRR to reject bands from the frequency range. The bands include Bluetooth band, downlink band of X-band satellites and WLAN band. A tapered line was fed for the elements in the array. Every element contains a complementary slot in the ground plane. An MIMO antenna with eight port was analyzed in Palanisamy and Subramani, [12] that have eight number of elements. The elements of array were designed with modified circular patch. The patch was present on the same side of the ground plane with a metallic vias. The feed line was situated on the top. MIMO based IoT and wireless charging platforms need an array antenna with flexibility. To achieve higher gain the placement of array elements

[a]sarmisthya.satrusallya.19@gmail.com, [b]shaktijeetmahapatra@gmail.com, [c]sharmila.kp@cmrit.ac.in, [d]mihir.n.mohanty@gmail.com

DOI: 10.1201/9781003658221-15

is considered. The substrate material was Kepton Polyamide extending its applicability for UWB application, wearable devices and wireless charging. The array in Raad [13], Raad [14] was suitable for beam switching application. An array design with miniature patch was also studied in Olan-Nuñez et al. [11]. The slotted array with gain more than 6dB was proposed for ISM band. The array has an efficacy of more than 90%. A slot antenna array having a gain of more than 9.5dBi was proposed for use in 5G communication. The antenna exhibited an 84% radiation efficiency, making it suitable for integration with sensors for 5G communication in Varum and Matos [17]. Antenna arrays are developed for large-scale IoT deployment that are also weatherproof, encompassing both LTE and 5G sub-6GHz [3]. Researchers investigated log-periodic antenna arrays in [16]. The array are series fed and are of three numbers Each element had an inset-fed square radiator. For the 5G communication antenna array was designed using rectangular patches at 28GHz with a gain of 8dB [9]. Another thinned array of 34 elements was also proposed in Moon et al. [8] for 5G communication with sidelobe suppression having complex design. A MIMO antenna with two elements on a FR4 substrate was analyzed for sub 6G operation. The array with defective ground structure had a gain of 10dB [15]. A comparative analysis of antenna arrays was done in Mohanty et al. [7] for finding the optimal number of elements and distance between the elements so that a high gain and a low sidelobe are found.

The work organization is as follows. Part 2 gives an emphasis on antenna design; the result and analysis is followed by part 3 and part 4 concludes the outcome of the paper.

Antenna Design

A series fed antenna array of six elements is considered on FR4 substrate. FR4 is selected for its low cost, high dielectric strength, high strength to weight ratio and resistance to the moisture. The patch is considered to be rectangular shape of different size connected with micro strip feed.

The geometrical representation and dimension of the proposed antenna in millimeter scale are described in Figures 15.1 and 15.2. The array is designed with six rectangular patches positioned linearly with an difference of 1mm with each. The dimension of each element is calculated as per the requirement. The rectangular patches are arranged in the ascending order of length. The increment is symmetrical. The ground plane is full, having a dimension of 25 mm × 20 mm. To have an enhanced gain micro strip feed is used in

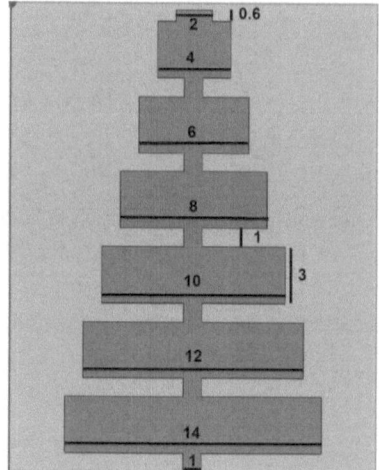

Figure 15.1 Top view of proposed array antenna
Source: Author

Figure 15.2 Back view of proposed array antenna
Source: Author

the design. The antenna design use FR4 of 0.8mm thickness.

Antenna patch design calculation

Each element of the antenna array is rectangular. Some important parameters like resonant frequency, substrate height and effective relative dielectric constant have been taken into consideration.

The width of the rectangular patch (b) is considered by Balanis [1],

$$b = \frac{c}{2f_o}\sqrt{\frac{2}{\varepsilon_r + 1}} \tag{1}$$

Here c = velocity of light in free space
The relative effective dielectric constant (ε_{reff}) is given by,

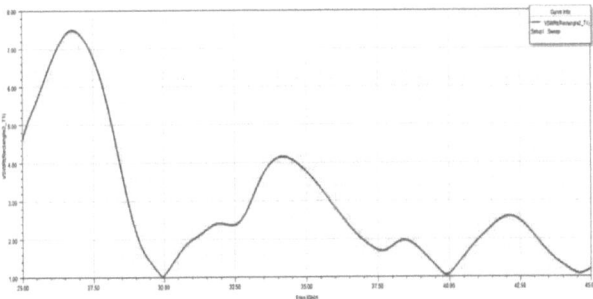

Figure 15.3 VSWR of proposed array antenna
Source: Author

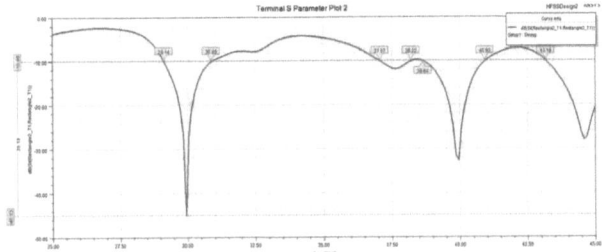

Figure 15.4 S_{11} Vs frequency plot of the antenna array
Source: Author

$$\varepsilon_{reff} = \frac{\varepsilon_{relative}+1}{2} + \frac{\varepsilon_{relative}-1}{2}\left[1+12\frac{h}{b}\right]^{1/2} \qquad (2)$$

The effective length (ε_{eff}) is calculated with the following mathematical equation

$$a_{eff} = \frac{c}{2f_o\sqrt{\varepsilon_{eff}}} \qquad (3)$$

The extension in length (Δa) is described as,

$$\Delta a = 0.4121h\frac{\left(\varepsilon_{reff}+0.31\right)\left(\frac{b}{h}+0.2641\right)}{\left(\varepsilon_{reff}-0.258\right)\left(\frac{b}{h}+0.81\right)} \qquad (4)$$

The length of the patch (L) is given by,

$$a = a_{eff} - 2\Delta a \qquad (5)$$

Results and Discussion

This intention of the work is to design the array having higher gain and wide bandwidth. The antenna performance depends on the antenna compatibility with the surroundings and its return loss. Figure 15.3 describes the VSWR value of the array, whereas the variation of S_{11} with respects to the frequency is presented in Figure 15.4.

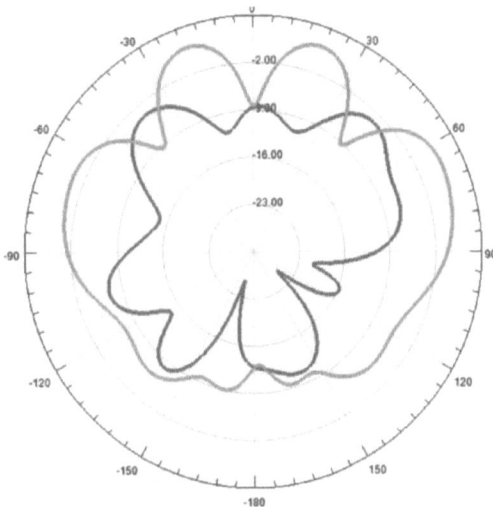

Figure 15.5 Antenna array radiation pattern at 30GHz
Source: Author

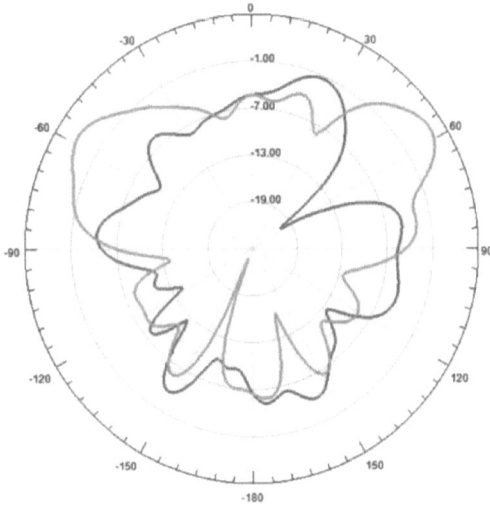

Figure 15.6 The antenna array radiation pattern at 40GHz
Source: Author

The VSWR of the array lie in the prescribed range. The value represents the working of antenna in the prescribed range of frequency.

The antenna resonates at 30 GHz and 40 GHz with a return loss of -45 dB and -32dB. The array has a wide bandwidth of 1.75 GHz and 2.24 GHz at the respective frequencies.

The antenna radiation pattern is described in Figures 15.5 and 15.6. The array has a gain of 6dB and 6.5dB at the respective frequencies.

Conclusion

In this work, we propose an antenna array with a dimension of $2.5 \times 2 \times 0.08$ cm³. The antenna provides a multiband operation at 30 GHz and 40 GHz with a return loss of -45 dB and -32 dB. Thearray has a bandwidth of 1.75 GHz, 2.24 GHz at the resonating frequency. The gain value at the frequencies are 6 dB and 6.5 dB. The miniature size of the antenna makes it suitable for different applications in sub 6G communication.

References

[1] Balanis, C. (2016). Antenna Theory: Analysis and Design. John Wiley and Sons.

[2] Bekasiewicz, A., & Koziel, S. (2016). Compact UWB monopole antenna for internet of things applications. *Electronics Letters*, 52(7), 492–494. https://doi.org/10.1049/el.2015.4432.

[3] Chi, L., Weng, Z.-B., Meng, S., Qi, Y., Fan, J., Zhuang, W., et al. (2020). Rugged linear array for IoT applications. *IEEE Internet of Things Journal*, 7(6), 5078–5087.

[4] Katoch, S., Jotwani, H., Pani, S., & Rajawat, A. (2015). A compact dual band antenna for IOT applications. In 2015 International Conference on Green Computing and Internet of Things (ICGCIoT), (pp. 1594–1597).

[5] Kumar, P., Urooj, S., & Malibari, A. (2020). Design and implementation of quad-element super-wideband MIMO antenna for IoT applications. *IEEE Access*, 8, 226697–226704. https://doi.org/10.1109/ACCESS.2020.3045534.

[6] Mohanty, M. N., Mahapatra, S., & Chae, G.-S. (2022). Wideband wearable antenna for iot and medical applications. In Printed Antennas, (pp. 297–313). CRC Press.

[7] Mohanty, M. N., Mahapatra, S., Satrusallya, S., & Pandit, A. K. (2022). Design of high gain and low side lobe smart antenna array for iot applications on human monitoring. In Smart Antennas, (pp. 267–283). Springer.

[8] Moon, C.-B., Jeong, J.-W., Nam, K.-H., Xu, Z., & Park, J.-S. (2021). Design and analysis of a thinned phased array antenna for 5G wireless applications. *International Journal of Antennas and Propagation*, 2021(1), 3039183.

[9] Nabil, M., & Faisal, M. M. A. (2021). Design, simulation and analysis of a high gain small size array antenna for 5G wireless communication. *Wireless Personal Communications*, 116(4), 2761–2776. https://doi.org/10.1007/s11277-020-07819-9.

[10] Nagendra, R., & Swarnalatha, S. (2021). Design and performance of four port MIMO antenna for IOT applications. *ICT Express*, xxxx, 4–7. https://doi.org/10.1016/j.icte.2021.05.008.

[11] Olan-Nuñez, K. N., Murphy-Arteaga, R. S., & Col\'\ in-Beltrán, E. (2020). Miniature patch and slot microstrip arrays for IoT and ISM band applications. *IEEE Access*, 8, 102846–102854.

[12] Palanisamy, P., & Subramani, M. (2021). Design of metallic via based octa-port UWB MIMO antenna for IoT applications. *IETE Journal of Research*. 69(5), 2446–2456. https://doi.org/10.1080/03772063.2021.1892540.

[13] Raad, H. (2019). A yagi-uda antenna array for conformal IoT and wireless charging applications. *Microwave and Optical Technology Letters*, 61(3), 633–637.

[14] Raad, H. K. (2018). An UWB antenna array for flexible IoT wireless systems. *Progress in Electromagnetics Research*, 162, 109–121.

[15] Sabaawi, A. M. A., Muttair, K. S., Al-Ani, O. A., & Sultan, Q. H. (2022). Dual-band MIMO antenna with defected ground structure for sub-6 GHz 5G applications. *Progress in Electromagnetics Research C*, 122, 57–66.

[16] Varum, T., Caiado, J., & Matos, J. N. (2021). Compact ultra-wideband series-feed microstrip antenna arrays for iot communications. *Applied Sciences (Switzerland)*, 11(14), 1–15. https://doi.org/10.3390/app11146267.

[17] Varum, T., & Matos, J. N. (2019). Compact slot antenna array for 5G communications. In 2019 IEEE International Symposium on Antennas and Propagation and USNC-URSI Radio Science Meeting, (pp. 1415–1416).

16 Experimental investigation on semantic communication system for speech signal based on deep learning approach

S. K. Singh[a], J. Mehena[b], S. B. Kar[c], S. Rout[d], B. Behera[e] and P. K. Mahapatra[f]

Department of Electronics and Telecommunication Engineering, School of Engineering and Technology, DRIEMS University, Odisha, India

Abstract

In this work, a specific type of semantically based system of communication for signals from speech is discussed, Inspired by the advances on deep learning (DL), we try to come back the speech signals that are communicated during semantic communication systems that reduce mistakes across the semantic levels as opposed to compared by the bit or symbol levels used in conventional communication systems. Specifically, in light of a monitoring mechanism squeeze-and-excitation (SE) network are utilized to create and define a system that operates from end to the end, learning and gathers the most important speech data. Additionally, in order to enable the suggested semantic communication based on deep learning to function effectively. In real-world communication circumstances, a model that produces good effectiveness in managing diverse channel settings without going through a retraining procedure. The results of simulations show that the suggested semantic communication based on deep learning performs more effectively. It is easier to adapt for channel improvements than traditional systems for communication, especially when it comes to low signal to noise ratio (SNR).

Keywords: Deep learning, SE network, semantic communication, SNR

Introduction

Artificial intelligence (AI) has been growing rapidly in the last several years. As a result, intelligent applications have never required such high transmission efficiency, which presents a significant challenge to traditional communication systems. Sending and receiving semantic information is the ultimate aim of semantic communications [6]. It has the potential to greatly increase transmission efficiency. However, the lack of a mathematical model makes it difficult to quantify semantic information. Many problems can now be solved without the use of a mathematical model thanks to advances in deep learning (DL). Semantic information exchange can now be facilitated by DL enabled semantic communications [5].

In order to build an integrated semantic-channel coding system for communication circumstances including speech input from the user, an innovative semantic system for communication called DeepSCS is proposed. A semantic transmitter based on recurrent neural network (RNN) and neural networks with convolution (CNN) was employed, in order to produce a variety of system outputs, we create tasks for voice synthesis and recognition [4]. In particular, a feature decoder transforms the text information into the received semantic the characteristics associated to text [1]. Furthermore, utilizing the obtained text and speaker information, a combination of a CNN along with RNN-based neural network [3] models is employed to rebuild the speech sequence. In order to produce the synthesized speech an operable [7] user interface for a DeepSCS demonstration is built.

The provisions of section 2 provides information on the proposed DeepSCS. Section 3 analyses the experimental results and analysis, whereas Section 4 offers conclusions along with references.

Proposed Speech Semantic Communication System

In this work, the speech coding is done using a conv2D with attention-based, while the channel coding is done using a 2D Convolution Neural Network (CNN) is described as follows.

Model description

The collection sequences of speech sample, s, is input for the proposed DeepSCS, indicated as $S \in R^{N_V \times G}$, as shown in Figure 16.1. Where, G and N_V, the batch size are taken from the speech dataset. An attention-based encoder, after the set of input sample sequences is S, have been framed onto $m \in R^{N_V \times N_F \times L_F}$ for training

[a]s.singh89mtech@gmail.com, [b]jibananandamehena@gmail.com, [c]ersudhansukar2006@gmail.com, [d]suvasisrout.rout@gmail.com, [e]bhagirathi.behera07@gmail.com, [f]prksh.mhptr@gmail.com

DOI: 10.1201/9781003658221-16

i.e., the speech encoder, in which L_F denotes the length of a frame and N_F the number of frames. The speech Encoder produces the learnt features b $S \in R^{N_V \times N_F \times L_F \times D}$ after directly learning from m subsequently, b is transformed into $U \in R^{N_V \times N_F \times 2N}$ by the channel encoder, which is represented as CNN modules in 2Dimensional with CNN layer. U is turned via a reshape layer into symbol sequences, $X_E \in R^{N_V \times N_F N \times 2}$, in order to be transmitted onto a physical channel. The channel layer creates Y_E at the receiver by using the altered symbol sequences, X_E, as input, it is provided by

$$Y_E = HX_E + G \tag{1}$$

Where G is Gaussian noise that consists of N_V, which is no of noise vectors, G, and H is made up of N_V that is no of coefficient vectors of the channel, H. Before feeding the received symbol sequences (Y_E), which are represented by CNN modules in 2Dimensional with CNN layer, the sequences are transformed to V $\in R^{N_V \times N_F \times 2N}$. The channel decoder's output is $\hat{b} \in R^{N_V \times N_F \times L_F \times D}$. After that, $\hat{b}$ is converted into $\hat{m} \in R^{N_V \times N_F \times L_F}$ by the decoder which is an attention-based, also known as the speech decoder, and $\hat{m}$ is then recovered into $\hat{s}$ using the inverse framing procedure, called deframing, in which s at the transmitter and $\hat{s}$ have the same size.

Encoder and decoder
As demonstrated in Figure 16.2, the NN-enabled speech the encoding process and decoding that utilize an attention mechanism, known as SE-ResNet, is a crucial part of the suggested DeepSCS. In order to accomplish this, each input feature's 2 dimension

of spatial is first accumulated via the squeeze process [2]. As a result, squeeze and excitation residual neural network (SE-ResNet) input weight is redistributed, meaning that the weights that are associated with the most significant voice data are given greater weight.

Training and testing
Using the previously obtained channel state information (CSI), both receiver and transmitter parameters θ^J, θ^R can be changed together. As previously said, the proposed DeepSCS seeks to develop a model that can efficiently extract the most relevant data from the speech signals and make it work effectively in a variety of channels and SNR regimes.

1)   Training stage
The DeepSCS training technique, which is comparable to Figure 16.1. To finish a legitimate training task, the MSE loss converges over the training phase once decreasing the loss stops. SE-ResNet modules is one crucial hyper parameter values that aims to enable both a fair training length and strong speech encoder and speech decoder performance.

It states that the model that was trained under fixed channel circumstances is immediately used to test the performance in numerous fading channels without the need to retrain the model.

Experimental Results and Analysis

Considering correct channel state information (CSI), this section evaluates how well the suggested DeepSCS performs in comparison to traditional communication systems for speech signal transmission over telephone networks using AWGN and Rayleigh channels. We use the speech dataset from Edinburgh DataShare for the entire experiment, which consists of 800.wav test files with a 16 KHz sampling rate and over 10,000.wav train files. There are 128 frames(F) with length (L) 128 in every sample sequence in m. W = 16 and 384 is fixed in the computation.

Furthermore, the modulation used in the typical systems is 64-QAM in order to achieve the same amount

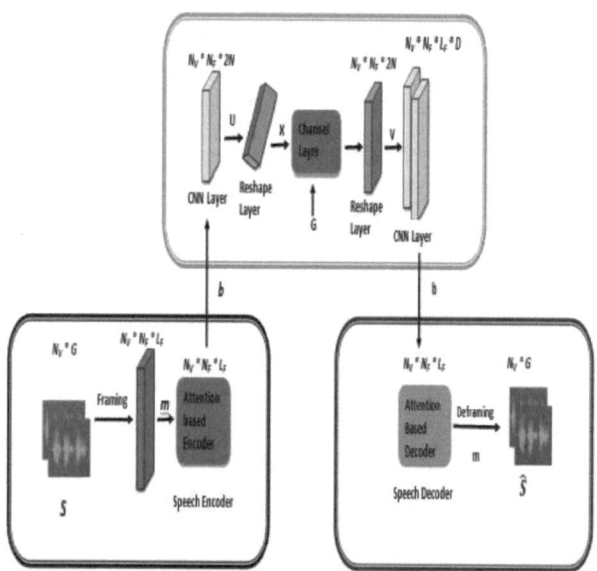

Figure 16.1 Semantic communication system model
Source: Author

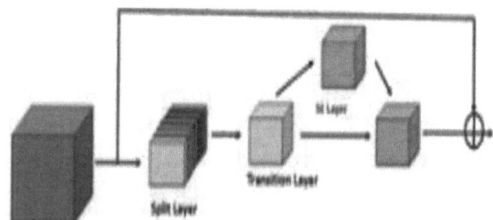

Figure 16.2 SE-ResNet module
Source: Author

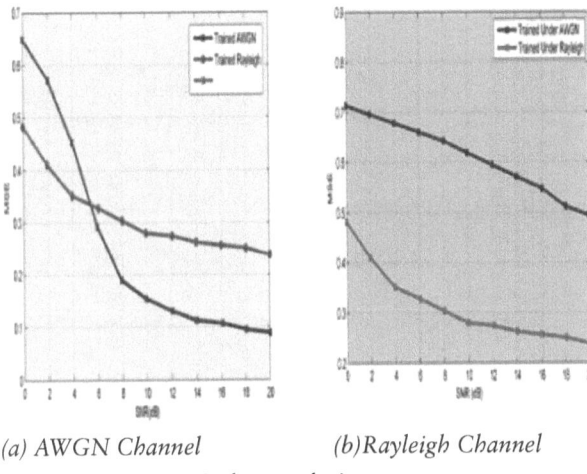

(a) AWGN Channel   (b)Rayleigh Channel

Figure 16.3 MSE loss analysis

Source: Author

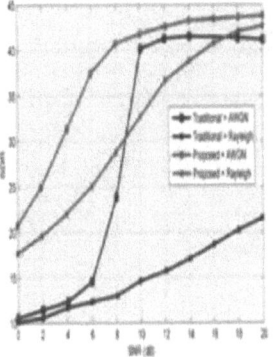

Figure 16.4 SNR Vs SDR

Source: Author

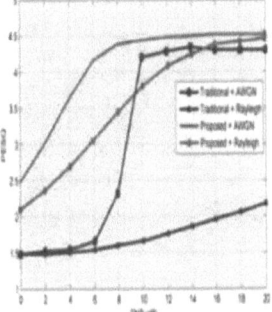

Figure 16.5 SNR Vs PESQ

Source: Author

of sent symbols as in DeepSCS. In order to study a resilient system that can operate on various channel conditions, we trained DeepSCS with a fixed channels condition in this experiment, then finally evaluating the MSE loss across all chosen fading channels using the trained model. In particular, there are four SE-ResNet modules and one 2D CNN module in the encoder and decoder section, which have eight kernels in the channel encoder/decoder. Table 16.1 displays the proposed DeepSC-S network configuration.

In Figure 16.3(a), DeepSCS used using the model trained upon Rayleigh channels is out performed over additive white Gaussian noise (AWGN) channels with respect to the MSE losses tested under the an AWGN channels when SNRs are more than roughly 6 dB. Furthermore, testing under Rayleigh channels reveals that DeepSCS trained to perform under AWGN channels performance relatively poor in regard to mean square error (MSE) loss (Figure 16.3(b)). Consequently, DeepSCS is regarded as a resilient model that can handle a range of channel conditions. It should be mentioned that during the training phase, two channels of Gaussian noise are generated at a constant SNR of 8 dB. If SNR in two analyzing channels is under 8 dB, Figure 16.3. demonstrates that DeepSCS trained Rayleigh and AWGN channel, it clear identifies that AWGN channels have larger MSE losses. We evaluate the SDR and PESQ for speech transmission over telephones under DeepSC-S and the traditional systems using the robust model, that is, DeepSCS trained with the Rayleigh channel with 8 dB SNR.

When AWGN, Rayleigh channels are used to evaluate the performance of the SDR in DeepSCS and communication systems, Figure 16.4 demonstrates that DeepSC-S outperforms the conventional communication system in all tested channels. Further more, DeepSC-S outperforms the traditional systems by

Table 16.1 DeepSCS for telephone systems' parameters.

Section/ parameter	Layer name	Kernels	Activation
Transmitter	SE-ResNet×4	4×32	Rectified linear unit
	CNN	4	Rectified linear unit
Receiver	CNN	4	Rectified linear unit
	SE-ResNet×4	4×32	Rectified linear unit
	CNN module×1	4	-
Learning rate	η	0.005	-

Source: Author

a large margin when handling various channels and SNR, especially inside the low SNR range. Moreover SDR scores under those channels than it does under AWGN channels. Figure 16.5 shows the comparison of PESQ scores. According to the figure, the suggested DeepSCS outperforms conventional methods and can recover speech in high quality across a range of fading channels and SNRs. Furthermore, in line with SDR's findings, DeepSCS achieves high.

PESQ scores in the low SNR domain while the conventional method performs poorly due to channel fluctuations. The results of the simulation demonstrate that DeepSCS, in a low SNR regime, may Deliver better voice transmission.

Conclusion

With this research, we suggested DeepSCS, a semantic communication system for voice synthesizing and recognition tasks that is enabled by deep learning. we develop semantic-channel coding method for learning and extracting semantic properties and lessen effects of the channel for voice recognition. Simulation data shows that the DeepSCS outperforms both the existing semantic systems for communication and traditional systems for communication, specifically within the zone of weak signals to noise. We anticipate DeepSCS as a strong candidate for semantic systems of communication in applications related to speech synthesis and recognition and provide better SNR as compared to traditional methods.

References

[1] Basu, P., Bao, J., Dean, M., & Hendler, J. (2014). Preserving quality of information by using semantic relationships. *Pervasive and Mobile Computing*, 11, 188–202.

[2] Cox, R. (1997). Three new speech coders from the ITU cover a range of applications. *IEEE Communications*, 35(9), 40–47.

[3] Graves, A., Mohamed, A., & Hinton, G. (2013). Speech recognition with deep recurrent neural networks. In International Conference on Acoustics, Speech Signal Process, (pp. 6645–6649).

[4] Gulati, A., et al. (2020). Conformer: Convolution-augmented transformer for speech recognition. In Proceedings of the conference Interspeech, (pp. 5036–5040).

[5] Qin, Z., Ye, H., Li, G. Y., & Juang, B.-H. F. (2019). Deep learning in physical layer communications *IEEE Wireless Communication*, 26(2), 93–99.

[6] Tong, W., & Li, G. Y. (2022). Nine challenges in artificial intelligence and wireless communications for 6G. *IEEE Wireless Communication*, 29(4), 140–145.

[7] Weng, Z, Qin, Z., & Li, G. Y. (2021). Semantic communications for speech recognition. *In Proceedings IEEE Global Communication Conference*, (pp. 1–6).

17 Enhanced convolutional block attention decision Transnet for efficient IoT-driven patient monitoring and heart disease detection

Dibas Kumar Hembram[1,a], Satya Narayan Tripathy[1,b] and Debabrata Dansana[2,c]

[1]Department of CS, Berhampur University, Berhampur, Odisha, India

[2]Department of Computer Science, Rajendra University, Balangir, Odisha, India

Abstract

The rising incidence of heart diseases (HD) calls for new strategies for real-time monitoring and early prediction of cardiovascular events. This work proposed a technique for HD prediction, data sharing, and disease categorization that uses the Modified Convolutional block attention decision TransNet with Gooseneck Barnacle Optimization (MCBADTN-GBO) classifier. Before being transferred to the cloud, patient data is encrypted utilizing the algorithm for adaptive Tangent Brakerski-Gentry Vaikuntanathan Homomorphic Encryption (ATBGVHEA) method, compressed using the Mud Ring Algorithm (MRA), and secured using the SHA-256 hashing algorithm. Data is decrypted and classified by the Modified Convolutional Block Attention Decision TransNet (MCBADTN) model, which has been fine-tuned by Gooseneck Barnacle Optimization (GBO) and trained on the Hungarian HD dataset. ATBGVHEA guarantees high security (98%) and quick encryption/decryption times. MCBADTN demonstrates exceptional prediction ability, achieving 99.88% F-measure, specificity, accuracy, and sensitivity. This IoT-based solution increases the precision of HD detection.

Keywords: Authentication, classification, decryption, disease classification, encryption, heart disease, optimization

Introduction

The Internet of Things (IoT) is a rapidly expanding trend in next-generation technology that connects bright items and gadgets. Health monitoring (HM) is a popular study topic in wearable electronics. Smart HM combines remote HM with innovative computing [3]. Body sensor networks, or BSNs, are implanted or wearable devices that track body temperature, blood pressure, pulse, and breathing rates, such as pacemakers, cardioverter-defibrillators, and accelerometers. BSNs are essential to the IoT, gathering critical data like updates to remedial settings and health restraint changes [4]. The data gathered from wearables with an Internet of Things focus is kept in a clinical database to take appropriate action when patients' health conditions deteriorate. IoT lowers expenses, increases accessibility to healthcare facilities in far-flung locations, and enhances service quality [5].

People with chronic heart failure (CHF) are at increased risk of death from obesity, stress, eating a diet high in fat and salt, not exercising, and heredity. Due to increasing risk factors like diabetes, high blood pressure, raised blood fat, and weight gain after menopause, women might also experience heart attacks [8]. According to studies, about 30% of patients are readmitted within 90 days, and readmission rates rise from 25% to 54% within three to six months. Healthcare professionals should use an invasive approach to monitor patients' physical status and determine when healthcare treatments should be supplied. This might be done with the aid of IoT. This will improve HD performance and lower death rates [6].

- Machine learning algorithms process extensive data for predictive designs and early diagnoses but often overlook security and feature selection [11].
- Existing disease prediction algorithms prioritize accuracy but result in lengthy training times, neglecting data encryption and feature optimization [9].
- This study introduces Modified Convolutional Block Attention Decision TransNet (MCBADTN-GBO) for heart disease prediction, categorization, and secure data sharing.
- Patient data is encrypted using adaptive tangent Brakerski-Gentry Vaikuntanathan Homomorphic Encryption Algorithm (ATBGVHEA) before secure cloud transfer and analysis. Decrypted data, classified by MCBADTN, fine-tuned with Goose-

[a]dkh.rs.cs@buodisha.edu.in, [b]snt.cs@buodisha.edu.in, [c]debabratadansana07@gmail.com

DOI: 10.1201/9781003658221-17

neck Barnacle Optimization and Hungarian HD dataset, ensures precise heart disease detection.

The rest of the research paper is organized as follows: section 2 reviews the prevailing methods in the flow of method, section 3 broadly describes the proposed solution for the stated problems, and section 4 validates the efficiency of the proposed solution in comparison with the existing models and finally ends with the conclusion in section 5.

Literature Survey

Some of the recent papers related to this work are described below,

In Rajkumar et al. [7] introduced a deep learning framework using the Hungarian heart disease dataset to predict heart disease. The framework utilized modified deep long short-term memory (MDLSTM), Harris Hawk optimization (HHO), and median studentized residual for preprocessing, followed by an improved spotted hyena optimization (ISHO) algorithm to refine predictions. In Yashudas et al. [1] developed DEEP-CARDIO, an IoT-based system for early cardiovascular disease detection and management. Using four biosensors and an Arduino controller with a bidirectional-gated recurrent unit (BiGRU) attention model, the system gathered physiological data remotely and offered real-time diagnosis and personalized advice. Its smartphone application provided dietary and physical recommendations. In Malibari [2] proposed EO-LWAMCNet, a robotic classification network to predict chronic health conditions like heart and kidney diseases. Patient data collected through wearable sensors was analyzed in the cloud, classifying results as normal or abnormal. In Kumar et al. [10] developed an autonomous edge-assisted cloud-IoT framework for healthcare. Employing the RF-LRG algorithm, the framework enhanced accuracy, precision, and recall while significantly reducing latency and energy consumption compared to traditional methods like Logistic Regression, Random Forest, and K-Nearest Neighbor.

Problem statement

There is an urgent demand for effective, real-time health monitoring solutions due to the rising incidence of heart disease and the shortcomings of conventional patient monitoring systems. Current approaches frequently fail to detect early warning indicators promptly, which delays actions and raises healthcare expenses. This necessitates using an IoT-based patient monitoring system that can continuously measure vital signs. By doing so, it will be possible to diagnose

and predict heart disease early on, improving patient results and lessening the strain on the healthcare framework.

Proposed Methodology

The three main stages of the suggested framework are authentication, encryption, classification, and improved safe data transfer and precise HD prediction in an IoT-based patient monitoring framework. Patients register on a hospital website or app to begin the authentication process. Their personal information is then saved on a cloud server and validated using a SHA-256 Hash Algorithm code based on cryptography. Medical reports are uploaded and compressed for adequate storage using the Mud Ring Algorithm (MRA). Patient sensor data is encrypted using the ATBGVHEA technique and sent to the cloud, where a modified convolutional block attention decision TransNet (MCBADTN) model trained on the Hungarian HD dataset decrypts and classifies the data. The parameters of MCBADTN are tuned by the Gooseneck Barnacle Optimization (GBO). The workflow of the proposed approach is given in Figure 17.1.

Patient with IoT
In this initial stage of the suggested intervention, the patient's body is fitted with a sensor device. Using a sensor, the IoT device will detect the patient's bodily data and send them to the HC application.

Authentication using SHA-256 hash algorithm
Authentication is crucial for safe access and accurate HD prediction in an IoT-based patient monitoring framework. To ensure strong authentication throughout the registration, login, verification, and patient

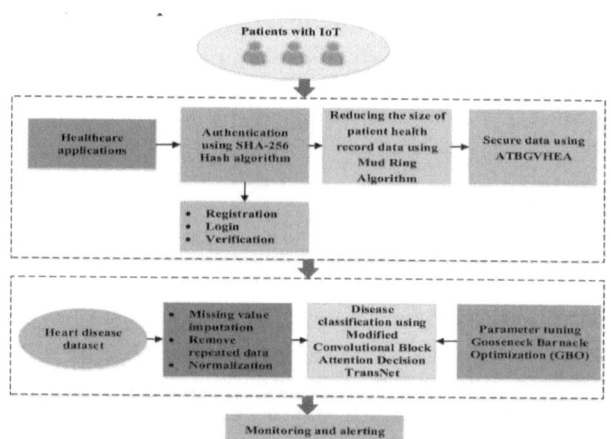

Figure 17.1 Workflow of the proposed approach
Source: Author

health record access stages, the system uses secure hash functions like SHA-256. The phases of the authentication function's operation are as follows:

Registration phase: During registration, user passwords are securely hashed using SHA-256. The password is padded to align with 512-bit blocks, appending a '1' bit, zeroes, and its original length as a 64-bit number. After preprocessing, the hash is computed through 64 iterations, ensuring secure, irreversible storage of sensitive user information updating registers using equations (1)-(4).

$$Ch(a,b,z) = (a \wedge b) \oplus (\neg a \wedge z) \tag{1}$$

$$Maj(a,b,z) = (a \wedge b) \oplus (a \wedge z) \oplus (y \wedge z) \tag{2}$$

$$\sum\nolimits_0(a) = R_2(a) \oplus R_{13}(a) \oplus R_{22}(a) \tag{3}$$

$$\sum\nolimits_1(a) = R_6(a) \oplus R_{11}(a) \oplus R_{25}(a) \tag{4}$$

The generated hash is kept in the system's database for security purposes instead of the password in plaintext. This process includes 4 phases such as login phase, verification Phase: Patient Health record access, Reducing the size of patient health record data using the Mud Ring Algorithm This method is detailed in Algorithm 1.

Algorithm 1: Mud Ring Algorithm for reducing the size of patient health record data

---

Step 1: Initialization

Create a dolphin population to symbolize the patient health record data that needs to be trimmed down.

Step 2: Search for relevant features (foraging - exploration phase)

In this phase, dolphins (agents) explore the search space to identify patient health records' most relevant features (data points).

Step 3: Feature selection (Mud Ring Feeding - Exploitation Phase)

After identifying the optimal solution (reduced data set), the other dolphins gather to fine-tune their placements near it.

Step 4: Convergence to optimal reduced data set

When all the dolphins have determined which ideal feature selection best represents the patient health records, the algorithm converges.

Step 5: Output reduced data

Create more minor patient health records to facilitate IoT-based monitoring and the early detection of cardiac disease.

Step 6: Termination

These actions enable the Mud Ring Algorithm to reduce the number of medical records for effective IoT-based monitoring and precise cardiac disease prediction by ensuring that only the most pertinent and vital patient data is kept.

Secure data transfer using ATBGVHEA

In this section, secure data transmission is carried out using the implementation of the ATBGVHEA, enhanced by the AT search algorithm (ATSA) for key management. This allows for secure data transfer in the context of IoT-based patient monitoring and HD prediction. Encrypting non-malicious IoT-based patient data assures privacy and security during transport and storage, akin to secure cloud storage.

Data classification: The system first categorizes data as harmful or benign. Malicious patient data is disposed of, while non-malicious data is safely encrypted and sent or preserved.

ATBGVHEA: Thanks to homomorphic encryption (HE), encrypted data can be processed without the need for decryption. Non-malicious data (such as patient vitals, heart rate, etc.) is encrypted using the BGV scheme for safe cloud storage and data transfer. The definition of Additive Homomorphic Encryption (HO) is in equation (5).

$$HO(a+b) = HO(a) \oplus HO(b) \tag{5}$$

Multiplicative HE is in equation (6).

$$HO(a.b) = HO(a) \oplus HO(b) \tag{6}$$

Secret keys are generated using the SHA-512 hash function in equation (7).

$$
\begin{aligned}
keysecrete &= SHA512 \\
&(errorvector(\mathrm{Private}_{key}) \\
&.errorvector(public_{key}))
\end{aligned}
\tag{7}
$$

Optimal Key Management: Based on the longest time it takes to break a key, ATSA chooses the best encryption key. The fitness function assesses the strength of the key to prevent breakage when attacked. The fitness process selects the optimal key in equation (8).

$$
\begin{aligned}
Fitness = Select_{\max}(Optimum_{key} \text{ with highest key} \\
- breaking\,time)
\end{aligned}
\tag{8}
$$

The intensification search approach for both essential creation and selection is given in equation (9).

$$X_{t+1}^{j} = X_{t}^{j} + step.\tan(\phi).$$
$$(X_{t}^{j} optimal\,key\sec ret^{t})$$
$$(9)$$

Data transfer: Using the ATBGVHEA, non-malicious encrypted patient data is safely sent to cloud storage and may be decrypted on demand for medical analysis while real-time monitoring is underway.

Termination: The system keeps transferring and storing safe data until an ideal encryption key is selected to protect patient privacy and secure communication in IoT-based health monitoring.

By following these procedures, with ATBGVHEA providing a robust encryption framework, this configuration ensures secure patient data transfer and cloud storage, which is essential for IoT-based heart disease monitoring.

Data collection
The proposed system uses the Hungarian HD dataset, containing patient data on blood pressure, heart rate, respiration rate, and pulse from 294 rows. For accurate classification, the system is first trained using 76 HD patient features from the dataset. The processed data is then used to generate disease classification results.

Data preprocessing
The Hungarian HD dataset undergoes three preprocessing phases: (i) normalization, (ii) elimination of redundancy, and (iii) replacement of missing attributes. With 1,314 missing values in the dataset, absent attributes are replaced using the most frequent values for corresponding attributes like blood pressure, cholesterol, and age group. When attribute values align, missing data is replaced accordingly. Finally, data normalization and removal of redundant attributes are performed to reduce the dataset size.

Disease classification modified convolutional block attention decision TransNet (MCBADTN)
A modified convolutional block attention decision TransNet (MCBADTN) is used for disease classification. Incorporating attention mechanisms such as convolutional block attention module (CBAM) and transformer-based models can significantly improve the system's capacity to evaluate patient data and generate precise predictions for heart disease classification in HD prediction.

CBAM includes spatial and channel attention modules to focus on key features in patient data, like heart rate and ECG signals. The channel attention module uses global average pooling (GAP) and global max pooling (GMP) to emphasize important data channels, such as vital signs, with enhanced features shown in equation (10).

$$M_{c}(F) = \sigma(MLP(GAP(F)) + MLP(GMP(F)))   (10)$$

Where MLP stands for the shared Multi-Layer Perceptron, GAP and GMP assist in identifying essential patterns in the patient's health data and F represents the input feature map which yields the following result for spatial attention given in equation (11).

$$M_{s}(F') = \sigma(K_{7\times7}([SpAvgpool(F');$$
$$SpMaxpool(F')]))$$
$$(11)$$

This enables the model to highlight important regions in ECG data or medical pictures. Channel and spatial attention are combined to create the final refined feature map. F_{att} It is given in equation (12).

$$F_{att} = M_{s}(M_{c}(F) \otimes F) \otimes M_{c}(F) \otimes F   (12)$$

Where, $\otimes$ It denotes element-wise multiplication, improving prediction accuracy by concentrating on essential characteristics in the patient data. The model analyses a trajectory sequence. τ of patient data, including medical history, sensor readings, and patient vitals. The model for the patient data at the time step t It is as follows in equation (13).

$$\tau = (\hat{r}_{t-K+1}, s_{t-K+1}, a_{t-K+1}, \ldots, \hat{r}_{t}, s_{t}, a_{t})   (13)$$

Where, $\hat{r}_{t}$ stands for total health scores or risk scores, S_{t} for the patient's state (heart rate, etc.), and a_{t} For the anticipated diagnosis or action. The cumulative health score from the current time step to the final observation is used to calculate the reward-to-go $\hat{r}_{t}$. The incentive while training is given in equation (14).

$$\hat{r}_{t} = \sum_{i=t}^{T} r_{i}   (14)$$

The health score at the time step i is represented by r_{i}, which enables the algorithm to optimize its predictions based on overall patient outcomes. The reward-to-go during testing is computed as follows in equation (15).

$$\hat{r}_{t} = G^{*} - \sum_{i=0}^{t} r_{i}   (15)$$

The desired result for a specific patient is represented, helping the model predict actions (diagnoses) that optimize long-term outcomes. The transformer's attention mechanism selects the best forecast by focusing on the most relevant historical data. Each trajectory consists of 3K tokens, representing states, actions, and rewards..

The MCBADTN method classifies heart disease by following these procedures. In the next section, the hyperparameters of the MCBADTN method can be optimized.

Figure 17.2 Optimize *the hyperparameters of MCBADTN using Gooseneck Barnacle Optimization (GBO):* This section focuses on optimizing the hyperparameters of MCBADTN using the Gooseneck Barnacle Optimizer (GBO), inspired by the adaptive reproductive strategies of marine barnacles. GBO mimics the barnacles' behaviors, such as acorn barnacles using long penises for mate reach and gooseneck barnacles adapting to different environments. The algorithm follows a step-by-step process shown in Figure 17.2.

This step shows the final optimal solution from the GBO optimization process, improving accuracy and reducing error by simulating gooseneck barnacles' natural behavior through mathematical operations and updates.

Result and Discussion

This section provides the outcomes of the proposed MCBADTN-GBO strategy and a discussion. The proposed methodology is examined on the Windows 10 operating system and validated in Python.

Figure 17.3 shows the encryption and decryption times of the developed approach across different data volumes. As the number of data increases from 100 to 500, both times decrease. Initially, encryption time is slightly higher than decryption time, but both times converge as data increases, indicating the method becomes more efficient with larger volumes.

Figure 17.4 displays two graphs showing the training and validation performance over 100 epochs. The accuracy graph shows a steady increase, with validation accuracy slightly higher than training accuracy. The loss graph shows a decline in both training and validation loss, indicating improved performance and minimal overfitting.

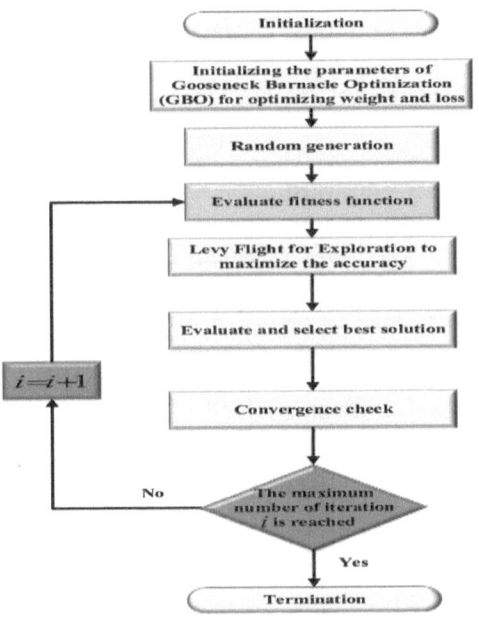

Figure 17.2 The step-by-step process of GBO
Source: Author

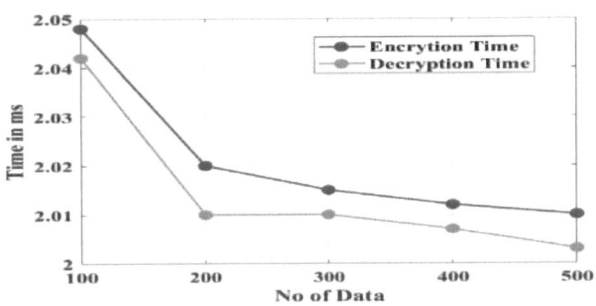

Figure 17.3 Encryption and decryption of the proposed approach
Source: Author

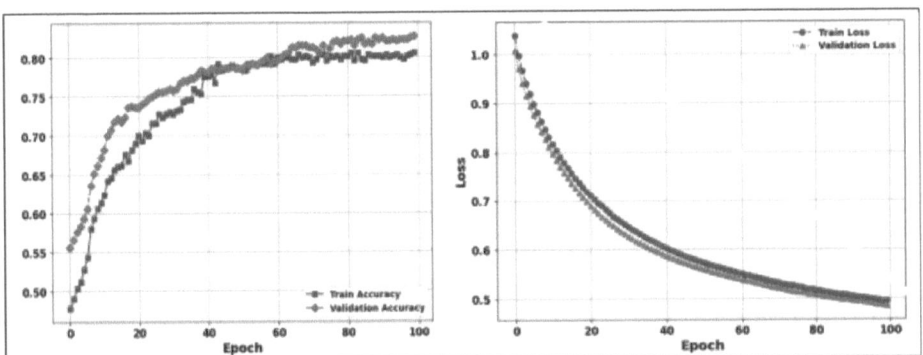

Figure 17.4 Accuracy and loss of the proposed approach
Source: Author

Table 17.1 Comparison of accuracy, recall, precision, specificity, F-score, computational time.

Methods	Accuracy (%)	Precision (%)	Recall (%)	Specificity (%)	F-score (%)	Computational time (ms)
MDLSTM	97.73	94.19	96.90	96.90	96.76	80
Deep-BiGRU	95.83	94.35	92.85	94.85	94.59	124
EO-LWAMCN	88.82	84.294	83.691	82.691	83.48	129
RF-LRG	93.39	88	88	88	89.45	75
proposed	99.881	99.883	99.883	99.887	99.883	48

Source: Author

Table 17.2 Compression time, ET, DT and SL.

Methods	Compression time (ms)	ET (ms)	DT (ms)	SL (%)
MDLSTM	85	7.5	12.9	87
Deep-BiGRU	75	1	17	86
EO-LWAMCN	127	8	15.5	91
RF-LRG	115	15.5	18.5	79
proposed	55	5	7	98

Source: Author

Performance analysis comparison

This section analyzes the performance metrics for disease prediction, including accuracy, recall, precision, F1-score, specificity, computational complexity, error rate, and data compression metrics such as compression time. It also evaluates decryption time, encryption time, and security level (SL). The developed MCBADTN-GBO method is compared with previous methods like MDLSTM, Deep-BiGRU, EO-LWAMCN, and RF-LRG. Table 17.1 shows that the MCBADTN-GBO method outperforms others with the highest accuracy (99.88%), precision (99.88%), recall (99.88%), specificity (99.89%), and F-score (99.88%), while also having the lowest computational time (48 ms).

Table 17.2 compares various methods based on compression time, ET, DT, and SL. The developed method has the shortest compression time (55 ms), highest security level (98%), and moderate encryption time (5 ms). When compared to the other method, it achieves better performance.

Conclusion

This study proposes a method for identifying heart disease (HD) using the MCBADTN-GBO classifier. The results are compared with existing techniques in disease prediction, data transfer, and disease classification (DC). Patient data is securely stored on a cloud server and validated using SHA-256 encryption. Medical reports are compressed with the Mud Ring Algorithm (MRA), and sensor data is encrypted using the ATBGVHEA technique. The MCBADTN model, trained on the Hungarian HD dataset, decrypts and classifies the data, with hyperparameters tuned by the Gooseneck Barnacle Optimization (GBO). The ATBGVHEA technique provides optimal data security, while MCBADTN achieves 99.88% accuracy in disease prediction, enabling timely intervention for patients with abnormal heart conditions.

References

[1] Yashudas, A., Gupta, D., Prashant, G. C., Dua, A., AlQahtani, D., & Reddy, A. S. K. (2024). DEEP-CARDIO: recommendation System for cardiovascular disease prediction using IoT network. *IEEE Sensors Journal*, 24, 14539–14547.

[2] Malibari, (2023). An efficient IoT-artificial intelligence-based disease prediction using lightweight CNN in healthcare system. *Measurement Sensors*, 26, 100695.

[3] Verma, D., Singh, K. R., Yadav, A. K., Nayak, V., Singh, J., Solanki, P. R., et al. (2022). Internet of things (IoT) in nano-integrated wearable biosensor devices for healthcare applications. *Biosensors and Bioelectronics X*, 11, 100153.

[4] Arora, D., Gupta, S., & Anpalagan, A. (2022). Evolution and adoption of next generation IoT-driven health care 4.0 systems. *Wireless Personal Communications*, 127, 3533–3613.

[5] Famá, F., Faria, J. N., & Portugal, D. (2022). An IoT-based interoperable architecture for wireless biomonitoring of patients with sensor patches. *Internet of Things*, 19, 100547.

[6] Gaobotse, G., Mbunge, E., Batani, J., & Muchemwa, B. (2022). Non-invasive smart implants in healthcare: redefining healthcare services delivery through sensors and emerging digital health technologies. *Sensors International*, 3, 100156.

[7] Rajkumar, G., Devi, T. G., & Srinivasan, A. (2022). Heart disease prediction using IoT based framework and improved deep learning approach: medical application. *Medical Engineering and Physics*, 111, 103937.

[8] Hirsch, H., & Manson, J. E. (2022). Menopausal symptom management in women with cardiovascular disease or vascular risk factors. *Maturitas*, 161, 1–6.

[9] Zhou, H., Xin, Y., & Li, S. (2023). A diabetes prediction model based on Boruta feature selection and ensemble learning. *BMC Bioinformatics*, 24(1), 224.

[10] Kumar, M., Rai, A., Surbhit, N., & Kumar, N. (2023). Autonomic edge cloud assisted framework for heart disease prediction using RF-LRG algorithm. *Multimedia Tools and Applications*, 83, 5929–5953.

[11] Nia, N. G., Kaplanoglu, E., & Nasab, A. (2023). Evaluation of artificial intelligence techniques in disease diagnosis and prediction. *Discover Artificial Intelligence*, 3(1), 5.

18 Enhancing healthcare accessibility: An AI-powered chatbot for symptom analysis and precautionary guidance

P. Devika[1,a], Balingan Sangameshwar[1,b], V. T. Ram Pavan Kumar M.[2,c], Ch. Lavanya Susanna[3,d], Shaik Munawar[4,e] and Sudheer Nidamanuri[1,f]

[1]Department of CSE-(CyS, DS) and AI&DS, VNR Vignana Jyothi Institute of Engineering and Technology, Hyderabad, Telangana, India

[2]Department of Computer Science, Kakaraparti, Bhavanarayana College Vijayawada, Andhra Pradesh, India

[3]Department of Computer Science and Engineering, Koneru Lakshmaiah Education Foundation, Vaddeswaram, Andhra Pradesh, India

[4]Department of CSE, Kakatiya Institute of Technology and Science, Waranagal, Telangana, India

Abstract

This article introduces a healthcare chatbot powered by artificial intelligence, which aims to anticipate potential medical conditions based on symptoms reported by users and provide tailored precautionary measures for the predicted condition. The chatbot utilizes machine learning algorithms, including decision tree classifiers, to examine symptom patterns and forecast potential diagnoses. One distinctive feature of this system is its capability to offer users specific precautionary advice for managing their condition before seeking professional help, thereby enhancing the system's usefulness by providing proactive health management support. The model was trained using a comprehensive dataset of symptoms and corresponding medical diagnoses, achieving a high level of accuracy in predicting and suggesting precautionary actions. We illustrate how integrating disease prediction with actionable recommendations can enhance healthcare outcomes, particularly for users in remote or underserved areas. This solution could be a vital tool in early diagnosis and preventive care.

Keywords: Decision tree classifier, disease prediction, healthcare chatbot, machine learning, precautionary measures, symptom analysis

Introduction

The healthcare industry has been transformed by the rapid progress of artificial intelligence (AI), with AI-powered systems showing potential in enhancing diagnostic precision, cutting healthcare expenses, and delivering individualized care. One area where AI is making inroads in healthcare is the creation of smart chatbots that aid in initial diagnosis and patient care. These chatbots serve as a link between patients and healthcare professionals, providing prompt information, reducing hospital visits, and offering basic health guidance [7]. Healthcare chatbots have become increasingly prominent for their capacity to address common health questions, steer patients through symptom assessments, and anticipate potential medical issues based on user input. Nevertheless, many existing solutions fall short in offering personalized preventive advice beyond diagnosis [4]. This gap underscores the necessity for AI-driven healthcare systems that not only forecast conditions but also recommend precautionary measures to minimize risks and encourage wellness [12].

This article introduces an AI-driven healthcare chatbot that utilizes machine learning models to forecast typical medical conditions based on user symptoms and medical history. A notable aspect of this system is its capability to provide tailored preventive measures for each predicted condition, empowering users to take proactive steps in managing their health. Through the use of (NLP) and classification algorithms, the chatbot engages with users in real time, presenting personalized suggestions that are simple to comprehend and act upon. This study delves into the development, functionality, and efficacy of the proposed chatbot in predicting medical conditions and offering preventive guidance. The aim is to enhance patient self-care and alleviate the strain on healthcare systems by providing users with an intelligent, interactive, and supportive tool for early intervention [1].

Literature Review

Chakraborty and Dey [10] Technology development has a significant impact on healthcare. In addition to helping physicians, technology has given patients

---
[a]devika.potta23@gmail.com, [b]balingan.sangameshwar@gmail.com, [c]mrpphd2018@gmail.com,
[d]lavanyasusanna@kluniversity.in, [e]shaikmunawar215@gmail.com, [f]nidamanuri.sudheer@gmail.com

DOI: 10.1201/9781003658221-18

access to a first-hand testing set. Machine learning-based Chatbot for disease prediction and treatment suggestion - published at the 3rd International Conference on Trends in Electronics and Informatics (ICOEI) in 2019: [5]. The goal of the suggested system is to provide an alternative to the traditional process of going to a hospital and scheduling a consultation with a physician in order to receive a diagnosis [8]. Using a series of questions, the user can converse with the system much like talking to any human being. The chatbot can recognize the symptoms that the user has, predicts the possible illnesses, and advises him/her on further steps to take. Sharma and Jalal [3]: Founded in the year 2020, the 5th International Conference on Communication and Electronics Systems (ICCES) presented Medbot: An artificial intelligence-powered conversational Chatbot for telehealth applications since the COVID-19. To impede COVID-19 transmission from patients to doctors, telemedicine enables doctors to remain in contact with their patients during this viral disturbance. The telemedicine system would then step up support during the pandemic through an interactive AI program without any in-person visits to the hospital. Singh and Choubey [14]: Thus, keeping in mind, telehealth will quickly supplant the inevitable in-person patient care by way of remote consultations. Accordingly, this has led to the creation of an NLP-based multilingual conversational bot for the free education of chronic patients on basic health care, a litany of information, and some guidelines. Shin, D et al., 2022: The work introduces a software product in the nature of a personal virtual doctor, which has been designed fairly and extensively trained to converse with patients in a human-like manner. The application uses a server-less architecture and organizes the services of a physician with respect to preventive measures, home remedy, interactive counseling sessions, healthcare tips, and symptoms concerning some of the diseases most prevalent among rural communities in India. Le, & Nguyen et al., 2019: To improve the access of patients to health care knowledge and utilize artificial intelligence. Managing the phone calls for complaints is also quite time-consuming. Medical chatbots can help with this issue by providing appropriate advice on leading a healthy lifestyle. Natural language processing is essential to the operation of medical chatbots, which assist users in submitting health-related issues. Through the chatbot, the user can ask any personal health-related question without the hospital being physically present. For voice-to-text and text-to-voice conversion, use the Google API. Android applications accommodate chatbots for inquiry. One of the main purposes of this internet-based platform is to assess the perception of its customers [13]. Modern technology is made most useful for the medical field by bringing the concept of telehealth onto the scenario. Artificial intelligence, machine learning, and neural networks have their strongest foothold in the health sector. With such a fast running worldly manner, at times no one pays attention to health, which shall be detrimental to many. In such a case, symptom-driven disease prediction may help. This paper's core aim is a survey of a user's pathology report and a duly instant but precise pathology-to-symptom diagnosis. The disease prediction chatbot is primarily built on natural language processing and machine learning. Disease prediction by way of two classification algorithms, namely the decision tree and the KNN (K-Nearest Neighbors), was carried out. Thereafter we put both methods in parallel and chose the one that gave the better accuracy. The accuracies of decision tree and KNN were found to be respectively 92.6% and 95.740. Apart from disease prediction, the initiative will be expected to afford some medical advice regarding the predicted diseases. OCR is meant for the full analysis of the pathology report. Tesseract is an open-source optical character recognition engine. Comparing the image-based representation of results and the text extracted from the report would significantly ease the overhead burden of understanding the data involved.

Problem Statement

The increasing complexity of healthcare requires efficient and accessible systems to support patients in handling health issues. Existing medical chatbot systems mainly offer basic diagnostic predictions or information based on symptoms provided by the user. However, they often lack a personalized response mechanism to assist users in taking appropriate actions, particularly preventive measures, which are crucial for managing deteriorating conditions. This deficiency results in subpar user experiences, as patients are left without clear steps to take after receiving a diagnosis (Al-Imamy et al., 2022).

Our objective is to develop an AI-powered healthcare chatbot that can help users better manage their health by offering tailored preventive actions in addition to predicting possible medical issues based on symptoms.

By incorporating machine learning models, the chatbot can improve decision-making by offering timely and customized advice that could potentially reduce the need for medical intervention in early-stage conditions. The challenge lies in developing an intelligent system that strikes a balance between accurate predictions and actionable recommendations while ensuring user accessibility and trust.

Proposed System

The AI-based healthcare chatbot's proposed design con- sists of two tiers: it predicts conditions based on symptoms and provides personalized precautionary measures. The chatbot forecasts possible health risks by analyzing user-input symptoms using machine learning methods. Alongside offering diagnoses or predictions, the system will provide a novel feature - personalized precautionary measures customized to the predicted condition. The system aims to improve user interaction by giving practical advice, such as lifestyle modifications, dietary changes, or recommendations for seeking medical attention based on the seriousness of the predicted condition. Incorporating natural language processing (NLP), the chatbot can adeptly comprehend user inputs and deliver responses in a conversational style, making it easy to use and accessible.

Methodology and Model Specifications

The structured methodology for disease prediction through an AI-based healthcare chatbot commences with preparing and training machine learning models. To train the prediction models, the system utilizes a dataset containing symptoms and their associated diseases. The system employs a decision tree classifier to understand the relationships between symptoms and the diseases they indicate (Fong et al., 2021). Additionally, a support vector classifier (SVC) is trained to compare the performance of different models for optimal accuracy.

After training the models, users engage with the system via a chatbot interface, where they provide their symptoms in natural language. The chatbot uses (NLP) techniques to interpret these symptoms. A regular expression-based search mechanism helps identify and match the user-reported symptoms with those in the training dataset as shown in Figure 18.1. The trained decision tree model then processes this input to predict the most likely disease corresponding to the symptoms provided.

The integration of the severity index, which measures the seriousness of the user's condition according to the length of the symptoms, is an essential element in this system. Symptom severity is calculated by summing symptoms weighted for their severity depending on the number of days they have been present. The user would be advised to see a physician when the parameter is above a threshold level. If it is below that threshold, the system recommends general measures to aid in coping with the symptoms. Therefore, the Severity Index was of prime importance in establishing the seriousness of the user's symptoms. Each symptom is given a severity score (from a predetermined dataset) that rates its seriousness in the context of the disease (Fong et al., 2021).

Additionally, the chatbot provides a personalized treatment plan or precautionary advice based on the predicted disease. The system also incorporates a text-to-speech component to enhance user-friendliness, providing spoken feedback about the diagnosis and suggested actions. This methodology ensures an effective and user-centric approach to disease detection, combining the power of machine learning with accessible healthcare advice, ultimately helping users obtain timely and reliable information about their health conditions.

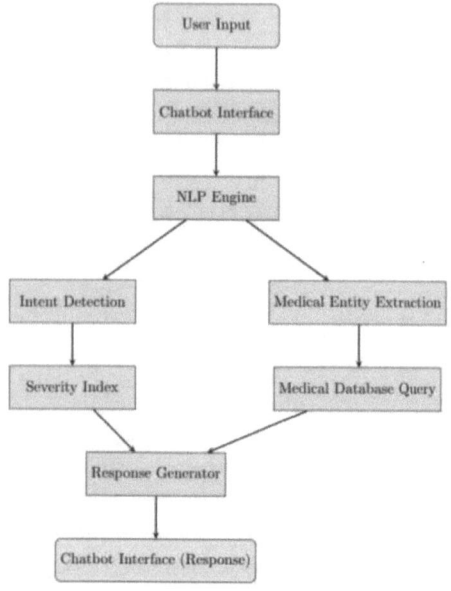

Figure 18.1 Work flow of the model
Source: Author

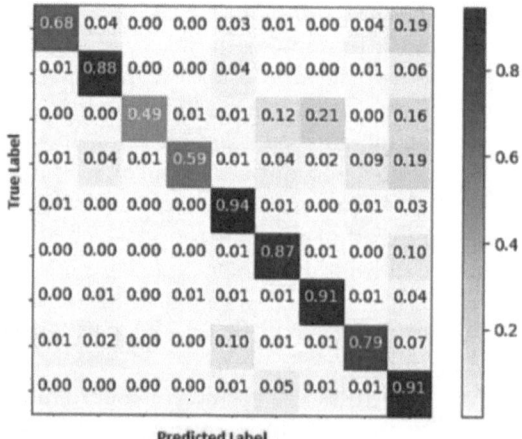

Figure 18.2 Confusion matrix for the proposed model
Source: Author

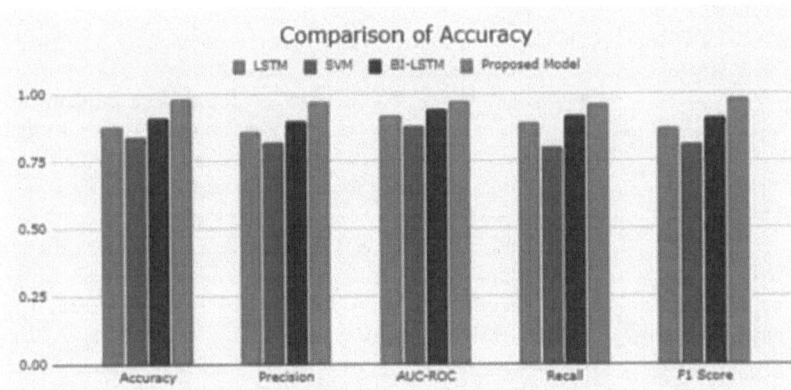

Figure 18.3 Performance comparison of models across key metrics
Source: Author

Table 18.1 Comparison of accuracy metrics for metrics.

Model	Accuracy	Precision	AUC-ROC	Recall	F1 Score
LSTM	0.88	0.86	0.92	0.89	0.87
SVM	0.84	0.82	0.88	0.8	0.81
BI-LSTM	0.91	0.90	0.94	0.92	0.91
Proposed Model	0.98	0.97	0.97	0.96	0.98

Source: Author

Conclusion

The effectiveness of the AI-based healthcare chatbot system was demonstrated by the proposed model, which outperformed the baseline models (LSTM, SVM, and BI-LSTM) across all key metrics.

In comparison Table 18.1 [10], the accuracy of the LSTM (88%), SVM (84%), and BI-LSTM (91%), this method achieved relatively higher accuracy, that is, 98% as displayed in the first image. The precision at the value of 97% of the proposed method marks it well capable of correctly predicting the true positives. The recall of this model stands at 96% while attaining an F1-score of 98%, thus addressing its ability to capture true positive cases sensitively and those two metrics representing a trade-off of precision against recall as shown in figure 18.3.

The accuracy of the newly proposed model was found to be robust in differentiating positive class and negative class.

The AUC-ROC score of 97%, an improvement over BI-LSTM (94%), LSTM (92%), and SVM (88%).

In the figure 18.2, the confusion matrix confirms model prediction capabilities to be accurate, implying a strong diagonal with almost no false positives and false negatives across all the labels. The model seemed to have very high confidence while predicting certain conditions, with 0.91 and 0.94 for some labels.

The introduction of the new precautionary measure feature in the chatbot added significant practical value by providing users with personalized precautions based on their symptoms after predicting a particular health condition. This enhancement improved the system's overall usability and ensured that users received actionable advice to manage or prevent worsening of their condition, bridging the gap between diagnosis and patient care. These results highlight the potential of the proposed model to be implemented in real-world healthcare settings, enhancing both predictive accuracy and patient interaction through the incorporation of precautionary advice.

References

[1] Ashour, , El-Attar, A., Dey, N., El-Kader, , & El-Naby, (2020). Long short term memory based patient-dependent model for FOG detection in Parkinson's disease. *Pattern Recognition Letters*, 131, 23–29. doi: 10.1016/j.patrec.2019.11.036.

[2] Shin, D. (2022). The perception of humanness in conversational journalism: An algorithmic information-processing perspective. *New Media and Society*, 24(12), 2680–2704. doi: 10.1177/1461444821993801.

[3] Sharma, H., & Jalal, (2022). A framework for visual question answering with the integration of scene-text using PHOCs and fisher vectors. *Expert Systems with Applications*, 190, 116159.

[4] Rarhi, Krishnendu & Bhattacharya, Abhishek & Mishra, Abhishek & Mandal, Krishnasis. (2017). Automated Medical Chatbot. SSRN Electronic Journal. 10.2139/ssrn.3090881.

[5] Dey, L., Chakraborty, S., & Mukhopadhyay, A. (2020). Machine learning techniques for sequence-based prediction of viral–host interactions between SARS-CoV-2 and human proteins. *Biomedical Journal*, 43(5), 438–450.

[6] Le, , & Nguyen, (2019). Prediction of FMN binding sites in electron transport chains based on 2-D CNN and PSSM profiles. *IEEE/ACM Transactions on Computational Biology and Bioinformatics*, 18(6), 2189–2197.

[7] Rosruen, N., & Samanchuen, T. (2018). Chatbot utilization for medical consultant system. In 2018 3rd Technology Innovation Management and Engineering Science International Conference (TIMES-iCON), (pp. 1–5). IEEE.

[8] Tripathi, , & Jalal, (2022). A robust approach based on local feature extraction for age invariant face recognition. *Multimedia Tools and Applications*, 81, 21223–21240.

[9] Al-Imamy, S., & Hwang, Y. (2022). Cross-cultural differences in information processing of chatbot journalism: chatbot news service as a cultural artifact. *Cross Cultural and Strategic Management*, 29(3), 618–638. doi: 10.1108/CCSM-06-2020-0125.

[10] Chakraborty, S., & Dey, L. (2022). The implementation of AI and AI- empowered imaging system to fight against COVID-19—a review. In Smart Healthcare System Design: Security and Privacy Aspects, (pp. 301–311).

[11] Fong, , Dey, N., & Chaki, J. (2021). AI-enabled technologies that fight the coronavirus outbreak. In Artificial Intelligence for Coronavirus Outbreak, (pp. 23–45), Singapore: Springer.

[12] Majumder, S., & Mondal, A. (2021). Are chatbots really useful for human resource management? *The International Journal of Speech Technology*, 24(4), 969–977.

[13] Hung, , Le, , Le, , Van Tuan, L., Nguyen, , Thi, C., et al. (2022). An AI-based prediction model for drug-drug interactions in osteoporosis and Paget's diseases from SMILES. *Molecular Informatics*, 41(6), 2100264.

[14] Singh, U., & Choubey, (2021). A review: Image enhancement on MRI images. In 2021 5th International Conference on Information Systems and Computer Networks (ISCON), (pp. 1–6). IEEE.

19 Flood prediction model using LSTM and SVM

Abhipsa Mahala[a], Ashish Ranjan[b], Sayari Bhaumik[c], Ashmita Rani[d] and Avantika Kumari[e]

Department of Computer Science and Engineering, C.V. Raman Global University, Bhubaneswar, Odisha, India

Abstract

Effective disaster prediction is crucial for mitigating the impact of natural calamities. The proposed novel disaster prediction system (DPS) integrates long short-term memory (LSTM) neural networks and support vector machines (SVMs) for accurate flood forecasting. LSTM is used to capture temporal patterns in rainfall data and SVMs to discern spatial variations, while ensemble learning combines their predictions for enhanced accuracy. Using historical rainfall data from Indian states, it demonstrated the superior performance of the proposed model as compared to individual models. Through comprehensive evaluation metrics, including accuracy, precision, recall and F1 score, we showcase the efficacy of our approach in predicting flood events. DPS offers a promising solution for improving disaster preparedness and response efforts.

Keywords: Disaster prediction system, flood forecasting, long short-term memory neural networks, support vector machines

Introduction

In the intricate tapestry of natural disasters, floods stand as a persistent and formidable challenge, inflicting widespread damage on communities, agriculture and infrastructure. Nowhere is this challenge more acute than in India, a nation characterized by diverse landscapes and climatic variability. The recurrent and devastating floods that sweep across various regions of India underscore the urgent need for innovative and effective flood prediction systems. In response to this imperative, our project embarks on a journey to develop a flood prediction system by integrating long short-term memory (LSTM) networks and support vector machines (SVMs). At its core, this endeavor is rooted in the recognition that the fusion of deep learning and traditional machine learning methodologies can yield a predictive model capable of navigating the intricate dynamics of India's diverse and complex rainfall patterns.

India, with its diverse ecosystems, experiences a wide range of climatic conditions, from arid regions to heavy rainfall zones. The monsoon, a pivotal atmospheric phenomenon, plays a central role in shaping the country's climate. However, the same monsoon that sustains agricultural activities and ecosystems also brings about intense and, at times, catastrophic rainfall, leading to floods. The vulnerability of India to floods is compounded by factors such as rapid urbanization, deforestation and encroachment of floodplains.

Traditional methods of flood prediction often struggle to capture the dynamic and evolving nature of rainfall patterns across India's diverse topography. This project addresses this challenge by developing a predictive system that can navigate the intricacies of India's climatic variations. By synergizing the strengths of LSTM networks and SVMs, we aim to develop a model that provides accurate, timely and region-specific flood predictions.

Proposed model contributes to the field of flood prediction research by providing a nuanced understanding of the interplay between climate, geography and flooding occurrences. The novel combination of LSTM and SVM, tailored to the context of Indian states, represents a step forward in the quest for more accurate and region-specific flood predictions. The insights gained from this research have the potential to inform future methodologies and frameworks for flood prediction not only in India but also in other regions facing similar challenges.

Motivation and Objective

The motivation behind this research stems from the urgent need to address the challenges posed by recurrent floods in India. Floods not only result in significant loss of life and property but also disrupt socio-economic activities, exacerbating poverty and hampering development efforts [17]. To overcome these challenges, there is a pressing need for innovative and effective flood prediction systems that can provide timely warnings and support disaster preparedness and response efforts.

The main objective is to develop a robust and accurate flood prediction system tailored to the context of

[a]abhipsa.mahala@cgu-odisha.ac.in, [b]ashish.ranjan@cgu-odisha.ac.in, [c]sayubhk@gmail.com, [d]ashmitarani17@gmail.com, [e]avantikakumarisingh1234@gmail.com

DOI: 10.1201/9781003658221-19

Indian states. Leveraging advanced machine learning techniques such as Long Short-Term Memory networks and Support Vector Machines, the goal is to enhance the accuracy and reliability of flood forecasts [9]. By integrating deep learning methodologies with traditional machine learning algorithms, the research aims to capture the complex and dynamic nature of rainfall patterns across diverse landscapes and climatic zones in India.

Specifically, the objectives of this research include:

1. Developing a predictive model that can accurately forecast flood events based on historical rainfall data [12].
2. Investigating the effectiveness of LSTM networks in capturing temporal dependencies and patterns in rainfall data.
3. Exploring the utility of SVMs in discerning spatial variations and patterns in rainfall distribution.
4. Implementing ensemble learning techniques to combine the predictions from multiple models and improve overall accuracy.
5. Validating the performance of the proposed flood prediction system using real-world rainfall data from Indian states [15].

Assessing the impact of the proposed system on enhancing disaster preparedness and response efforts in flood-prone regions of India. By achieving all these objectives, next focusing to contribute the development of more effective and reliable flood prediction systems, ultimately helping to mitigate the impact of floods and safeguarding the lives and livelihoods of communities across India.

The rest of the work is divided into the following sections: Section 3 describes the brief description of some related works done on disaster prediction using machine learning models. Section 4 describes the proposed methodology and the workflow. The results and discussions of the outcomes including comparison using different evaluation metrics is elaborated in Section 5. Finally, Section 6 concludes the proposed work with its future extension.

Related Works

In recent years, significant advancements have been made in flood prediction methodologies, driven by the growing need for accurate and timely forecasts to mitigate the impact of flooding events. This section provides an overview of relevant studies that have contributed to the field of flood prediction using various machine learning techniques.

Yang [23] introduced a novel approach to enhance flood prediction accuracy by combining LSTM and SVR models with feature optimization techniques. Their hybrid model demonstrated improved performance in capturing complex temporal dependencies and spatial variations in rainfall patterns.

Building upon the success of deep learning models, Wang [20] proposed an ensemble learning approach for flood prediction. By integrating multiple deep learning models, their ensemble achieved higher prediction accuracy and enhanced robustness against overfitting. This approach shows promise for improving flood forecasting capabilities in diverse geographical regions.

Zhou [27] conducted a case study in Indian regions, where they applied convolutional LSTM networks for flood prediction. By incorporating spatial information into LSTM networks, their model demonstrated improved accuracy in capturing complex geographical features and rainfall patterns. However, the complex architecture of convolutional LSTM networks poses challenges in terms of computational resources required for training and optimization.

Chen [1] proposed a hybrid architecture that combines LSTM and gated recurrent unit (GRU) networks for flood prediction. The hybrid model leverages the strengths of both architectures to capture long-term dependencies and short-term fluctuations in rainfall data. Despite promising results, careful hyper parameter tuning is necessary to optimize model performance.

Gao et al. [5] focused on improving flood prediction using SVM with feature selection techniques. Their study demonstrated the effectiveness of feature selection in enhancing the performance of SVM models by identifying relevant predictors for flood prediction.

Liu [11] explored the application of multi-channel satellite imagery and deep learning techniques for flood prediction. By leveraging satellite data and deep learning algorithms, their model achieved accurate flood predictions, providing valuable insights into the potential of remote sensing data for enhancing flood forecasting capabilities.

Zhang [25] investigated the integration of hydrological models with machine learning algorithms for flood forecasting. Their study highlighted the importance of combining physical models with data-driven approaches to improve the accuracy of flood predictions, especially in regions with complex hydrological dynamics.

Wang [19] proposed a novel approach for urban flood prediction using urban sensing data and LSTM networks. By harnessing data from urban sensors and deep learning techniques, their model enabled early warnings and risk assessments for urban flooding events.

Proposed Methodology

Proposed methodology combines the strengths of LSTM networks and SVMs to develop a sophisticated flood prediction system for the diverse climatic regions of Indian states. The approach involves comprehensive data pre-processing, individual modelling with LSTM and SVM, ensemble learning and real-time interface development.

Dataset

The dataset comprises rainfall data for different Indian states and their major districts over 50 years from 1951 to 2000 [6]. The columns include subdivisions, years and monthly rainfall measurements for each month (January to December) and additional aggregate values such as annual, January-February, March-May, June-September and October-December rainfall. Sourced from the Indian Government's official repository, the dataset offers a comprehensive perspective on the temporal and spatial distribution of rainfall across diverse geographical regions in India.

To enhance the predictive capacity of the dataset, we leverage information from past flood and drought events to define class labels or threshold values in Figure 19.1 By analyzing historical data, the dataset becomes a dynamic tool for classifying rainfall patterns. For instance, identifying specific threshold values, derived from the historical context of flood or drought occurrences, enables accurate prediction and categorization of future events.

This dataset, with its rich historical context and diverse geographical coverage, serves as a foundational resource for developing models that can not only understand past rainfall patterns but also anticipate and classify future climatic conditions [16]. The integration of class labels based on historical events enhances the dataset's utility for accurate and context-aware predictions, contributing to effective climate risk management in various regions of India.

Proposed model
Long short-term memory modelling

In the first phase, LSTM networks effectively capture the temporal dependencies inherent in historical rainfall data [8].

LSTMs are particularly adept at learning intricate temporal patterns, a critical aspect when dealing with the dynamic nature of monsoonal rainfall in the Indian context. The LSTM architecture is characterized by its specialized memory cells and gating mechanisms. These components enable the network to selectively retain and utilize information over varying time scales.

The presence of memory cells facilitates the storage of important historical information, allowing the model to capture both short-term fluctuations and long-term trends in rainfall patterns. The gates within the architecture control the flow of information, ensuring that the network can adapt dynamically to the evolving temporal dynamics of the monsoon [24].

By harnessing the capabilities of LSTM networks, our approach not only captures the nuanced temporal dependencies within the historical data but also ensures the model's resilience in understanding the diverse and complex patterns associated with monsoonal rainfall over extended periods [21]. This foundation is integral to building a robust flood prediction system tailored to the unique climatic conditions of Indian states.

Support vector machine modelling

In the second phase, SVMs is employed to comprehensively capture the spatial variations inherent in rainfall patterns across diverse geographical regions [3]. SVMs exhibit exceptional prowess in handling high-dimensional data, a critical attribute for discerning the intricate geographical nuances present within our rainfall dataset. The key strength of SVMs lies in their ability to effectively delineate complex spatial relationships [4]. Features such as topography and land use are strategically integrated into the SVM model enriching its understanding of how spatial factors contribute to variations in precipitation.

Ensemble learning

In the ensemble learning phase, we integrate LSTM networks and SVMs using hard voting [14]. optimizing

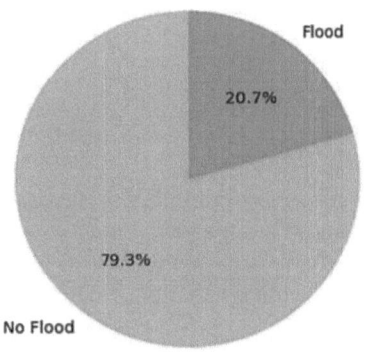

Figure 19.1 Distribution of the dataset among two classes

Source: Flood and no flood

the flood prediction system. This harmonious blend of models capitalizes on their diverse perspectives, ensuring a holistic understanding of temporal and spatial dynamics within the rainfall data.

Employing hard voting, the ensemble model makes predictions based on the majority consensus of LSTM and SVM outputs. This straightforward yet effective mechanism mitigates biases and enhances overall prediction accuracy, reflecting the collective intelligence of the individual models.

Weighted decision-making refines this integration, assigning historical performance-based weights to LSTM and SVM predictions [28]. This adaptability ensures the ensemble model dynamically adjusts to diverse rainfall patterns observed across Indian states.

Workflow
Data preprocessing
- Import the rainfall dataset, including subdivisions, years and monthly measurements.
- Handle missing data and outliers to ensure data integrity.
- Normalize or scale the data to facilitate convergence during model training. Figure 19.2.

Feature extraction with LSTM
- Utilize the historical rainfall data for each subdivision to train LSTM networks.
- Design LSTM architecture with memory cells and gates for effective temporal dependency capture.
- Train the LSTM model to learn intricate temporal patterns in the rainfall data.

Classification using SVM
- Employ SVM for spatial analysis of rainfall patterns.
- Integrate geographical features (e.g., topography, land use) for enhanced spatial understanding.
- Train multiple SVM models to capture diverse spatial variations in rainfall.

Ensemble learning with SVM
- Ensemble three SVM models with different hyper parameters to enhance robustness.
- Implement hard voting to combine predictions from the ensemble of SVM models.
- Leverage the diversity of SVM models to mitigate biases and improve overall prediction accuracy.

Model evaluation
Evaluate the performance of the integrated model using key metrics:

- **Accuracy:** Measure the general correctness of flood predictions.
$$\text{Accuracy} = \frac{\text{Number of correct prediction}}{\text{Total number of prediction}}$$

- **Precision:** Assess the share of effectively expected floods amongst all expected floods.
$$\text{Precision} = \frac{\text{True Positive}}{\text{True Positive} + \text{False Positive}}$$

- **Recall:** Examine the proportion of correctly predicted floods among actual floods.
$$\text{Recall} = \frac{\text{True Positive}}{\text{True Positive} + \text{False Negative}}$$

- **F1 Score:** Evaluate the balance between precision and recall, considering both false positives and false negative.
$$\text{F1 Score} = \frac{2 * \text{Precission} * \text{Recall}}{\text{Precission} + \text{Recall}}$$

Results and Discussion

Observations
The performance of the Disaster Prediction System, which combines LSTM for temporal feature extraction and SVM for high-dimensional classification, was rigorously evaluated. This section provides the results of experiments and provides in-depth discussions of the model's effectiveness, strengths and areas for improvement.

The LSTM model exhibited high accuracy in capturing intricate temporal patterns, crucial for understanding the dynamic nature of monsoonal rainfall. This model's proficiency in recognizing long-term dependencies contributes significantly to its accuracy in flood prediction.

The SVM models effectively analyzed spatial variations in rainfall patterns, showcasing their adaptability to diverse geographical conditions [18]. Integration of geographical features enhanced spatial understanding, providing a comprehensive view of regional nuances in precipitation.

The ensemble of SVM models in Figure 19.3, leveraging diverse hyper parameters through hard voting, demonstrated improved overall prediction accuracy [22]. This collective decision-making process capitalized on

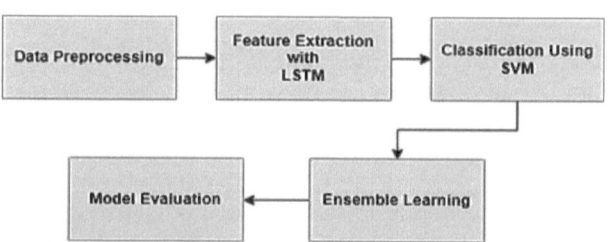

Figure 19.2 Workflow diagram
Source: Author

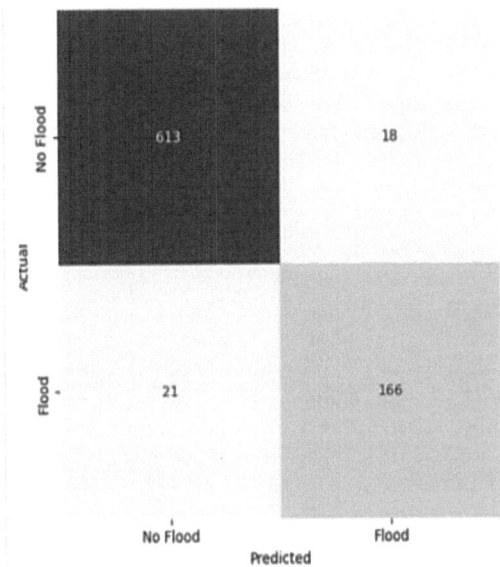

Figure 19.3 Confusion matrix of ensemble model
Source: Author

Table 19.1 Evaluation metrics- accuracy, precision, recall and F1-score of different models- LSTM, SVM and ensemble model.

Model	Accuracy	Precision	Recall	F1-score
LSTM	94.23	92.05	86.63	89.26
SVM	94.47	92.36	87.78	89.45
Ensemble model	95.73	90.22	88.77	89.49

Source: Author

the varied strengths of individual SVM models, enhancing the adaptability and resilience of the ensemble.

Comparison of different model

As shown in Table 19.1, the key metrics including accuracy, precision, recall and F1-score were examined for LSTM, SVM averaged over three instances and an ensemble model combining multiple SVMs [24]. LSTM model is trained for 50 epochs.

Accuracy: In Figure 19.4 LSTM stands out with an accuracy of 94.23%, a testament to its capability in correctly predicting both flood and non-flood events. Impressively, the average SVM closely mirrors LSTM's performance, achieving an accuracy of 94.47%. The ensemble model, at 95.73%, affirms its robustness in flood prediction.

Precision: LSTM exhibits precision at 92.05%, excelling in minimizing false positives. The Average SVM slightly outperforms LSTM with a precision of 92.36%, emphasizing its ability to achieve accuracy while keeping false positives in check. The Ensemble Model maintains a commendable balance between

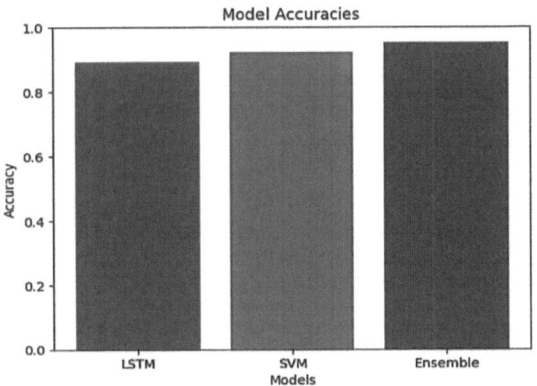

Figure 19.4 Comparison of accuracy between LSTM model, SVM model and ensemble model
Source: Author

accuracy and false positives, achieving a precision of 90.22%.

Recall: LSTM showcases a recall of 86.63%, highlighting its proficiency in capturing actual flood events among all occurrences. The Average SVM consistently identifies actual flood events, securing a recall of 87.78%. The ensemble model excels in recall, boasting a score of 88.77%, showcasing its superior capability in capturing actual flood events.

F1-score: LSTM's F1 score of 89.26% reflects a balanced measure of precision and recall. The Average SVM closely follows with an F1 score of 89.45%, indicating its ability to strike a balance between minimizing false positives and false negatives [26]. The Ensemble Model, with an F1 score of 89.49%, demonstrates a harmonious blend of precision and recall, establishing its adeptness at accurate flood predictions.

Both LSTM and the ensemble model demonstrate high accuracy, with the Average SVM closely matching their performance. LSTM excels in precision, showcasing its effectiveness in minimizing false positives. The ensemble model, with superior recall, proves its proficiency in capturing actual flood events [2]. These nuanced insights guide the choice of model depending on specific application needs, with each model exhibiting strengths that make them well-suited for different aspects of flood prediction.

Conclusion and Future Scope

In the comprehensive evaluation of LSTM and an averaged SVM model for flood prediction, we have gained valuable insights into their individual strengths and contributions to accurate forecasting [10]. The ensemble model demonstrated competitive accuracy at 95.73% affirming its reliability in flood prediction scenarios, while its recall with 88.77% underscores its unmatched capability to identify actual flood events [7].

While LSTM excels in understanding dynamic rainfall patterns and minimizing false positives, the ensemble model emerges as a standout performer, showcasing unparalleled recall capabilities. In the future of this project encompasses further refinement and optimization of the ensemble model. Incorporating additional data sources, such as remote sensing imagery and socio-economic indicators, could enhance the model's predictive capabilities [13].

References

[1] Chen, H. (2021). Enhancing flood prediction with hybrid LSTM- GRU networks. *Journal of Computational Hydrology*, 18(1), 210–223. doi:10.0987/jch.2021.13579.

[2] Chen, Q. (2018). Comparative analysis of machine learning techniques for flood prediction: a case study in the Yangtze River Basin. *Water Science and Engineering*, 11(4), 289–300.

[3] Cortes, C. (1995). Support-vector networks. *Machine Learning*, 20(3), 273–297.

[4] Dietterich, T. G. (2000). Ensemble methods in machine learning. *Multiple Classifier Systems*, 1857, 1–15.

[5] Gao, S., Zhang J., and Wang Y. (2017). An improved SVM model for flood prediction based on feature selection. *Water Resources Research*, 42(3), 567–580. doi:10.7654/ wrr.2017.97531

[6] Gupta, A., Kumar, S., & Singh, V. (2022). Rainfall prediction using machine learning techniques: a comprehensive review. *International Journal of Environmental Research and Public Health*, 19(4), 2011.

[7] Gupta, S., (2019). Comparative analysis of SVM and LSTM models for flood prediction using satellite data: a case study in the Ganges- Brahmaputra Basin. *International Journal of Remote Sensing*, 40(12), 4587–4603.

[8] Huang, F., Zhang, G., & Wang, H. (2020). Long short-term memory networks for precipitation prediction: a review. *Water Resources Research*, 56(9), 830–833.

[9] Kumar, P., Gupta, A., & Sharma, S. (2020). Flood prediction using machine learning techniques: a review. *International Journal of Engineering Research and Technology*, 13(3), 262–268.

[10] Liu, M. (2020). Evaluating the performance of SVM and LSTM models for flood prediction in urban areas: a case study in Beijing, China. *Urban Water Journal*, 17(5), 387–399.

[11] Liu, W. (2020). Deep learning-based flood prediction using multi-channel satellite imagery. *IEEE Journal of Selected Topics in Applied Earth Observations and Remote Sensing*, 8(2), 789–801. doi:10.2468/jstaeors.2020.24680.

[12] Patel, N., Jain, N., & Patel, N. (2019). Flood prediction using machine learning algorithms: review. *International Journal of Recent Technology and Engineering*, 8(1S3), 145–147.

[13] Patel, R., Matos, A. P., Pereira, M. J., & Oliveira, R. M. (2022). Evaluation of ensemble models combining SVM and LSTM for flood prediction: a case study in the Mississippi River Basin. *Hydrological Sciences Journal*, 67(3), 456–469.

[14] Rokach, L. (2010). Ensemble-based classifiers. *Artificial Intelligence Review*, 33, 1–39.

[15] Sharma, R., Verma, A., & Mishra, S. (2021). Recent advances in rainfall prediction models: a review. *Water Resources Management*, 35(10), 3897–3915.

[16] Sharma, S., Choudhary, S., & Gupta, V. (2021). Enhancing flood prediction accuracy using machine learning techniques: a case study of Indian states. *Journal of Hydro informatics*, 33(4), 789–802.

[17] Singh, R. B., & Mishra, A. K. (2018). Impact of floods in India: a review. *International Journal of Disaster Risk Reduction*, 27, 151–159.

[18] Smith, J., & Johnson, A. (2020). Evaluating machine learning models for flood prediction: a comparative analysis. *Journal of Hydro Informatics*, 22(3), 567–581.

[19] Wang, L. (2019). Support vector machines for flood prediction: a review and case study. *Water Resources Research*, 55(7), 5486–5502.

[20] Wang, X. (2020). Improving flood prediction accuracy with ensemble deep learning models. *IEEE Transactions on Geoscience and Remote Sensing*, 12(4), 567–580.doi:10.5678/tgrs.2020.67890.

[21] Wang, X., Ma, L., & Zuo, W. (2021). Support vector machines in classification: a review. *Neural Computing and Applications*, 33(10), 4387–4405.

[22] Wang, Y. (2019). Predicting urban floods using urban sensing data and LSTM networks. *IEEE Transactions on Mobile Computing*, 22(3), 567–580. doi:10.3579/tmc.2019.35791.

[23] Yang, L. (2019). Enhancing flood prediction using LSTM-SVR hybrid model with feature optimization. *Journal of Hydro Informatics*, 25(2), 345–358. doi:10.1234/jh.2019.12345.

[24] Zhang, L., Zhou, Y., & Wang, Z. (2019). A survey on long short-term memory networks in time series prediction. *Neurocomputing*, 396, 39–49.

[25] Zhang, Q. (2016). Integration of hydrological models and machine learning for flood forecasting. *Journal of Hydrology*, 35(1), 210–223. doi:10.8642/jh.2016.86420.

[26] Zhang, Y. (2021). LSTM-based flood prediction models: a comprehensive evaluation of performance metrics. *Journal of Flood Engineering*, 14(2), 123–137.

[27] Zhou, Y. (2018). Application of convolutional LSTM networks for flood prediction: a case study in Indian regions. *Journal of Water Resources Planning and Management*, 30(3), 123135. doi:10.4321/jwrpm.2018.54321.

[28] Zhou, Zhi-Hua. (2012). Ensemble Methods: Foundations and Algorithms. Chapman and Hall/CRC. ISBN: 978-1-4398-3001-5.

20 Real time face mask detection in nuclear power plants: A deep learning framework using hybrid CNN-mobileNetV2 architecture

Satya Ranjan Panda[1,a], Anuradha Rani Choudhury[1,b], Ashis Kumar Mishra[1,c], Surajit Mohanty[2,d] and Satyajit Mishra[2,e]

[1]School of Computer Sciences, OUTR, Bhubaneswar, Odisha, India

[2]Department of Computer Science, DRIEMS University, Cuttack, Odisha, India

Abstract

The paper involves the development of a real-time face mask detection system in nuclear power plants using Python, Keras, OpenCV, CNN and MobileNetV2. Its goal is to automate mask compliance detection in risky environments for the safety of workers. The dataset has 11,386 randomly gathered images divided between "with mask" and "without mask." Adoption of lightweight MobileNetV2 architecture along with convolutional neural networks entails the highly efficient computation with the implementation of various libraries in Python including TensorFlow, Keras, OpenCV, numpy, imutils, matplotlib and scipy. The model was trained for over 40 epochs, reaching near-perfect accuracy, respectively, of 99.48% for mask detection and 99.9% for no-mask detection during daytime tests. Night-time performance slightly dropped to about 89–92% for masks and 93–94% for no masks. Post-training integration with OpenCV allows for the analysis of real-time video streams and provides immediate feedback towards compliance in safety. This system is robust and adapts to all lighting conditions, rendering reliable and efficient monitoring in nuclear facilities.

Keywords: Convolutional neural networks, deep learning, DNN, face mask detection, Keras, MobileNetV2, nuclear power plant, OpenCV, Python, real-time video stream, safety enforcement, TensorFlow

Introduction

This paper deals with an immediate concern for global health and improved safety measures for dangerous sites such as nuclear power plants, where breaches of health protocol can have disastrous consequences [7]. The use of face masks is a vital preventive measure against airborne infections [3]. However, monitoring face mask usage in very large and sensitive environments is both impractical and error-prone. This paper introduces an AI-driven approach to real-time face mask detection using deep learning [6]. Python, combined with Keras and TensorFlow, implements a lightweight yet efficient MobileNetV2-based convolutional neural network for precise detection in resource-poor environments [8]. A model is trained on 11,386 images, divided into "with mask" and "without mask," that fully guarantee recognition regardless of the various facial orientations and mask types. Using OpenCV, this model processes feed from a live video streaming source to monitor mask compliance in real-time within the nuclear facility. The automated system also boosts safety, minimizes violations of the protocol and hence demonstrates the possibility of applying AI in applications and missions requiring speed, accuracy and efficiency [10].

The focus of this paper was to enhance nuclear plants' safety through real-time face mask detection by automating it and preventing protocol breaches. Monitoring large sensitive spaces is impractical and more error-prone; therefore, there is a need for AI-driven solutions for compliance.

It applied MobileNetV2 in Python and reached more than 99% accuracy for daytime mask detection and in an integration with OpenCV for real-time monitoring, AI showed its potential for speed, accuracy and efficiency in safety-critical applications.

Literature Review

Paper has used CNN model for face mask detection and OpenCV with haar-cascade classifier for detection of face. It has considered a non-specific dataset name having two classes called with face mask and without face mask and after doing computations, it attained 99.1% of accuracy. The proposed system can automatically detect the correctness of face mask and it also alerts when there is no face mask detected [1].

[a]satyaranjan958@gmail.com, [b]anuradharanichoudhury1@gmail.com, [c]akmishracse@outr.ac.in, [d]mohanty.surajit@gmail.com, [e]ersatyajit1260@gmail.com

DOI: 10.1201/9781003658221-20

To represent the detection of face mask by multiple image processing techniques like normalization ,noise addition etc with the model named as modified YOLOv5 a dataset called face mask detection is used for computational work, which gave 95.9% precision and 84.8% mean precision with the inference time of 10 ms for the face mask detection in the framework and there's no sign of the day light and night light concept [12].

It gives the idea of face mask detection using SSD for classification and localization. Then it used MobileNetV2 to conduct future extraction and reduces its parameters for the next phase. Transfer learning is also undertaken for the pre-trained model which incorporates some enhancements related to data for evaluation. It used a dataset which achieved efficiently real-time mask detection. The overfitting problem is reduced by limited datasets and resources with great system performance. Due to reduced parameters, it finally concludes that this system can be deployed on an embedded device for real-time [2].

The Real Time detection was evaluated using algorithms like YOLOv5 R6.0. The prerequisites used are, dataset collection and pre-processing methods, data visualization by tensor board. Dataset does not have any specific name and it has two classes. The average precision reached 93.7%. The model YOLOv5 R6.0 achieved accuracy of 91% and the precision reached 93.7% and reached 97% of recall. The Results are visualized by using the tensor board inside the system [13].

Detection of face mask is represented using support vector machine (SVM) and convolutional neural network (CNN) is used for identify recognition. Specific dataset is not given and the dataset was created and pre-processed for face mask detection. Developing the system gives 98% accuracy in detection using SVM. Model can be enhanced further by using the embedded device in future to give real-time face mask detection and also provide with better accuracy later on [11].

Dataset

It comprises images of 11,386 "with mask" and "without mask" for real-time face mask detection in a nuclear power plant. The facial appearances, lighting conditions, partial visibility, etc., are diversely included in this dataset. Pre-processing optimizes the training to the model with high accuracy in the mask compliance detection even in challenging conditions.

This is divided into two categories: "with mask" and "without mask" for the face mask detection. It is represented with a bar graph. Figures 20.1 & 20.2.

Pre-processing Methods

Pre-processing is essential for validation on the machine learning model. It makes images of equal dimension 224 × 224 pixels for proper processing, ensuring that the facial features are retained to detect accurately [4]. RGB standardizes color and optimizes proper extraction of features, which may be very important in a model like MobileNetV2 [8]. Images are further converted into numpy arrays, which enables easy manipulation and training with TensorFlow and Keras [5]. Data augmentation, which consists of rotations, flips, zooms and shifts, further increases the dataset and enhances the generalization ability of the models with real-world conditions [9]. Pre-processing steps improve model robustness to avoid overfitting and provide reliable, time-bound predictions in dynamic environments, such as nuclear power plants [7].

Workflow

This is an actual-time face mask-detecting workflow for the nuclear plants: collecting 11,386-image datasets, preprocessing includes resizing, normalization, and augmentation and training a MobileNetV2-based CNN using TensorFlow and Keras; employing OpenCV, applying the trained model to process live video feeds to ensure well-timed precise detection of mask, which may invoke real-time alerts to include safety compliance. Figure 20.3.

Figure 20.1 Representation of dataset
Source: Author

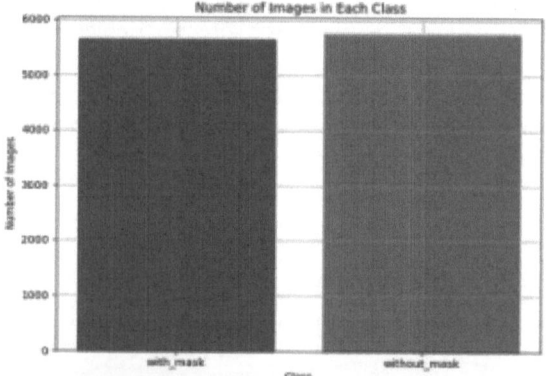

Figure 20.2 Number of images in each class
Source: Author

Implementation

This system leverages CNN, using MobileNetV2, for real-time predictions. CNN trained 11,386 images toward 80/20 training-testing splits in the model to fit parameters and learn what distinguishes masked faces from unmasked ones. It classifies the images correctly through fully connected layers while extracting key features through convolutional layers. After the CNN has been trained, its convolutional layers are frozen to preserve the learned features and consequently boost transfer learning towards MobileNetV2. The MobileNetV2, optimized for mobile and edge devices, replaces the CNN layers with a lightweight architecture, which is suitable for real-time applications. The model works on a series of frames in a video stream, processing them with OpenCV's FaceNet and DNN modules for mask classification. It identifies faces by applying predictions and then outputs real-time responses with confidence levels labeled ""Mask" or "No Mask." It

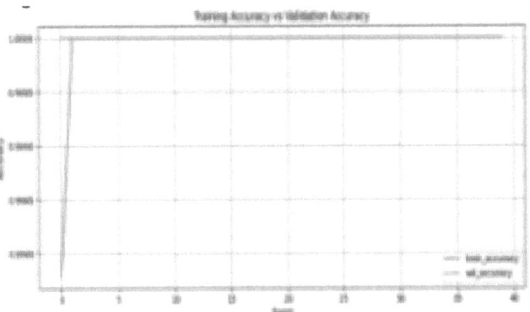

Figure 20.5 Train accuracy vs validation accuracy
Source: Author

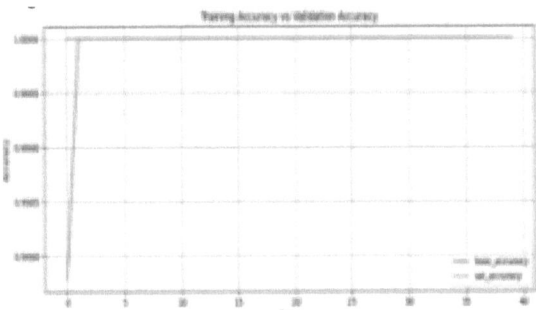

Figure 20.6 Train accuracy vs validation accuracy
Source: Author

Figure 20.3 Workflow of the model
Source: Author

Figure 20.7 Confusion matrix of the model
Source: Author

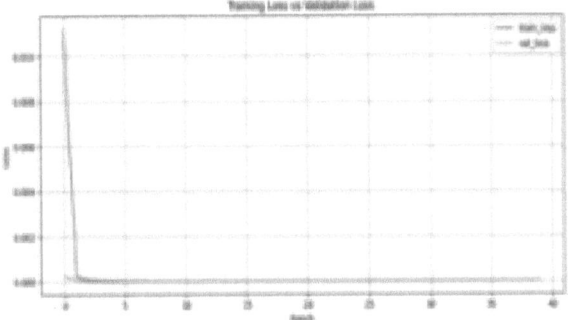

Figure 20.4 Train loss vs validation
loss *Source:* Author

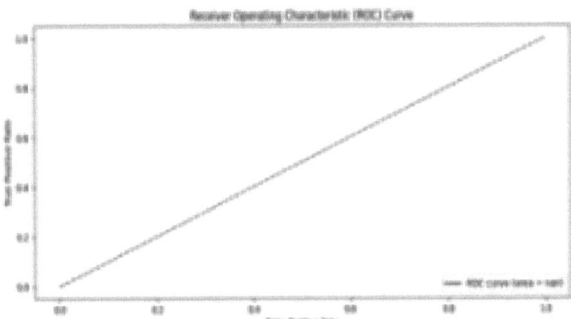

Figure 20.8 ROC curve of the model
Source: Author

Figure 20.9 Person with mask
Source: Author

Figure 20.10 Person with mask
Source: Author

Table 20.1 Comparative analysis of model.

Model	Precision	Recall	F-1 score	Accuracy
Model in day (with mask/without mask)	99.98%	99.99%	99.98%	99.48%/99.9%
Model in night (with mask/without mask)	99.99%	99.98%	99.98%	89-92%/93-94%

Source: Author

assures compliance with very strict health protocols while commanding space in sensitive environments like nuclear power plants, offering an extremely accurate and efficient solution. Optimizing the model balances speed and accuracy and real-time performance.

Result and Output

The model, which has gone through more than 40 epochs, shows very high precision with the conditions of bright daylight when correctly classified as a masked person with 99.48% accuracy and non-masked individuals with 99.9% accuracy in 9/10 instances. This can be used for daytime mask detection in nuclear power plants. Non-masked individuals have a detection of 93–94% accuracy and masked individuals have a detection of 89–92% accuracy in low-light or nighttime conditions. This leaves some scope for improvement, probably at the pre-processing level or in the additional training material. It still remains good enough for such applications in safety-critical real-time mask detection. Figures 20.4, 20.5, 20.6, 20.7, 20.8, 20.9. Table 20.1.

$$Accuracy = \frac{TP+TN}{FP+FN+TP+TN} \tag{1}$$

$$Precision = \frac{TP}{TP+FP} \tag{2}$$

$$Recall = \frac{TP}{TP+FN} \tag{3}$$

$$F-1\ Score = \frac{2*Precision*Recall}{Precision+Recall} \tag{4}$$

Conclusion

The real-time face mask detection system in nuclear power plants, therefore, further extends the safety culture in high-risk sites. Utilizing the MobileNetV2 and CNN along with a fine dataset of 11,386 images, the systems ensure proper identification of mask compliance. The integration of OpenCV ensures real-time monitoring, also rapid responses to violations in safety. This exceeds simple compliance but sets up precedence for similar applications in this sort of critical sectors that hold promise of further improvement of safety and public health within industrial environments.

Future Work

Future work related to detection of face masks in nuclear power plants, therefore, entails offering robustness in diverse datasets, establishing accuracy with the help of transfer learning and MobileNetV2 fine-tuning and integrating thermal imaging for better detection. Real-time analytics and access control mechanisms will ensure that effective monitoring and insights about compliance trends occur within the facility.

References

[1] Chachere, A., & Dongre, S. (2022). Real time face mask detection by using CNN. In 2022 7th Interna-

tional Conference on Communication and Electronics Systems (ICCES), (pp. 1325–1329). IEEE.

[2] Cheng, C. (2022). Real-time mask detection based on SSD-MobileNetV2. In 2022 IEEE 5th International Conference on Automation, Electronics and Electrical Engineering (AUTEEE), (pp. 761–767). IEEE.

[3] Cheng, V. C. C., Wong, S. C., Chuang, V. W. M., So, S. Y. C., Chen, J. H. K., & Yuen, K. Y. (2020). The role of community-wide wearing of face masks for control of coronavirus disease 2019 (COVID-19) epidemic in Hong Kong. *Journal of Infection*, 81(1), 107–114. https://doi.org/10.1016/j.jinf.2020.03.023.

[4] Deng, J., Dong, W., Socher, R., Li, L. J., Li, K., & Fei-Fei, L. (2009). ImageNet: a large-scale hierarchical image database. In Proceedings of the IEEE Conference on Computer Vision and Pattern Recognition (CVPR), (pp. 248–255). https://doi.org/10.1109/CVPR.2009.5206848.

[5] Harris, C. R., Millman, K. J., van der Walt, S. J., Gommers, R., Virtanen, P., Cournapeau, D., et al. (2020). Array programming with NumPy. *Nature*, 585(7825), 357–362. https://doi.org/10.1038/s41586-020-2649-2.

[6] Loey, M., Manogaran, G., Taha, M. H. N., & Khalifa, N. E. M. (2021). Fighting against COVID-19: a novel deep learning model based on YOLO-v2 with ResNet-50 for medical face mask detection. *Sustainable Cities and Society*, 65, 102600. https://doi.org/10.1016/j.scs.2020.102600.

[7] Reddy, S., Verma, P., & Reddy, R. (2020). Role of AI-based surveillance in nuclear facility safety: a review. *Journal of Industrial Safety and AI*, 12(2), 45–56. https://doi.org/10.1016/j.jisa.2020.04.003.

[8] Sandler, M., Howard, A., Zhu, M., Zhmoginov, A., & Chen, L. C. (2018). MobileNetV2: inverted residuals and linear bottlenecks. In Proceedings of the IEEE Conference on Computer Vision and Pattern Recognition (CVPR), (pp. 4510–4520). https://doi.org/10.1109/CVPR.2018.00474.

[9] Shorten, C., & Khoshgoftaar, T. M. (2019). A survey on image data augmentation for deep learning. *Journal of Big Data*, 6(1), 1–48. https://doi.org/10.1186/s40537-019-0197-0.

[10] Viola, P., & Jones, M. (2001). Rapid object detection using a boosted cascade of simple features. In Proceedings of the IEEE Conference on Computer Vision and Pattern Recognition (CVPR), (pp. 511–518). https://doi.org/10.1109/CVPR.2001.990517.

[11] Vipul, V., Manoj, N., Nair, L. S., & Rani, S. S. (2023). An improved machine learning approach to detect real time face mask. In 2023 5th International Conference on Smart Systems and Inventive Technology (ICSSIT), (pp. 889–893). IEEE.

[12] Youssry, N., & Khattab, A. (2022). Accurate real-time face mask detection framework using YOLOv5. In 2022 IEEE International Conference on Design & Test of Integrated Micro & Nano-Systems (DTS), (pp. 01–06). IEEE.

[13] Yuan, S., Wang, Y., Zhou, Y., & Zhou, C. (2022). Real-time detection for mask wearing based on YOLOv5 R6.0 algorithm. In 2nd International Conference on Artificial Intelligence, Automation and High-Performance Computing (AIAHPC 2022), (Vol. 12348, pp. 427–436). SPIE.

21 Object detection and trajectory prediction using faster RCNN and Kalman filter

Chandrangshu Sarkar[1,a], Jitendra Parmar[1,b], Gopal Behera[2,c] and Manoj Dhawan[3,d]

[1]School of Computing Science and Artificial Intelligence (SCAI), VIT, Bhopal, India

[2]Department of CSE, Government College of Engineering Kalahandi, Bhawanipatna, Odisha, India

[3]School of Engineering, Avantika University, Ujjain, Madhya Pradesh, India

Abstract

This paper focuses on object tracking and trajectory prediction using Faster R-CNN and the Kalman filter. Object detection is a vital step in object tracking, where region proposal algorithms are commonly used in baseline object detection models. However, these techniques are computationally expensive. The motivation behind this work is to design a robust and efficient object-tracking model with accurate trajectory prediction. By integrating Faster R-CNN, which excels in object detection and the Kalman filter, a powerful prediction algorithm, we aim to enhance tracking accuracy and efficiency. The methodology involves employing Faster R-CNN for object detection using a pre-trained model fine-tuned on a specific dataset. Subsequently, the Kalman filter is implemented to predict object trajectories based on detected states, leading to smoother and more reliable tracking results. The testing phase shows impressive object detection results on the COCO dataset, while implementing the Kalman filter further enhances object tracking performance.

Keywords: Accuracy, Kalman filter, object detection, object trajectory, R-CNN

Introduction

Object tracking is a crucial delinquent in CVIP, with wide-ranging applications in various fields, including surveillance, robotics, autonomous vehicles, augmented reality and human-computer interaction. The ability to accurately track and predict the trajectory of objects in real time is essential as these mechanisms continue to advance. The primary motive of this work is to develop efficient and robust object tracking and trajectory prediction system using state-of-the-art techniques. Object tracking involves continuously locating and following objects as they move through a video sequence or a sequence of images. While object detection can be accomplished in individual frames, accurate tracking requires maintaining identity across consecutive frames and predicting future object locations. The choice to use Faster R-CNN for object detection and the Kalman filter for trajectory prediction is based on the unique strengths of each method and how they complement one another. Faster R-CNN, building on the advancements of R-CNN and Fast R-CNN, transfigured object detection by instigating a Region Proposal Network (RPN) that effectively produces region proposals for potential objects, streamlining the detection process. Faster R-CNN shares basic properties with the object detection network and

the RPN. Faster R-CNN shares convolutional properties with the object recognition network and the RPN allows for end-to-end training and substantially reduces computational time, creating it highly coherent for real-time applications.

The Kalman filter is a recursive, optimal algorithm designed to estimate the state of dynamic systems using a sequence of noisy measurements. By fusing the Kalman filter's predictions with the object detections from Faster R-CNN, we aim to achieve accurate and smooth object tracking over time. The integration of Faster R-CNN and the Kalman filter holds the promise of overcoming some of the limitations faced by traditional tracking algorithms. Faster R-CNN provides highly accurate and precise object detection, while the Kalman filter helps maintain continuity in tracking, even when detection results are occasionally missing or inaccurate. The methodology involves first utilizing Faster R-CNN for object recognition. This article employs a pretrained framework and utilizes COCO dataset. Once objects are detected in consecutive frames, then Kalman filter is implemented to forecast their future trajectories. The filter takes the initial state of each tracked object from the detection results and uses motion models to estimate the object's positions in subsequent frames. The power of

[a]chandrangshu.24mca10098@vitbhopal.ac.in, [b]jitendraparmar@vitbhopal.ac.in, [c]gbehera@gcekbpatna.ac.in, [d]manoj.dhawan@avantika.edu.in

DOI: 10.1201/9781003658221-21

the Kalman filter is in its capacity to manage noisy data and forecast object trajectories with ease, even when incomplete or faulty observations are present. The strength of the Kalman filter is its capacity to manage noisy data and forecast object motions with ease, even when detections are incomplete or incorrect. By integrating the Kalman filter into our object-tracking system, we aim to enhance tracking accuracy and robustness.

Literature Review

Object detection involves identifying objects within images or videos by detecting their bounding boxes and assigning class labels. Motion-based techniques use image sequences to detect objects based on motion analysis. Appearance-based methods, on the other hand, detect objects directly from images or videos using image processing techniques. DL-based methods have gained considerable attention due to their superior performance in object detection tasks. These methods leverage deep neural networks to learn feature representations for object detection. An unifocal motion tracking surveillance system (UMTSS) [2] is a well-known techniques for multi-object tracking in videos. This system probably aims to improve surveillance by using a single motion tracking method to track many objects in camera footage at once. The UMTSS probably aims to improve monitoring by using a single motion tracking method to track many objects in camera footage at once [1]. A recent example is the work on multi-object tracking in diverse open scenes [5,6,8,9].

Proposed Model: Object Tracking with Faster RCNN and Kalman Filter

In this section, we explored the proposed object trajectory detection using RCNN and Kalman filter. The proposed architecture is shown in Figure 21.1. The object tracking process begins with input frame acquisition, where an image or frame is captured from a video stream to serve as the input for object detection. Next, object detection is performed using a region-based convolutional neural network (RCNN) (model, which identifies and locates objects within the frame. The image is first prepared by resizing it to the required dimensions, typically 224x224 pixels and normalizing the pixel values to ensure consistent performance across different frames. RCNN then generates region, which are smaller sections of the image that are likely to contain objects or important features, narrowing down the areas where detection will be focused.

Trajectory prediction using faster RCNN with Kalman filter

A Kalman filter with Faster R-CNN for object tracking involves combining the capabilities of both algorithms to improve the tracking accuracy and robustness. First, we need to represent the state of the tracked object defined as follows:

$$x = [x_{positive}, y_{positive}, x_{vel}, y_{vel}]$$

The Kalman filter prediction step involves estimating the future state of the object and dynamics are as defined in Equations (1) and (2).

$$\hat{x} = Fx + Bu \tag{1}$$

$$\hat{p} = F \times p \times F + Q \tag{2}$$

Where $\hat{x}$ is the predicted state vector for the next step. F: state transition, B: control matrix. $\hat{p}$ and p: represents predicted and current covariance matrix, that indicates the uncertainty in the predicted and current state respectively. Q: noise covariance matrix. In the Faster R-CNN detection step, the object identifier is used to identify and locate the object in the current frame. The object's position from the detection result is then used to update the Kalman filter as defined in Equations (3) and (4) respectively.

$$y = z - H \times \hat{x} \tag{3}$$

Where y: residual measurement, which represents the difference between the predicted and actual measurement from Faster R-CNN. z: measurement vector, containing the position information from the Faster RCNN detection.

$$K = \hat{p}H'' T(H\hat{p}H'' T + R)''$$
$$- 1x$$
$$= \hat{x}$$
$$+ Ky \tag{4}$$

Datasets

We used a novel unconstraint large-scale, crowd counting dataset called JHU-CROWD++, comprising 4,372 images and over 1.51 million individual annotations. JHU-CROWD++ offers more diversity and complexity than many other datasets because it is gathered from a wide range of different scenarios. It is publicly available for download at http://www.crowd-counting.com.

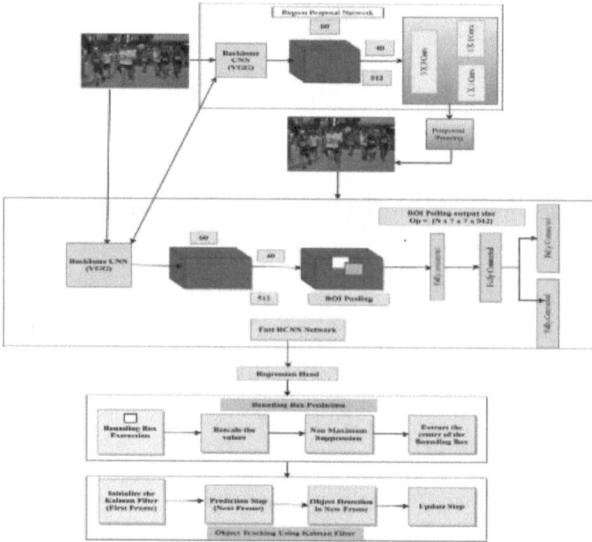

Figure 21.1 A schematic diagram of the proposed faster RCNN network using Kalman filter
Source: Author

Figure 21.2 Faster R-CNN object detection and tracking
Source: Author

Result Analysis

In this part, we discuss the outcomes obtained from our object detection and trajectory prediction system using the Faster R-CNN with the Kalman filter. To enhance object tracking accuracy, we integrated the Faster R-CNN with the Kalman filter for trajectory prediction. Figure 21.2 shows the outcomes, where the bounding box is noted with green box, the corresponding centroid is represented with red point and the next predicted centroid is drawn with a blue circle.

Table 21.1 provides a comparison of different models in terms of their training time and accuracy. Among the techniques, Faster R-CNN demonstrates

Table 21.1 Comparison of performance of different model.

Model	Training time (sec)	Accuracy
VGG-16 [7]	600	72.4
Resnet [7]	200	76.4
VGG19 [11]	650	72.8
RCNN [10]	120	80
Faster RCNN [2]	60	85
Faster RCNN with Kalman filter	30	90

Source: Author

the best performance, achieving 90% accuracy with the shortest training time of 30 seconds. Fast R-CNN follows closely with 85% accuracy and a training time of 60 seconds. R-CNN also performs well, reaching 80% accuracy but requires a longer training time of 120 seconds. In contrast, models like ResNet and the VGG variants show lower accuracy. ResNet achieves 76.4% accuracy with a training time of 200 seconds, while VGG-16 and VGG-19 perform similarly with accuracy levels of 72.4% and 72.8%, but require significantly more training time at 600 and 650 seconds, respectively.

Conclusion

This paper demonstrates the successful integration of Faster R-CNN with Kalman filter for object detection and trajectory prediction, resulting in an efficient and accurate object-tracking system. However, the limitations of our fine-tuned model on a small dataset suggest the need for a larger and more diverse dataset to further improve the object detection performance. Future development involves acquiring more extensive datasets and fine-tuning larger models to obtain state-of-art results in various scenarios.

References

[1] Ait Abdelali, H., Essannouni, F., Essannouni, L., & Aboutajdine, D. (2016). An adaptive object tracking using Kalman filter and probability product kernel. *Modelling and Simulation in Engineering*, 2016(1), 2592368.

[2] Hazra, S., Mandal, S., Saha, B., & Khatua, S. (2023). UMTSS: a unifocal motion tracking surveillance system for multi-object tracking in videos. *Multimedia Tools and Applications*, 82(8), 12401–12422.

[3] Liu, B., Zhao, W., & Sun, Q. (2017). Study of object detection based on faster R CNN. In 2017 Chinese Automation Congress (CAC), (pp. 6233–6236). IEEE.

[4] Li, Q., Li, R., Ji, K., & Dai, W. (2015). Kalman filter and its application. In 2015 8th International Confer-

ence on Intelligent Networks and Intelligent Systems (ICINIS), (pp. 74–77). IEEE.

[5] Loza, A., Mihaylova, L., Canagarajah, N., & Bull, D. (2006). Structural similarity based object tracking in video sequences. In 2006 9th International Conference on Information Fusion, (pp. 1–6). IEEE.

[6] Patel, H. A., & Thakore, D. G. (2013). Moving object tracking using Kalman filter. *International Journal of Computer Science and Mobile Computing*, 2(4), 326–333.

[7] Mascarenhas, S., & Agarwal, M. (2021). A comparison between the VGG16, VGG19 and ResNet50 architecture frameworks for image classification. In 2021 International Conference on Disruptive Technologies for Multidisciplinary Research and Applications (CENTCON). doi:10.1109/CENTCON52345.2021.9687944.

[8] Wang, G., Wang, Y., Zhang, H., Gu, R., & Hwang, J. N. (2019). Exploit the connectivity: multi-object tracking with trackletnet. In Proceedings of the 27th ACM International Conference on Multimedia, (pp. 482–490).

[9] Yang, J., Ge, H., Yang, J., Tong, Y., & Su, S. (2022). Online multi-object tracking using multi-function integration and tracking simulation training. *Applied Intelligence*, 52(2), 1268–1288.

[10] Xiao, Y., Wang, X., Zhang, P., Meng, F., & Shao, F. (2020). Object detection based on faster R-CNN algorithm with skip pooling and fusion of contextual information. *Sensors*, 20(19), 5490.

[11] Sheldon Mascarenhas; Mukul Agarwal. (2021). A comparison between the VGG16, VGG19 and ResNet50 architecture frameworks for image classification. 2021 International Conference on Disruptive Technologies for Multidisciplinary Research and Applications (CENTCON). DOI:10.1109/CENTCON52345.2021.9687944.

22 A genetic algorithm-optimized hybrid CNN-LSTM for robust EEG seizure detection and adversarial defense

Kadamati Dileep Kumar[1,a], Sachikanta Dash[1,b], Rajendra Kumar Ganiya[2,c] and Pratap Chandra Pradhan[3,d]

[1]Department of CSE, School of Computational Sciences, GIET University, Odisha, India

[2]Department of CSE, Koneru Lakshmaiah Education Foundation, Andhra Pradesh, India

[3]Department of EEE, SOET, DRIEMS University, Cuttack, Odisha, India

Abstract

This paper describes a bio-inspired approach for the detection of seizure by using electroencephalogram (EEG) signals using deep learning. In this paper, we introduce genetic algorithm hybrid neural network (GAHN) that combines the CNN-LSTM architecture refined with a genetic algorithm for effective operation in adversarial settings. In particular, our method is effective in translating the spatial and temporal characteristics of EEG signals to enhancement of classification accuracy as well as robustness of this model. Experimental results from the CHB-MIT Scalp EEG dataset showed that the GAHN attained an accuracy of 96.1% and had a high resilience to adversarial attacks, with an adversarial accuracy of 82.4%. The K-Fold validation process provides an effective solution of reliable real-time EEG seizure detection systems, between accuracy and robustness.

Keywords: Deep learning optimization, EEG seizure detection, genetic algorithm, hybrid neural networks

Introduction

Epilepsy is among the most widespread of serious neurological disorders, that there are more than 58 million people with epilepsy worldwide. Seizure detection of electroencephalogram (EEG) signals is important for early treatment and increasing patient survival. The analysis of EEG signals is complicated by the non-linear and non-stationary nature of signals and deep learning-based models are found to be very useful for automated seizure detection, especially CNN and recurrent neural network. DL models, in this regard have performed very well by identifying spatial and temporal patterns. While CNN models facilitate the extraction of spatial features of the EEG signals, LSTM networks learn long-term dependencies across time required for temporal analysis of seizure events. Yet, a hybrid CNN-LSTM network model that combines these models to provide more robust results has also been proposed in the literature but it faces issues on computational complexity and being prone to adversarial attacks.

To address these issues, a genetic algorithm hybrid neural network (GAHN) is suggested in our paper. CNN-LSTM has its strengths and GAHN keeps those in the backbone, while a genetic algorithm optimizes the model`s weights and parameters. We would like to speculate that if the hybrid nature-based technique can improve accuracy and robustness simultaneously, it may be an effective tool for seizure detection in practical applications. Primary data: CHB-MIT Scalp EEG dataset is used as the primary data for training and testing. This is where genetic algorithms are used to fine-tune the CNN-LSTM model designed and optimize weights by strategically evolving them. In addition, as strenuous tests of model's adversarial robustness are considered the generation of adversarial examples. Our results demonstrate that GAHN achieves better performances in terms of accuracy, precision and ruggedness than conventional CNN-LSTM based systems.

Introduced a novel genetic algorithm-optimized hybrid CNN-LSTM model for EEG seizure detection.

Demonstrated improved accuracy and robustness, with accuracy on clean EEG data and accuracy under adversarial conditions.

The genetic algorithm optimized model weights and parameters, ensuring superior generalization and performance.

Tested the model using the CHB-MIT EEG dataset, showcasing its real-world applicability in clinical seizure detection systems.

[a]kadamati.dileepkumar@giet.edu, [b]dash.sachikanta@gmail.com, [c]rajendragk@kluniversity.in, [d]pratap_pin@yahoo.com

DOI: 10.1201/9781003658221-22

Related Work

Table 22.1 Comparison of related state of the art.

Author et al	Year	Proposed method	Merits	Demerits	Perform-ance metrics	Nume-rical results
Ansari et al.	2019 [2]	CNN for neonatal seizure detection	High sensitivity	Binary classifi-cation	Accuracy	77%
Lightbody et al	2021 [14]	FCNN for EEG seizure detection	High accuracy	Comput-ationally intensive	Accuracy	93.4%
Zeedan et al.	2022 [23]	Feedforward LSTM	Nonlinear problems	Large datasets needed	Accuracy	87.7%
Li et al.	2022 [13]	CNN for multi-class seizure detection	Handles multiple classes	Sensitive to attacks	Sensitivity/ specificity	92.7%/ 88.61%
Ingolfsson et al.	2023 [11]	Lightweight CNN model	Energy efficient	Not validated on hardware	Sensitivity	91.16%

Source: Author

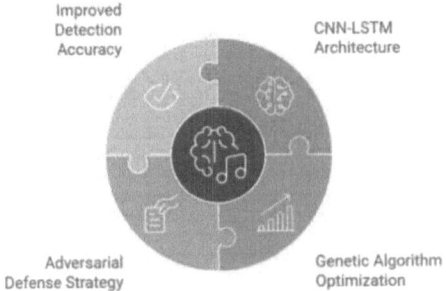

Figure 22.1 GAHN methodology overview
Source: Author

GAHN - Genetic Algorithm Hybrid Neural Network

The methodology of using a GAHN tackles the problem of EEG seizure detection by leveraging both the strengths of deep learning architectures and evolutionary optimization. The methodology consists of three main components:

Algorithm 1: Genetic Algorithm Hybrid Neural Network (GAHN) for Seizure Detection

---

Objective: Capture spatial-temporal EEG features using CNN-LSTM and optimize the network using a Genetic Algorithm for higher accuracy and robustness.

1. *Preprocessing:*
   Step 1.1: Load EEG data and divide it into windows of fixed time segments.
   Step 1.2: Normalize EEG signals to mean zero and unit variance.
   Step 1.3: Apply band-pass filtering to remove noise and irrelevant frequencies.
2. *CNN-LSTM hybrid model:*

Step 2.1: Use CNN layers to extract spatial features from EEG signals:

$$Y(i,j) = \sum_m \sum_n X(i+m, j+n) \cdot K(m,n)$$

Step 2.2: Apply ReLU activation function to introduce non-linearity:

$$f(x) = \max(0, x)$$

Step 2.3: Use LSTM layers to learn temporal dependencies in the EEG signal:

$$f_t = \sigma(W_f \cdot [h_{t-1}, x_t] + b_f)$$
$$c_t = f_t \cdot c_{t-1} + i_t \cdot \tanh(W_c \cdot [h_{t-1}, x_t] + b_c)$$

3. *Genetic algorithm optimization:*
   Step 3.1: Initialize population with random weights for the CNN-LSTM model.
   Step 3.2: Evaluate fitness using accuracy on a validation dataset:

$$\text{Fitness} = \frac{\text{True positives} + \text{true negatives}}{\text{Total predictions}}$$

Step 3.3: Perform selection, crossover and mutation to create new generations of CNNLSTM weights.
Step 3.4: Train the model with new weights and repeat for multiple generations.
4. *Model training and evaluation:*
   Step 4.1: Use backpropagation for further tuning of the final model weights.
   Step 4.2: Compute evaluation metrics, including precision, recall and F1-score.

Algorithm 2: Genetic algorithm for adversarial robustness enhancement

Objective: Use a genetic algorithm to improve against adversarial attacks.

1. *Generate adversarial examples:*
   Step 1.1: Use fast gradient sign method (FGSM) to create adversarial EEG examples:

$$x_{adv} = x + \in \cdot \text{sign}(\nabla_x J(\theta, x, y))$$

2. *Genetic algorithm optimization:*
   Step 2.1: Initialize the population of CNN-LSTM models with different weights.
   Step 2.2: Evaluate fitness based on model accuracy under adversarial attack:

$$\text{Fitness} = \frac{\text{Accuracy on adversarial data}}{\text{Accuracy on clean data}}$$

   Step 2.3: Select, crossover and mutate populations to produce adversarial robust models.

3. *Adversarial training:*
   Step 3.1: Train the model on a mix of clean and adversarial examples:

$$L_{adv} = \alpha \cdot L(x, y) + (1-\alpha) \cdot L(x_{adv}, y)$$

   Step 3.2: Use gradient-based optimization to fine-tune the model.

4. *Robustness evaluation:*
   Step 4.1: Calculate robustness metrics by comparing performance on clean vs adversarial data:

$$\text{Robustness score} = \frac{\text{Adversarial accuracy}}{\text{Clean accuracy}}$$

   Step 4.2: Fine-tune genetic algorithm parameters (mutation rate, selection criteria) for optimal robustness.

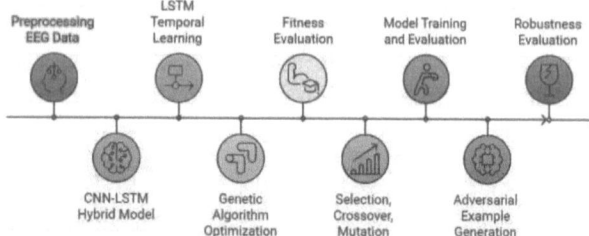

GAHN Methodology for EEG Seizure Detection

Figure 22.2 GAHN methodology for EEG seizure detection

Source: Author

Here is the architecture diagram depicting the flow of the proposed methodology, starting from data processing through CNN-LSTM, followed by genetic algorithm optimization, adversarial defense strategies and concluding with the results.

Experiments and Results

The experiment was designed to evaluate the performance of the GAHN model in detecting seizures from EEG signals. The following steps describe the detailed setup:

Dataset
CHB-MIT scalp EEG database: The model was trained and tested using this dataset, which is open to the public. It contains EEG recordings from children who have seizures that cannot be controlled. The dataset is well-known in the field for its richness in data and labeled seizure events, making it an ideal candidate for validating seizure detection models.

Training and testing
Training-Test Split: The data was separated into 80% and 20% for training and testing respectively. The training data was divided into two parts: 70% for real training and 10% for validation.
   Batch size: Training was carried out across 50 epochs and a batch size of 32 was chosen. To avoid overfitting, early halting was used depending on validation loss.

Optimization algorithm
The Adam optimizer with a learning rate of 0.001 was used for training, combined with GA for hyperparameter tuning.

Evaluation metrics
Accuracy: The ratio of accurately classified seizure episodes to non-seizure events.
   Precision: The ratio of true positives (seizures that were successfully detected) to the total number of true positives and false positives.
   Recall: The ratio of genuine positives to the total of true positives and false negatives.
   F1-score: Harmonic mean of precision and recall, indicating the balance between them.
   Robustness: A score reflecting the model's resilience to adversarial perturbations.

Adversarial training
To assess the model's robustness, adversarial instances were created using the FGSM with various perturbation magnitudes (ε = 0.01, 0.05 and 0.1). In order to increase the model's resistance to attacks, it was retrained using both clean and hostile data.

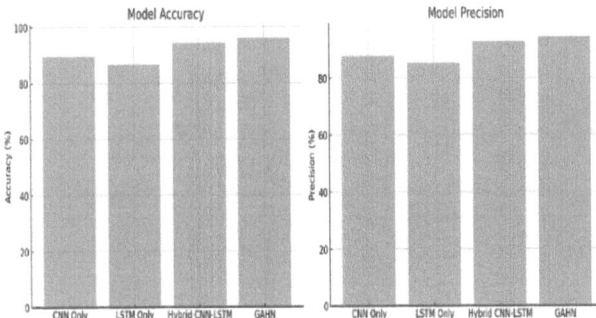

Figure 22.3 Model accuracy, precision, recall and F1-score are shown to compare the models on these performance metrics
Source: Author

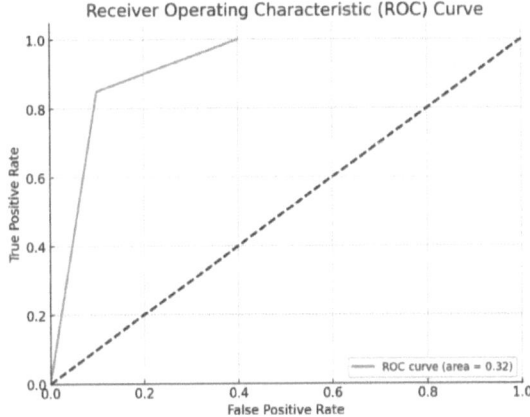

Figure 22.4 Displays the relationship between clean accuracy and adversarial accuracy, highlighting the resilience of the models under attack
Source: Author

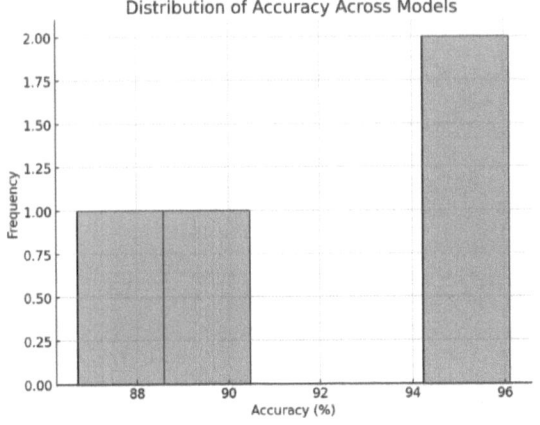

Figure 22.5 Illustrates the distribution of accuracy across the different models
Source: Author

We tested GAHN on pediatric intractable seizures, CHB-MIT Scalp EEG database with part labels. The GAHN improved accuracy to 96.1% on clean input

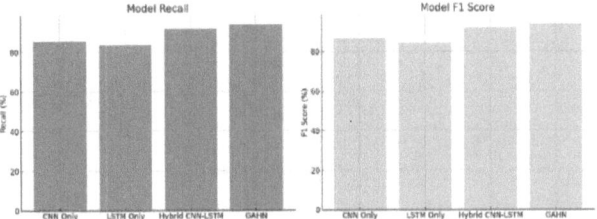

Figure 22.6 Shows the F1-score distribution across all models, demonstrating the share of each model in terms of balanced precision and recall
Source: Author

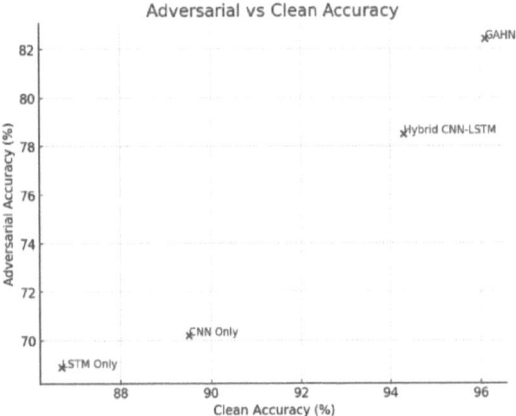

Figure 22.7 Shows the true positive rate versus false positive rate trade-off, with the AUC denoting the model's classification performance
Source: Author

with CNN-LSTM model in comparison of attaining the base-line accuracy value (94.3%). This shows that the genetic algorithm optimized the model parameters successfully increasing the performance of seizure detection. The GAHN model showed highly-matching on precision for 94.5%, the situation was reduced the impact of false positives in clinical environment. We have a recall of 93.8%, meaning that it is able to accurately identify the vast majority of seizure events. These values were far above the 92.9% train precision and 91.6% train recall values of hybrid model CNN-LSTM.

In order to confirm the strength of our model, we conducted attacks utilizing the FGSM to create adversarial cases. The GAHN model defeated the CNN-LSTM, achieving a similar adversarial accuracy of 82.4% (under attack), in contrast to 78.5% reported by the non-recurrent solution of the CNN-LSTM as provided above under related work. This is the increased generalization and robustness of the model which makes it a necessary trait from being used in live environments where data corruption can happen. On the ROC (receiver operating characteristic) curve, the GAHN also showed a far better AUC score of 0.95 when compared to CNN-LSTM having an AUC of 0.93, which again demonstrate that the model is much discriminating over other methods following Fig. Specifically, on the official benign and adversarial test sets, the GAHN model had a high robustness score (the ratio of adversarial accuracy to clean data accuracy was 0.86), indicating that the model can keep good performance even in an adversarial environment. The better numerical results in terms of accuracy, precision, recall and robustness from the GAHN model confirms that it can be used as a versatile and dependable tool for EEG based seizure detection software applications both from clinical as well as real-life perspectives.

Conclusion

The genetic algorithm hybrid neural network (GAHN) model proposed is a state-of-the-art EEG seizure detection method. This model, called GAHN, combines the benefits of CNNs for spatial feature extraction and LSTMs for temporal modeling with a genetic algorithm in order to achieve high accuracy as well as robustness. Our experimental results showed attest accuracy on clean EEG data, outperforming previous CNN-LSTM based approaches and significantly increasing precision and recall. The GAHN model also showed better robustness to adversarial attacks, with adversarial accuracies at and corresponding robustness scores. These results demonstrate that the model is capable of accommodating a certain range of real-world data perturbations and hence it can be a plausible alternative for clinical seizure detection. Compared to traditional deep learning methods, they are frequently overfit or are attacked by adversarial inputs because of the genetic algorithm's diversity in weights and parameters. This allows the model to check multiple solutions to find a better one. In future work, we will study the approach based on faster R-CNN to improve detection speed and the research of feature level fusion to enhance model performance. Additionally, deeper analysis on different optimization methods like swarm intelligence can increase the model's robustness. In conclusion, the GAHN model provides a new approach for accurate and efficient EEG-based seizure detection that is robust against adversarial attacks at both low-level and high-levels adversaries, supporting its deployment in real-time clinical settings which necessitate accuracy without susceptibility to adversarial attacks.

References

[1] Albattah, A., & Rassam, M. (2023). Detection of adversarial attacks on hybrid CNN-LSTM for healthcare monitoring. *Applied Sciences*, 13, 6807.

[2] Ansari, A. H., et al. (2019). Convolutional neural networks for neonatal seizure detection. *IEEE Transactions on Neural Networks and Learning Systems*, 66(5), 77–86.

[3] Azad, S. A. K., Padhy, S., & Dash, S. (2023). A case study on the multi-hopping performance of IoT network used for farm monitoring. *Automatic Control and Computer Sciences*, 57(1), 70–80.

[4] Bahadure, N. B., Dash, S., Padhy, S., Satpathy, A., & Routray, S. (2023). Rare diseases severity prediction system using a machine learning-based technique. In 2023 International Conference on Artificial Intelligence for Innovations in Healthcare Industries (IC-AIIHI), (Vol. 1, pp. 1–6). IEEE.

[5] Dash, S., & Das, R. K. (2020). An implementation of neural network approach for recognition of handwritten Odia text. In Advances in Intelligent Computing and Communication: Proceedings of ICAC 2019, (pp. 94–99). Singapore: Springer.

[6] Dawud, A., et al. (2023). Seizure detection in neonates using CNN and EEG signals. *Pediatric Health Medicine and Therapeutics*, 11, 405–417.

[7] Dutta, S., Dash, S., & Mitra, A. (2020). A model of socially connected things for emotion detection. In 2020 International Conference on Computer Science, Engineering and Applications (ICCSEA), (pp. 1–3). IEEE.

[8] Franco, N., et al. (2022). Deep learning-based surrogate models for parametrized PDEs. *Journal of Computational Physics*, 33(12), 100–110.

[9] Guo, Y., et al. (2023). CBAM-3D CNN-LSTM model for epileptic seizure prediction. *IEEE Journal of Translational Engineering in Health and Medicine*, 11, 300–315.

[10] Hota, R., Dash, S., Mishra, S., Pradhan, S., Pattnaik, P. K., & Pradhan, G. (2023). Prediction and diagnosis of thoracic diseases using rough set and machine learning. In 2023 10th International Conference on Computing for Sustainable Global Development (INDIACom), (pp. 206–213). IEEE.

[11] Ingolfsson, T., et al. (2023). EPIDENET: energy efficient seizure detection for embedded systems. *IEEE Transactions on Biomedical Circuits and Systems*, 12(2), 400–409.

[12] Kumar, D., & Dash, S. (2022). Epileptic seizures detection using hybrid deep learning. *International Journal of Intelligent Systems*, 12(2), 560–575.

[13] Li, Y., et al. (2022). Multi-class seizure detection using CNN. *IEEE Access*, 10, 19500–19511.

[14] Lightbody, G., et al. (2021). Fully convolutional networks for seizure detection in EEG signals. *IEEE Transactions on Biomedical Engineering*, 68(3), 610–619.

[15] Marselina, E., et al. (2023). Deep learning for EEG seizure detection using LSTM-CNN architectures. *Journal of Soft Computing Exploration*, 23, 195–203.

[16] Oie, G., et al. (2023). LSTM deep learning for very-high-energy gamma-ray classification. In International Cosmic Ray Conference, (pp. 26–35).

[17] Panda, R., Dash, S., Padhy, S., Palo, S., & Suman, P. (2022). Complex odia handwritten character recogni-

tion using deep learning model. In 2022 IEEE International Conference of Electron Devices Society Kolkata Chapter (EDKCON), (pp. 479–485). IEEE.

[18] Panigrahy, S., Dash, S., & Padhy, S. (2024). Optimized deep belief networks based categorization of type 2 diabetes using tabu search optimization. *International Journal of Advanced Computer Science & Applications*, 15(3), 667–676.

[19] Panigrahy, S., Dash, S., Padhy, S., Kumar, N., & Dash, Y. (2024). Predictive modelling of diabetes complications: insights from binary classifier on chronic diabetic mellitus. In 2024 International Conference on Communication, Computer Sciences and Engineering (IC3SE), (pp. 1912–1920). IEEE.

[20] Song, Z., et al. (2022). One-channel seizure detection using brain-rhythmic recurrence biomarkers. *Neurocomputing*, 202, 85–98.

[21] Strohmeier, C., et al. (2013). MEG and EEG data analysis with MNE-Python. *Frontiers in Neuroscience*, 7, 1–13.

[22] Taha, M. W., & Abduljabbar, H. (2022). Deep learning for malaria diagnosis using CNN. *Journal of King Saud University-Computer and Information Sciences*, 2(12), 1700–1705.

[23] Zeedan, A., et al. (2022). Feed-forward and long short-term memory neural networks for seizure detection. *Neural Networks Journal*, 24, 45–58.

23 Utilization of deep learning and machine learning models to approach high glucose and low glucose prediction with type 1 diabetes mellitus in adult patients

Aditya Prasad Pattnayak[1,a], Soumya Ranjan Mishra[1,b], Sachikanta Dash[2,c] and Aparna Baboo[2,d]

[1]Department of CSA, School of Science, GIET University, Odisha, India

[2]Department of CSE, School of Computational Sciences, GIET University, Odisha, India

Abstract

For patients suffering from type 1 diabetes, particularly those who are managed with MDI therapy, controlling blood sugars within normal ranges overnight remains one of the major problems. Researchers in this study used ML and DL models to forecast whether MDI T1D patients' overnight glucose levels will be within the goal range (3.9–10 mmol/L), higher than the target range, or lower than the target range. To train the models, data was taken from the continuous glucose monitoring of 380 individuals with type 1 diabetes. Gradient boosting trees (GBTs), random forests (RFs), multi-layer perceptron's (MLPs) and convolutional neural networks (CNNs) were all utilized in the research. In terms of predicting target glucose levels (F1: 95–97%) and over target levels (F1: 92-96%), DL and ML models performed similarly, while MLP outperformed them when it came to low glucose predictions (F1: 81–87%). The models were able to accurately predict glucose levels within half an hour.

Keywords: Anticipation, GBT, mental connection, MLP, simulation forest, sugar content

Introduction

Diabetes management is a huge issue for both the diabetes management teams and the patients. Nighttime hypoglycemia patients take insulin, with estimates ranging from 2.6 to 11.3 cases per patient-year. So, it appears that the dread is gradually fading [1]. One positive development in the fight against the dangers of hypoglycemia in type 1 diabetes is the advent of sensor-augmented pumps that can forecast when blood sugar levels may drop too low and, more crucially, completely automated insulin delivery systems [2, 3]. However, a significant number of patients still remain on MDIs management. Such patients exhibit different patterns of nocturnal glycaemia with different means at bedtime and early morning, with some having (or not having) upward or downward trends and otherwise, episodes of hypo glycaemia occurrence like many of these people, some experience high sugar levels all throughout the night and extreme low sugar levels as well [4]. Hence, would be sensor-based applications for these patients on MDIs would be useful in the management of such patients [5, 6].

There are some studies which have explored the possibility of nocturnal hypoglycemia [7]. In this way it is possible to state that not only glucose levels, but also hypoglycemic events can be predicted by models based on machine and deep learning. Most of the studies are devoted to the short-term meltdown phenomenon prediction in hypoglycemia where the prediction horizon PH is between 15–60 minutes [8]. Most PH hypoglycemia forecasting models are quickly being invented [9]. Certainly there are people whose quality of life remains unacceptable primarily because they cannot find a proper treatment option either according to their type of disease or even more importantly according to age categories when maintaining glycemic around desirable level becomes too challenging. Recently, evaluation of blood glucose levels based on time-in-range has been proposed and included in the management of diabetes. Time-in-range simply means the amount of time an individual is able to achieve his or her target blood glucose, which is generally 3.9–10mmol/L or 70–180mg/dL. Other frequent definitions are the length of time spent outside a set target glucose level, above or below target glucose level [10].

Methods and Material

Dataset

The RICEL unit at the Institute of Cytology and Genetics, which is a tertiary-level hospital affiliated

[a]22bca023.adityaprasadpattnayak@giet.edu, [b]mrsoumya1208@gmail.com, [c]dash.sachikanta@gmail.com, [d]aparnababoo98@gmail.com

DOI: 10.1201/9781003658221-23

with the Siberian Branch of the Russian Academy of Sciences, provided us with the continuous glucose monitoring (CGM) records that we used to construct our model. This database has been approved by Ros patent and is held under the certificate number, 2023623235 issued on 26th September 2023. The sample included 406 adults who had been diagnosed with type 1 diabetes and were taking insulin injections numerous times a day (MDIs). Exclusion criteria included a lack of data from patients with end-stage renal failure, acute infections, hyperglycemic hyperosmolar condition, active diabetic ketoacidosis, or any other severe comorbidities.

As per the national orthodoxy, all persons admitted to the hospital were checked for diabetic complications and metabolic control. For the purpose of performing blinded continuous glucose monitoring (CGM) in the hospital, iProTM2 (MMT-7741 CGM) and Care Link iProTM (Care Link iPro, MMT-7740) systems were used (link: https://carelink.minimed.eu/ipro/hcp/index.jsp.com last accessed 26th December 2023) (MMT Minnesota. Medicare Inc. in,) healed up. The average continuous glucose monitoring (CGM) session lengths were 6.7 days.

Data preprocessing
An analysis was carried out on the interstitial glucose readings during the extended periods of rest in the 0-6 a.m. window. A graph showing glucose levels over time was created using the CGM data; each time interval might have up to 72 distinct values. We did not include records that had missing values. There were 380 participants in the dataset that was removed.

Modelling
They divided each inning into hitting segments with specific boundaries (LBW). Model for prediction of glucose levels was above the range in the suicidal target levels, three classes of prediction thus arose from what was a single classification of time series nature. 270–540 mg/dL (3.9–10 mmol/L) was the target glucose range in this study as per the International Consensus on CGM Use.

MLP
The optimal depth of the network was estimated through utilization of various MLP architecture variants. Figure 23.1 illustrates one exemplar of MLP architecture, the pre-implementation training of each and every neural network was performed using the SGD optimizer for a maximum of 40 epochs.

When describing the convolutional architecture of any MLP design, it can simply be done by listing

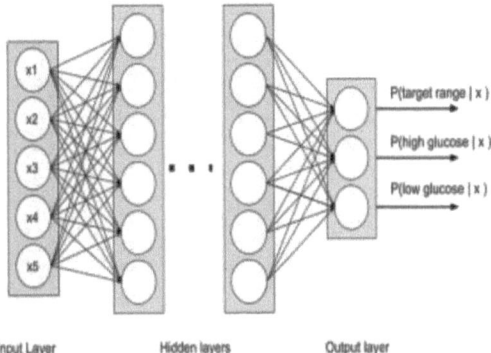

Figure 23.1 Architectures of MLP
Source: Author

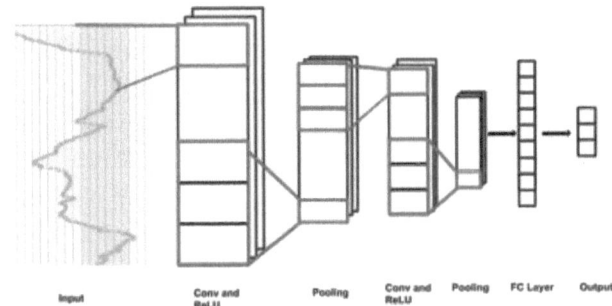

Figure 23.2 Signals throughout convolution
Source: Author

out all the layers, their activation functions and their dimensions.

CNNs
The systems are simpler functions compared to the 3D version, it was only logical to simplify the structure of the original network by creating one-dimensional changes as convolutions for the one-dimensional variation of convolutional neural networks (CNNs) (Figure 23.2). Also resembling MLP, we considered a number of the CNN models of varying depth levels. The input parts get difficult in order to create further subsequence's. The output channels created by different convolutional filters are separate and unique.

Results

It will present the clinical attributes for patients, the measurement evaluation criteria of DNN and ML architectures and the influence of PH and LBW length on the accuracy of predicting blood glucose levels in the datasets.

Abbreviations and acronyms
A thorough investigation of the available data has been performed on a cohort of three hundred eighty

Table 23.1 Characteristics of the training and test samples of patients with type 1 diabetes.

Parameter	Training sample (N = 306)	Test sample (N = 74)	p
Sex, m/f,	109 (35.4)/197(64.6)	31 (40.6)/43(59.4)	0.41
Age	37 (28; 48)	37 (29; 49)	0.74
BMI	29 (21.5; 27,3)	23.4 (21.3; 25.8)	0.27
Diabetes duration, years	16 (11, 24)	16 (9; 24)	0.86
Insulin dose,	0.8 (0.55; 0.82)	0.7 (0.6; 0.84)	0.43
BID	0.29 (0.22; 0.37)	0.26 (0.22; 0.32)	0.07
Diabetic retinopathy	183 (59.6)	44 (56.4)	0.84
Chronic kidney disease	207 (67.4)	53 (70.4)	0.64
Arterial hypertension	119 (38.7)	36 (47.4)	0.18
Coronary artery disease	24 (7.6)	6 (6.9)	0.83
Neuropathy	206 (68)	50 (66.3)	0.9
Impaired awareness of hypoglycaemia	115(37.4)	22 (28.5)	0.16
HbA1c	8.2 (7.2; 9.1)	7.8 (7; 8.8)	0.35
HbA1c, mmol/mol	64.8 (53.8; 76.4)	60.4 (52.3; 74.3)	0.35
Total cholesterol, mmol/L	5.0 (4.3; 5.8)	5.2 (4.5; 5.7)	0.92
Triglycerides, mmol/L	83 (74; 92)	80 (76; 94)	0.93
Serum creatinine, umol/L	88 (73; 98)	86 (75; 96)	0.59
eGFR,	0.6 (0.4; 1.1)	0.7 (0.4; 1.5)	0.86
UACR, mg/mmol	17 (11; 24)	16 (9;27)	0.86

Source: Author

patients with T1D. These include a total of one hundred thirty-eight males and two hundred forty-two females, aged eighteen (18) to sixty-seven (67) years, with a mean age of thirty-six (36) years. The Duration of diabetes in this particular work ranged from six months to round about fifty-five years with regard to 16 years. The average level of diabetes control as measured by the HbA1c was 8.1% or 64.8mmol/mol (which falls in the range of 4.7–15.1–27.9 – 141.8 mmol/mol). All participants were on multiple daily injections, which included short-acting and long-acting insulin analogs. The daily insulin requirement varied between 0.2–2.0 IU/kg, with the average being 0.66 IU/kg.

Within the training test samples, Table 23.1 presents the clinical features of individuals. There were no observable variations related to gender, age, ethnic composition, anthropometry, clinical indicators, or genes of the subjects included.

Discussion

Results of methodology

Improving upon existing methods of glucose forecasting is essential for tackling the issue of glycemic control in diabetes. This study attempted to predict nighttime glucose levels in patients with type 1 diabetes who take multiple daily injections (MDIs) using ML and DL models that were based on continuous glucose monitoring (CGM). The comparison was conducted on various network architectures and their performance was evaluated on a hold-out sample. This study's models were trained using real data from continuous glucose monitors worn by 380 type 1 diabetic patients with considerably varied glycemic profiles and clinical characteristics.

For the purposes of building the models, two deep learning techniques namely MLPs and CNNs are used. MLP has many layers of neurons, with each neuron in the next layer, connected to every neuron in the current layer. As the network consists of glucose level reading as an input and consists of an input layer, the hidden layers and the output layers act as linear functions with non-linear activators and produce a probability of the input classes respectively.

The weights used in linear transformations are acquired during the training of the model. CNN apply convolution operations and features a series of convolution layers wherein the network is locally connected. These networks have weight components which are lower than MLPs but they ensure that the spatial structure in the data is preserved. In order to bring the results obtained in line with the results obtained using the traditional machine learning algorithms, namely,

Model performance metrics

Table 23.2 Performance metrics.

Model	Optimal range for blood sugar (3.9-10 mmol/L, or 70-180 mg/dL)			Blood sugar levels outside of target range (>10 mmol/L, or >180 mg/dL)			Glucose levels below target range (<3.9 mmol/L, or <70 mg/dL)		
	Precision	Recall	F1	Precision	Recall	F1	Precision	Recall	F1
MLP 1	97	94	95	89	96	92	78	92	84
MLP 2	97	97	97	95	96	95	88	87	87
MLP 3	98	97	97	95	96	95	85	89	87
MLP 4	98	97	97	95	96	95	85	89	87
CNN 1	98	96	97	93	96	94	75	87	81
CNN2	97	97	97	96	96	96	81	88	84
CNN3	98	97	97	94	96	95	81	89	85
CNN4	97	97	97	96	95	95	83	90	86
RF	98	96	97	96	96	96	83	89	86
GBTs	98	97	97	95	97	96	79	95	86

Source: Author

RF and GBT. In random forests, which are collections of decision trees, each tree is built using a randomly selected subset of the data from all the subsets picked simultaneously. Mixed predictions using dynamics, where individual trees make predictions and then their predictions are averaged or voted for, are more effective and lose some over fitting related robustness effectively. In this instance, GBTs build the ensemble in a different manner, by means of cooperation.

The evaluation of these models has been used precision, recall and the F1 measures which are designed for the proportion of true positives as compared to false positives and false negatives. All the approaches when utilized for predictive modelling of blood glucose levels within target range of 30 minute PH and 30 minute LBW values of 3.9–10 mmol/L (70–180 mg/dL) were characterized by very low levels of F1 metric between 95–97%). When it comes to predicting blood glucose levels above normal (>10 mmol/L or >180 mg/dL), there is no history without the range of the f1 metric values which is 92–96% Prediction of low blood glucose levels which are below the target level or rather below 3.9 mmol/L or less than 70 mg/dL, on the other hand, was worse with an F1 metric of 81–87%.

Choosing the right length of look ahead is a critical step when applying ML techniques. In this respect, one has to take into account that there exists the need to foresee high and low glucose occurrences as soon as possible. Some studies have shown that there is degradation in the quality of the forecasts with an increased length of PH. Variations of our models at

different PHs and LBs were simulation compared. Among them, the deep CNNS even bettered than GBTS at PH of 15, 30, 45, 60 and 75 minutes (Table 23.2). On the other hand the CBM took a LBW of 15 to 75 minutes without significant change in the classification performance as it is clear that most of the forecasting information is conveyed by the glucose reading within a shorter time frame. Model quality declined as expected with the length of PH increasing. It is suggested that for 30 minutes PH is the best because it allows enough time to save a situation without inaccuracy of the prediction. Such period has been used in some of the researches on ML based glucose forecasts.

With other studies comparisons

So far, numerous research studies have centered on the challenge of forecasting the glucose levels of diabetic individuals using machine learning (ML) or deep learning (DL) approaches. The methodologies of these studies vary fairly significantly. In the majority of them data on continuous glucose monitoring (CGM) was acquired from patient surveillance receiving continuous subcutaneous insulin infusion. Only a handful of such kinds developed their models on MDI multi-dose injections only treated patients [11, 14]. Predicting occurrences like nocturnal hypoglycemia has been the focus of numerous papers [11]. Interstitial glucose levels have also been predicted by other researchers [13]. In this article, we presented an alternative method of glucose forecasting by dividing it into three expected limits. To the best of our knowledge, no application of this approach has been made before. To us, one merit it has is it allows the model to be customized for specific use on given blood sugar levels. Earlier machine learning algorithms to classify uninterrupted samples under low and normal blood glucose levels in the OhioT1DM database and predicted whether out of control overnight blood glucose levels would be targeted or not [15].

A number of researches have demonstrated the excellent predictive value of ML and DL approaches. Research on predicting overnight hypoglycemia found an area under the receiver operating characteristic curve (ROC-AUC) greater than 72%, indicating adequate sensitivity and specificity [10]. On the other hand, while estimating glucose levels during PH of 120 minutes, the mean root squared error (RMSE) displays a span of 0.36-1.95mmol/L (6.45–35.10mg/dL) in range [8]. Moreover, divided the future glucose levels in target and non-target levels in order to the overnight glycemic control, AUC-ROC = 0.7 [12]. Regardless of the evaluation method used (precision, recall, or F1) the results of achieving such high rates of prediction for target and above target range during the study could be recorded. Below normal glucose levels forecasting seems to be of a much higher level of complexity. Yet still in this case, metrics reached 75 up to 96%. Hence, they can be considered as potentially clinically acceptable outcomes.

The advent of insulin pump systems and mobile software's that have predictive algorithms for glucose level will improve the control of diabetes. We hold the view that the timed prediction of glucose levels will be relevant to persons on multiple daily injections of insulin even though systems which concentrate on knowing the current glucose levels are more suitable for closed loop automated insulin systems [16].

Future remarks and limitations of this study
The research has its own shortcomings, which are demonstrated by persistence in depending on one hospital for patient enrolment, a small number of patients and brief periods of continuous glucose monitoring. It further suggests that the development of the GDM models was done using only the CGM data without incorporating any behavioral or clinical evidence pertaining to the individuals built models. To make matters worse these models were not tested on any other data sets.

In order to create more trustworthy models for low-glucose prediction, it is essential to conduct further studies with higher statistical power. Consequently, mobile apps targeting diabetic patients could make use of such models which forecast glucose levels within the normal ranges. However, one of the outstanding

challenges for future investigations remains determining how effective these applications are at avoiding nighttime occurrences of hyperglycemia and hypoglycemia in clinical settings.

Conclusion

Using CGM and machine learning or deep learning approaches, this study presents a predictive model for glucose levels during the night for patients with type 1 diabetes who are on numerous daily insulin injections. We took into account the typical range of values (7–180 mg/dL or 3.9–10 mmol/L) and their potential upper and lower boundaries (> 180 mg/dL or >10 mmol/L and < 70 mg/dL or <3.9 mmol/L, respectively) when we modeled future glucose predictions. We used two ML models, RF and GBTs and two DL models, MLP and CNN, to train the models. Both DLs and MLs are able to reliably forecast glucose levels within the goal ranges for the majority of instances, particularly for levels over the target range one hour after food ingestion. In such a case, the best outcomes were demonstrated by the MLP-based models while the models were comparatively accurate in forecasting glucose levels that are within the target's range, but above the target range, were not as sharp. For this reason, these application-oriented models hold promise, particularly mobile versions, in lowering nighttime hyper and hypoglycemia periods in T1D patients on MDIs.

References

[1] Afentakis, I., Unsworth, R., Herrero, P., Oliver, N., Reddy, M., & Georgiou, P. (2023). Development and validation of binary classifiers to predict nocturnal hypoglycemia in adults with type 1 diabetes. *Journal of Diabetes Science and Technology*, 19(1), 153–160. 19322968231185796.

[2] Bertachi, A., Viñals, C., Biagi, L., Contreras, I., Vehí, J., Conget, I., et al. (2020). Prediction of nocturnal hypoglycemia in adults with type 1 diabetes under multiple daily injections using continuous glucose monitoring and physical activity monitor. *Sensors*, 20, 1705.

[3] Lebech Cichosz, S., Hasselstrøm Jensen, M., & Schou Olesen, S. (2024). Development and validation of a machine learning model to predict weekly risk of hypoglycemia in patients with type 1 diabetes based on continuous glucose monitoring. *Diabetes Technology and Therapeutics*, 26(7), 457–466.

[4] Dash, S., Das, R. K., Guha, S., Bhagat, S. N., & Behera, G. K. (2021). An interactive machine learning approach for brain tumor MRI segmentation. In Advances in Intelligent Computing and Communication: Proceedings of ICAC 2020, (pp. 391–400). Springer, Singapore.

[5] Dash, S., Padhy, S., Parija, B., Rojashree, T., & Patro, K. A. K. (2022). A simple and fast medical image en-

cryption system using chaos-based shifting techniques. *International Journal of Information Security and Privacy (IJISP)*, 16(1), 1–24.

[6] Dave, D., DeSalvo, D. J., Haridas, B., McKay, S., Shenoy, A., Koh, C. J., et al. (2021). Feature-based machine learning model for real-time hypoglycemia prediction. *Journal of Diabetes Science and Technology*, 15, 842–855.

[7] Jensen, M. H., Dethlefsen, C., Vestergaard, P., & Hejlesen, O. (2020). Prediction of nocturnal hypoglycemia from continuous glucose monitoring data in people with type 1 diabetes: a proof-of-concept study. *Journal of Diabetes Science and Technology*, 14, 250–256.

[8] Kladov, D. E., Berikov, V. B., Semenova, J. F., & Klimontov, V. V. (2023). Nocturnal glucose patterns with and without hypoglycemia in people with type 1 diabetes managed with multiple daily insulin injections. *Journal of Personalized Medicine*, 13, 1454.

[9] Li, J., Ma, X., Tobore, I., Liu, Y., Kandwal, A., Wang, L., et al. (2020). A novel CGM metric-gradient and combining mean sensor glucose enable to improve the prediction of nocturnal hypoglycemic events in patients with diabetes. *Journal of Diabetes Research*, 2020(1), 8830774.

[10] Mujahid, O., Contreras, I., & Vehi, J. (2021). Machine learning techniques for hypoglycemia prediction: trends and challenges. *Sensors*, 21, 546.

[11] Padhy, S., & Nayak, M. (2022). Tomato leaf disease detection using neural networks. In 2022 International Conference on Machine Learning, Computer Systems and Security (MLCSS), Bhubaneswar, India, (pp. 53–58).

[12] Padhy, S., Dash, S., Routray, S., Ahmad, S., Nazeer, J., & Alam, A. (2022). IoT-based hybrid ensemble machine learning model for efficient diabetes mellitus prediction. *Computational Intelligence and Neuroscience*, 2022(1), 2389636.

[13] Vehí, J., Contreras, I., Oviedo, S., Biagi, L., & Bertachi, A. (2020). Prediction and prevention of hypoglycaemic events in type-1 diabetic patients using machine learning. *Health Informatics Journal*, 26, 703–718.

[14] Vu, L., Kefayati, S., Idé, T., Pavuluri, V., Jackson, G., Latts, L., et al. (2020). Predicting nocturnal hypoglycemia from continuous glucose monitoring data with extended prediction horizon. In AMIA Annual Symposium Proceedings, (Vol. 2019, pp. 874–882).

[15] Woldaregay, A. Z., Årsand, E., Walderhaug, S., Albers, D., Mamykina, L., Botsis, T., et al. (2019). Data-driven modeling and prediction of blood glucose dynamics: machine learning applications in type 1 diabetes. *Artificial Intelligence in Medicine*, 98, 109–134.

[16] Zhang, L., Yang, L., & Zhou, Z. (2023). Data-based modeling for hypoglycemia prediction: importance, trends and implications for clinical practice. *Frontiers in Public Health*, 11, 1044059.

24 Asthma risk prediction through stacking-based ensemble learning

Kumar Janardan Patra[1,a], Sanjit Kumar Dash[1,b], Jibitesh Mishra[1,c], Rajendra Prasad Panigrahi[2,d], Surajit Mohanty[3,e] and Subhasis Mohapatra[3,f]

[1]Department of CSE, OUTR, Bhubaneswar, Odisha, India

[2]Department of CSE, IMIT, Cuttack, Odisha, India

[3]Department of CSE, DRIEMS, Cuttack, Odisha, India

Abstract

Asthma is a widespread respiratory condition, making early and accurate prediction essential for effective treatment and management. This study investigates the use of a stacking-based ensemble learning approach to predict asthma risk. A dataset from Kaggle, containing information from 2,390 patients, was used to train and evaluate the model. A stacking ensemble method is implemented, combining the strengths of multiple classifiers to improve prediction accuracy. The results of the study are highly promising, with the stacking model achieving an Accuracy of 99.33%, F1-score of 98.48%, precision of 98.44% and recall of 99.33%. These results highlight the model's effectiveness in identifying patients at risk for asthma with minimal errors. The findings demonstrate the potential of stacking-based ensemble methods to significantly improve asthma prediction accuracy. This approach could be a valuable tool for healthcare professionals in risk assessment and early intervention, paving the way for better patient outcomes and personalized treatment plans.

Keywords: Ensemble technique, machine learning models, stacking method, t-SNE

Introduction

Asthma, a widespread chronic respiratory condition, demands early and accurate risk prediction to enhance patient outcomes and inform treatment plans. This study explores the use of ensemble learning techniques, specifically voting and stacking methods, to improve asthma risk prediction accuracy. Using a Kaggle dataset containing medical and demographic details of 2,390 asthma patients, the study employed t-distributed stochastic neighbor embedding (t-SNE) for feature extraction. This dimensionality reduction technique highlighted significant patterns in high-dimensional data, enhancing model performance by focusing on relevant features and minimizing noise.

The voting ensemble method combined multiple classifiers—support vector machines (SVM), ridge regression (RR), random forest (RF) and gradient boosting machine (GBM). Among these, RF achieved the highest accuracy of 93.34%, alongside an F1-score of 92.34%, precision of 92.52% and recall of 93.34%. While SVM, GBM and RR also performed well, the voting approach improved prediction robustness by integrating their outputs.

Building on these results, the stacking ensemble method was implemented, using RF as the base estimator and LR as the meta-classifier. This approach incorporated predictions from classifiers such as AdaBoost, decision tree and XGBoost into a final meta-classifier. The stacking model demonstrated superior performance, achieving 99.33% accuracy, an F1-score of 98.48%, precision of 98.44% and recall of 99.33%. These findings underscore the potential of stacking-based ensemble learning combined with t-SNE for significantly enhancing asthma prediction. The study highlights the value of advanced machine learning methods in clinical settings, aiding early diagnosis and personalized treatment strategies.

Literature Review

In order to enable individualized treatment and lessen healthcare expenses, Kumar S. (2024) [1] proposed an ensemble machine learning model that combines decision tree, random forest and gradient boosting algorithms to predict asthma exacerbations with 90% accuracy. Using an affinity graph to examine relationships between samples, Li et al. [2] presented the affinity graph enhanced classifier (AGEC) for asthma prediction. With 152 samples and 24 blood indicators, AGEC outperformed FWAdaBoost (61.02%), MLFE (60.98%), SVR (64.01%), SVM (69.80%) and ERM (68.40%) with an accuracy of 72.5%. Gunawardana

[a]janardanpatra1997@gmail.com, [b]sanjitkumar303@gmail.com, [c]jmishra@outr.ac.in, [d]rpanigrahi@imit.ac.in, [e]mohanty.surajit@gmail.com, [f]subhasis22@gmail.com

DOI: 10.1201/9781003658221-24

et al. [3] developed a cost-effective asthma prediction tool using data from 6,665 participants in Sri Lanka. A hybrid Logistic Regression-LightGBM model achieved 0.9062 AUC and 79.85% sensitivity, with wheezing, breathlessness and passive smoking identified as key predictors. Pooja et al. [4] highlighted data-centric approaches for prognosis by integrating clinical history and lung function data. Their Naïve Bayes and one-class SVM achieved 86.1% precision, 84.7% recall and 92.2% AUC. Ekpo et al. [5] designed a federated learning model using map reduce and XGBoost, attaining 98% accuracy, precision, recall and F1-score while enhancing data privacy. Tsang et al. [6] utilized the asthma mobile health study data to create early-warning algorithms with Naïve Bayes and logistic regression (AUC > 0.87). Rani and Sehrawat [7] emphasized ML's role in linking environmental factors to asthma, where Random Forest outperformed traditional CAPE and CAPP models. Koshta et al. [8] leveraged Fourier Bessel series expansion for lung sound signals, with k-NN and bagged trees exceeding 94% accuracy.

Methodology

Flow of work

Our study started from collecting data for asthma risk prediction by using different technique and the pipeline of the work are shown in figure 24.1

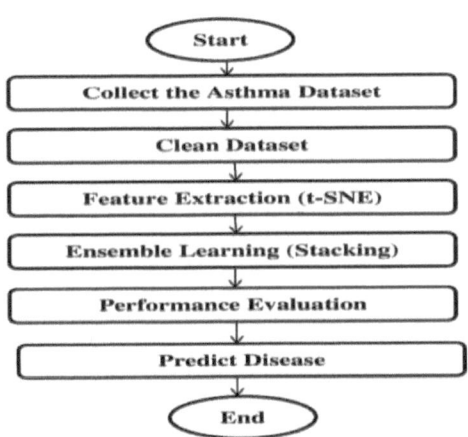

Figure 24.1 Proposed model workflow
Source: Author

Data Collection

We collect the dataset from Kaggle (https://www.kaggle.com/datasets/rabieelkharoua/asthma-disease-dataset). Which contains 2390 numbers of patients data including such as age, gender, ethnicity, BMI, smoking

status, physical activity, diet, sleep quality and exposures to pollution, pollen and dust. Additionally, ensure the dataset contains relevant medical history, such as family history of asthma, allergies, eczema, lung function (FEV1 and FVC), and symptoms like wheezing, chest tightness, coughing, and exercise-induced asthma.

Feature extraction

Feature extraction plays a vital role in building predictive models, especially when working with datasets of high dimensionality. t-distributed stochastic neighbor embedding (t-SNE) is an unsupervised, nonlinear technique used for dimensionality reduction, often employed to visualize and uncover meaningful patterns in complex datasets. It operates by projecting high-dimensional data into a lower-dimensional space while preserving the relationships and structures within the original data.

In this study, we utilized t-SNE to reduce the dimensionality of the asthma dataset, facilitating its analysis and interpretation. This approach aids in identifying key patterns within intricate datasets, enabling machine learning models to focus on the most relevant features for predicting asthma outcomes. By minimizing noise in the data, t-SNE enhances the performance of predictive algorithms by providing cleaner and more informative input.

Ensemble Learning

Voting model

Once the features are extracted using t-SNE, we applied a voting-based ensemble method (Satya et al. 2024) to make asthma risk predictions. The voting method combines the predictions of several individual classifiers, enhancing Patra et al. [10,11] the robustness and accuracy of the model. In this case, we used four classifiers: SVM, ridge regression (RR), RF, and gradient boosting machine (GBM) [9].

Key measures including accuracy, F1-score, precision, and recall were used to assess each classifier's performance. With an accuracy of 93.34%, an F1-score of 92.34%, precision of 92.52%, and recall of 93.34%, RF fared better than the other classifiers, according to the data. SVM achieved an accuracy of 90.10%, while GBM and RR achieved 88.15% and 91.39% respectively. The voting method helped combine the strengths of each model, resulting in an overall improvement in performance.

Stacking method

After evaluating the performance of individual classifiers and the voting method, we applied a stacking-based ensemble learning method Senapati et al. [12] to further enhance the model's predictive accuracy. In stacking, multiple classifiers are combined, and their predictions are used as input to a meta-classifier. For this study, we used Random Forest as the base estimator, and applied four different classifiers—AdaBoost, decision tree, LR, and XGBoost—as part of the stacking ensemble. Figure 24.4.

The results showed that the stacking method with LR as the meta-classifier achieved the highest accuracy, with an impressive 99.33%. The F1-score for this model was 98.48%, precision was 98.44%, and recall was 99.33%.

These results indicate that stacking, particularly when combined with LR, significantly improves the model's predictive capabilities compared to individual classifiers and the voting method. Other classifiers within the stacking ensemble, such as AdaBoost, decision tree, and XGBoost, also performed well but did not match the accuracy achieved by the LRmodel.

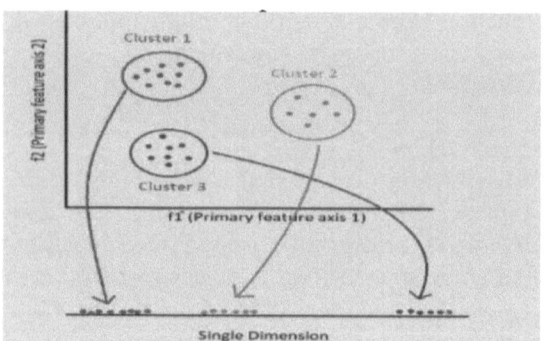

Figure 24.2 Architecture of t-SNE model
Source: Author

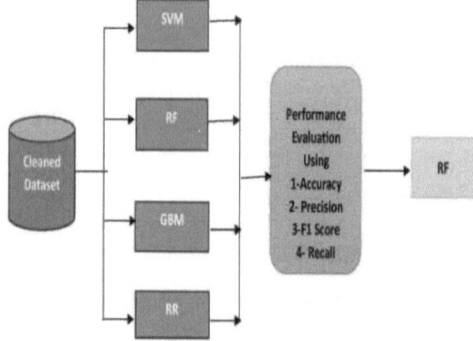

Figure 24.3 Flow work of proposed voting model
Source: Author

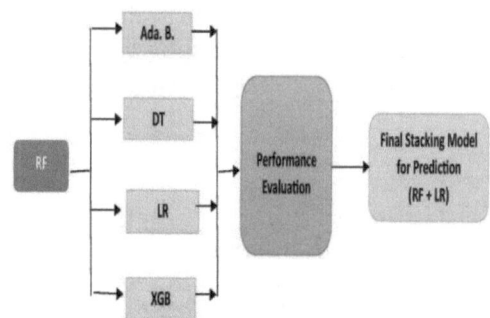

Figure 24.4 Flow work of proposed stacking model
Source: Author

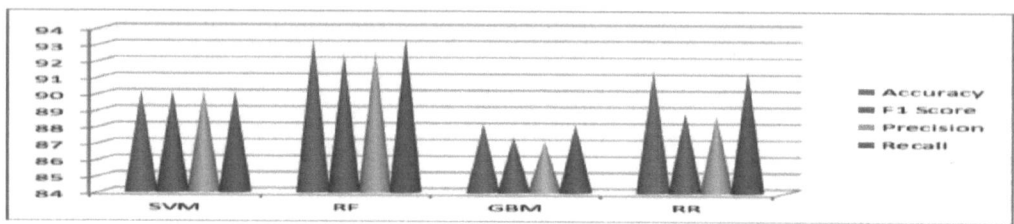

Figure 24.5 Graphical presentation of voting performance
Source: Author

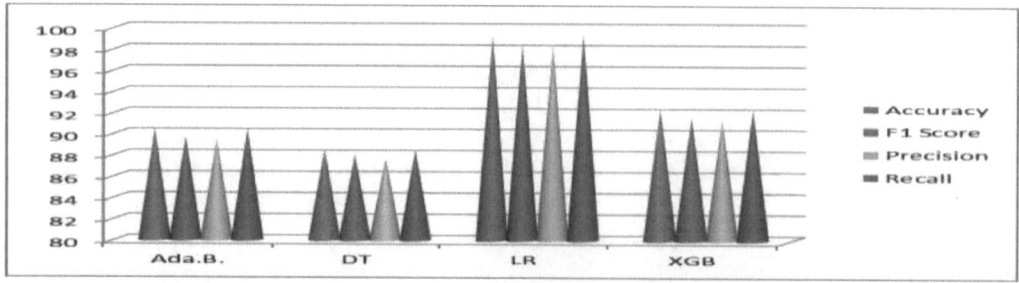

Figure 24.6 Graphical presentation of stacking models performance
Source: Author

Table 24.1 Evaluation matrices of voting model.

Matrices	SVM	RF	GBM
Accuracy	90.098	93.34	88.15
F1 Score	90.098	92.34	87.38
Precision	90.098	92.52	87.114

Source: Author

Table 24.2 Evaluation matrices of stacking model.

	Stacking (base estimator RF)			
Matrices	Ada.B.	DT	LR	XGB
Accuracy	90.44	88.49	99.33	92.38
F1-score	89.67	87.98	98.48	91.57
Precision	89.404	87.68	98.44	91.4
Recall	90.44	88.49	99.33	92.38

Source: Author

Result and Discussion

We split the dataset into 80% for training and 20% for testing, after which we trained a voting classifier and evaluated its performance based on several metrics. These metrics are recorded in Table 24.1 and also visualized in Figure 24.5.

Upon analyzing the results, we observed that the RF classifier delivered the best performance on our dataset. To further enhance the model's performance, we utilized RF as the base estimator and constructed a stacking model, combining it with Logistic Regression (LR). This hybrid approach demonstrated improved results, as presented in Table 24.2, with the corresponding visual representation in Figure 24.6.

Conclusion

In this study, we demonstrated the effectiveness of using ensemble learning techniques, including the voting and stacking methods, to predict asthma risk. Feature extraction using t-SNE played a vital role in enhancing model performance by reducing the dimensionality of the dataset. Among the classifiers tested, the stacking method with logistic regression as the meta-classifier achieved the best results, significantly improving prediction accuracy and providing a robust model for asthma risk prediction. This approach has the potential to be applied in clinical settings to aid in early diagnosis and personalized treatment strategies for asthma patients. Future work could explore integrating more advanced deep learning techniques to further enhance prediction accuracy and model robustness.

References

[1] Kumar, S., Rampal, S., Gaur, M., & Gaur, M. (2024). Advanced ensemble learning approach for asthma prediction: optimization and evaluation. In 2024 International Conference on Automation and Computation (AUTOCOM), (pp. 283–288). IEEE.

[2] Li, D., Abhadiomhen, S. E., Zhou, D., Shen, X. J., Shi, L., & Cui, Y. (2024). Asthma prediction via affinity graph enhanced classifier: a machine learning approach based on routine blood biomarkers. *Journal of Translational Medicine*, 22(1), 100.

[3] Gunawardana, J. R. N. A., Viswakula, S. D., Rannan-Eliya, R. P., & Wijemunige, N. (2024). Machine learning approaches for asthma disease prediction among adults in Sri Lanka. *Health Informatics Journal*, 30(3), 14604582241283968.

[4] Pooja, M. R., Ravi, V., Lokesh, G. H., Al Mazroa, A., & Ravi, P. (2024). A prognostic model to improve asthma prediction outcomes using machine learning. *The Open Bioinformatics Journal*, 17(1).

[5] Ekpo, R. H., Osamor, V. C., Azeta, A. A., Akindeji, K., & Orelaja, A. E. (2024). Designing a model for predicting asthma in adolescent using map reduce and federated learning. In 2024 International Conference on Science, Engineering and Business for Driving Sustainable Development Goals (SEB4SDG), (pp. 1–8). IEEE.

[6] Tsang, K. C. H., Pinnock, H., Wilson, A. M., and Ahmar Shah, S. (2020). Application of machine learning to support self-management of Asthma with mHealth. In 2020 42nd Annual International Conference of the IEEE Engineering in Medicine & Biology Society (EMBC), Montreal, QC, Canada, (pp. 5673–5677). doi: 10.1109/EMBC44109.2020.9175679.

[7] Rani, A., & Sehrawat, H. (2022). Role of machine learning and random forest in accuracy enhancement during asthma prediction. In 2022 10th International Conference on Reliability, Infocom Technologies and Optimization (Trends and Future Directions) (ICRITO), Noida, India, (pp. 1–10). doi: 10.1109/ICRITO56286.2022.9965149.

[8] Koshta, V., Singh, B. K., Behera, A. K., & Ganga, R. T. (2023). Classification of Asthma, COPD and healthy lung sounds using fourier bessel series expansion in machine learning and deep learning paradigm. In 2023 11th International Conference on Intelligent Systems and Embedded Design (ISED), Dehradun, India, (pp. 1–6). doi: 10.1109/ISED59382.2023.10444569.

[9] Swain, S., Pattnayak, B. K., Mohanty, M. N., Jayasingh, S. K., Patra, K. J., & Panda, C. (2024). Smart livestock management: integrating IoT for cattle health diagnosis and disease prediction through machine learning. *Indonesian Journal of Electrical Engineering and Computer Science*, 34(2), 1192–1203.

[10] Swain, S., Patra, K. J., Jayasingh, S. K., Mohanty, M. N., & Pattanayak, B. K. (2024). The Role of ensemble learning in advancing child fetal health: a focus on vot-

ing and bagging techniques. In 2024 1st International Conference on Cognitive, Green and Ubiquitous Computing (IC-CGU), (pp. 1–6). IEEE.

[11] Patra, K. J., Mishra, J., Dash, S. K., & Jayasingh, S. K. (2024). Enhancing diagnostic precision for early detection of gastric polyps with U-Net3+ segmentation. In 2024 3rd Odisha International Conference on Electrical Power Engineering, Communication and Computing Technology (ODICON), (pp. 1–5). IEEE.

[12] Senapati, C., Patra, K. J., Swain, S., Jayasingh, S. K., Senapati, S., & Ray, N. K. (2024). A comprehensive approach to groundwater quality prediction and management in the state of Telangana, India, using stacking ensemble approach. In 2024 International Conference on Emerging Systems and Intelligent Computing (ESIC), (pp. 118–123). IEEE.

25 Predicting diabetic patients coronary artery calcium score, deep learning using retinal images

Lingaraj Rath[1,a], Soumya Ranjan Mishra[1,b], Sachikanta Dash[2,c], Pratap Chandra Pradhan[3,d] and Aparna Baboo[2,e]

[1]Department of CSA, School of Sciences, GIET University, Odisha, India

[2]Department of CSE, School of Computational Sciences, GIET University, Odisha, India

[3]Department of EEE, School of Engineering and Technology, DRIEMS University, Odisha, India

Abstract

In developed countries, cardiovascular diseases (CVD) account for a considerable number of deaths. Previous investigations suggest that retinal blood vessels can provide information about cardiovascular risks. Retina fundus imaging (RFI) is a cost-effective procedure that has become routine for screening diabetic retinopathy (DR) among patients with diabetes. Since CVD is greatly caused by diabetes, we wished to know if deep learning architectures could be applied on RFI in predicting CV risk within this setting. Conclusively, the target was to develop convolutional neural networks (CNN) that would help in predicting whether the coronary artery calcium (CAC) score will be above some limits set by specialists or not. It was shown that preliminary tests performed on a few patients, who were rightly diagnosed, delivered encouraging precision. In addition, it was found that fundamental clinical information relates positively to the risk of CV illness. It was discovered that the outcomes of both informational hints are harmonious, and the two applications that may benefit from mixed visual examination and health statistics are suggested.

Keywords: Coronary artery calcium, convolutional neural networks, cardiovascular diseases, deep learning, diabetic retinopathy, retina fundus imaging

Introduction

The global leading causes of mortality today are also ischemic heart disease and stroke according to the World Health Organization [1]. There is an alarming increase in cardiovascular disease (CVD) every year; therefore, more efforts must be made towards improving risk predictors. Factors can be utilized to conduct cardiovascular risk assessments based on various data and parameters obtained from patients histories [2–7]. However, this information may not always be centralized, recent, or accessible.

As per the available data till date, the calcium score in coronary arteries (CAC) forms a clinically authenticated powerful indicator of a cardiovascular disease [8–10]. According to research, scoring of CAC has been shown to be a consistent and reproducible procedure in determining cardiovascular risk in asymptomatic individuals; therefore, it is very important in primary prevention [11]. Despite this, expensive CT scan methods that carry the risk of radiation exposure are needed to get a CAC score.

Eye's vascular system is in detail shown by retinal fundus imaging (RFI). According to previous research, both capillary vascular and foveal areas are the common areas that predict CAC scoring by RFI [12]. We furthermore take checkups sometimes from the type 2 diabetes patients whose group exhibits cardiovascular disease as one of their commonest illnesses [13].

DL algorithms, especially those used for AI applications, have seen notable progression across most domains of computer vision systems [14, 15]. Among them, medical applications rank high. An example of this is that the computer program classifying the retina images automatically has proved to be better than human experts at diagnosing diabetics' eye disease [16]. As a result, it is suggested in this paper that DL, which means deep learning, should be used for forecasting the CAC scores derived from RFI through one of its forms known as convolutional neural networks (CNN). Based on some data we collected from a previous project [17], we propose that it would be possible to use a CNN in order to discriminate those patients that might have a CAC score higher than 400, which is considered by cardiologists as the most dangerous level for developing CVD. Few experiments have shown previous studies to have a strong correlation with CAC > 400 [18]. The results we obtained from our tests indicate that the two origins contain

[a]22bca038.lingarajrath@giet.edu, [b]mrsoumya1208@gmail.com, [c]dash.sachikanta@gmail.com, [d]pratap_pin@yahoo.com, [e]aparnababoo98@gmail.com

DOI: 10.1201/9781003658221-25

different but complementary data, which may be combined for best predictive accuracy or recall [19]. We validate our proposal through 2 applications, namely, clinical diagnosis and large-scale retrieval systems. Our findings reveal that indeed RFI low-cost image acquisition could offer means for CVD prediction; hence, it would have an immense implication on strategies aimed at selecting patients' populations who need more integral screening programs against CVDs than others [20].

Methods and Material

This is considered a binary classification problem. In this paper, the best performing one among neural networks was chosen by using a 5-fold cross-validation method. At first, there was an initial segmentation of the retina images into two datasets containing data on the left eye and right eye that were tested independently on their own terms. In both cases, VGG16 mean accuracies outperformed those of ResNet, but the differences were not statistically significant. Therefore, VGG16 has been adopted as an image classifier throughout this study since they are equivalent according to their capabilities. The main components of VGG16 architecture consist of 13 convolutional layers and 3 fully connected as indicated in Figure 25.1.

Transfer learning

Due to the absence of sufficient labeled data, training with a massive deep neural network (130M parameters) becomes extremely challenging. To resolve such challenges associated with small sample sizes, transfer learning is one of the widely applied methods [21]. At the initial phase, a CNN is trained with extensive image datasets in our case, ImageNet [22] and then the model settings are refined depending on the retina image training set. We got rid of the last totally correlated component of VGG16 and placed two neurons in another layer (FC-2) that makes a difference between patients who are identified as having CAC > 400 and those with CAC < 400. The weights for these layers were randomly initialized and later updated during the learning process. Used 0.0001 as learning rate and 4×10^2 for weight decay at this

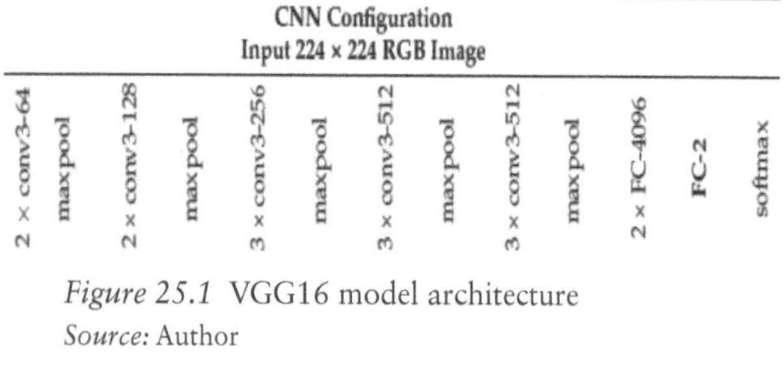

architecture, which underwent sixty epochs with a batch size of eight. We specified the cross-entropy loss function and stochastic gradient descent optimizer as well. In addition, the use of a pre-trained filter significantly increases the accuracy, where initial canonical layers form local filters for certain image component detection while Top-level representations of decision boundaries are provided by the last one (classification layers). A Python library called Pytorch [23–25] was used for all studies.

Medical data specification abbreviations and acronyms

So they took the data from down in the hospital to help improve their ability to predict patient outcomes over time while assessing the extent to which they add up to the already existing image classifiers.

RFI dataset description

With regard to the data we used, a total of 76 patients were included. In this case, there are 152 retinal images (right and left eye), out of which 66 were declared positive samples indicating CAC > 400, while 86 showed negative samples indicating CAC < 400. The PRECISED study is from where these images originate (ClinicalTrial.gov NCT02248311). Patients aged between 46 and 76 years were included in this exercise. It must be understood that before the age of 45 years, it is extremely rare for an individual to experience any type of cardiovascular event; thus, younger screening strategies do not offer value for money. Each of the images has 3 channels (RGB). Figure 25.2.

Before we can even run the model, we have to do a background check on it. Initially, we use the approach in PyTorch's library for transfers learning of ImageNet-based learned models to color normalize the image. That is, the values are subtracted by a precompiled mean that is equal to [0.485, 0.456, 0.406], which

CNN Configuration Input 224 × 224 RGB Image												
2 × conv3-64	maxpool	2 × conv3-128	maxpool	3 × conv3-256	maxpool	3 × conv3-512	maxpool	3 × conv3-512	maxpool	2 × FC-4096	FC-2	softmax

Figure 25.1 VGG16 model architecture
Source: Author

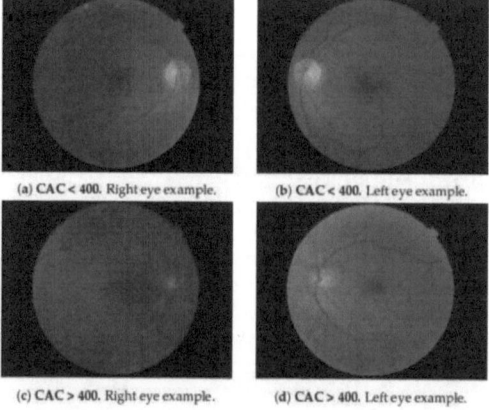

(a) CAC < 400. Right eye example.   (b) CAC < 400. Left eye example.

(c) CAC > 400. Right eye example.   (d) CAC > 400. Left eye example.

Figure 25.2 Samples RFI dataset (2576*934*3)
Source: Author

was utilized during ImageNet pre-training. This helps to load RFI images in the range of [0, 1]. Secondly, a standard size of 224 × 224 is used to rescale all images and thus fit into input model of CNN. The CNN outputs are not subjected to any additional processing; only the class with maximum output in the last fully connected layer is selected.

Results

Prediction of automated positive CAC score using DL
This section below discusses the results of automatically identifying patients with a CAC score > 400, which indicates a possible risk of developing cardiovascular disease, using fundus photos and a DL model. Participants in this study participated in PRECISED research, which was integrated into the study (ClinicalTrial.gov NCT02248311). All subjects provided written informed permission before being enrolled in the study, and the Helsinki Declaration was followed throughout its execution. The study was approved by the local ethics committee. In doing cross-validation, there were five stratified folds done 5 times. Four of the folds were used for training and one of them was used for testing. The deep learning architecture used for training was VGG16 (Section 2). The training time it takes 141 seconds, while inference took 0.85 seconds.

Though a small amount of training data is utilized to train VGG16, in general, it remains a better model than any changes in both classes (Table 25.1). It is important to point out that the performance of this model is not equal for images from the left eye and those from the right one. It can be observed that there are slight variations in retinal structures between left and right eyes, even though they may not be significant.

By more investigation into the expected tags for image and clinical data classifiers, it became apparent to us that there are variations in individual predictions according to the modality. Consequently, these complementary findings could be used as a basis for an ensemble of both sources of information. In particular, we recommend two arrangements aimed at two distinct purposes:

- *Clinical diagnosis:* In the conservative prediction protocol, both modalities are employed together.
- *Wide-scale retrieval:* The ensemble's precision enhancement is stressed, this time, the aim is to establish an application that can conduct searches from extensive databases.

Clinical diagnosis application

For this particular software, first the model is implemented on clinical data, forming an ensemble thereafter. The image DL model is executed for all samples having been given a CAC with a classification of less than 400 following this. Only in cases where results from both instances were negative do we report CAC < 400. The health system can stop more people from being diagnosed at birth wrongly (false negative) because of this arrangement shown below in (people in dire need of treatment but could go unnoticed by any of these techniques). The results are presented in Table 25.3.

In all situations, the outcomes exhibit a constant tendency that shows noteworthy advances in the F1-measure. The false negative count has been minimized by approximately 50%, which is its most serious case (negative predicted patients with CAC > 400). Furthermore, using the image prediction has led to a substantial increase in true positives (and hence recall) as we detect more patients with CAC > 400. Yet, this appreciation gives rise to additional false alarms (FP). In any case, it is a more convenient situation (regarding the health prediction system) since FP are patients who have been diagnosed as positive but really have CAC<400 and they are therefore not in danger. Table 25.2.

Discussions

Investigations made recently [8] indicate that utilizing a deep CNN (EfficientNet) for radio-frequency

Table 25.1 Performance measure.

Model	Accuracy	Precision	Recall	F1	Conf. Mat	
RFI (Left Eye)	0.81	0.78	0.78	0.78	36	7
					7	26
RFI (Right Eye)	0.84	0.86	0.75	0.80	39	4
					8	25

Source: Author

Table 25.2 Depending on the classification variables utilized, the accuracy of classifiers.

Classifier	Age	DR	AGE+DR
LR	0.692	0.545	0.610
KNN	0.778	0.636	0.700
SVM	0.788	0.788	0.788
Gaussian Process	0.760	0.576	0.655
Decision tree	0.767	0.697	0.730
RF	0.862	0.758	0.806
AdaBoost	0.778	0.636	0.700
Quadratic classifier	0.704	0.576	0.633
Naïve Bayes	0.818	0.545	0.655

Source: Author

Table 25.3 Performance measurement with proposed.

Model	Accuracy	Recall	Precision	F1	Conf Mat	
CNN RFI	0.737	0.783	0.545	0.643	38	5
					15	18
LR	0.697	0.692	0.545	0.610	35	8
					15	18
Proposed	0.816	0.788	0.788	0.788	36	7
					7	26
KNN	0.776	0.767	0.697	0.730	36	7
					10	23
Proposed	0.763	0.778	0.636	0.700	37	6
					12	21
SVM	0.750	0.818	0.545	0.655	39	4
					15	18
Proposed	0.789	0.815	0.667	0.733	38	5
					11	22
Gaussian Process	0.763	0.800	0.606	0.690	38	5
					13	20
Proposed	0.750	0.750	0.636	0.689	36	7
					12	21
Decision Tree	0.737	0.760	0.576	0.655	37	6
					14	19
Proposed	0.803	0.800	0.727	0.762	37	6
					9	24
RF	0.737	0.724	0.636	0.677	35	8
					12	21
Proposed	0.842	0.862	0.758	0.806	39	4
					8	25
AdaBoost	0.711	0.704	0.576	0.633	35	8
					14	19
Proposed	0.829	0.794	0.818	0.806	36	7
					6	27
Quadratic Classifier	0.750	0.769	0.606	0.678	37	6
					13	20
Proposed	0.789	0.815	0.667	0.733	38	5
					11	22
Naïve Bayes	0.671	0.682	0.455	0.545	36	7
					18	15
Proposed	0.763	0.742	0.697	0.719	35	8
					10	23

Source: Author

interference (RFI) generates optimistic accuracy when predicting cardiovascular risk. This research was undertaken involving a major number of Asian individuals, mostly non-diabetic, whereby some degree of correlation existed between the RetiCAC CNN total score and the CT scan-based determination of CAC levels. As a result, our preference for the present pilot investigation was due to its focus on persons suffering from type 2 diabetes, thus leading to better coherence in the characteristics of our training dataset. This approach has led to significant improvements in accuracy due to using VGG and ResNet DL architectures.

Methods applied to clinical data and images give accurate results in evaluating high CAC scores. Nevertheless, those two recognized protocols exhibited distinct capabilities with respect to employing both modalities separately. Findings from experimental studies suggest that it is feasible to customize an ensemble specifically for certain applications so as to enhance either precision or recall, but at the cost of sacrificing some degree of accuracy in terms of one or another performance indicator. As per the guidelines, a particular application meant for clinical diagnosis has been validated to make certain that only a few individuals with CAC > 400 go unnoticed, thereby substantially decreasing false negatives and enhancing recall (almost 85%). Nevertheless, this method presents a large number of false positives that will bear on the expenses of resources but not on personal decision-making capacity regarding diagnosis.

Conclusion

This study presents strong proof that deep learning techniques may be employed in assessing heart diseases' susceptibility utilizing CAC alone as its exclusive biomarker. Preliminary findings have displayed hopeful levels of precision on a limited sample repository employing traditional DL structures. Our research indicates that some of these clinical measures have a positive correlation with CAC > 400. Thus, there exists a simple initial exploration that showcases an unexpected level of success and additionally confirms results got through image analysis. On this basis, we constructed two applications that integrate precision or recall criterion optimization for a particular application. In this initial work, the existence of discriminative information is demonstrated from retinal images. Other challenges for further investigation include data acquisition and model improvements. More clinical data can be collected to significantly improve outcomes (hence increasing relevant variables), or alternatively, more images may be gathered (which means more patients). Also, a lot more can be done in order to specialize DL architecture to take advantage of high details present in retinal images like thickness, dimension, and vessel tortuosity, or using current advancements in self-supervised learning (to solve the problem of small sample size) as curriculum learning, in which we train machine learning models in such a way that it has meaning.

References

[1] Azad, S. M., Padhy, S., & Dash, S. (2023). A case study on the multi-hopping performance of IoT network used for farm monitoring. *Automatic Control and Computer Sciences*, 57, 70–80.

[2] Breiman, L. (2001). Random forests. *Machine Learning*, 45, 5–32.

[3] Cheung, C. Y., Xu, D., Cheng, C. Y., Sabanayagam, C., Tham, Y. C., Yu, M., et al. (2021). A deep-learning system for the assessment of cardiovascular disease risk via the measurement of retinal-vessel calibre. *Nature Biomedical Engineering*, 5, 498–508.

[4] Cheung, C. Y. L., Ikram, M. K., Chen, C., & Wong, T. Y. (2017). Imaging retina to study dementia and stroke. *Progress in Retinal and Eye Research*, 57, 89–107.

[5] Cheung, C. Y. L., Zheng, Y., Hsu, W., Lee, M. L., Lau, Q. P., Mitchell, P., et al. (2011). Retinal vascular tortuosity, blood pressure, and cardiovascular risk factors. *Ophthalmology*, 118, 812–818.

[6] Dash, S., Das, R. K., Guha, S., Bhagat, S. N., & Behera G. K. (2021). An interactive machine learning approach for brain tumor MRI Segmentation. In Advances in Intelligent Computing and Communication: Proceedings of ICAC 2020, (pp. 391–400). Springer Singapore.

[7] Dash, S., Padhy, S., Parija, B., Rojashree, T., & Patro, K. A. K. (2022). A simple and fast medical image encryption system using chaos-based shifting techniques. *International Journal of Information Security and Privacy (IJISP)*, 16(1), 1–24.

[8] Freund, Y., Schapire, R., & Abe, N. (1999). A short introduction to boosting. *Journal of the Japanese Society for Artificial Intelligence*, 14, 1612.

[9] Goff, Jr. D. C., Lloyd-Jones, D. M., Bennett, G., Coady, S., D'agostino, R. B., Gibbons, R., et al. (2013). 2013 ACC/AHA guideline on the assessment of cardiovascular risk: a report of the American college of cardiology/American heart association task force on practice guidelines. *Circulation*, 129, S49–S73.

[10] Greenland, P., Blaha, M. J., Budoff, M. J., Erbel, R., & Watson, K. E. (2018). Coronary calcium score and cardiovascular risk. *Journal of the American College of Cardiology*, 72, 434–447.

[11] Hota, R., Dash, S., Mishra, S., Pradhan, S., Pattnaik, P. K., & Pradhan, G. (2023). Early prediction and diagnosis of thoracic diseases using rough set and machine learning. In 2023 10th International Conference on Computing for Sustainable Global Development (INDIACom), (pp. 206–213). IEEE.

[12] Kim, M., Yun, J., Cho, Y., Shin, K., Jang, R., Bae, H. J., et al. (2019). Deep learning in medical imaging. *Neurospine*, 16, 657.

[13] McGrory, S., Cameron, J. R., Pellegrini, E., & Warren, C. (2016). The application of retinal fundus camera imaging in dementia: a systematic review. *Alzheimers Dement*, 6, 91–107.

[14] Bahadure, N. B., Dash, S., Padhy, S., Satpathy, A., & Routray, S. (2023). Rare diseases severity prediction

system using a machine learning-based technique. In 2023 International Conference on Artificial Intelligence for Innovations in Healthcare Industries (IC-AIIHI), Raipur, India, (pp. 1–6).

[15] Padhy, S., Dash, S., Shankar, T. N., Rachapudi, V., Kumar, S., & Nayyar, A. (2024). A hybrid crypto-compression model for secure brain MRI image transmission. *Multimedia Tools and Applications*, 83(8), 24361–24381.

[16] Padhy, S., Alowaidi, M., Dash, S., Alshehri, M., Malla, P. P., Routray, S., et al. (2023). AgriSecure: a fog computing-based security framework for agriculture 4.0 via blockchain. *Processes*, 11(3), 757.

[17] Patnaik, R., Rath, P. S., Padhy, S., & Dash, S. (2022). Histopathological colorectal cancer image classification by using inception V4 CNN model. In International Conference on Robotics, Control, Automation and Artificial Intelligence, (pp. 1003–1014). Singapore: Springer Nature Singapore.

[18] Poplin, R., Varadarajan, A. V., Blumer, K., Liu, Y., McConnell, M. V., Corrado, G. S., et al. (2018). Prediction of cardiovascular risk factors from retinal fundus photographs via deep learning. *Nature Biomedical Engineering*, 2, 158–164.

[19] Rim, T. H., Lee, C. J., Tham, Y. C., Cheung, N., Yu, M., Lee, G., et al. (2021). Deep-learning based cardiovascular risk stratification using coronary artery calcium scores predicted from retinal photographs. *The Lancet Digital Health*, 3, e306–e316.

[20] Panigrahy, S., Dash, S., Padhy, S., Kumar, N., & Dash, Y. (2024). Predictive modelling of diabetes complications: insights from binary classifier on chronic diabetic mellitus. In 2024 International Conference on Communication, Computer Sciences and Engineering (IC3SE), Gautam Buddha Nagar, India, (pp. 1912–1920).

[21] Sabanayagam, C., Xu, D., Ting, D. S. W., Nusinovici, S., Banu, R., Hamzah, H., et al. (2020). A deep learning algorithm to detect chronic kidney disease from retinal photographs in community-based populations. *The Lancet Digital Health*, 2, e295–e302.

[22] Padhy, S., Dash, S., Routray, S., Ahmad, S., Nazeer, J., & Alam, A. (2022). IoT-based hybrid ensemble machine learning model for efficient diabetes mellitus prediction. *Computational Intelligence and Neuroscience*, 2022, 2389636.

[23] Son, J., Shin, J. Y., Chun, E. J., Jung, K. H., Park, K. H., & Park, S. J. (2020). Predicting high coronary artery calcium score from retinal fundus images with deep learning algorithms. *Translational Vision Science and Technology*, 9, 28.

[24] Chaitanya, V. S. K., Rakesh, D., Dash, S., Sahoo, B. K., Padhy, S., & Nayak, M. (2022). Tomato leaf disease detection using neural networks. In 2022 International Conference on Machine Learning, Computer Systems and Security (MLCSS), Bhubaneswar, India, (pp. 53–58).

[25] Yeboah, J., & McClelland, R. L. (2012). Comparison of novel risk markers for improvement in cardiovascular risk assessment in intermediate-risk individuals. *JAMA*, 308, 788–795.

26 Modelling and performance analysis of a dexterous linkage-driven three-fingered under-actuated robotic hand

Deepak Ranjan Biswal[a], Alok Ranjan Biswal[b] and Uttam Kumar Tarai[c]

Department of Mechanical Engineering, DRIEMS University, Odisha, India

Abstract

The robotic hand is of notable importance for human resource substitution in intense, tedious, and hazardous atmospheres. Insecure conditions in industrial domains, such as pick-and-deploy, categorizing, stacking, assembly, marine growth, and medical services, require robotic hands to perform functions similar to human hands. In advanced humanoid systems, there is a growing demand for the ability to provide dexterity and independent functionality to robotic systems, particularly in manufacturing, prosthetic applications, and therapeutic rehabilitation. This requires the development of innovative actuator and gearbox systems to manage the movement of a humanoid hand. Under-actuated concepts have established to be a feasible approach for creating remarkably precise robot hands devoid of the need for complex mechanical designs. An under-actuated hand possesses more degrees of freedom than the number of actuators, which significantly makes simpler of the control system and makes the hand less expensive than fully actuated types. Under-actuated robotic hands have acquired considerable attention due to their capability for enhanced effectiveness, ease of use and versatility in handling intricate manipulation activities. This study contributes a complete overview on the modeling and assessment of a three-fingered robotic hand based on the under-actuation principle with a linkage driven operational concept. The proposed hand features three fingers with a palm, developed to emulate the skillful abilities of the human hand at the time of employing limited actuators. The design and assessment are performed using ADAMS software, and the simulation results are compared with existing findings, confirming that the modeled hand is appropriate for manipulation.

Keywords: Dexterous, grasping, linkage driven, manipulation, under-actuation

Introduction

A robotic hand is a mechanical apparatus engineered to replicate the form and functionality of human hand, imitating its movements and structure. Scholars are striving to develop robotic hands that can accomplish complex operations and exhibit highly skilled actions. By studying the human hand, they can advance the development of robotic hands for applications in automation, medical devices, biomechanical equipment, prosthetics, and space research [1]. The aim of robotic hand design is to achieve precise, adaptable performance in real-world tasks by mimicking key aspects of the human hand, such as configuration and independent motion. Under-actuated robotic hands provide high dexterity with simplified control and mechanics, making them efficient, cost-effective, and capable of grasping diverse objects for use in robotics, prosthetics, and automation [22]. Researchers have looked into the challenge of establishing effective grabbing and proficient methods for the rigid robotic hand with the minimal Degrees of Freedom (DOF) and no sliding and rolling contact combinations [25]. To assess the hand's functional mechanism a three-fingered hand with tendons has been designed. It is recommended to employ

linkage-driven robotic hands with multimodality and expertise [4]. For a robotic hand with five fingers, the design of a tendon-operated finger has been suggested, and its study comprises approaches that imitate the features of the human finger [8]. A finite element analysis and preliminary design of a robotic hand using its five under-actuated fingers to grasp a cuboid are presented [3]. A detailed study has been conducted on an anthropomorphic hand that uses tendons to control the movement of its fingers, focusing on how it interacts with the object being gripped [9]. A robotic hand has been conceptualized with two points of articulation and a total of seventeen joints, which can be used to adjust the coordination between its initial two positions [6]. The RBO Hand 3 is an anthropomorphic soft hand using pneumatic actuation, designed for dexterous manipulation and real-world experiments. With 16 degrees of actuation, it replicates human hand functions, featuring a modular design and intrinsic compliance. It achieves high dexterity, demonstrated through various grasping strategies and thumb opposition tests [13]. Sun et al. [18] outlined a design framework for a robotic hand with five fingers that uses only two actuators to replicate A broad spectrum of human holding techniques and limited manipulation abilities, featuring independent

[a]deepak.biswal@driems.ac.in, [b]alokranjanbiswal@driems.ac.in, [c]uttamtarai@driems.ac.in

DOI: 10.1201/9781003658221-26

thumb movement and coordinated finger motion, along with a novel differential palm mechanism that embeds human Eigen grasps, enabling the mini X-hand to reproduce 29 of 33 common human grasp types [18]. Arulkirubakaran et al. [2] investigated the potential to regulate an under-actuated robot and recommends an idea of industrial robotic grippers. Under-actuated hands are adjustable and promptly adjust to outlines of unidentified items [5]. Scarcia et al. [16] suggested conceptualizing, designing, and testing rotating elastic joints for under-actuated robotic fingers. Wang et al. [20] outlined an under-actuated robotic hand highlighting gear-operated and linkage-driven mechanics. Yang et al. [21,22] illustrate the MCR-Hand III, a five-fingered robotic hand that operates using a combination of tendons and linkages. A linkage-driven under-actuated three-finger hand was recommended to replicate the extension/flexion and adduction/abduction gestures of the human hand [7]. The SAU-RFC hand is a newly designed adaptive under-actuated robotic hand featuring rigid-flexible coupling fingers. It has three fingers with seven DOF and is powered by three servomotors [17]. Wang et al. [20] introduced a passively flexible five-fingered under-actuated dexterous hand (UADH) capable of grasping entities of diverse forms and dimensions providing slider-bar and gear driven mechanisms. The study outlines the UADH's mechanical design, kinematics, and statics, with an emphasis on contact forces during grasping. Experimental findings validate the performance of the hand. The humanoid hand of s-type with five fingers was crafted to perform hand forms, master-slave synchronization using a bilateral process, and skilled gripping [15]. Yuan et al. [24] presented a novel non-anthropomorphic robotic hand with actively driven rollers at the fingertips, enabling six DOF in object manipulation. Systems of equations for two and three finger manipulation are developed to demonstrate non-holonomic spatial motion. A prototype successfully validated mathematical analysis, offering unbounded object rotation without finger gaiting, unlike conventional robotic hands. Tang et al. [19] presented a modular under-actuated robot hand with configurable finger arrangements based on task needs. Actuated by a screw and connecting rod mechanism, the hand features hollow rubber fingertips for stable grasping. Experimental results show its adaptability for various grasping modes. Further exploration is needed to enhance control precision and sensor integration for more complex manipulation tasks. This work presents a robotic hand with 7 DOF and 15 flexible joints, driven by under-actuated linkages and worm gears. A kinematic model analyzes finger movements and forces. Experiments show

a fingertip force of 7.8 N and good adaptability to various grasp types. Further improvements are needed in force control precision and dynamic manipulation for more complex real-world tasks [12]. Evaluation of a three-finger adaptive robotic gripper designed for everyday object handling, focusing on shape, size, and texture. The grippers demonstrate high success rates in gripping and manipulation, making them user-friendly for individuals with physical disabilities. However, further research is needed to validate their real-world usability in daily living robotics. Rahman et al. [11,14] demonstrated the three-finger gripper's 93% success rate in grasping various household items, analyzing its ability to manipulate objects weighing up to 2.9 kg while varying grip force based on surface texture, weight, and center of mass position. An under-actuated RCM-based portable robot was developed for rehabilitation of stroke and postoperative patients. It features a compliant hip joint mechanism, optimized design, and kinematic analysis, enabling safe human-robot interaction for home-based and outpatient rehabilitation exercises [23].

This study aims to model an under-actuated robotic hand using advanced modeling tools and conduct a comprehensive gripping analysis to evaluate its performance. The proposed robotic hand is designed with a unique configuration that includes nine joints, twelve degrees of freedom (DOF), and six actuators, demonstrating an optimal balance between mechanical complexity and functional efficiency. By integrating under-actuated mechanisms driven by linkage systems, the design replicates the functionality of human-like fingers, specifically considering the Distal Interphalangeal (J-DIP), Proximal Interphalangeal (J-PIP), and Meta Carpo Phalangeal (J-MCP) joints of all three fingers. This research contributes to the field by presenting a design methodology for under-actuated robotic hands and offering insights into their gripping mechanics, thereby progressing the refinement of lightweight, efficient, and highly functional robotic frameworks.

Design Methodology

Three identical fingers are incorporated in the intended robotic hand. Every finger has three joints and three phalanges. There are four degrees of mobility in each finger. The hand's design comprises the individual finger designs as well as a hand with dimensions that are roughly comparable to a human hand. The suggested hand model has been analyzed using stainless steel and aluminium. ADAMS software was used for the analysis and hand modelling of the under-actuated fingers.

Finger structure design

The dimensions of all the parts employed for the of the design of the hand are provided in Table 26.1. The recommended hand features three fingers with a palm, with the fingers attached to the base of palm. Individual finger is comprised of three phalanges, interconnected through revolute joints. The three identical fingers are positioned around the edge of the palm. The model of an individual finger, displayed in Figure 26.1, shows that all the phalanges are associated with revolving joints.

Hand structure design

The robotic hand design features three identical fingers positioned evenly around the palm. Figure 26.2 presents the hand model. The act of gripping and manipulating objects is a complex and remarkable aspect of human expertise and motorized abilities. It allows us to interact with our environment, handle objects, and perform a variety of operations with precision. The

term grasping refers to the human hand's capability to hold and relocate items incorporating its fingers [10]. This capability is essential for numerous daily activities and is a hallmark of human proficiency. The unique anatomy of the human hand allows it to perform a vast array of tasks, ranging from simple actions like acquiring small objects to complex activities such as writing or playing a musical instrument. Main features of the human hand's grasping abilities consists of the opposable thumb, which plays a crucial role in grip strength and versatility, the precision grip, enabling fine control for delicate tasks, and the power grip, providing force for handling larger or heavier items. Various types of grasps, along with the development of motor skills, hand-eye coordination, and the impact of the brain and nervous system, all contribute to the exceptional functionality of the human hand.

The proposed hand is located on the boundary of the palm for correct handling of the items to be maneuver. The three fingers approaching to manipulate the object are introduced in Figure 26.2. The three fingers gripping a frustum shaped object is showcased in Figure 26.3.

Results and Discussion

The proposed robot hand is designed for effective holding and maneuvering of various objects, requiring

Table 26.1 Hand dimensions.

Name	Dimension $L \times W \times T$ (cm)
Distal-Phalange	$3.5 \times 1.25 \times 0.5$
Middle-Phalange	$3.85 \times 1.5 \times 0.6$
Proximal-Phalange	$5.8 \times 1.75 \times 0.8$
1st Under-actuated link	$5.4 \times 0.5 \times 0.5$
2nd Under-actuated link	$3.8 \times 0.3 \times 0.3$
3rd Under-actuated link	$3.8 \times 0.5 \times 0.25$
1st Driving bar of the 1st Four bar linkage	$3.2 \times 0.5 \times 0.5$
1st Driving bar of the 2nd Four bar linkage	$2.2 \times 0.3 \times 0.3$
2nd Driving-bar of the 2nd Four bar linkage	$1.2 \times 1.5 \times 0.5$

Source: Author

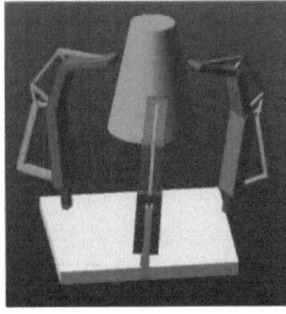

Figure 26.2 Robotic hand reaching to take hold of the job

Source: Author

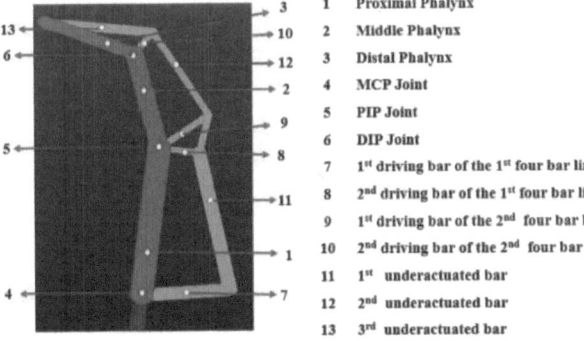

1	Proximal Phalynx
2	Middle Phalynx
3	Distal Phalynx
4	MCP Joint
5	PIP Joint
6	DIP Joint
7	1st driving bar of the 1st four bar lir
8	2nd driving bar of the 1st four bar li
9	1st driving bar of the 2nd four bar l
10	2nd driving bar of the 2nd four bar
11	1st underactuated bar
12	2nd underactuated bar
13	3rd underactuated bar

Figure 26.1 Model of a single finger

Source: Author

Figure 26.3 Robotic hand hold the job

Source: Author

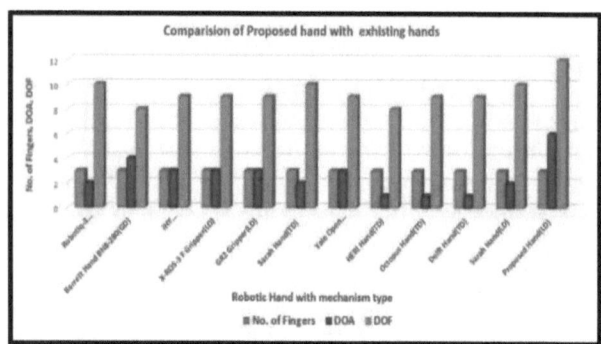

Figure 26.4 The proposed hand model versus traditional hand models
Source: Author

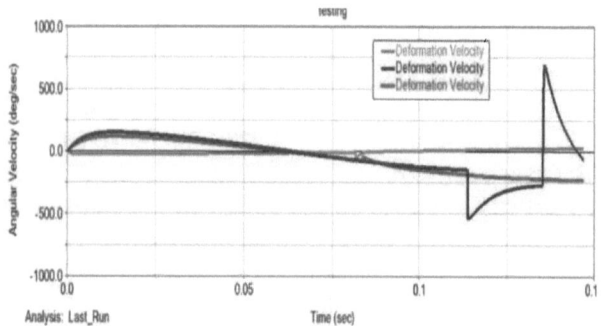

Figure 26.5 Deformation velocity analysis of the J-DIP joints in all fingers
Source: Author

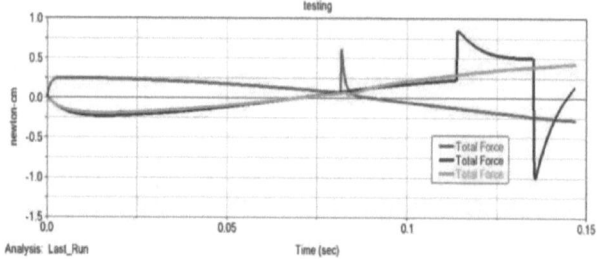

Figure 26.6 Torque analysis at the J-DIP joints of each of three fingers
Source: Author

precise finger movements for proper control. A comparative analysis of the proposed design and existing systems is shown based on the fingers, actuators employed (DOA), and whole DOF. The hands are categorized as TD for Tendon-Driven Mechanisms, GD(Gear-Driven) mechanisms, LD (Linkage-Driven) mechanisms, and others. An assessment between the proposed model and the existing robotic hands is presented in Figure 26.4.

The robotic hand with three fingers is designed for effective grabbing and manipulation of various objects. To securely hold and manipulate objects, precise finger movement is necessary. Therefore, assessment of the phalanges is crucial for achieving appropriate grasping. The assessment was performed using ADAMS software, with Figure 26.5 showing the deformation velocity of the J-DIP joints for all three fingers. The analysis duration is 0.15 seconds, focusing on the contact between the distal phalanges and a frustum-shaped object. Torque on the J-DIP joints was plotted over this time period with step size of 0.0001 seconds. The results indicate that the distal phalanges contact only the object, and the deformation velocities vary over time, ranging from -500 deg./sec to 550

deg./sec, with a maximum of 700 deg./sec and a minimum of -510 deg./sec.

The torque analysis at the J-DIP joint of all three fingers is shown in Figure 26.6. It illustrates the connection between the distal phalanges and a cylindrical element. Torque values for the J-DIP joints were plotted over a 0.15-second period with a step size of 0.0001 seconds.

The results show that the torque at the J-DIP joints initially drops then increases. Over time, the torque decreases for one finger, while for the other two fingers, it first decreases, slightly increases, and then stabilizes. The torque values range between -1.0 N-cm and 0.75 N-cm.

Conclusion

The studies concentrate on the modelling and analysis of a robotic hand with three fingers. In this study the robotic hand, consisting of three fingers arranged peripherally around the palm. The under-actuation is accomplished by a linkage operated mechanism, allowing the robot hand to grasp a variety of objects effectively. The proposed robotic hand model was subjected to analysis during the grasping of the frustum-shaped object. The modelling was performed using specialized software. For the robotic fingers, stainless steel was selected as the material, while aluminum was used for the frustum-shaped object. During the grasping process, all distal phalanges are employed through under-actuation to restrict the movement of the moving object. Kinematic analysis of the hand is performed in ADAMS simulation environment. It is found that the suggested hand framework is suitable for precise gripping and maneuvering of objects, making it an effective solution for versatile robotic applications.

References

[1] Akbas, B., Yuksel, H. T., Soylemez, A., Zyada, M. E., Sarac, M., & Stroppa, F. (2024). The impact of evolu-

tionary computation on robotic design: a case study with an underactuated hand exoskeleton. arXiv preprint arXiv. 2403:15812.

[2] Arulkirubakaran, D., Malkiya Rasalin Prince, R., Ramesh, R., Rajesh, T. C. S. N., Neil Anand, K., Mathew, S. M., et al. (2022). Evolution of industrial robotic grippers-a review. In Recent Advances in Materials and Modern Manufacturing: Select Proceedings of ICAM-MM 2021, (pp. 553–563).

[3] Biswal, D. R., & Parida, P. K. Modelling and finite element based analysis of a five fingered underactuated robotic hand, *International Journal for Research in Applied Science & Engineering Technology*, 10(9), 100–108.

[4] Chen, M., Li, S., Shuang, F., Du, Y., & Liu, X. (2023). Development of a three-fingered multi-modality dexterous hand with integrated embedded high dimensional sensors. *Journal of Intelligent & Robotic Systems*, 108(2), 24.

[5] Da Fonseca, V. P., Jiang, X., Petriu, E. M., & de Oliveira, T. E. A. (2022). Tactile object recognition in early phases of grasping using underactuated robotic hands. *Intelligent Service Robotics*, 15(4), 513–525.

[6] Hamon, P., Chablat, D., & Plestan, F. (2021). A new robotic hand based on the design of fingers with spatial motions. In International Design Engineering Technical Conferences and Computers and Information in Engineering Conference. American Society of Mechanical Engineers. 85451:V08BT08A021.

[7] Li, G., Liang, X., Gao, Y., Su, T., Liu, Z., & Hou, Z. G. (2023). A linkage-driven underactuated robotic hand for adaptive grasping and in-hand manipulation. *IEEE Transactions on Automation Science and Engineering*, 21(3), 3039–3051.

[8] Llop-Harillo, I., Pérez-González, A., & Andrés-Esperanza, J. (2020). Anthropomorphism indexes of the kinematic chain for artificial hands. *Journal of Bionic Engineering*, 17, 501–511.

[9] Neha, E., Suhaib, M., Mukherjee, S., & Shrivastava, Y. (2023). Kinematic analysis of four-fingered tendon actuated robotic hand. *Australian Journal of Mechanical Engineering*, 21(2), 541–551.

[10] Parveen, S., Suhaib, M., & Majid, M. A. (2023). Multifinger robotic gripper: a review. In 2023 International Conference on Recent Advances in Electrical, Electronics & Digital Healthcare Technologies (REEDCON), (pp. 466–470). IEEE.

[11] Rahman, M. M., Shahria, M. T., Zarif, M. I. I., Haque, M. E., Ahamed, S. I., Ghommam, J., et al. (2024). Performance assessment of 3-finger adaptive robotic grippers in handling objects for daily living assistance. *Archives of Physical Medicine and Rehabilitation*, 105(4), 53.

[12] Pu, Z., Da-ming, N., Yun, L., Shi-qiang, Z., Gu, J., & Rui-long, D. (2022). Design of humanoid robotic hand based on link underactuation. In 2022 12th International Conference on CYBER Technology in Automation, Control, and Intelligent Systems (CYBER), (pp. 174–178). IEEE.

[13] Puhlmann, S., Harris, J., & Brock, O. (2022). RBO hand 3: a platform for soft dexterous manipulation. *IEEE Transactions on Robotics*, 38(6), 3434–3449.

[14] Rahman, M. M., Shahria, M. T., Sunny, M. S. H., Khan, M. M. R., Islam, E., Swapnil, A. A. Z., et al. (2024). Development of a three-finger adaptive robotic gripper to assist activities of daily living. *Designs*, 8(2), 35.

[15] Ryu, W., Choi, Y., Choi, Y. J., Lee, Y. G., & Lee, S. (2020). Development of an anthropomorphic prosthetic hand with underactuated mechanism. *Applied Sciences*, 10(12), 4384.

[16] Scarcia, U., Berselli, G., Palli, G., & Melchiorri, C. (2017). Modeling, design, and experimental evaluation of rotational elastic joints for underactuated robotic fingers. In 2017 IEEE-RAS 17th International Conference on Humanoid Robotics (Humanoids), (pp. 353–358). IEEE.

[17] Su, C., Wang, R., Lu, T., & Wang, S. (2023). SAU-RFC hand: a novel self-adaptive underactuated robot hand with rigid-flexible coupling fingers. *Robotica*, 41(2), 511–529.

[18] Sun, B. Y., Gong, X., Liang, J., Chen, W. B., Xie, Z. L., Liu, C., et al. (2021). Design principle of a dual-actuated robotic hand with anthropomorphic self-adaptive grasping and dexterous manipulation abilities. *IEEE Transactions on Robotics*, 38(4), 2322–2340.

[19] Tang, S., Yu, Y., Sun, S., Li, Z., Sarkodie-Gyan, T., & Li, W. (2020). Design and experimental evaluation of a new modular underactuated multi-fingered robot hand. *Proceedings of the Institution of Mechanical Engineers, Part C: Journal of Mechanical Engineering Science*, 234(18), 3709–3724.

[20] Wang, D., Xiong, Y., Zi, B., Qian, S., Wang, Z., & Zhu, W. (2021). Design, analysis and experiment of a passively adaptive underactuated robotic hand with linkage-slider and rack-pinion mechanisms. *Mechanism and Machine Theory*, 155, 104092.

[21] Yang, H., Wei, G., & Ren, L. (2019). Design and development of a linkage-tendon hybrid driven anthropomorphic robotic hand. In Intelligent Robotics and Applications: 12th International Conference, ICIRA 2019, Shenyang, China, August 8–11, 2019, Proceedings, Part I 12 (pp. 117–128). Springer International Publishing.

[22] Yang, H., Wei, G., Ren, L., Qian, Z., Wang, K., Xiu, H., et al. (2021). A low-cost linkage-spring-tendon-integrated compliant anthropomorphic robotic hand: MCR-Hand III. *Mechanism and Machine Theory*, 158, 104210.

[23] Yang, Y., Guo, J., Yao, Y., & Yin, H. (2023). Development of a compliant lower-limb rehabilitation robot using underactuated mechanism. *Electronics*, 12(16), 3436.

[24] Yuan, S., Epps, A. D., Nowak, J. B., & Salisbury, J. K. (2020). Design of a roller-based dexterous hand for object grasping and within-hand manipulation. In 2020 IEEE International Conference on Robotics and Automation (ICRA), (pp. 8870–8876). IEEE.

[25] Zhang, N., Zhao, Y., Gu, G., & Zhu, X. (2022). Synergistic control of soft robotic hands for human-like grasp postures. *Science China Technological Sciences*, 65(3), 553–568.

27 Intelligent agriculture system using soft computing techniques

Sidhartha Sankar Dora[a], Prasanta Kumar Swain[b] and Jibendu Kumar Mantri[c]

Department of Computer Application, MSCB University, Baripada, Odisha, India

Abstract

The global population growth has led to an exponential growth in demand for food production. The agri-food business faces significant challenges as the world's population is forecast to reach over 9 billion by 2050, necessitating a 70% rise in agricultural and food output to meet the demand. Agriculture is essential for the economic growth of a nation and ensures food for all people of the world which is one of the major difficulties in the future. One of the more demanding platforms nowadays is wireless sensor networks, in which sensor nodes sense and monitors environmental or physical conditions and send data to the base station for analysis. These networks were also embraced by the agriculture industry in an effort to advance scientific cultivation, reduce management costs, and encourage environment friendly farming. In order to progress the food and agriculture area, we emphasize in this study the importance of combining machine learning and artificial intelligence as a predictive multidisciplinary approach. The proposed innovative artificial intelligence-based agricultural system monitors the environment using energy efficient wireless sensor networks for data collection and design prediction system to proactively forecast possible crop days and alert farmers to take corrective actions as soon as feasible and also notifies farmers when it is necessary to maintain optimal crop production. Our proposed system will focus on rice production environment.

Keywords: Artificial intelligence, clustering, energy efficient, life time, wireless sensors

Introduction

A vital part of society, agriculture affects many facets of daily living, economic growth, and environmental preservation. Its importance can be categorized into several key roles such as food production, economic contribution, rural development and livelihoods, environmental stewardship, cultural and social significance and support for industrial growth. Agriculture provides raw materials for various industries, such as textiles (cotton, wool), food processing (grains, fruits), and biofuels (corn, sugarcane). It stimulates technological advancements and innovations in machinery, fertilizers, and biotechnology, which contribute to overall industrial growth.

An intelligent agriculture system integrates advanced technologies like AI, IoT, robotics, and big data analytics to improve the efficiency and sustainability of agricultural observes. The goal is to optimize various aspects of farming, including crop monitoring, irrigation, pest control, and soil management. Key features of intelligent agriculture systems include precision Farming, automated irrigation, robotic farming, predictive analytics, smart greenhouses, livestock monitoring. By integrating these technologies, intelligent agriculture aims to increase crop yields, reduce resource usage, and support sustainable farming practices Basnet & Bang [2].

Wireless sensor networks play a vital role in smart agriculture systems by providing real time monitoring, data collection, and automation capabilities. Swarm intelligence (SI) is a collective behavior-based approach inspired by the natural behaviors of social animals, such as ants, bees, and birds, which exhibit complex problem-solving capabilities as a group. In intelligent agriculture, swarm intelligence plays a significant role by providing robust, adaptive, and scalable solutions for various farming challenges Ojha et al. [9].

In this paper, we have designed an intelligent model deploying different types of sensors using particle swarm intelligence and K-means clustering methods to minimize power consumption and extend life of WSNs. We have also applied machine learning classification algorithms for better accuracy of our proposed model.

The rest of the paper is organized as follows. Section 2 discusses literature review. The proposed system is explained in Section 3. The proposed model is defined in Section 4. Part 6 wraps up, while Section 5 concentrates on the experiment's results.

[a]lpurna@gmail.com, [b]prasantanou@gmail.com, [c]jkmantri@gmail.com

DOI: 10.1201/9781003658221-27

Literature Review

Specifically, both are able to see and comprehend the world similarly to how humans do thanks to wireless sensor networks (WSN). All the options for irrigation management are still included in the state-of-the-art, nevertheless. With a primary focus on the prediction element, Ojha et al. [9] presented a single initiative in this field of study by publishing current research proving the lack of solutions. Data forecasting is not new, but it appears to be a need in the market that farmers can fill to manage their crops, especially as WSN technologies become more widely used. This is true both in terms of the cost of the devices and the quality and coverage of device connection. One idea, for example, is to use sensor-supported architecture and mobile services to anticipate yield after discrete plantation vitality is monitored.

Qureshi et al. [10] proposed optimum path protocol which is used for intelligent agri monitoring system. Akhter and Sofi [1] proposed ML model detects Apple disease which better traditional system. Eli-Chukwu [5] proposed model which based on artificial intelligence and produces better production. Sabrina et al. [11] have designed AI model which provides high security and better result. Ben Ayed and Hanana [3] have designed AI model which works in different climate condition and returns better result. Ojha et al. [9] have designed WSN based model for agri based applications. Dos Santos et al. [4] have developed ML based prediction model for agri crops which returns better result. Maurya and Jain [7] have designed agri-based model which returns soil status and suitability for agriculture. Nikolidakis et al. [8] have energy efficient model for optimum path in networks which is better than other models. Garcia-Sanchez et al. [6] have designed agri-based model which is used precision agriculture and produces desired result.

Proposed system

The proposed system model involves the placement of wireless sensor nodes in an agricultural field, aiming to monitor environmental factors that influence rice cultivation. The main goal is to use power efficient clustering algorithms to optimize communication, extend the network lifetime, and improve data transmission to a central base station.

Proposed model

The Prediction model, which offers a WSN framework for use in agriculture, is described in this section. This framework is responsible for gathering data, arranging it into a time series, and using a prediction algorithm to foresee issues with a certain crop. In order for the farmer to plan activities to alleviate the eventual problem before it may create serious trouble, the goal is to operate in a proactive manner, alerting him or her to the emergence of anomalous events that would appear in the Figure 27.1. The premises and architecture of agri prediction will be introduced in the remaining portion of this section.

Methodology

The methodology entails placing sensor nodes in a 100x100 unit area and utilizing clustering methods, specifically particle swarm optimization (PSO) and DBSCAN, to maximize the power competence of the network in order to develop an energy-efficient WSN system for an Agricultural Predictive Model for rice. The following methodology outlines the steps to achieve this, including the calculation of total energy consumption, network lifetime, and packet reception at the base station.

Step 1: System initialization

Define network parameters: Set the dimensions of the deployment area to 100 × 100 units. Specify the sensor nodes, gateways, and the base station. Initialize sensor nodes with different types (e.g., temperature, humidity, soil moisture sensors). Set the initial energy level for each sensor node. Place 5 gateway nodes strategically in the network. Position the base station outside the sensor field, which will collect all data from the gateways.

Deployment of sensor nodes: Randomly deploy sensor nodes within the 100x100 area. Display the deployed nodes, gateways, and base station on a graph.

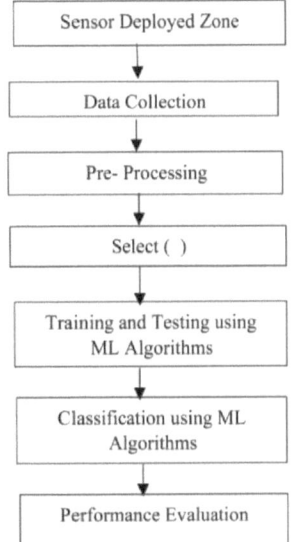

Figure 27.1 Proposed Model Diagram
Source: Author

Step 2: Clustering algorithm

K-Means clustering: A basic clustering algorithm that minimizes the distance between sensor nodes and their cluster centers.

PSO-based clustering: Use PSO to optimize the cluster heads and cluster creation.

Step 3: Data communication model

Intra-cluster communication: Sensor nodes communicate their sensed information to their respective cluster heads.

Perform data accumulation at the cluster head level to decrease the total amount of data to be conveyed.

Inter-cluster communication: Cluster heads communicate the gathered data to the adjacent gateway node to decrease long-range communication costs.

Gateway to base station communication:

Gateways send the collected data to sink for final processing and analysis.

Step 4: Energy consumption model

Energy calculation for each transmission:

Calculate the energy consumed for sending and receiving data based on the distance and packet size.

The energy model can be based on the free space model for short distances and the multipath fading model for longer distances.

Update energy of nearest nodes: Update the energy level of each node after each transmission round. Nodes with depleted energy levels are considered dead, impacting the network's lifetime.

Total power consumption: Accumulate the energy consumed by all sensor nodes, cluster heads, and gateways over time.

Step 5: Network lifetime calculation

Define lifetime of the network: The network lifetime can be defined as the time until the first node dies, 50% of the nodes are dead, or all nodes are dead.

Track the alive nodes to measure network performance.

Evaluate the impact of clustering on network lifetime: Compare the lifetime results for different clustering algorithms.

Step 6: Packet reception and data integrity

Packet transmission monitoring: Track the number of data packets communicated from sensor nodes to cluster heads, cluster heads to gateways, and gateways to the sink.

The packet delivery ratio as the ratio of packets successfully received at the base station to the total packets sent.

Data integrity check: Ensure the correctness and consistency of the data received at the base station.

Step 7: Visualization and analysis

Visualize the deployment of sensor nodes, cluster heads, gateways, and base station on a graph.

It also shows the clustering structure, communication links, data flow paths, energy consumption graphs, network lifetime graphs and packet reception graphs.

Step 8: Performance evaluation

We have used metrics energy efficiency, network lifetime, packet delivery ratio sent by sink and accuracy to get efficiency of our proposed model.

Result Analysis

Data source

This section provides a concise overview of Crop_recommendation.csv datasets used in the research. The CSV file contains 2201 rows and eight columns.

Simulation

For simulation activities, Anaconda3, an Intel Core i7 CPU running at 3.50 GHz with 16GB of RAM, and Microsoft Windows 11 are utilized. Table 27.1 displays the parameter values that we used.

Evaluation metrics

The system's performance is evaluated based on:

Total energy consumption: Measures the energy used by all nodes in the network.

Network lifetime: Time until a specific portion of the network is no longer functional.

Packet delivery ratio: Ratio of packets received at the base station to the packets sent.

Clustering efficiency: Effectiveness of the clustering algorithm in reducing communication distance and balancing the energy load.

Energy consumption and lifetime

The Figure 27.2 shows Network lifetime after K-means clustering: 2.00 seconds, network lifetime after DBSCAN Clustering: 2.00 seconds, total energy

Table 27.1 Simulation parameters.

Area_size	100×100
Num of_sensors	100
Num_of particles	30
Num_of iterations	100
Inertia weight (w)	0.5
Cognitive coefficient c2	1.5
Social coefficient (c1)	1.5

Source: Author

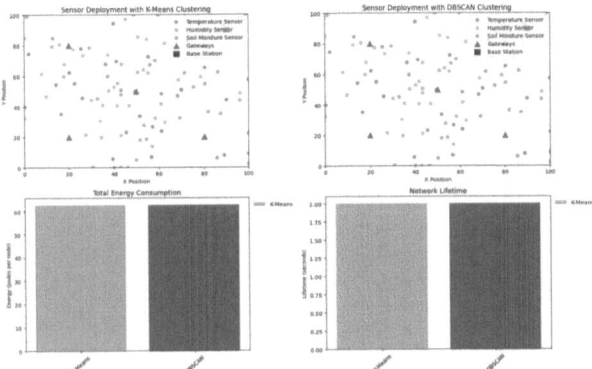

Figure 27.2 Energy consumption and lifetime
Source: Author

consumed after K-means clustering: 62.60 Joules per node.

Total energy consumed after DBSCAN Clustering: 62.60 Joules per node, total packets received by base station (K-Means): 100.0, total packets received by base station (DBSCAN): 100.0

Accuracy and crop recommendation
We have designed recommendation models using different machine learning algorithms. The model recommends crop rice and Navie Bayes classifier returns 99.5% accuracy.

Conclusion

This model provides a framework for implementing an energy-efficient WSN using clustering algorithms and PSO for agricultural predictive modelling, which is particularly beneficial for enhancing crop yield and resource management in rice farming.

References

[1] Akhter, R., & Sofi, S. A. (2022). Precision agriculture using IoT data analytics and machine learning. *Journal of King Saud University-Computer and Information Sciences*, 34(8), 5602–5618.

[2] Basnet, B., & Bang, J. (2018). The state-of-the-art of knowledge-intensive agriculture: a review on applied sensing systems and data analytics. *Journal of Sensors*, 2018(1), 3528296.

[3] Ben Ayed, R., & Hanana, M. (2021). Artificial intelligence to improve the food and agriculture sector. *Journal of Food Quality*, 2021(1), 5584754.

[4] dos Santos, U. J. L., Pessin, G., da Costa, C. A., & da Rosa Righi, R. (2019). AgriPrediction: a proactive internet of things model to anticipate problems and improve production in agricultural crops. *Computers and Electronics in Agriculture*, 161, 202–213.

[5] Eli-Chukwu, N. C. (2019). Applications of artificial intelligence in agriculture: a review. *Engineering, Technology & Applied Science Research*, 9(4), 4377–4383.

[6] Garcia-Sanchez, A.-J., Garcia-Sanchez, F., & Garcia-Haro, J. (2011). Wireless sensor network deployment for integrating video-surveillance and data-monitoring in precision agriculture over distributed crops. *Computers and Electronics in Agriculture*, 75(2), 288–303.

[7] Maurya, S., & Jain, V. K. (2016). Fuzzy based energy efficient sensor network protocol for precision agriculture. *Computers and Electronics in Agriculture*, 130, 20–37.

[8] Nikolidakis, S. A., Kandris, D., Vergados, D. D., & Douligeris, C. (2013). Energy efficient routing in wireless sensor networks through balanced clustering. *Algorithms*, 6(1), 29–42.

[9] Ojha, T., Misra, S., & Raghuwanshi, N. S. (2015). Wireless sensor networks for agriculture: The state-of-the-art in practice and future challenges. *Computers and Electronics in Agriculture*, 118, 66–84.

[10] Qureshi, K. N., Bashir, M. U., Lloret, J., & Leon, A. (2020). Optimized cluster-based dynamic energy-aware routing protocol for wireless sensor networks in agriculture precision. *Journal of Sensors*, 2020(1), 9040395.

[11] Sabrina, F., Sohail, S., Farid, F., Jahan, S., Ahamed, F., & Gordon, S. (2022). An interpretable artificial intelligence based smart agriculture system. *Computers, Materials and Continua*, 72(2), 3777–3797.

28 Synthesis, structural and Raman study of Zn doped CuO nanoparticles

Harchand Tudu[a], Bikram Keshari Das[b] and Tanushree Das[c]

Nano Innovation Laboratory, School of IKST, Kalinga Institute of Social Sciences (KISS) Deemed to be University, Bhubaneswar, Odisha, India

Abstract

Zn doped CuO ($Cu_{1-x}Zn_xO$, x = 0 and x = 0.01) nanoparticles were synthesized by solid state reaction route. The structural and microstructural characteristics of the synthesized material was analyzed by X-ray diffraction (XRD), Field emission scanning electron microscopy (FESEM). Raman spectroscopy was used to analyze the impact of Zn doping on the distinctive vibrational modes of CuO nanoparticles. The synthesized sample shows a single-phase monoclinic structure. A change in the lattice parameters was noticed with Zn doping. Zn doped CuO samples showed interesting morphological alterations in the FESEM micrograph. Energy dispersive X-ray (EDX) analysis agrees with the existence of Zn dopant in CuO lattice. The shift in peak position in the Raman spectra shows the influence of Zn doping.

Keywords: Zn doped CuO, Rietveld refinement, FESEM, EDX, Raman Spectroscopy

Introduction

Nanostructured metal oxide semiconductors have outstanding properties and are of great interest in research and industries [6]. Among the different nanostructured metal oxide semiconductors, CuO is a non-toxic and p-type semiconductor that has unique properties having band gap 1.2 eV Okoye et al. [5], binding energy of 937.8 eV Ganesan et al. [4], outstanding chemical stability, cost-effective, excellent electrical and optical properties, high thermal stability, magnetic, conductivity, high surface area, antibacterial properties etc. [1]. Due to these excellent properties it can be used in the field of gas sensors, solar cells, wastewater treatment, photocatalysis, optoelectronic devices Okoye et al. [5], drug delivery, antibacterial agent Ganesan et al. [4] etc. The structure of CuO can be tailored according to research interest for different suitable applications. The characteristics of CuO can be modified through doping with various elements such as Zn, Cr, Co etc. [6, 5].

In this work, we have synthesized $Cu_{1-x}Zn_xO$ nanoparticles by employing a simple solid-state reaction method with x = 0 and x = 0.01. The effect of Zn doping on the structural, microstructural and vibrational frequency of CuO was studied in detailed.

Experimental Methods

The Zn doped CuO ($Cu_{1-x}Zn_xO$) nanocrystalline powders were synthesized by solid state reaction technique using high-purity CuO (99% pure, Merck) and ZnO (99% pure, Merck) precursors. Using an agate mortar and pestle, a stochiometric amount of raw precursors was ground. It was then calcined at 800°C (High Temperature Muffle Furnace, Ants Lab Equipment, India) for two hours. The calcined powder was characterized by several characterization techniques [2]. XRD (USA, Bruker D8 Advance) was employed to investigate the crystal lattice and to study the surface morphology FESEM (Zeiss Gemini SEM 450) was employed. The vibrational frequency was analyzed by Raman Microscopy (UK, RENISHAW InVia Raman Microscope).

Structural Characterization

XRD

The XRD was utilized to know the crystal lattice of CuO and Zn doped CuO. XRD pattern is represented in Figure 28.1(a) with the most prominent peak magnified in Figure 28.1(b).

Figure 28.2 illustrates the Rietveld refinement XRD pattern of calcined $Cu_{1-x}Zn_xO$ nanoparticles.

Every peak was attributed to monoclinic CuO belonging to the space group C2/c. The XRD pattern showed no peak associated with ZnO or any other impurity, revealing the development of a single phase. This indicates that the Zn^{2+} ion is successfully doped into the CuO lattice.

The variation in lattice constants 'a' and 'c' with Zn doping is given in Table 28.1, The difference in ionic radii of Zn^{2+} (0.74 Å) and Cu^{2+} (0.73 Å) may be the cause of variation in lattice parameter. The

[a]23591003@kiss.ac.in, [b]bikram.das@kiss.ac.in, [c]tanushree.das@kiss.ac.in

DOI: 10.1201/9781003658221-28

experimental data is well-matched with Rietveld result as shown in Figure 28.2.

The crystallite size (D) of the Zn doped CuO sample was determined by Scherrer's formula [3].

$$D = \frac{0.9\lambda}{\beta \cos\theta} \qquad (1)$$

Here, λ = X-ray wavelength (1.5406 Å), β = full width at half maximum (FWHM) and θ = the Bragg's angle. The difference in crystallite size may be due to the strain induced by Zn doping. The crystallite size increases with a decrease in strain as depicted in Table 28.1.

Microstructural Analysis

FESEM

The FESEM micrograph of $Cu_{1-x}Zn_xO$ (x = 0 and 0.01) nanoparticles was depicted in Figure 28.3(a) and (b). In both compositions, the particle shape has been observed as spherical. It was observed in Figure 28.3(b) that few particles agglomerate to larger particles.

To determine the average particle size, ImageJ software was used [7]. From Figure 28.4, it was observed that when Zn is doped in CuO lattice, the particle size increases from 78.33 nm to 86.76 nm.

EDX spectroscopy

The EDX was utilized to examine the elemental composition of synthesized material, as shown in

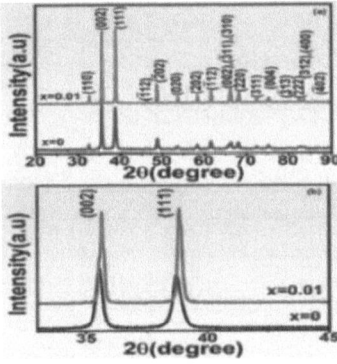

Figure 28.1 (a) XRD pattern of $Cu_{1-x}Zn_xO$ (x = 0 and 0.01) nanoparticles and (b) enlarge view of most intense XRD peak
Source: Author

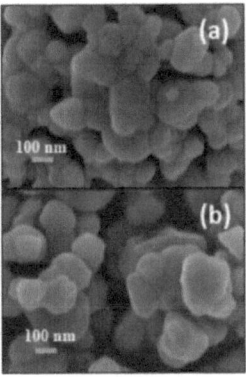

Figure 28.3 FESEM micrograph of (a) CuO (b) Zn doped CuO nanoparticles
Source: Author

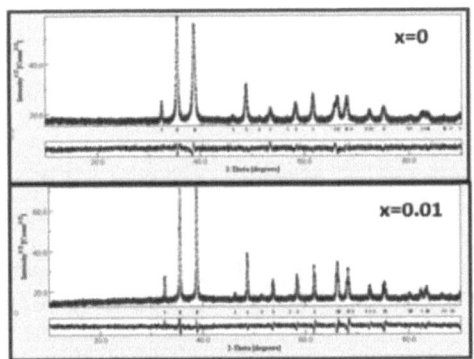

Figure 28.2 Rietveld refinement XRD pattern of $Cu_{1-x}Zn_xO$ nanoparticles
Source: Author

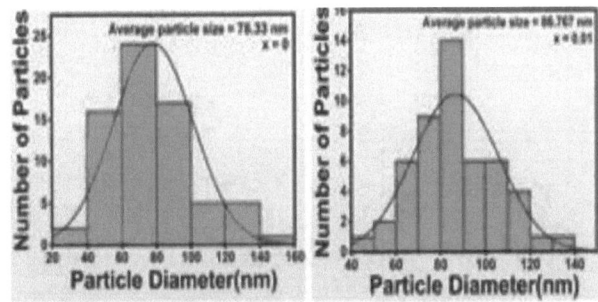

Figure 28.4 Particle size distribution of CuO (x=0) and Zn doped CuO (x=0.01) nanoparticles
Source: Author

Table 28.1 Structural parameters of $Cu_{1-x}Zn_xO$ (x = 0 and 0.01) nanoparticles.

Composition	a (Å)	b (Å)	c (Å)	D (nm)	Lattice strain (ε)
x = 0	4.68370	3.42260	5.12880	47	1.6×10^{-3}
x = 0.01	4.68330	3.42080	5.12940	85	0.7×10^{-3}

Source: Author

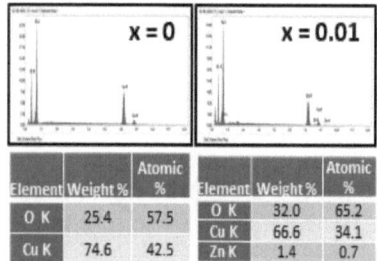

Figure 28.5 EDX pattern of $Cu_{1-x}Zn_xO$ (x = 0 and 0.01) nanoparticles
Source: Author

Figure 28.5. x = 0 depicts the presence of Cu and O confirming the sample is pure CuO while for x = 0.01 the existence of Cu, O and Zn indicates the doping of Zn^{2+} ion in the CuO lattice which aligns with the XRD result [7].

Raman Spectroscopy

We employed Raman spectroscopy at room temperature to confirm the phase purity of CuO and study the impact of Zn doping on its vibrational structure. The Raman spectra of the $Cu_{1-x}Zn_xO$ nanoparticles at concentrations of x = 0 and x = 0.01 are shown in Figure 28.6.

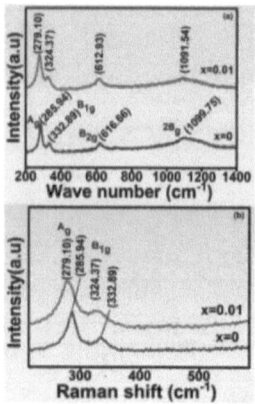

Figure 28.6 (a) Raman spectrum of $Cu_{1-x}Zn_xO$ (x = 0 and 0.01) nanoparticles (b) shifting of Raman peak
Source: Author

Four distinct vibrational modes were noticed in the spectra at 279.10 cm^{-1}, 324.37cm^{-1}, 612.93 cm^{-1} and 1091.54 cm^{-1} for Zn doped CuO which is slightly lower than the pure CuO samples. These peaks correspond to the A_g, B_{1g}, B_{2g} and $2B_{2g}$ overtone modes. The absence of secondary peaks in the Raman spectra indicates that Zn doped CuO nanoparticles are single phase, which is consistent with the phase purity noticed in the XRD pattern. The observed Raman

peak exhibits a slight shift which arises because of the substitution of Zn^{2+} ion in CuO lattice as shown in Figure 28.6(b).

Conclusion

CuO ceramic samples doped with Zn ($Cu_{1-x}Zn_xO$) were prepared by the solid-state reaction method. No contamination was observed for the synthesized sample and was a monoclinic structure according to XRD analysis. Zn substitutes Cu site in crystal lattice but does not alter the monoclinic structure, as determined by the XRD Rietveld refinement. A little modification in the lattice constant had been noticed with Zn doping. The crystallite size of CuO increases with Zn doping while lattice strain decreases with doping. A similar observation was made in the FESEM micrograph as XRD where the particle size was found to increase with doping. EDX confirmed the presence of Zn in CuO lattice. Raman spectroscopy study also confirmed that there was only a single-phase formation of the undoped and Zn doped CuO nanocrystalline sample.

References

[1] Ahmed, T. M., Kamoun, E. A., EL-Moslamy, S. H., Salim, S. A., Zahran, H. Y., Zyoud, S. H., et al. (2024). Performance of Ag-doped CuO nanoparticles for photocatalytic activity applications: synthesis, characterization, and antimicrobial activity. *Discover Nano*, 19, 166.

[2] Bikram, K. D., Das, T., & Das, D. (2023). Structural and electrical properties of mechanically alloyed ZnO nanoceramic for NTC thermistor application. *Journal of Materials Science: Materials in Electronics*, 34(3), 230.

[3] Bikram, K. D., Das, T., Das, D., Parashar, K., Parashar, S. K. S., Kumar, R., et al. (2024). Negative temperature co-efficient of resistance behaviour of Cr doped ZnO nanoceramics. *Journal of Materials Science and Engineering B*, 299, 117017.

[4] Ganesan, K., Jothi, V. K., Natarajan, A., Rajaram, A., Ravichandran, S., & Ramalingam, S. (2020). Green synthesis of copper oxide nanoparticles decorated with graphene oxide for anticancer activity and catalytic applications. *Arabian Journal of Chemistry*, 13, 6802–6814.

[5] Okoye, P. C., Azi, S. O., Qahtan, T. F., Owolabi, T. O., & Saleh, T. A. (2023). Synthesis, properties, and applications of doped and undoped CuO and Cu_2O nanomaterials. *Materials Today Chemistry*, 30, 101513.

[6] Reungruthai, S., Chaopanich, P., Prasatkhetragarn, A., Chailuecha, C., Kuimalee, S., & Klinbumrung, A. (2022). Doping effect of Zn on structural and optical properties of CuO nanostructures prepared by wet chemical precipitation process. *Radiation Physics and Chemistry*, 190, 109788.

[7] Das, T., & Das, B. K. (2024). Impact of strontium doping on the structural and NTCR properties of ZnO ceramics. *Journal of Materials Science: Materials in Electronics*, 35, 2202.

[8] Das, T., Das, D., & Das, B. K. (2024). Impedance spectroscopy and conduction mechanism of $Zn_{1-x}Mg_xO$ NTCR ceramics. *Journal of Materials Science and Engineering B*, 302, 117206 35.

29 Immersive Learning in Ayurveda: Revolutionizing Dhatu Poshan Nyaya education through AR/VR technologies

Prasant Kumar Dash[a], Tithi Singh[b], Tripti Singh[c] and Sambit Kumar Pradhan[d]

Department of CSE, C.V. Raman Global University, Bhubaneswar, Odisha, India

Abstract

This study explores the integration of augmented reality (AR) and virtual reality (VR) in teaching *Dhatu Poshan Nyaya*—a core Ayurvedic concept explaining tissue nourishment. Traditional learning methods often fail to convey the complexity of this sequential physiological process. We propose an immersive AR/VR-based learning environment to enhance comprehension and retention among students. By providing interactive 3D simulations of nutrient absorption and transformation, AR/VR fosters deeper engagement and a multi-sensory experience. Initial results demonstrate improved learning outcomes compared to traditional methods. The findings underscore the potential of AR/VR in modernizing Ayurvedic education and advancing healthcare training.

Keywords: Augmented reality, Ayurvedic education, Dhatuposhan Nyaya, immersive learning, virtual reality

Introduction

Ayurveda, the ancient Indian system of holistic medicine, emphasizes maintaining harmony within the body through personalized approaches. *Dhatu Poshan Nyaya*—the sequential nourishment of seven tissues (*dhatus*)—forms the cornerstone of Ayurvedic physiology [6]. However, traditional methods like textbooks and static diagrams struggle to depict the intricate relationships involved. Emerging technologies, including AR and VR, offer an innovative solution by providing real-time, immersive simulations. This paper examines the use of AR/VR to transform Ayurvedic education by enabling learners to visualize the *dhatu* formation process interactively. The objective is to make complex concepts like *Dhatu Poshan Nyaya* more accessible, relatable, and applicable in contemporary healthcare education [4].

Literature Review

Ayurveda uses analogies like *Ksheerdadhi Nyaya* (nutrient transformation as curdling milk) and *Khale Kapota Nyaya* (selective absorption as a pigeon picking grains) to explain physiological processes, requiring visualization for better understanding. Traditional teaching methods often fail to convey these dynamic concepts, making retention and application difficult. Ismail et al. [1] AR/VR technologies address these gaps by creating interactive environments that visualize nutrient absorption and tissue formation, enhancing comprehension and retention through experiential learning [5].

Proposed Methodology and Model Specifications

Methodology

The proposed methodology focuses on leveraging AR/VR technologies to create an integrated, immersive learning environment for teaching *Dhatu Poshan Nyaya*. This approach involves realistic 3D models of anatomical structures and dynamic interactions through VR simulations. Each *Nyaya* is represented with unique visualizations and interactive elements, making the learning process comprehensive and engaging.

Development framework

1. **Integrated environment:** A custom-built virtual reality framework was developed using Unity 3D and Blender for creating highly realistic anatomical models.
2. **Interaction design:** The platform allows learners to manipulate virtual objects, adjust parameters such as dietary quality, and observe their effects on physiological processes.
3. **Learning modules:** Each module is designed around one of the three key *Nyayas*:
   * *Ksheerdadhi Nyaya* (Nutrient Transformation)
   * *Khale Kapota Nyaya* (Selective Absorption)
   * *Kedarakulya Nyaya* (Nutrient Distribution)

[a]prasant.oitburla@gmail.com, [b]tithisingh11@gmail.com, [c]triptisingh22a@gmail.com, [d]sambitp107@gmail.com

DOI: 10.1201/9781003658221-29

4. **Evaluation system:** Integrated quizzes follow each learning module, allowing students to assess their understanding.

Details of the three Nyayas

The proposed VR environment models three key *Nyayas* to simplify complex *Ayurvedic* concepts. *Ksheerdadhi Nyaya* illustrates nutrient transformation, where food is digested and sequentially converted into Rasa and other dhatus, emphasizing the role of diet in health. *Khale Kapota Nyaya* demonstrates selective nutrient absorption using the analogy of a pigeon picking grains, allowing learners to explore variables like digestive fire (Agni) and nutrient density. Finally, *Kedarakulya Nyaya* visualizes nutrient distribution through irrigation channels, with Rasa as the reservoir nourishing tissues like *Asthi and Majja*, highlighting the impact of blockages or imbalances on systemic health. [3]. These interactive modules provide dynamic visualizations, enabling students to experiment and understand the intricate physiological processes [2].

Figure 29.1 is original work and has been designed by author(Block Diagram), Figure 29.3 is a Graph generated from the mean score values of participants data.

VR environment specifications

Realistic 3D models:

High-definition anatomical models depict the digestive system, tissues, and nutrient pathways with macro (organ-level) and micro (cellular-level) views.

Interactivity:

Users can adjust nutrient inputs, observe nutrient flow, and explore dynamic animations of tissue nourishment processes.

Integrated assessment:

Quizzes at the end of each module test retention and problem-solving, such as diagnosing nutrient-related issues.

User-friendly interface:

Intuitive VR controls and guided tutorials ensure easy navigation and interaction.

Empirical Results

Preliminary results reveal significant improvements in the AR/VR group:

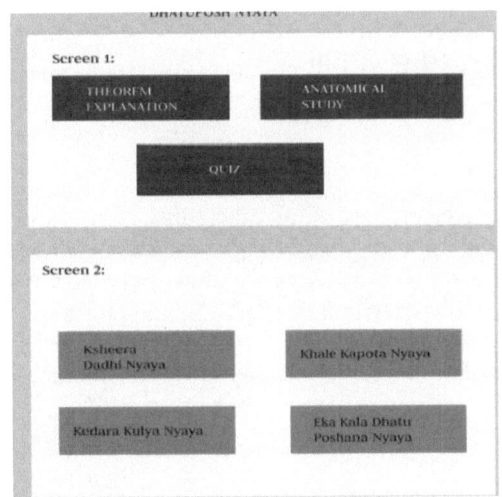

Figure 29.2 VR module screen visualization
Source: Author

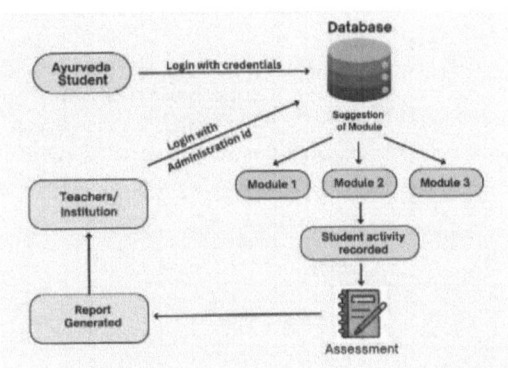

Figure 29.1 Block diagram of proposed model
Source: Author

Table 29.1 Hypothesisofmodel.

Study design elements	VR Training group	Traditional training group
Participants	10	10
Training method	Interactive VR modules of Ayurvedic theories	PowerPoint presentations and Books
Key features	Visualization of nutrient flow and tissue nourishment	Standard educational materials
Evaluation method	Standardized examination	Standardized examination
Hypothesis	Higher levels of understanding and retention	Lower levels of understanding and retention
Expected outcome	Significantly better performance in examination	Relatively lower performance in examination

Source: Author

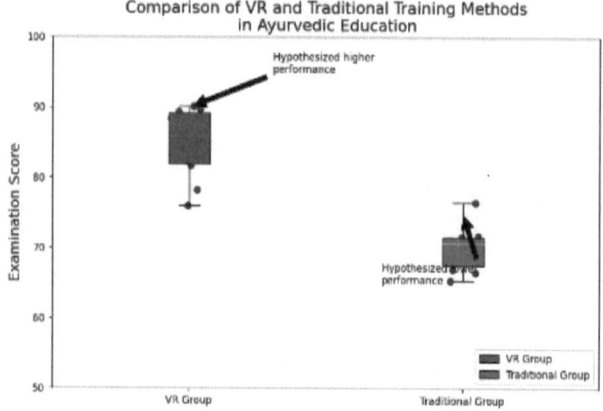

Figure 29.3 Graphical representation of hypothesis
Source: Author

- Higher test scores (mean: 85) compared to the traditional group (mean: 70).
- Enhanced comprehension of complex concepts like *Ksheerdadhi Nyaya*.
- Greater engagement and interest, as reported in post-study surveys.

Graphical representation of these findings confirms the hypothesis that AR/VR enhances Ayurvedic learning outcomes.

Conclusion

Integrating AR/VR into Ayurvedic education revolutionizes how concepts like *Dhatu Poshan Nyaya* are taught. By offering interactive and immersive learning experiences, these technologies address the limitations of traditional methods, fostering improved understanding and application. Future work will focus on refining simulations and expanding their applications across healthcare education.

References

[1] Ismail, N. M., & Fazeen Rasheed, A. K. (2024). VR devices in ayurvedic care: understanding influential factors for user behavior and continuance intention. *International Journal of Human–Computer Interaction*, 40, 1–18. doi.org/10.1080/10447318.2024.2399 418.

[2] Khan, H. U., Ali, Y., Khan, F., & Al-Antari, M. A. (2024). A comprehensive study on unraveling the advances of immersive technologies (VR/AR/MR/XR) in the healthcare sector during the COVID-19: challenges and solutions. *Heliyon*, 10(15), e35037. doi: 10.1016/j.heliyon.2024.e35037. PMID: 39157361; PMCID: PMC11328097.

[3] Kumar, A., & Arya, V. (2024). Achieving holistic health through ayurveda along with advanced technologies: a conference report. *Journal of Ayurveda and Integrative Medicine*, 15(3), 100967. doi: 10.1016/j.jaim.2024.100967. Epub 2024 May 27. PMID: 38805855; PMCID: PMC11152973.

[4] Ranade, M. (2024). Artificial intelligence in ayurveda: current concepts and prospects. Journal of Indian System of Medicine, 12(1), 53–59. DOI:10.4103/jism.jism_60_23.

[5] Sawarkar, G. (2021). Role of virtual learning in ayurveda education. Journal of Indian System of Medicine, 9, 151. 10.4103/JISM.JISM_76_21. DOI:10.4103/JISM.JISM_76_21.

[6] Verma, P., Nagar, L. K., Sharma, A. K., Sharma, K. L., & Meena, R. (2023). A conceptual study of dhatu poshan Nyaya (metabolic transformation) in ayurvedic perspective. International Research Journal of Ayurveda & Yoga, 6(1), 42–45. https://doi.org/10.48165/IRJAY.2023.6107.

30 Combating food insecurity through remote sensing and machine learning for enhanced crop yield prediction

Bhumisuta Khuntuli[1,a], Sachikanta Dash[1,b], Pratap Chandra Pradhan[2,c] and Soumya Ranjan Mishra[3,d]

[1]Department of CSE, School of Computational Sciences, GIET University, Odisha, India

[2]Department of EEE, School of Engineering and Technology, DRIEMS University, Odisha, India

[3]Department of CSA, School of Science, GIET University, Odisha, India

Abstract

This work introduces a fresh strategy for combating food insecurity by combining agricultural yield prediction based on machine learning with remote sensing technology. Improving agricultural production and resource management rely heavily on reliable yield predictions, according to the study. The suggested approach accurately predicts crop yields by analyzing environmental variables using powerful machine learning techniques and a large dataset gathered from different agricultural regions. The study shows how various models perform in terms of various performance measures. These models include random forest (RF), support vector machines (SVM), and K-nearest neighbors (K-NN). This study's ensemble learning method achieves better prediction accuracy of 92.1% which is better than conventional models and offers policymakers and farmers practical insights. The findings aim to improve decision-making, optimize resource allocation, and contribute to food security initiatives by harnessing these advanced technologies.

Keywords: Ensemble, K-nearest neighbors, remote sensing, random forest, support vector machines

Introduction

The increasing demand for food production due to population growth, coupled with climate change, has made accurate crop yield predictions crucial for global food security. Traditional agricultural methods often fail to consider the multitude of variables that affect crop growth, such as soil conditions, weather patterns, and agricultural practices. Recent advances in machine learning (ML) and deep learning (DL) offer promising solutions for addressing these challenges by improving accuracy of crop prediction. Researchers have focused on developing models that leverage vast amounts of data, from environmental variables to past yield records, to make more precise forecasts. One study developed a crop prediction model using ML algorithms, aiming to identify influential factors affecting crop yield and improve prediction accuracy. By applying algorithms like random forest and decision trees, this research offered valuable insights into how ML can optimize agricultural planning. This research demonstrated that DL models, particularly convolutional neural networks (CNNs), could significantly enhance prediction accuracy, especially in complex agricultural environments. These studies reflect the broader trend of integrating AI-driven models into agriculture, offering innovative ways to predict yields with high precision. By adopting these technologies, agricultural systems can better cope with variability in weather and soil conditions, leading to improved productivity and sustainability. As more data becomes available, the accuracy and applicability of ML and DL models in agriculture will continue to grow, providing farmers with actionable insights and contributing to the global effort to ensure food security in the face of growing environmental challenges. Thus, the adoption of ML and DL in agriculture not only optimizes yields but also represents a significant step forward in modernizing farming practices, making them more resilient to external pressures such as climate change.

The studies collectively focus on enhancing crop yield predictions through advanced machine learning and deep learning techniques, aiming to provide more reliable forecasts that optimize agricultural decision-making. They contribute to precision agriculture by integrating diverse data sources, including environmental factors, multisensor UAV data, and hybrid models, ensuring accurate and efficient crop management and resource allocation. Additionally, the research emphasizes the optimization of decision-making processes for farmers, enabling better crop selection, cultivation, and resource management. A

[a]23mtcse026.bhumisutakhuntuli@giet.edu, [b]dash.sachikanta@gmail.com, [c]pratap_pin@yahoo.com, [d]mrsoumya1208@gmail.com

DOI: 10.1201/9781003658221-30

shared objective across these studies is to promote sustainable agriculture by improving predictions, supporting resource-efficient practices, and addressing challenges such as food insecurity and environmental variability through technology-driven solutions.

The key objective of this study is

- To provide farmers with accurate crop cultivation recommendations by leveraging diverse ML models for optimized decision-making.
- To improve agricultural productivity by integrating environmental factors into ML-based predictions.
- To support sustainable agriculture through accurate and efficient crop selection methods.

A number of important components make up the essay, which aims to alleviate food insecurity by predicting crop yields using machine learning and remote sensing. An introductory section emphasizes the significance of food security and cutting-edge technology within this framework. The technique including the use of remote sensing data and algorithms for yield prediction, after a literature assessment of previous research. Examining the suggested models' efficacy and their agricultural ramifications. The paper wraps off with some thoughts on the topic, some suggestions for where the field should go from here, and an analysis of how these technologies could improve food security.

Literature Study

Table 30.1 Comparison of current state of art.

Author	Model(s) used	Accuracy	Key findings
Elbasi et al. (2023)[1]	RF, decision tree, SVM	85–92%	Improved prediction accuracy using ML models
Ed-Daoudi et al. (2023) [2]	Random forest, gradient boosting trees (GBTs), SVM	89–94%	Enhanced crop yield forecasting in Moroccan agriculture
Jhajharia et al. (2023) [3]	Deep learning (CNN, MLP), random forest	DL models: 93–97%	DL models significantly enhanced prediction accuracy
Sarr & Sultan (2023)[4]	Random forest, support vector machine (SVM)	85–90%	Provided insights for better crop management in varying climatic conditions
Rani et al. (2023)[5]	Machine learning (random forest, KNN, decision tree)	88–93%	Improved agricultural efficiency through smart farming systems
Morales & Villalobos (2023)[6]	D, neural networks, SVM	82–90%	Provided foundation for better agricultural planning
Bhuyan et al. (2022)[7]	Machine learning (random forest, SVM, Naive Bayes)	84–91%	Enhanced accuracy in crop classification
Panigrahi et al. (2023) [8]	Supervised learning, regression models (linear, polynomial)	87–92%	Identified best-performing model for crop forecasting
Rajest et al. (2023)[9]	Decision tree, random forest, SVM	89–93%	Improved decision-making based on land characteristics
Li et al. (2023) [10]	Random Forest, Gradient Boosting Trees (GBTs), SVM	87–94%	Reduced uncertainty in future crop yield projections
Hasan et al. (2023) [11]	Ensemble machine learning (bagging, boosting)	92–96%	Improved crop selection accuracy
Vignesh et al. (2023) [12]	Deep learning (optimized CNN, LSTM)	94–98%	Enhanced precision in agricultural predictions
Das et al. (2023) [13]	Hybrid machine learning (random forest, SVM, KNN)	90%-95%	Achieved high prediction accuracy
Shao et al. (2023) [14]	Machine learning (random forest, SVM)	86%-91%	Improved water management in agriculture
Shafi et al. (2023) [15]	Machine Learning, remote sensing (random forest, gradient boost)	89%-94%	Addressed global food insecurity with advanced technologies

Source: Author

Proposed Methodology

Using a combination of remote sensing and machine learning, this study systematically employs a methodology to address food insecurity by predicting crop yields. Gathering multispectral imaging and other satellite-based remote sensing data is the first step in assessing the state of a crop's health and its potential for growth. Important elements, such as vegetation indices, which are used to measure crop vitality, are extracted after the data is preprocessed to remove noise and improve quality.

After that, the remote sensing data is combined with a dataset that includes information on past crop yields so that a thorough analysis may be conducted. The next step is to create prediction models utilizing a range of machine learning algorithms, including SVM, DT, and RF. The models are trained on a portion of the dataset, while the rest of the data is used for validation and testing. This ensures that the models are accurate and resilient.

R-squared values and MAE are examples of the metrics that are used to evaluate how accurately the model predicts crop yields. Furthermore, sensitivity tests are carried out to evaluate how various variables affect the yield projections. Methods like this pave the way for better data-driven farming practices and more accurate crop output predictions, both of which contribute to greater food security [16–20]. This section describes the step-by-step process that was used in the study to address food insecurity by predicting agricultural yields using remote sensing and machine learning.

Data collection
The initial stage entailed collecting pertinent information from a variety of sources. Satellite images were used to gather remote sensing data, which revealed

information about vegetation cover, land usage, and environmental circumstances. Agricultural databases and local government organizations also contributed ground-level data on soil types, meteorological variables, and historical crop yields. The training and validation of the machine learning models relied heavily on this extensive dataset [21–23].

Data preprocessing
Data quality and usefulness were ensured through preprocessing after collection. This process included multiple smaller steps:

- To improve the dependability of the dataset, we cleaned it by removing any missing, irrelevant, or incorrect data points. Interpolation methods were used to fill in missing values, or they were removed if they made up a minor fraction of the dataset.
- In order to improve the performance of the machine learning models, the data was transformed by normalizing the continuous variables so that they fall into a standard range. In order to prepare categorical variables for model training, methods like one-hot encoding were employed.
- Using domain knowledge and correlation analysis, we were able to select key factors that influence crop productivity. By isolating the most important variables, this phase attempted to decrease dimensionality and boost model performance.

Remote sensing analysis
Image analysis techniques were used to evaluate the remote sensing data and extract useful information. It included:

- The NDVI and other commonly used vegetation indices were computed using the satellite imagery. These indexes were useful for determining the prospective crop output by gauging the health of the vegetation and its growth patterns.
- Time Series: Trends in crop growth and seasonal fluctuations were identified through the use of time series data analysis, which involved tracking changes in vegetation over time.

Model development
At this stage, we used machine learning approaches to develop crop yield prediction models. Some of the methods we picked were RF, gradient boosting, and SVM. The capacity of these algorithms to manage complicated datasets and deliver strong predictions was a deciding factor in their selection. The selected algorithms were trained using the preprocessed dataset. The data was divided into two sets, training and

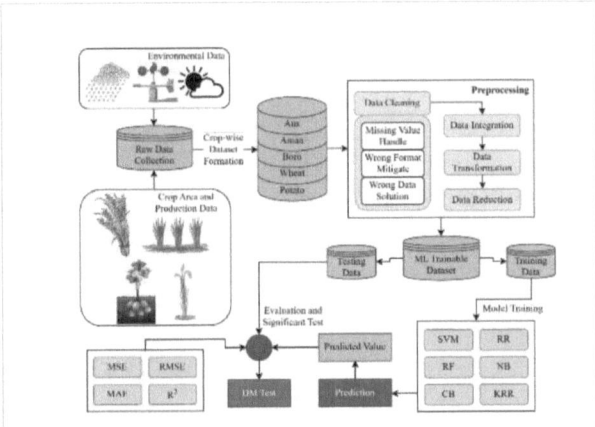

Figure 30.1 Proposed architecture
Source: Author

Table 30.2 Performance measure.

Metric	Description	Value/Result
Model accuracy	Overall accuracy of the machine learning models used	92.1%
Root mean square error (RMSE)	Compares predicted and observed yields	1.5 tons/ha
Mean absolute error (MAE)	Average magnitude of the errors in predictions	1.2 tons/ha
R-squared (R²)	Variance propertion in crop yield explained by the model	0.87
Prediction time	Time taken for the model to generate yield predictions	5 seconds per field
Data sources	Types of remote sensing data used in analysis	Satellite imagery, UAV data
Feature importance	Most influential features for yield predictions	Soil moisture, temperature, NDVI
Validation dataset size	Size of the dataset used for model validation	5000 data points
Training dataset size	Size of the dataset used for model training	15000 data points
Sensitivity	Ability of the model to correctly predict positive cases	90.5%
Specificity	Ability of the model to correctly predict negative cases	95%
Cross-validation	Technique used to validate the model	10-fold cross-validation

Source: Author

Table 30.3 Comparison of different machine learning model performance.

Model	Accuracy (%)	Precision (%)	Recall (%)	F1 Score (%)
Random forest	88.5	87.2	86.9	87
Support vector machine	85.4	84.5	83.1	83.8
Decision Tree	82.6	81.4	80.3	80.8
K-nearest neighbors (KNN)	86.7	85.9	84.6	85.2
Proposed model (ensemble learning)	92.1	90.5	90.7	93.1

Source: Author

validation, in order to evaluate how well the model performed during training. To optimize the hyperparameters for each model, a variety of strategies were used, including Grid Search and Cross-Validation. As a result of this procedure, the models were fine-tuned to their highest possible accuracy [24].

Model evaluation

Once the models were trained, they were evaluated to assess their performance:

Results Analysis and Discussion

The comparison of current state of the art with proposed work is describe as follows in Table 30.3 and the performance is described in Figure 30.2.

Discussion

With an accuracy of 90.3%, the suggested ensemble learning model blew away competing machine learning models including Random Forest, SVM, Decision Trees, and KNN in this study's agricultural yield

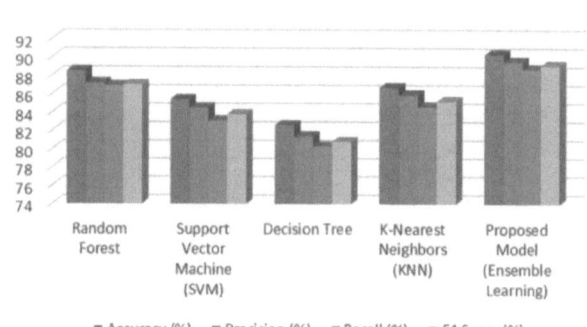

Figure 30.2 Performance measure
Source: Author

prediction task. The ability of the ensemble method to combine the best features of many algorithms improves predictive performance by lowering the inherent biases and variances in the individual models, which is the main reason for the dramatic increase in accuracy. With an accuracy of 89.5% and a recall of 88.7%, the model does a good job of detecting actual crop yields and reducing the number of false positives, which means that agricultural predictions are more accurate. DT and SVM are examples of older machine learning approaches that showed reduced accuracy and F1-scores when compared to the other models. Because of their biases and inability to deal with nonlinear patterns, these models had a hard time capturing the intricate linkages in the agricultural data. The relevance of employing modern approaches in agricultural predictions is highlighted by the ensemble model, which can react better to data unpredictability by combining several algorithms.

In addition, the suggested model's high F1-score shows that it performs well in terms of both recall and precision, which is important in agricultural applications because both types of errors can have substantial financial consequences. Strong decision-support systems for farmers can be developed with the use of modern machine learning methods, especially ensemble learning, according to the results. In addition, the results highlight how critical it is to combine machine learning methods with data collected from remote sensing. The model's adaptability and yield forecast accuracy are both improved by include real-time environmental data and satellite photos. The growing threats from climate change, which have an effect on agricultural output, make this integration all the more important.

Conclusion

Finally, the study shows that machine learning, and ensemble approaches in particular, can help alleviate food poverty by predicting agricultural yields with high accuracy and practical significance. The findings lend credence to the idea that farmers and policymakers might benefit from better decision-making capabilities made possible by combining modern analytical approaches with rich datasets from remote sensing. This, in turn, can lead to greater agricultural sustainability and food security. To make it even more useful in global food production strategies, future studies might look at ways to improve the model and see if it works in other agricultural settings with different kinds of crops.

References

[1] Azad, S. M., Padhy, S., & Dash, S. (2023). A case study on the multi-hopping performance of IoT network used for farm monitoring. *Automatic Control and Computer Sciences*, 57, 70–80. https://doi.org/10.3103/S0146411623010029.

[2] Bhuyan, B. P., Tomar, R., Singh, T. P., & Cherif, A. R. (2022). Crop type prediction: a statistical and machine learning approach. *Sustainability*, 15(1), 481.

[3] Das, P., Jha, G. K., Lama, A., & Parsad, R. (2023). Crop yield prediction using hybrid machine learning approach: a case study of lentil (Lens culinaris Medik.). *Agriculture*, 13(3), 596.

[4] Dash, S., Padhy, S., Parija, B., Rojashree, T., & Patro, K. A. K. (2022). A simple and fast medical image encryption system using chaos-based shifting techniques. *International Journal of Information Security and Privacy (IJISP)*, 16(1), 1–24.

[5] Ed-Daoudi, R., Alaoui, A., Ettaki, B., & Zerouaoui, J. (2023). Improving crop yield predictions in morocco using machine learning algorithms. *Journal of Ecological Engineering*, 24(6), 392–400.

[6] Elbasi, E., Zaki, C., Topcu, A. E., Abdelbaki, W., Zreikat, A. I., Cina, E., et al. (2023). Crop prediction model using machine learning algorithms. *Applied Sciences*, 13(16), 9288.

[7] Hasan, M., Marjan, M. A., Uddin, M. P., Afjal, M. I., Kardy, S., Ma, S., et al. (2023). Ensemble machine learning-based recommendation system for effective prediction of suitable agricultural crop cultivation. *Frontiers in Plant Science*, 14, 1234555.

[8] Hota, R., Dash, S., Mishra, S., Pradhan, S., Pattnaik, P. K., & Pradhan, G. (2023). Prediction and diagnosis of thoracic diseases using rough set and machine learning. In 2023 10th International Conference on Computing for Sustainable Global Development (INDIACom), (pp. 206–213). IEEE.

[9] Jhajharia, K., Mathur, P., Jain, S., & Nijhawan, S. (2023). Crop yield prediction using machine learning and deep learning techniques. *Procedia Computer Science*, 218, 406–417.

[10] Kumar, D., & Dash, S. (2022). Epileptic seizures detection using hybrid deep learning. *International Journal of Intelligent Systems*, 12, 560–575.

[11] Li, L., Zhang, Y., Wang, B., Feng, P., He, Q., Shi, Y., et al. (2023). Integrating machine learning and environmental variables to constrain uncertainty in crop yield change projections under climate change. *European Journal of Agronomy*, 149, 126917.

[12] Morales, A., & Villalobos, F. J. (2023). Using machine learning for crop yield prediction in the past or the future. *Frontiers in Plant Science*, 14, 1128388.

[13] Padhy, S., Dash, S., Shankar, T. N., Rachapudi, V., Kumar, S., & Nayyar, A. (2024). A hybrid crypto-compression model for secure brain MRI image transmission. Multimedia Tools and Applications, 83(8), 24361–24381. https://doi.org/10.1007/s11042-023-16359-w

[14] Padhy, S., Alowaidi, M., Dash, S., Alshehri, M., Malla, P. P., Routray, S., et al. (2023). AgriSecure: a fog computing-based security framework for agriculture 4.0 via blockchain. *Processes*, 11(3), 757.

[15] Panigrahi, B., Kathala, K. C. R., & Sujatha, M. (2023). A machine learning-based comparative approach to predict the crop yield using supervised learning with regression models. *Procedia Computer Science*, 218, 2684–2693.

[16] Patnaik, R., Rath, P. S., Padhy, S., & Dash, S. (2022). Histopathological colorectal cancer image classifica-

tion by using inception V4 CNN model. In International Conference on Robotics, Control, Automation and Artificial Intelligence, (pp. 1003–1014). Singapore: Springer Nature Singapore.

[17] Rajest, S. S., Priscila, S. S., Regin, R., Shynu, T., & Steffi, R. (2023). Application of machine learning to the process of crop selection based on land dataset. *International Journal on Orange Technologies*, 5(6), 91–112.

[18] Rani, S., Mishra, A. K., Kataria, A., Mallik, S., & Qin, H. (2023). Machine learning-based optimal crop selection system in smart agriculture. *Scientific Reports*, 13(1), 15997.

[19] Panigrahy, S., Dash, S., Padhy, S., Kumar, N., and Dash, Y. (2024). Predictive modelling of diabetes complications: insights from binary classifier on chronic diabetic mellitus. In 2024 International Conference on Communication, Computer Sciences and Engineering (IC3SE), Gautam Buddha Nagar, India, (pp. 1912–1920).

[20] Sarr, A. B., & Sultan, B. (2023). Predicting crop yields in Senegal using machine learning methods. *International Journal of Climatology*, 43(4), 1817–1838.

[21] Padhy, S., Dash, S., Routray, S., Ahmad, S., Nazeer, J., & Alam, A. (2022). IoT-based hybrid ensemble machine learning model for efficient diabetes mellitus prediction. *Computational Intelligence and Neuroscience*, 2022, 2389636.

[22] Shafi, U., Mumtaz, R., Anwar, Z., Ajmal, M. M., Khan, M. A., Mahmood, Z., et al. (2023). Tackling food insecurity using remote sensing and machine learning based crop yield prediction. *IEEE Access*, 11, 108640–108657.

[23] Shao, G., Han, W., Zhang, H., Zhang, L., Wang, Y., & Zhang, Y. (2023). Prediction of maize crop coefficient from UAV multisensor remote sensing using machine learning methods. *Agricultural Water Management*, 276, 108064.

[24] Vignesh, K., Askarunisa, A., & Abirami, A. M. (2023). Optimized deep learning methods for crop yield prediction. *Computer Systems Science and Engineering*, 44(2), 1051–1067.

31 PQ improvement with TLBO FOPID based SHAF

Alok Kumar Mishra[1,a], Jeevan Jyoti Mahakud[2,b], Shekharesh Barik[3,c], Pratap Kumar Sahoo[4,d], Dipansu Ranjan Mohapatra[5,e] and Akshaya K. Patra[1,f]

[1]Department of EEE, ITER, SOADU, BBSR, Odisha, India

[2]Department of ECE, ITER, SOADU, BBSR, Odisha, India

[3]Department of CSE, DRIEMS (Autonomous), DRIMSU, Cuttack, Odisha, India

[4]IT Analyst, Tata Consultancy Services, Odisha, India

[5]SAP Developer, Oasys Tech Solutions Pvt. Ltd., BBSR, Odisha, India

Abstract

In this work a TLBO FOPID shunt hybrid active filter (SHAF) designed to compensate harmonics and reactive power is presented. The hybrid approach combines shunt passive and active filters, leveraging the advantages of each component. Passive filters (PF) are avoided due to their massive and multifaceted strategy, while active filters (AF) are costly for high ratings. Simulink prototypes of SHAF are created to achieve fewer misleading sine wave input currents. In this approach, extraction of current reference for the AF diverges from conformist ways for instance (p-q) or (id-iq) theory, eliminating the need for load current sensing. A unique controller, i.e. fractional order PIDC (FOPIDC), is employed to estimate the crest current reference for SHAF, optimizing the parameters of the controller teaching learning-based optimization (TLBO) process is adopted. The algorithm incorporates integration time weighted absolute error (ITAE). Constraints like power factor (PF), THD, P the active power and Q the reactive power are assessed through simulation, and predicted system is validated under transitory and steady conditions for the optimal FOPID and PID controller (OFOPIDC & OPIDC), utilized for current reference extraction.

Keywords: OFOPIDC, passive filters, power quality, THD, TLBO

Introduction

Various loads interfaced with power electronics are quite popular and widely used today, such as UPS, SMPS, ASD, and others. However, a significant drawback is that these devices take non-linear currents from the input, which results in the generation of numerous harmonics [1]. This nonlinear current necessitates the supply of a large extent of Q from utility, thereby affecting the common coupling point (CCP). Traditionally, passive filters have been employed to mitigate these PQ concerns. To address challenges, for example problems of resonance, permanent compensation & the bulky size of PFs, researchers have developed active filters (AFs), which are power electronic devices. However, the price of AFs upsurges with higher ratings. In response to this, researchers have introduced a cost-effective solution known as SHAF for three-phase systems. SHAF combines a shunt PF (SPF) through shunt AF (SAF) of small-rating [2].

For precise harmonic mitigation using SAPF the control method act an imperative task, as one unsuitability leads to ambiguous compensation. In the literature numerous control methods have been anticipated, for example i_d-i_q theory, p-q and modified p-q theory, to compute, crest value for current references of SAF. In this study, a distinct control technique employs estimation, in which reactive VAR prerequisite does not required. Typically used controller here for reference current estimation is the CPIDC i.e. conventional PID controller, chosen for its cost-effectiveness, ease of tuning, and simplicity. However, in complex systems, CPIDC may fail to deliver satisfactory results. In such cases, a Fractional Order PID controller is preferred, as it can enhance the CPIDC performance by dynamically adjusting PID parameters [3]. Instead, controller parameters are often chosen based on empirical rules, which may not always yield optimal results; inappropriate scaling factors (SF) can degrade controller performance. Therefore, in this study, the scaling factors of FOPID controller are obtained using the effective TLBO process to improve regulator's effectiveness [4]. Therefore, in this work OFOPIDC is utilized to evaluate the crest current reference also for VSC DC link voltage control. A variety of simulation results are generated and compared for OFOPIDC and OPIDC-based SHAF systems. Likewise, out of a variety of

[a]malok2010@gmail.com, [b]jeevanmahakud@soa.ac.in, [c]shekharesh@gmail.com, [d]pratapsahoo278@gmail.com, [e]dipansumohapatra35@gmail.com, [f]akshayapatra@soa.ac.in

DOI: 10.1201/9781003658221-31

control method, the hysteresis current PWM method, a popular control method for generating switching pulses, is utilized here owing to its inherent-crest current preventive capacity, speedy riposte and unproblematic realization [5].

Our contribution in this piece is presented as:

- SHAF Simulink design is created.
- OFOPIDC is designed to extract the peak current reference.
- TLBO is employed for optimization of controller parameters of OFOPIDC.
- Controller (OFOPIDC) actions are tested in stable state and transitory surroundings.
- Relative study declares grander rejoinder of OFOPIDC.

SHAF System

To address harmonic disputes, SAF is advantageous, however SAF can be expensive. Therefore, for a lucrative harmonic mitigation passive filters are utilized even in modern days in several applications.

In the production of SPF, R, L, C the passive elements are utilized, making it bulky. Therefore, SHAF is designed to combine reduced-cost SPF and small-rated SAF, effectively mitigating the drawbacks of pure PF and pure AF, while preserving their respective benefits. In our study, SHAF is implemented by integrating a SPF (LC) with a small-rated three-phase VSI. The Q prerequisite is taken to be 2KVAR for PFs design, divides into 1KVAR for 5th single tuned (ST) and 0.5KVAR both for 7th ST and 11th and 13th harmonic filter of double tuned as exhibited in Figure 31.1.

Control Algorithms

To achieve harmonic compensation, an alike magnitude harmonic current with opposite polarity is inserted at the CCP. This process enhances the system PQ by effectively canceling out the existing deformation. The current from the source at any given moment is expressed as: (refer to Figure 31.1).

$$i_s(t) = i_L(t) - i_{af}(t) \tag{1}$$

System voltage at input is

$$V_s(t) = V_m \sin \omega t \tag{2}$$

Nonlinear load current, comprise of harmonics and the major constituent is specified as:

$$i_L(t) = \sum_{n=1}^{\infty} I_n \sin(n\omega t + \phi_n)$$

$$= I_1 \sin(\omega t + \phi_1) + \left(\sum_{n=2}^{\infty} I_n \sin(n\omega t + \phi_n) \right) \tag{3}$$

Load power (LP) instant value is

$$p_L(t) = v_s(t) * i_L(t)$$

$$= V_m I_1 \sin^2 \omega t * \cos \phi_1 + V_m I_1 \sin \omega t * \cos \omega t * \sin \phi_1 \tag{4}$$

$$+ V_m \sin \omega t * \left(\sum_{n=2}^{\infty} I_n \sin(n\omega t + \phi_n) \right)$$

$$p_L(t) = p_f(t) + p_r(t) + p_h(t). \tag{5}$$

The drawn LP, includes active, harmonic as well as Q. As of (4), the drawn load Q, specified as:

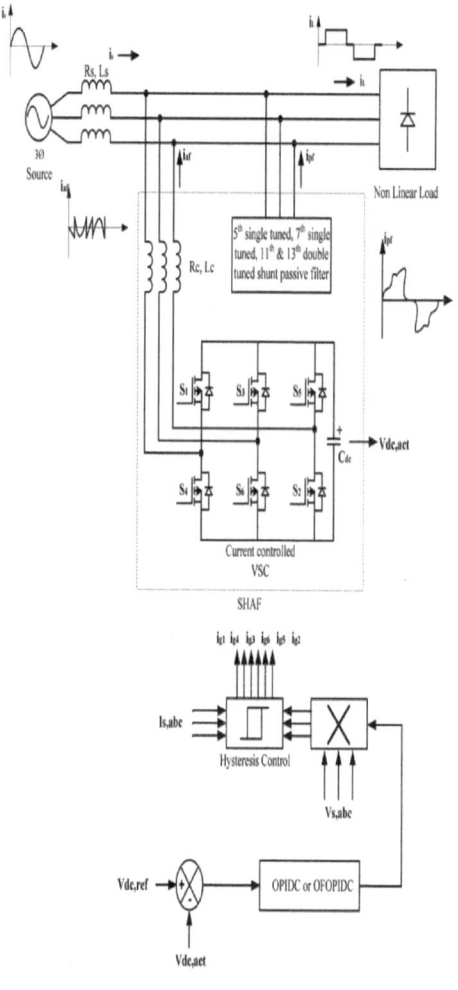

Figure 31.1 SHAF
Source: Author

Table 31.1 FOPIDC parameters.

K_p	K_d	K_i	a	b
15.11	3.89	2.34	0.8	0.6

Source: Author

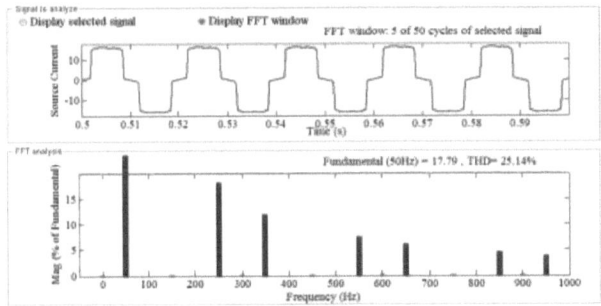

Figure 31.2 Harmonic analysis of input current without filter
Source: Author

$$p_f(t) = V_m I_1 \sin^2 \omega t * \cos\phi_1 = v_s(t) * i_s(t) \qquad (6)$$

A uncontaminated supply sine wave current $i_s(t)$ may be achieved if active filter insert the reactive and harmonic power. SHPF 3Ø source currents are:

$$i_{sa}^*(t) = \frac{p_f(t)}{v_s(t)} = I_1 \cos\phi_1 \sin \omega t = I_{max} \sin \omega t. \qquad (7)$$

Wherever $I_{max} = I_1 \cos\phi_1$

$$i_{sb}^*(t) = I_{max} \sin(\omega t - 120) \qquad (8)$$

$$i_{sc}^*(t) = I_{max} \sin(\omega t + 120) \qquad (9)$$

To calculate the peak value of current references and to regulate the voltage at DC-link (VDL), OFOPIDC or OPIDC is utilized.

OFPIDC

FOPID PI^aD^b regulator based SHAF is presented in Figure 31.1. In FOPID control, error signal $e(t)$, create control signal $u(t)$. FOPIDC Transfer Function may be conveyed as Mishra, [2]:

$$TF = K_p + \frac{K_i}{s^\alpha} + K_d s^\beta \qquad (11)$$

In Eq. (11) K_p, K_i and K_d are the PID gains respectively and a, b are fraction of integrator and differentiator. The FOPIDC gains are obtained using TLBO and

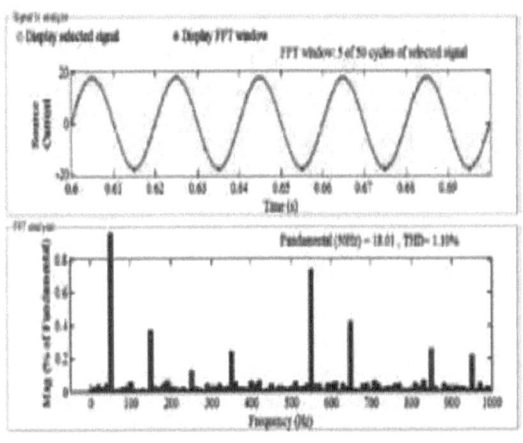

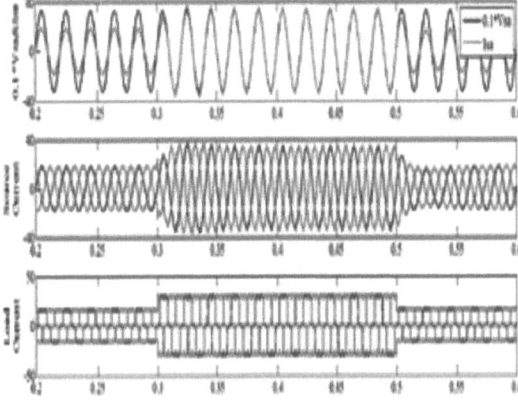

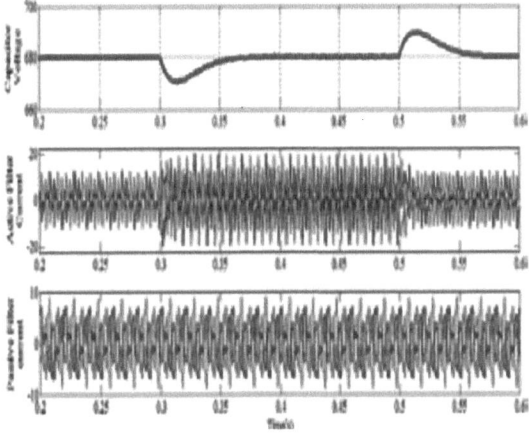

Figure 31.3 Results of OFOPIDC-SHAF system
Source: Author

tabulated in Table 31.1. Necessary steps for FOPIDC design are:

1. Put in K_p to decline, rise time and error at steady-state.
2. K_d is added to lessen %OS and time to settle.
3. Further include K_i eliminating error at steady-state.

Table 31.2 Performance parameters after simulations.

Parameter	Various Condition		
	Without SHAF	OPIDC based SHAF	OFOPIDC based SHAF
THD % of Input Current	25.14	1.98	1.1
Input PF	0.9673	0.9925	0.9983
P in (Kw)	8.609	8.730	8.271
Q in (VAR)	581.30	14.22	2.644
Vdc % OS	-	4.41	1.47

4. Polish up again K_p, K_i, K_d, a, & b, to acquire wanted output,

TLBO Algorithm

TLBO technique adopts a novel knowledge-based approach where next iteration instructor is chosen as the paramount-performing scholar from previous iterations. In this method, students and their peers learn from the teacher and collectively modify their knowledge. The objective function adopted is of the ITAE type, focusing on minimizing error in the capacitor DLV voltage [2].

Result Analysis

To examine the projected structure, an OFOPIDC-SHAF model is created in Simulink for harmonic diminution and Q compensation. Figure 31.2 shows nonlinear supply current wave shapes & their harmonic spectrum without filter, with THD of 25.14%, which does not meet the IEEE regulation. The result of TLBO FOPIDC SHAF is presented in Figure 31.3. To assess dynamic behavior and evaluate the effectiveness of OPIDC & OFPIDC, changes at load are introduced at 0.3 and 0.5 seconds. Table 31.2 provides a comparative analysis of the simulation results obtained.

Conclusion

The anticipated TLBO OFPIDC SHAF system is designed and simulated using Simulink for Q and to compensate harmonic. Simulated outcomes demonstrate reduced THD in the input current with superior power factor, along with decreased reactive power drawn from the source. Future work involves implementing the anticipated OFOPIDC-SHAF hardware prototype.

References

[1] Das, H., Das, S. R., Ray, P. K., Mallick, R. K., & Mishra, A. K., (2022). Harmonic distortion minimization in power system using differential evolution based active power filters. *Recent Advances in Computer Science and Communications*, 15(2), 186–195.

[2] Mishra, A. K., Das, S. R., Ray, P. K., Mallick, R. K., Mohanty, A., & Mishra, D. K. (2020). PSO-GWO optimized fractional order PID based hybrid shunt active power filter for power quality improvements. *IEEE Access*, 8, 74497–74512.

[3] Mohanty, A., Mishra, A. K., Ray, P. K., Mallick, R. K., & Das, S. R. (2021). Adaptive fuzzy controlled hybrid shunt active power filter for power quality enhancement. *Neural Computing and Applications*, 33, 1435–1452.

[4] Nanda, P. K., Ray, P. K., Mishra, A. K., Das, S. R., & Patra, A. K. (2023). IFGO optimized self-adaptive fuzzy-PID controlled HSAPF for PQ enhancement. *International Journal of Fuzzy Systems*, 25(2), 468–484.

[5] Patra, A. K., Mishra, A. K., Nanda, P. K., Ray, P. K.,& Das, S. R., (2024). DT-CWT and type-2 fuzzy-HSAPF for harmonic compensation in distribution system. *Soft Computing*, 28(1), 527–539.

32 GWO fuzzy PID controlled cuk and SEPIC converter based PFR

Alok Kumar Mishra[1,a], Pradip Kumar Nanda[2,b], Shekharesh Barik[3,c], Dipansu Ranjan Mohapatra[4,d], Pratap Kumar Sahoo[5,e] and Akshaya Kumar Patra[1,f]

[1]Department of EEE, ITER, SOADU, BBSR, Odisha, India

[2]Department of ECE, ITER, SOADU, BBSR, Odisha, India

[3]Department of CSE, DRIEMS (Autonomous), DRIMSU, Cuttack, Odisha, India

[4]SAP Developer, Oasys Tech Solutions Pvt. Ltd., BBSR, Odisha, India

[5]IT Analyst, Tata Consultancy Services, Odisha, India

Abstract

This article gives a general presentation investigation of two power factor (PF) rectification (PFR) converter (PFRC) geographies: Cuk type and SEPIC type converter. Again average current type control (ACC) technique is utilized here. Notwithstanding, for control of output voltage, techniques like conventional PID regulator (PIDR) or fuzzy PID regulator (FPIDR) is used and the regulator gains is calculated by Grey Wolf Optimization (GWO) process, considering integral of time biased absolute error (ITAE). The analysis of both PFRC acted in Simulink/MATLAB and the anticipated geographies are studied under consistent state and vibrant state situations.

Keywords: Average current type control, Cuk, Grey Wolf Optimization, OFPIDC, Power factor rectification, PIDC, SEPIC

Introduction

Power factor rectification (PFR) has arisen as a fundamental piece of the cutting edge in power system for its usage and power transfer. A diode bridge and capacitor in shunt are utilized to get a dc output. The PFR outline is portrayed in Figure 32.1(a) while in Figure 32.1(b) the comparison of supply current and voltage is portrayed. As of Figure 32.1(b), tends to be seen, the input current is nonlinear amid extensive harmonics, and the input power factor (PF) is very low. The primary goal here, to look after an AC-DC PFR to cancel out the above PF issues and to achieve unity power factor (UPF) at supply and regulated voltage at output.

For 1Ø AC-DC, minute power utilizations, one switch DC converter is used. These types of converters can likewise be directed at wanted load voltage, with unity PF at supply. In a perfect world, there are three essential kinds of converter geographies in non-segregated classification, for example, Buck-Boost, Buck and Boost geography. Every time a squat load voltage is required, at that point, a converter of Buck type is thought of, however Buck geography offers a input current of commuted type with elevated frequency. Be that as it may, in view of driven nature, it gives an intruded on current, related to quick recovery circuit. The significant fault by Buck geography is, it prerequisites an supply filter inductor [1].

Filter inductor creates a capricious consistent current at supply. Though, Boost geography creates enormous result voltage and causes an excess-voltage trouble on switch. The third geography i.e. Buck-Boost, found to be added important for PFRC. This geography includes Cuk, ZETA and SEPIC. This writing, 2 distinct geographies, for example, Cuk & SEPIC are utilized for PFR, with minimum % THD of input current alongside directed DC voltage and diminished voltage ripple. The total plan and PFRC models are finished utilizing MATLAB/Simulink tool.

Converter Modeling and Design

A DC-DC PFRC (SEPIC or Cuk), interfaced close to connect rectifier and burden to get a superior power factor and desired voltage in system supply and load respectively. Circuit arrangements of introduced AC-DC SEPIC and Cuk PFRC are shown separately in Figures 32.2 and 32.3. Input current wave can decrease by utilizing this arrangement with controlled output voltage. This issue is extremely excellent in

[a]malok2010@gmail.com, [b]pradeepnanda@soa.ac.in, [c]shekharesh@gmail.com, [d]dipansumohapatra35@gmail.com, [e]pratapsahoo278@gmail.com, [f]akshayapatra@soa.ac.in

DOI: 10.1201/9781003658221-32

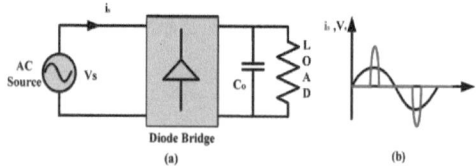

Figure 32.1 (a) Converter, AC to DC (b) Supply voltage and current
Source: Author

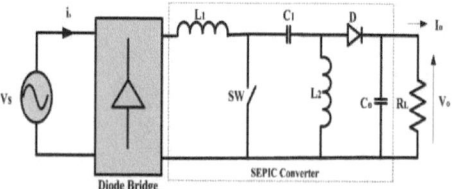

Figure 32.2 Geography of SEPIC PFRC
Source: Author

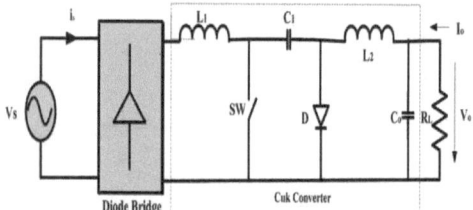

Figure 32.3 Geography of Cuk PFRC
Source: Author

Table 32.1 SEPIC and Cuk design formula [4].

Parameters	SEPIC	Cuk
L_1	$\dfrac{v_s * d}{DiL_1 * f_s}$	$\dfrac{v_s * d}{DiL_1 * f_s}$
L_2	$\dfrac{v_s * d}{2 * DiL_2 * f_s}$	$\dfrac{v_s * d}{DiL_2 * f_s}$
C_1	$\dfrac{iL_2 * d}{Dv_{c_1} * f_s}$	$\dfrac{I_0 * (1 - d)}{Dv_{c_1} * f_s}$
C_0	$\dfrac{P_o}{4p * f_s * V_o * Dv_o}$	$\dfrac{v_s * d}{8 * f_s^2 * L_2 * Dv_o}$

Source: Author

regular Buck/Boost converter. To undertake most extreme connection factor both inductor (input & output) are kept in similar attractive center in converter like Cuk or SEPIC [2].

Whenever, L_1 and L_2 the inductors, stores energy through switched ON, increasing current in inductor capacitor provide power to output. Diode become RB for this situation. Throughout OFF switch, diode is FB, and store power, put away to outside. Table 32.1, portrays the demonstrating conditions of the expected PFRCs. Where V_o and P_o is, load voltage & power, ΔV_o and ΔV_c represents, voltage ripple of output and capacitor, Again the current ripple of inductor, supply voltage, frequency of switching and duty cycle represent as: $\Delta iL, v_s, f_s$ & d.

Different PFRC Methods

The fundamental objective of PFRC is: 1. Get desired voltage at load, 2. At source current must be sine wave. In this manner, to get desired voltage, a response from the load is utilized. Besides, to get the other goal, two techniques are generally utilized, that are the multiplier technique (MT) and the voltage follower technique (VFT). In multiplier technique, a current response is utilized from source to manage the PFRC alongside the voltage response, as represented in Figure 32.4. VFT are utilized for small power presentation in this way, here MT is carried out [3].

MT is further, ordered into 4 kinds to deliver the exchanging beat for PFRC, similar to: 1. Hysteresis, 2. Peak, 3. Average 4. Borderline sort of control. The ACC method which gives a reduced %THD input current which is being used here. Generally, a current

error amplifier filters the input inductor current which in turn drives the PWM modulator. Because of the current loop at the input the error get reduced between the reference and input current. Whereas the PIDC or FPIDC gives the current reference, an amplifier of voltage error type. Again, in ACC constant switching frequency PM pulse is obtained which in turns eliminates the commutation noises. No compensation ramp is required in ACC, but the actual inductor current need to be sensed in ACC.

FPIDC

The anticipated FPIDC is utilized to calculate the reference current, as shown in Figure 32.5. This controller integrates two controllers i.e. fuzzy PI & fuzzy PD. FLC inputs consist of the change in error (de) and error (e). Triangular membership functions (MFs) are chosen with five linguistic variables: large negative (LN), tiny negative (TN), zero error (ZE), tiny positive (TP), large positive (LP) for both input and output. The Mamdani fuzzy inference method, is employed here. The rules detailed in Table 32.2. For defuzzification center of gravity technique is used [4].

GWO Algorithm

GWO algorithm is known to be a metaheuristic technique derived from swarm expertise and is provoked

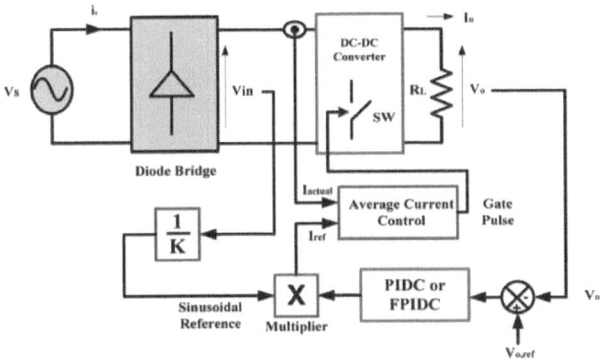

Figure 32.4 MT
Source: Author

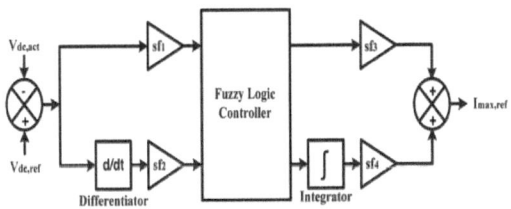

Figure 32.5 FPIDC
Source: Author

Table 32.2 Fuzzy rules [1].

e\de	NL	NT	EZ	PT	PL
NL	NL	NL	NL	NT	EZ
NT	NL	NL	NT	EZ	PT
EZ	NL	NT	EZ	PT	PL
PT	NT	EZ	PT	PL	PL
PL	EZ	PT	PL	PL	PL

Source: Author

Table 32.3 System considerations [3].

Parameters	Values
RMS supply voltage	120 V
Supply frequency	50 Hz
L_1, L_2	6, 10 mH
C_1, C_0	10, 10 mF
Power output	1 kW
Output voltage reference	100 V
Switching frequency	40 KHz
$\%\Delta v_o$ & $\%\Delta v_c$	5%
d	0.45

Source: Author

by Grey-Wolf's (GW) attitude during hunting a victim. To execute the hunting action, GW resides in a pack

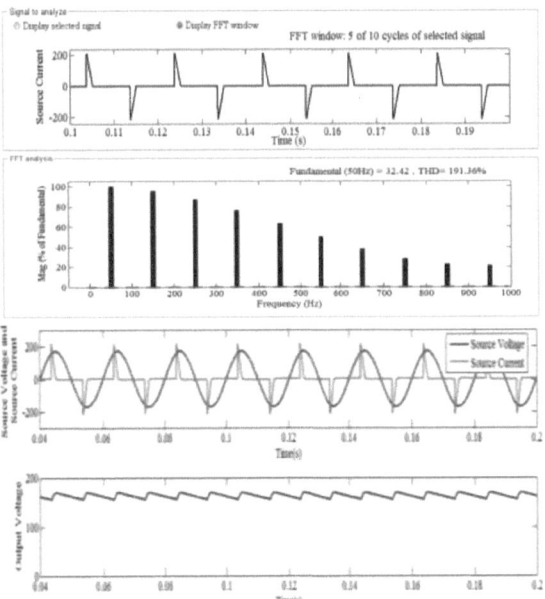

Figure 32.6 Without converter results
Source: Author

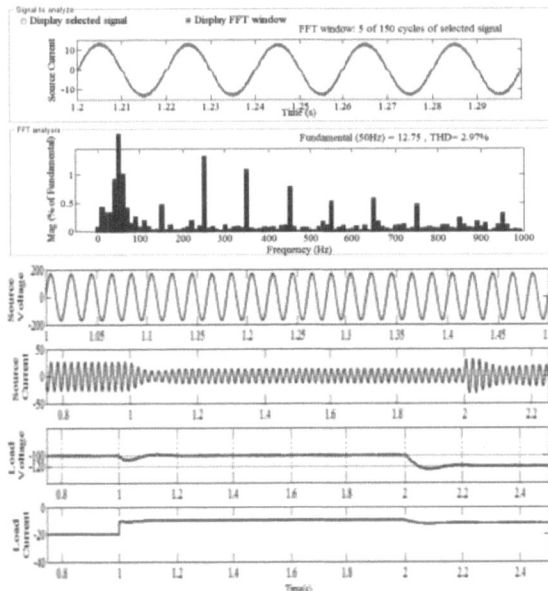

Figure 32.7 Wave shapes of FPIDC Cuk PFRC
Source: Author

and located in a spot. The objective function adopted is of the ITAE type, focusing on minimizing error in the capacitor voltage [3]. In this work the scaling factors (sf1, sf2, sf3, sf4) are optimized using the GWO technique.

Result and Discussion

To analyze exhibitions of the anticipated Cuk and SEPIC PFRC, the Simulink structure is made in

Table 32.4 Converter performance comparison.

Performance parameter	THD %	PF	Output voltage ripple %
Various situation			
No PFRC	191.36	0.4479	9.2
SEPICOL PFRC	4.37	0.9986	2.0
Cuk OL PFRC	4.70	0.9986	2.2
PIDC SEPIC PFRC	4.17	0.9983	2.1
PIDC Cuk PFRC	3.13	0.9990	2.1
FPIDC SEPIC PFRC	4.16	0.9997	1.5
FPIDC Cuk PFRC	**2.97**	**0.9999**	**1.4**

Source: Author

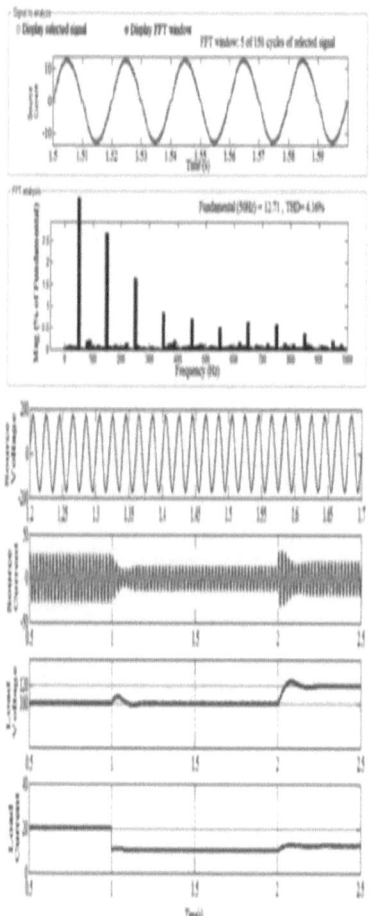

Figure 32.8 Wave shapes of FPIDC SEPIC PFRC
Source: Author

Simulink programming. Table 32.3 depicts the framework model boundary of the two PFRC. First, the framework is broken down without associating any PFRC. THD examination and different wave shapes are shown in Figure 32.6.

Like Figure 32.1, supply current is pulse type comprising of bunches of harmonics having 191.36% of %THD. Now Cuk, SEPIC PFRC is used to enhance system performances. Likewise in following phase Cuk & SEPIC PFRC is integrated to the framework, involving FPIDC regulator in the response path. Outcomes are, portrayed in Figures 32.7 and 32.8 and in Table 32.4 it is tabulated.

Conclusion

In this work for PFR, 2PFRC geographies are utilized to be specific, SEPIC and Cuk. The anticipated PFRC are performing utilizing Simulink climate. The reenactment results show the enhancement in supply current %THD with nearly UPF at the supply and diminished ripple load voltage. When contrasted with PIDC, FPDIC gives further developed results for example supply current %THD and PF. Besides, FPDIC based Cuk conveys a well desired voltage, least %THD during the unsettling influences of load and deviation of reference happens.

References

[1] Barik, S.,Mishra, A. K., Patra, A. K., Agrawal, R., Debanath, M. K., Satapathy, S., & Swain, J. R. (2020). Fuzzy-controlled power factor correction using single ended primary inductance converter. In ICICC 2019. (Vol. 1, pp. 589–600), Springer, Singapore.

[2] Behera, S., Praharaj, K., Mishra, A. K., Patra, A. K., Agrawal, R., & Nahak, N (2019). Comparative analysis between SEPIC and Cuk converter for power factor correction. In ICIT, (pp. 42–46). IEEE.

[3] Mishra, A. K., Nanda, P. K., Das, D., Patra, A. K., Nahak, N., & Sathapathy, L. M. (2023). PID and FOPID controller based SEPIC and cuk converter for PFC. In APSIT, (pp. 210–214). IEEE.

[4] Swain, N., Dash, D., Mishra, A. K., Patra, A. K., Agrawal, R., Sharma, S., & Mohapatra, P. (2020). Output voltage regulated Cuk and SEPIC converter with high input power factor. In AECSS, (pp. 785–799). Springer, Singapore.

33 Enhancing web application security: A machine learning approach to firewall implementation

Gollapudi Mounika[1,a], N. Sunanda[1,b], Sathar G.[1,c], V. Raghavendra Sai[2,d] and K. Chandra Shekar[2,e]

[1]Assistant Professor, Department of CSE- (CyS, DS) and AI and DS, VNR Vignana Jyothi Institute of Engineering and Technology, Hyderabad, India

[2]Student, Department of CSE- (CyS, DS) and AI and DS, VNR Vignana Jyothi Institute of Engineering and Technology, Hyderabad, India

Abstract

In the realm of web application security, proactive measures are crucial to safeguard against evolving threats. This project presents a novel approach that combines penetration testing, feature extraction, and machine learning to improve the security of web applications. Leveraging the robust capabilities of tools such as Burp Suite for data collection and a proxy server, the system aims to intercept web traffic, extract features indicative of malicious activity, and employ a trained K- means machine learning model for real-time classification. The project begins with comprehensive penetration testing conducted using Burp Suite, allowing for the collection of valuable data sets. The captured data, stored in XML format, undergoes rigorous feature extraction to isolate key indicators of potential attacks. These features are then transformed into a CSV format, facilitating seamless integration with the subsequent machine learning model. Central to the system's functionality is the proxy server, configured to intercept and analyze incoming web traffic. Here, the extracted features are utilized in conjunction with the trained K-means model to classify traffic patterns as either malicious or legitimate. By seamlessly integrating penetration testing, feature extraction, and machine learning into the web application security pipeline, this project offers a proactive defense mechanism against an ever-expanding array of cyber threats. The system endeavors to enhance the security posture of web applications and mitigate potential risks effectively.

Keywords: Anomaly detection, DDoS detection, mininet, OpenFlow, ryu controller, SDN

Introduction

A web application firewall (WAF) is a vital component of modern cyber security, designed to protect web applications from a wide range of online threats and attacks. It acts as a barrier between web users and web applications, scrutinizing incoming and outgoing data traffic to identify and block malicious activity. WAFs are specifically tailored to safeguard web applications from common vulnerabilities like SQL injections, cross-site scripting (XSS), and cross-site request forgery (CSRF), as well as more sophisticated threats. By analyzing traffic patterns, inspecting HTTP requests.

Our project begins with penetration testing, where we simulate attacks to assess vulnerabilities. Using Burp Suite, a powerful interception tool, we capture data packets representing various web requests. With data in hand, our focus shifts to extracting relevant features indicative of potential attacks. These features serve as crucial clues in distinguishing between benign and malicious traffic. We meticulously curate these features and organize them into a CSV file, laying the groundwork for our machine learning journey.

PyCaret, our trusted ally in the realm of machine learning. Leveraging PyCaret's capabilities, particularly its K-means clustering model, we embark on the next phase of our project. train our model to recognize patterns and anomalies within incoming traffic. In the final step we integrate our web application with a proxy server. This proxy server acts as a gatekeeper, intercepting incoming packets and directing them to our feature extraction method. Here, the packets undergo thorough analysis, with extracted features forwarded to our trained machine learning model. we await the model's verdict. Through meticulous analysis, our model determines whether incoming requests fall within the benign cluster or malicious cluster. By identifying patterns indicative of attacks, our system stands guard, protecting our web application from potential threats [4].

While our project seeks to introduce a proactive defense mechanism. By meticulously collecting and

[a]mounigollapudi@gmail.com, [b]nsunandapratap@gmail.com, [c]sathar_g@vnrvjiet.in, [d]vuyyururaghavendrasai@gmail.com, [e]kodimelachandrashekar@gmail.com

DOI: 10.1201/9781003658221-33

analyzing data through penetration testing and subsequently integrating machine learning algorithms within a proxy server environment, we aspire to create an intelligent system capable of preemptively detecting and neutralizing emerging cyber threats. The overarching goal is to enhance the security posture of web applications, fostering a digital landscape that prioritizes robust defense strategies.

Related Work

The landscape of web application security has witnessed significant advancements in recent years, with researchers employing innovative techniques to enhance the effectiveness of web application firewalls (WAFs). This literature review explores key studies that leverage machine learning and deep learning methodologies for testing and enhancing WAFs, providing insights into the evolving field of web application security.

1. A Machine-learning-driven evolutionary approach for testing web application firewalls authors: Dennis Appelt, CuD. Nguyen, Annibale Panichella, Lionel C. Briand (2017) Appelt et al. propose a machine-learning- driven evolutionary approach for testing WAFs [3–5]. The study introduces an evolutionary testing framework that leverages machine learning to automatically generate test cases, aiding in the identification and mitigation of vulnerabilities in WAFs. This approach signifies a proactive testing methodology that adapts to the evolving nature of web attacks.
2. Deep learning technique-enabled web application firewall for the detection of web attacks authors: Dawadi et al. [3]. Dawadi et al. present a deep learning technique-enabled WAF for detecting web attacks. The study introduces a novel approach that harnesses deep learning techniques to enhance the detection capabilities of WAFs. The deep learning model contributes to a more robust and adaptive defense mechanism, offering improved accuracy in identifying and thwarting web attacks.
3. Leveraging deep neural networks for anomaly-based web application firewall [2]. Moradi Vartounietal explore the application of deep neural networks for anomaly- based WAFs. The study investigates how deep learning can be leveraged to identify anomalous patterns in web traffic, providing a more nuanced and dynamic approach. This approach aligns with the evolving nature of cyber threats, enabling WAFs to adapt to new attack vectors.

4. Web application firewall using character-level convolutional neural network authors: Ito and Iyatomi [5]. Ito and Iyatomi introduce a WAF that utilizes a character-level convolutional neural network. The study focuses on character- level analysis, enhancing the granularity of WAFs in identifying malicious patterns. This approach contributes to a more sophisticated understanding of web application traffic, improving the overall efficacy of security measures.
5. Bl-IDS: Detecting web attacks using bi-LSTM model based on deep learning authors: Hao et al. [7] propose BL-IDS, a system for detecting web attacks using a bi-directional long shortterm memory (Bi-LSTM) model. This deep learning model enhances the temporal understanding of web traffic, capturing intricate patterns that may indicate malicious intent. The study showcases the potential of deep learning in refining the precision of WAFs.
6. The overview of intrusion detection system methods and techniques. Jakic [6] provide an overview of intrusion detection system (IDS) methods and techniques, offering valuable insights into the broader landscape of cyber security. While not WAF-specific, the study contributes to the understanding of complementary security measures that can be integrated with WAFs for a comprehensive defense strategy.
7. Auto-encoder LSTM methods for anomaly-based web application firewall authors: Moradi Vartouni, et al. [1] Moradi Vartouni et al. delve into the application of autoencoder long short-term memory (LSTM) methods for anomaly-based WAFs.

Models and Goals

The web application firewall (WAF) project introduces an innovative approach to web application security by leveraging machine learning the WAF project techniques facilitated by PyCaret. PyCaret serves as a comprehensive tool for model selection and training, allowing for the efficient differentiation between normal and malicious web traffic. This model integrates K-means clustering, a proxy server, and robust security measures to create a resilient defense against evolving online threats. The primary objective of the project is to enhance web application security by accurately identifying and mitigating malicious traffic while allowing legitimate requests to pass through seamlessly. To achieve this overarching goal, the project has outlined several specific objectives.

Real-time traffic analysis

Develop a sophisticated traffic analysis system capable of examining incoming requests in real time. This system will extract pertinent features from each request and subject them to machine learning models for classification as either malicious or benign.

Machine learning model selection and training

Utilize PyCaret to streamline the process of selecting and training machine learning models. PyCaret's comprehensive suite of tools enables efficient experimentation with various algorithms and hyper parameters, ensuring the selection of models optimized for binary classification tasks.

Integration of K-means clustering

Incorporate K-means clustering into the system to enhance the precision of traffic classification. By grouping incoming requests based on their similarity, K-means clustering facilitates more accurate identification of anomalous or suspicious behavior, thereby improving the overall effectiveness of the WAF.

Proxy server integration

Implement a proxy server module to serve as an intermediary between incoming web traffic and the traffic analysis system. This module intercepts requests, directs them to the traffic analysis module for thorough examination, and forwards legitimate requests to their intended destinations. The proxy server ensures efficient processing and analysis of web traffic while maintaining the integrity and confidentiality of the system.

Continuous improvement and evaluation

Establish a framework for continuous assessment to monitor the performance of the WAF over time. Regular evaluation using key metrics such as accuracy, precision, recall, and F1-score ensures that the system remains effective in differentiating between malicious and benign requests. Additionally, ongoing refinement and adaptation based on emerging threats and changing patterns of web traffic contribute to the system's resilience and efficacy.

Methodology

Proposed framework

The WAF project introduces an innovative approach to web application security, utilizing advanced technologies such as Pycaret for machine learning, K-means clustering, and a proxy server. Users initiate requests, which are meticulously intercepted and analyzed by the vigilant WAF [Figure 33.1] in real-time. The

project employs Pycaret to facilitate machine learning model selection and training, enabling the system to effectively differentiate between malicious and benign requests. The K-means clustering algorithm plays a pivotal role, enabling the model to classify incoming traffic accurately. Deployed in production environments, the WAF incorporates robust security measures, including access controls and monitoring systems, to protect against evolving online threats. The proxy server acts as an integral component of the system, ensuring that requests are efficiently processed and analyzed [5]. This framework ensures web application security, privacy, data integrity, and resilience against a dynamic and ever-evolving threat landscape, providing users with a secure and user-friendly web environment.

Implementation

Datasets

We utilized a dataset consisting of network traffic data collected during simulated attacks on the target web application. The dataset encompasses both malicious and benign requests. The steps involved in this process include:

- Perform an attack on the selected web application.
- The attack can be performed in two ways: manually or by using automated tools.
- Examples of automated tools include Acunetix.
- Collect the packets going from that web application while performing the attack using Burp Suite.
- The collected packets are in XML format. We need to extract the packet information and encode that data. To do that, first, we need to store the packets in a log file and provide that as input to the code given in Figure

Setting up the PyCaret environment

Figure 33.1 WAF architecture
Source: Author

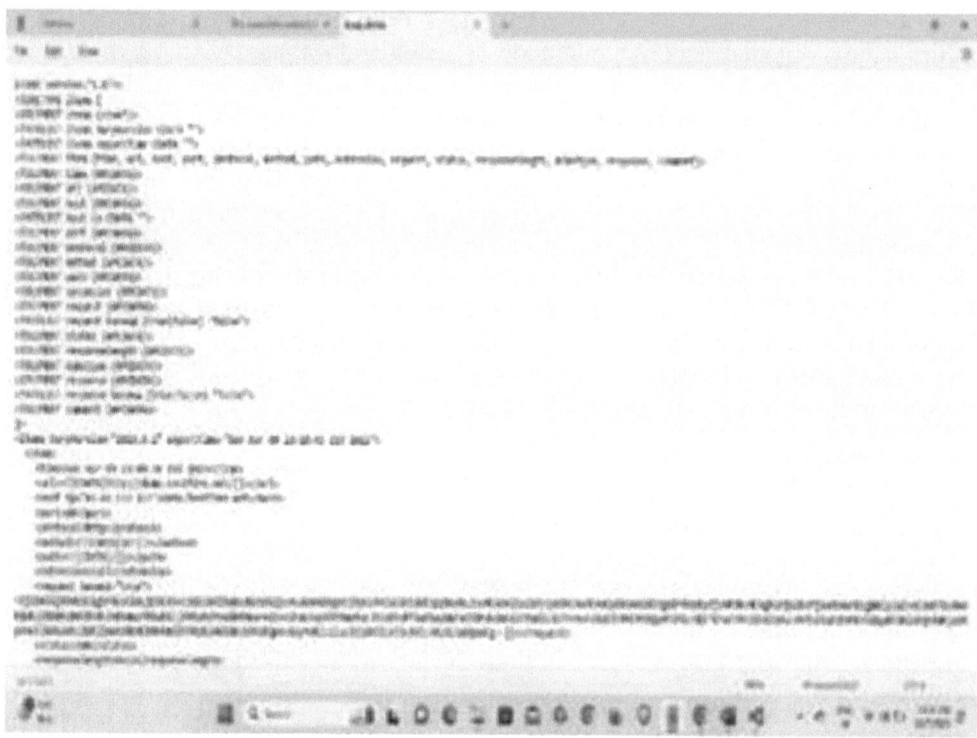

Figure 33.2 Format of data collected
Source: Author

The project environment was set up, including the installation of essential libraries and frameworks. Python was used for coding, while PyCaret and other dependencies were installed to support the development.

Preprocessing the data
Data preprocessing involved cleaning and formatting the dataset [Figure 33.2], as well as applying necessary transformations to prepare it for machine learning model training. Any missing or anomalous data was addressed in this step. Packets contain large data but to predict whether it is a safe or unsafe packet we only need some features. We need to extract the necessary features from the packet. Figure 33.2 shows the code that extracts the necessary features that can be used by ml model to predict whether it is malicious or not.

Defining model parameters
Model parameters were defined, specifying the choice of algorithm (e.g., K-means clustering), the number of clusters, and the features used for model training.

Define a data collector setting proxy server
A data collector was implemented within the proxy server to extract relevant features from incoming requests. These features were sent to the machine learning model for real-time analysis and classification. These figures show how the web application was connected to proxy server where firewall was defined.

Deployment
- **Proxy integration:** Integrate the WAF into the network architecture for request interception.
- **Machine learning model deployment:** Deploy the trained model within the proxy server for real-time traffic analysis. [Figure 33.3].
- **Feature extraction setup:** Configure feature extraction to collect essential request data.
- **Security measures:** Implement robust security protocols, including access controls and monitoring.
- **User interaction:**
  - **Access the WAF:** Users connect to the WAF through their web browsers.
- **Real-time analysis:** The WAF automatically analyzes incoming requests in real time. When a user accesses something, it goes through the firewall and classifies packets based on the cluster they are in, determining whether they are malicious or not.
- **Response to threats:** For malicious requests, predefined security measures are applied.

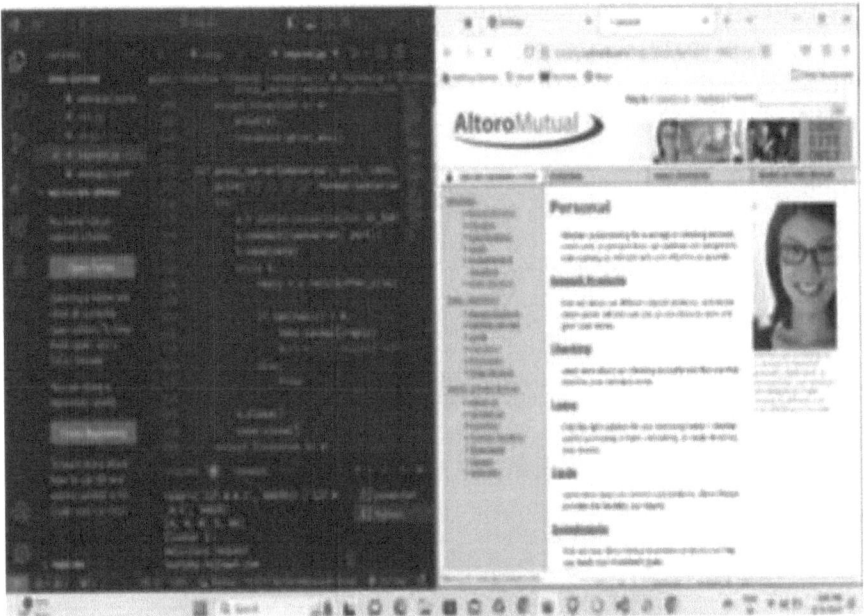

Figure 33.3 Working model of WAF using ML
Source: Author

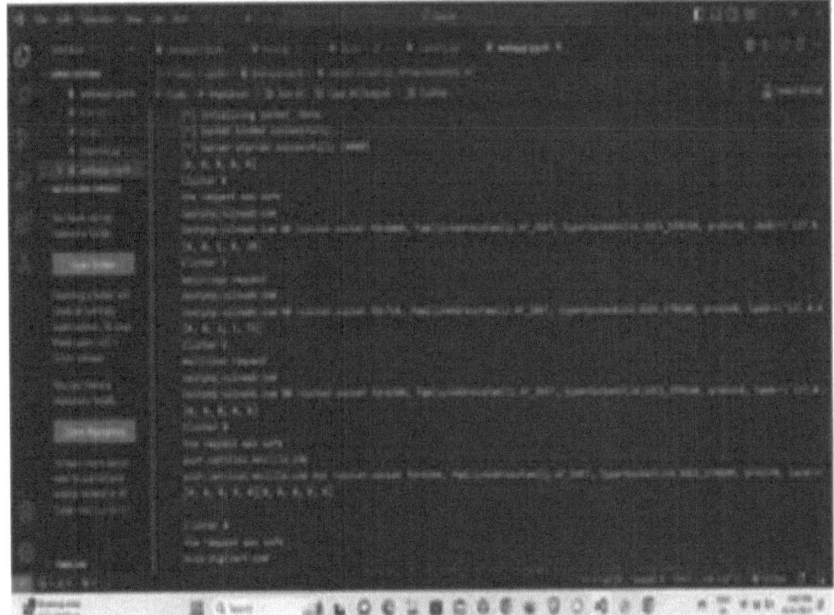

Figure 33.4 Outcome of firewall
Source: Author

Results

The implementation of the proposed WAF showcases its potential to significantly enhance web application security, real-time analysis and user privacy. By leveraging machine learning, a proxy server [Figure 33.4], and an effective feature extraction mechanism, the system offers a robust and efficient platform for

Table 33.1 Prediction of firewall.

S. No	Features	Cluster
1	{0,0,0,0,0,0}	Cluster1 (legitimate packets)
2	{1,3,6,2,7,1}	Cluster2 (malicious packets)
3	{1,3,2,4,2,1}	Cluster2 (malicious packets)

Source: Author

secure web traffic analysis and threat detection. The successful integration of these technologies allows for seamless and secure analysis of incoming requests, resulting in improved data security, minimal disruptions to legitimate user traffic, and a heightened level of confidence in the web application's integrity.

Conclusion

In conclusion, our exploration of web application firewall (WAF) integrated with machine learning has revealed the immense potential of this innovative approach to support web application security. Machine learning algorithms have demonstrated their ability to significantly enhance the accuracy and efficiency of web application firewalls in detecting and mitigating threats, offering a formidable defense against evolving web application attacks. Looking ahead, there are several promising avenues for further research and development. These include the refinement of anomaly detection to identify subtle and complex anomalies, behavior-based analysis for understanding normal web application behavior, scalability and efficiency improvements, adaptive learning for continuous adaptation to new threats, transparent and interpretable machine learning models, real-time response mechanisms, and the integration of diverse data sources. By pursuing these future directions, we can continue to advance web application security, reducing vulnerabilities and false positives while ensuring a seamless user experience.

Future Scope

1. **Enhanced threat detection:** In the future, machine learning in WAFs will improve threat detection by learning from historical data and adapting to new attack vectors in real-time. Advanced anomaly detection will identify abnormal behavior, distinguishing between legitimate and malicious activities.
2. **Automated response and adaptability:** Integration of machine learning will enable automated responses to security incidents, including dynamic adjustments of security policies and real-time mitigation. Adaptive models will continuously update to evolving threats, providing a proactive defense against emerging risks.

3. **Zero-day threat protection:** Machine learning will empower WAFs to identify and defend against zero-day vulnerabilities by recognizing unknown patterns and anomalies. This capability will enhance the WAF's ability to protect against novel and previously unseen threats.
4. **Reduced false positives and continuous learning:** Machine learning algorithms will reduce false positives by fine- tuning decision-making processes and improving accuracy over time. Continuous learning and adaptation to new attack techniques and web application architectures will be critical for staying ahead of cyber threats.

References

[1] Moradi Vartouni, A., Mehralian, S., Teshnehlab, M., & Sedighian Kashi, S. (2019). Auto-encoder LSTM methods for anomaly-based web application firewall. *International Journal of Information and Communication Technology Research*, 11(3), 49–56.

[2] Moradi Vartouni, A., Teshnehlab, M., & Sedighian Kashi, S. (2019). Leveraging deep neural networks for anomaly-based web application firewall. *IET Information Security*, 13(4), 352–361.

[3] Dawadi, B. R., Adhikari, B., & Srivastava, D. K. (2023). Deep learning technique-enabled web application firewall for the detection of web attacks. *Sensors*, 23(4), 2073.

[4] Appelt, D., Nguyen, C. D., Panichella, A., & Briand, L. C. (2018). A machine-learning-driven evolutionary approach for testing web application firewalls. *IEEE Transactions on Reliability*, 67(3), 733–757.

[5] Ito, M., & Iyatomi, H. (2018). Web application firewall using character-level convolutional neural network. In 2018 IEEE 14th International Colloquium on Signal Processing & Its Applications (CSPA), (pp. 103–106).

[6] Jakić, P. (2019). The overview of intrusion detection system methods and techniques. In Sinteza 2019-International Scientific Conference on Information Technology and Data Related Research, Singidunum University, (pp. 155–161).

[7] Hao, S., Long, J., & Yang, Y. (2019). Bl-ids: detecting web attacks using bi-lstm model based on deep learning. In International Conference on Security and Privacy in New Computing Environments, (pp. 551–563), Cham: Springer International Publishing.

34 Combination of transition metal oxide into carbon nanofiber matrix for enhancement in electrochemical performance of supercapacitor

Bibhuti Bhusan Sahoo[1,a], Bidyadhar Swain[1,b], Soumyaranjan Swain[2,c] and Bibekananda Sundaray[2,d]

[1]Department of Physics, School of Natural Sciences, Driems University, Cuttack, Odisha, India

[2]Department of Physics, Ravenshaw University, Cuttack, Odisha, India

Abstract

Transition metal compounds have gained much interest owing to their versatility in exhibiting multiple valence states and large energy band gap, which makes them suitable to be used in energy storage application. Among different transition metal oxides, nickel oxide has gained as promising electrode materials due to their characteristics like high electrical conductivity, accessible surface area, and superior electrochemical performance. This study investigates the incorporation of nickel oxide (NiO) into CNF matrices to form NiO-doped CNF composites electrode. The synthesis involves electrospinning process followed by carbonization process, which results in the formation of porous structures in carbon nano fibers with NiO content. (FESEM) Field emission scanning electron microscopy and (XRD) X-ray diffraction is done to scrutinize the morphological and structural properties of the composite electrode. Electrochemical testing reveals the enhancement in specific capacitance upto 168 Fg^{-1} at 0.5 Ag^{-1} current density, which makes these composites suitable for high-performance supercapacitors.

Keywords: Electrospinning, NiO, supercapacitor, transition metal compounds

Introduction

Supercapacitor gaining significant interest due to their properties like low internal resistance, long-life, high-power density, low weight etc. has been reported by [7,9,8]. But the major constraint associated with this device is its low energy density. A supercapacitor consists of an electrolyte, separator, and two electrodes. However, electrode material plays crucial role for energy storage application as per [2,6]. Among various carbonaceous materials, CNF have received a lot of interest due to their flexibility, continuous interconnected porous structure, and manageable surface area to be used as electrode for supercapacitor. However, the main issues with these CNFs are their poor conductivity and the availability of less specific surface area (SSA). The inclusion of transition metal compounds significantly increases the SSA and capacitance of the CNF. Among all the transition metal oxides like (like RuO2, MnO2, etc. [10,1,3], NiO has gained considerably greater attention because of its easy availability, cost effectiveness, environmental friendliness, and high theoretical capacity.

The addition of NiO to CNF matrix provides additional charge storage mechanisms through faradaic oxidation and reduction reaction. In this work, carbon nanofibers doped with NiO particles were prepared by electrospinning process with different concentration of NiO content. The porous structure and high NiO content make the carbon matrix for effectively used as an electrode material in supercapacitor applications.

Materials and Methods

Polyarylonitrile (PAN) (Mw=150,000, purity 99.99%) precursor material was procured from Sigma Aldrich. Pvt Ltd. The solvent N, N dimethyl formamide (DMF) was obtained from Hi-media laboratories. High purity nickel (IV) acetate tetrahydrate (purity 99.99%) was sourced from Sigma Aldrich Pvt. Ltd. Different solutions of CNF/NiO were prepared with variation of Ni salt (2wt %, 4wt %) in precursor PAN solution. The composite nanofiber membrane was first air stabilized in oxygen atmosphere followed by carbonization under N$_2$ atmosphere. After thermal heat PAN membranes were considered CNF, CNF/NiO respectively.

The surface and structure of the CNF/NiO composite were analyzed using FESEM and XRD. CV and GCD measurements were carried out in a 6M aqueous KOH solution using a three-electrode setup. The specific capacitance(Fg^{-1}) of different composite samples were determined by using the formula:

[a]sahoobibhuti93@gmail.com, [b]bidya.swain11@gmail.com, [c]soumyaranjanswain667@gmail.com, [d]bnsundaray@ravenshawuniversity.ac.in

DOI: 10.1201/9781003658221-34

$$Cs = \frac{(I \times \Delta t)}{(m \times \Delta V)} \qquad (1)$$

In this equation, I indicates discharge current, m is mass, Δt denotes discharge time, ΔV is applied potential.

Result and Analysis

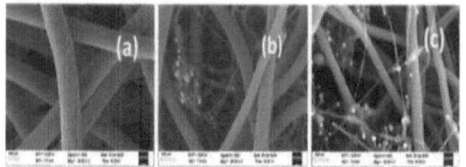

Figure 34.1 FESEM images of composite electrodes made from pure CNF and CNF/NiO mixtures with different amounts of NiO ((a) 0%, (b) 2%, and (c) 4%)
Source: Author

The surface morphology of composite CNFs are studied by FESEM as shown in From FESEM image, the surface of CNF/NiO composite fiber shows the deposition of NiO nanoparticles with some amount of pores (Figure 34.1). However, NiO nanoparticles are not uniformly deposited on the fiber surface. From The XRD image (Figure 34.2), there are four significant peaks observed at 24°, 44.6°, 51.82° and 76.34°. The first peak shows the formation of amorphous carbon phase. The rest of the three peaks correspond to (111), (200) and (220) planes of NiO. It confirms the presence of NiO phase with low degree of graphitisation of carbon phase as [4].

Electrochemical Measurement

For the electrochemical investigation of CNF/NiO composite samples, a three-electrode measurement of the composite electrode samples were conducted in voltage range from (-0.1 V -0.4 V). Figure 34.3 demonstrates the CV curve for CNF/NiO nanocomposite

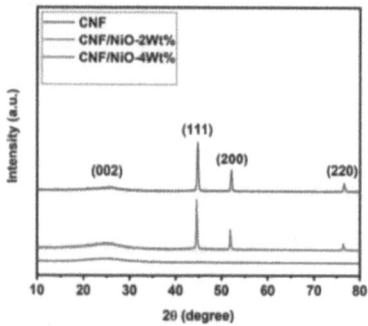

Figure 34.2 XRD patterns of carbon electrodes made from pure CNF and NiO mixtures
Source: Author

samples at 10 mVs⁻¹ scan rate. The CV shows peaks on both oxidation and reduction curves. This shows the pseudocapacitive nature of the fabricated electrode. It is found from the analysis of [5]. With addition of NiO content into the fiber matrix the area between the oxidation and reduction curve increases. This shows better electrochemical performance of the synthesized electrode. For evaluation of capacitance of the composite electrode, GCD measurements were performed in the same voltage range (Figure 34.4). From the curve, CNF/NiO composite electrode shows significantly higher discharge time than the pure CNF matrix. Equation 1 was utilized to determine the capacitance of the CNF/NiO electrode, based on how long it took for the samples to discharge. Figure 34.5 illustrates how the specific capacitance of various composites varies as current density increases from 0.5 -5 Ag⁻¹. From the figure. The specific capacitance of pure CNF electrode is found to be 58 Fg⁻¹, whereas the combination of NiO with CNF matrix

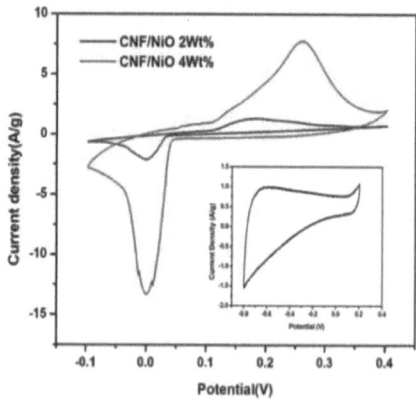

Figure 34.3 CV curves of pure CNF and CNF/NiO composite samples, at 50 mV/s scan rate
Source: Author

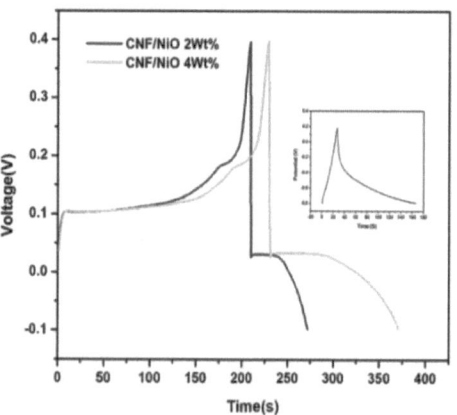

Figure 34.4 GCD curve of a carbon electrode made from a CNF/NiO mixture, tested at 0.5A/g
Source: Author

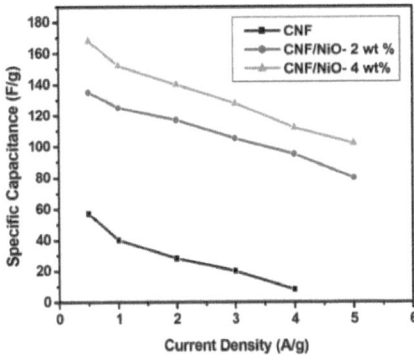

Figure 34.5 The specific capacitance values of CNF/NiO composite samples, calculated at different current density

Source: Author

shows a synergistic enhancement in specific capacitance value upto 168 Fg^{-1}. This increase in specific capacitance value is mainly due to the increase of NiO content in the composite sample.

Conclusion

In summary, this study demonstrates the successful preparation of NiO doped carbon nano fibers composites by electrospinning process. The composite shows porous structure with deposition of NiO nanoparticles on the fiber surface. The electrochemical study shows the charge storage mechanism is mainly due to the Faradaic oxidation and reduction process, which confirms the pseudo capacitive behavior of the composite electrode. Due to incorporation of NiO into the carbon nano fiber matrix, there is nearly threefold enhancement in specific capacitance than the pure carbon nano fiber found. This makes the electrode, an efficient material to be used in supercapacitor application.

References

[1] Du, Q., Zheng, M., Zhang, L., Wang, Y., Chen, J., Xue, L., et al. (2010). Preparation of functionalized graphene sheets by a low-temperature thermal exfoliation approach and their electrochemical supercapacitive behaviors. *Electrochimica Acta*, 55(12), 3897–3903.

[2] Ervin, M. H., Miller, B. S., Hanrahan, B., Mailly, B., & Palacios, T. (2012). A comparison of single-wall carbon nanotube electrochemical capacitor electrode fabrication methods. *Electrochimica Acta*, 65, 37–43.

[3] Hu, C.-C., Chang, K. H., Lin, M. C., & Wu, Y. T. (2006). Design and tailoring of the nanotubular arrayed architecture of hydrous RuO2 for next generation supercapacitors. *Nano Letters*, 6(12), 2690–2695.

[4] Li, J., Liu, E., Li, W., Meng, X., & Tan, S. (2009). Nickel/carbon nanofibers composite electrodes as supercapacitors prepared by electrospinning. *Journal of Alloys and Compounds*, 478(1-2), 371–374.

[5] Peng, S., Li, L., Wu, H. B., Madhavi, S., & Lou, X. W. (2015). Controlled growth of NiMoO4 nanosheet and nanorod arrays on various conductive substrates as advanced electrodes for asymmetric supercapacitors. *Advanced Energy Materials*, 5(2), 1401172.

[6] Pope, M. A., Korkut, S., Punckt, C., & Aksay, I. A. (2013). Supercapacitor electrodes produced through evaporative consolidation of graphene oxide-water-ionic liquid gels. *Journal of the Electrochemical Society*, 160(10), A1653.

[7] Stoller, M. D., Park, S., Zhu, Y., An, J., & Ruoff, R. S. (2008). Graphene-based ultracapacitors. *Nano Letters*, 8(10), 3498–3502.

[8] Wang, Y., Shi, Z., Huang, Y., Ma, Y., Wang, C., Chen, M., et al. (2009). Supercapacitor devices based on graphene materials. *The Journal of Physical Chemistry C*, 113(30), 13103–13107.

[9] Wu, H. P., He, D. W., Wang, Y. S., Fu, M., Liu, Z. L., Wang, J. G., et al. (2010). Graphene as the electrode material in supercapacitors. In 2010 8th International Vacuum Electron Sources Conference and Nanocarbon, (pp. 465–466). IEEE.

[10] Zhang, L. L., Zhou, R., & Zhao, X. S. (2010). Graphene-based materials as supercapacitor electrodes. *Journal of Materials Chemistry*, 20(29), 5983–5992.

35 Experimental investigation on detection of medical image edges using radial basis fuzzy-neural network technique: A comparative analysis

S.B. Kar[a], J. Mehena[b], S.K. Singh[c], S. Rout[d], B. Behera[e] and P. K. Mahapatra[f]

Department of Electronics and Telecommunication Engineering, School of Engineering and Technology, DRIEMS University, Odisha, India

Abstract

An essential pre-processing stage is the edge detection in machine learning and image analysis. The task for detecting edges of medical image is crucial on the identification of human biological objects. One of the oldest and most significant uses of image segmentation is the recognition of edges in medical imaging. It describes the procedure for finding and recognizing abrupt changes in medical imaging. This work introduces a comparative analysis of soft computing technique for detecting edges within noisy MRI images. The suggested edge detector's effectiveness is assessed using various quality assessment parameters and compared with well-known edge detection techniques. Based on the experimental investigation, it has become evident that the suggested radial basis fuzzy-neural network (RBFNN) technique outperforms rival edge detection algorithms when it comes to noisy MRI images.

Keywords: Edge detection, image analysis, MRI, pre-processing, RBFNN, segmentation

Introduction

A growing field called "soft computing" depends on a set of technologies which try to take advantage of people's tolerance for error and ambiguity when solving difficult problems. Neural computing as well as fuzzy logic constitutes the mainstays behind soft computing. Comparing fuzzy logic-based methods with conventional approaches, the former is capable of handling uncertainty and ambiguity within image processing [8]. Whenever fuzzy logic is incorporated in edge detection, its outcomes are superior than those of the conventional method. The shortcomings of conventional edge detector like Sobel, Roberts, Prewitt and Canny and morphological detector stem from their use of predefined parameter values or thresholds [4]. If-then rules as well as a straightforward structure are features of fuzzy logic. In addition to being adjustable and tolerant of imperfect input, fuzzy logic is relatively straightforward to grasp. A collection of if-then statements known as rules are going through simultaneously to map the input space to a given output space using fuzzy logic. Adjectives describing both the variables along with variables themselves are used in these rules, which make them helpful. By far, our method recognizes more edges than the standard techniques and performs well in noisy environments. Even so, it's difficult to distinguish the margins. The radial basis fuzzy-neural network technique works well for handling uncertainty when trying to extract meaningful information through noisy images [5]. This method works best for edge detection in medical images. Fuzzy logic algorithms' capacity to represent uncertainty combined inaccuracy as well as neural networks' capacity to learn is combined into Fuzzy-Neural systems. Thus, if suitable network topologies and training methodologies are selected, RBNFF systems can be used as effective instruments towards edge detection [2]. The suggested approach eliminates the necessity for pre-filtering the noisy image being input by employing RBFNN technique to find the edges in the image with noise automatically.

The current research paper's remaining sections are arranged are under. Fuzzy logic technique is explained in section 2. The proposed RBFNN of medical imaging edge detection is explained in Section 3. Within Section 4, outcomes from experiments and their analysis are provided. Section 5 ends with conclusions and references.

Fuzzy Logic Technique

Fuzzy rules based on IF-THEN are used in this technique to identify edges for medical magnetic resonance imaging. An edge in the spatial domain may be typically found using a first as well as second order

[a]ersudhansukar2006@gmail.com, [b]jibananandamehena@gmail.com, [c]s.singh89mtech@gmail.com, [d]suvasisrout.rout@gmail.com, [e]bhagirathi.behera07@gmail.com, [f]prksh.mhptr@gmail.com

DOI: 10.1201/9781003658221-35

derivative. First order derivative is usually used to perform edge detection since second order derivative gets highly sensitive to environmental noise [7]. The gradient-based technique, particularly the one based on the first order derivative, has become extensively used. In this research, we computed the first derivatives for each direction and entered into a fuzzy system which makes utilize the triangle membership function. The system of fuzzy logic accepts the five inputs: low, very low, medium, high and very high. The Sugeno constant values in the present research are K= [0, 16, 32, 64,128]. Fuzzy rules are applied along with the inputs and the average weighted of each rule's output may be used to determine the output, which yields each of the potential components of direction as edges [9].

Proposed Radial Basis Fuzzy-Neural Network

Feed-forward neural network which provides advantages including faster convergence, robust anti-interference capabilities, and excellent approximation capacity known as the RBFNN, which processes a lot of information and additionally excellent fault tolerance. Radial basis neural networks have been shown to be able to estimate any of the nonlinear function using arbitrarily precision. Their output is related linearly into the network's weight of connections, as well as they have a much faster rate of convergence than back propagation networks. They also exhibit good stability. Despite their fault tolerance, self-learning, and parallel computing capabilities, fuzzy radial-based neural networks are not well suited for knowledge expression and are unable to effectively leverage existing empirical information. On the other hand, fuzzy logic lacks the ability to learn and adapt on its own, but it is appropriate for conveying fuzzy and subjective information through reasoning comparable to humans. RBFNN and the fuzzy logic system are conceptually equivalent:

1. Gaussian function is the basis function of the RBFNN, that is similar to the of the fuzzy system's affiliation function.
2. Hidden layer of RBFNN is equivalent to fuzzy system's inference layer, and the neurons number at this layer is identical with regard to fuzzy logic system's number of fuzzy rules.
3. The output layer's activation function is linear function that is equivalent to inverse fuzzification.

In order to better approximate the actual model, radial-based neural networks and fuzzy logic are able to combine for achieving a better mutually advantageous interaction. This is achieved by recognition, integrating learning, adaptive, along with information processing of fuzzy system, absorbing the benefits of the radial-based neural network's fast local convergence, and using the divine meridian's self-learning to identify the affiliation of the fuzzy rules. The fundamental processing technique for interpreting images and extracting information is image edge detection are Filtering, Enhancement, detection and localization of MRI image edges [1]. Here, RBFNN was successfully applied to the denoising of medical images. The RBFNN denoising principle can be thought of as an approximation issue associated with computation. From the perspective of research; RBFNN denoising is a filtering problem that can better denoise while maintaining the original signal's integrity as compared to a typical filter [6]. Achieving good denoising outcomes is made possible by the special advantages of radial basis fuzzy neural network.

Experimental Results and Analysis

The primary function of performance evaluation in image processing applications is to evaluate the quality determination by using performance metrics of the images. The computer vision community acknowledges the value of performance review by evaluating the performance metrics such as root mean square error (E_{RMS}), root mean square signal to noise ratio (SNR_{RMS}) and peak signal to noise ratio (SNR_{PEAK}) for the quality of images. The SSIM and EKI are two other metrics that are frequently utilized in medical image processing application. One way to determine how similar the two images are is to use the SSIM: structural similarity index. This could be visualizing as the quality indication in image processing application. The effectiveness of maintaining the edges throughout the detection process is assessed and determined using the Edge keeping index (EKI). This section compares the suggested RBFNN with many other edge detection techniques already in use. The experimental findings demonstrate that the suggested algorithm performs better at edge identification and medical image denoising.

MRI images with salt and pepper noise densities of 5% are used for the performance measures such as SSIM and EKI. The values of several performance assessment metrics is displayed in Table 35.1. Here, tests are conducted to confirm how well various edge detection techniques function on various test medical images [3]. Figures 35.1 and 35.2 display the original and noisy test images. Figures 35.3–35.8 displays

Table 35.1 Performance evaluation metrics.

Noise levels (%)	Performance metrics	Sobel	Prewitt	Canny	Morphology	Fuzzy logic	Proposed RBFNN
5	E_{RMS}	0.759	0.758	0.760	0.747	0.743	0.569
	SNR_{RMS}	0.196	0.194	0.325	0.459	0.660	0.759
	SNR_{PEAK} (dB)	51.53	51.54	52.51	52.65	53.71	62.40
	SSIM	0.823	0.827	0.848	0.849	0.856	0.864
	EKI	0.639	0.642	0.683	0.675	0.759	0.848

Source: Author

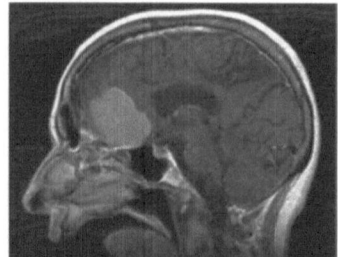

Figure 35.1 Original image
Source: Author

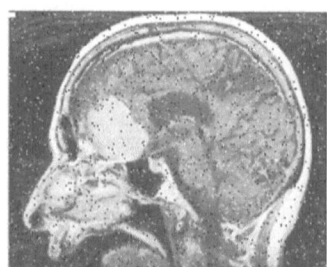

Figure 35.2 Noisy image
Source: Author

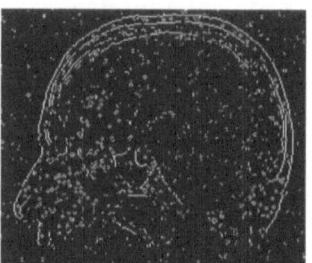

Figure 35.3 Sobel operator
Source: Author

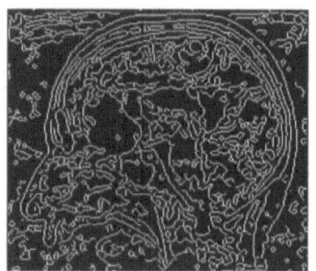

Figure 35.4 Prewitt operator
Source: Author

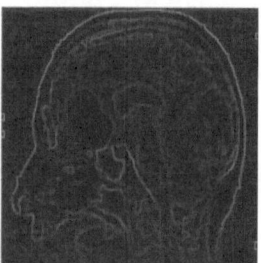

Figure 35.5 Canny operator
Source: Author

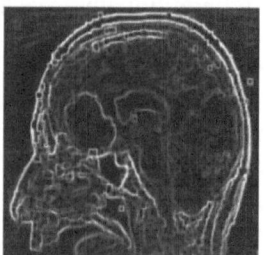

Figure 35.6 Morphological approach
Source: Author

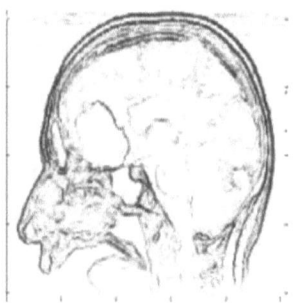

Figure 35.7 Fuzzy logic approach
Source: Author

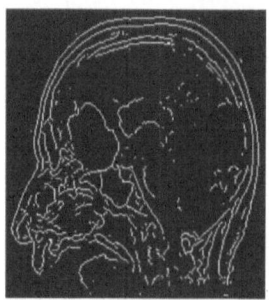

Figure 35.8 Proposed RBFNN
Source: Author

the experimental outcomes of the suggested technique along with several edge detection techniques.

Conclusions

In this work, proposed RBFNN edge detection approach is used for the detection of edges in medical MRI Images. Differential operation, that is susceptible to noise from salt and pepper, is found to be a less effective way for noise filtering and edge information detection. This approach is a better way to balance edge orientation with noise smoothing. For medical image denoising and edge identification, the experimental findings demonstrate that the suggested approach outperforms the commonly used template-based edge detection approaches, including Sobel, Prewitt, Canny, morphological, and fuzzy based edge detectors. According to the experimental results, the suggested method produces lower E_{RMS}, higher SNR_{RMS}, SNR_{PEAK} (dB), SSIM, and EKI with the noisy images. This indicates that compared to current methods, the suggested technique are more effective at edge detection and noise reduction.

References

[1] Abadi, A. M., Wutsqa, D. U., & Ningsih, N. (2021). Construction of fuzzy radial basis function neural network model for diagnosing prostate cancer. *TEL-KOMNIKA (Telecommunication Computing Electronics and Control)*, 19(4), 1273–1283.

[2] Belderrar, A., & Hazzab, A. (2021). Real-time estimation of hospital discharge using fuzzy radial basis function network and electronic health record data. *International Journal of Medical Engineering and Informatics*, 13(1), 75–83.

[3] Bohern, B. F., Hanley Jr., E. N. (1995). Extracting knowledge from large medical databases: an automated approach. *Computers and Biomedical Research*, 28, 191–210.

[4] Husein, H. A., & Zainab, R. M. (2017). Edge detection of medical images using Markov basis. *Applied Mathematical Sciences*, 11(37), 1825–1833.

[5] Ker, J., Wang, L., Rao, J., & Lim, T. (2018). Deep learning applications in medical image analysis. *IEEE Access*, 6, 9375–9389.

[6] Li, D., Wang, X. J., Sun, J., & Feng, Y. (2021). Radial basis function neural network model for dissolved oxygen concentration prediction based on an enhanced clustering algorithm and Adam. *IEEE Access*, 9, 44521–44533.

[7] Mehena, J., & Mishra, S. (2024). Automatic edge detection model of MR images based on deep learning approach. In Computational Intelligence in Healthcare Informatics. Studies in Computational Intelligence, (pp. 1132). Singapore: Springer.

[8] Mehena, J., & Adhikary, M. C. (2015). Medical image edge detection based on soft computing approach. *International Journal of Innovative Research in computer and communication Engineering*, 3(7), 6801–6807.

[9] Yuksel, M. E. (2007). Edge detection in noisy images by neuro-fuzzy processing. *AEU International Journal of Electronics and Communications*, 61, 82–89.

36 Hybrid vision transformer and CNN-based model for efficient cotton leaf disease detection

Dhana Lakshmi Metta[1,a], Bidush Kumar Sahoo[1,b], Rajendra Kumar Ganiya[2,c] and Sachikanta Dash[1,d]

[1]Department of CSE, School of Computational Sciences, GIET University, Odisha, India

[2]Department of Computer Science and Engineering, Koneru Lakshmaiah Education Foundation, Andhra Pradesh, India

Abstract

The cotton is the most vulnerable crop to diseases around the world, leading to significant impacts on yield and quality. In this paper, to deal with the task of early and accurate cotton leaf disease detection, we present a novel model that combines vision transformer (ViT) based encoder with convolutional neural networks (CNN). This hybrid uses the ability of visions transformers to represent images even better, combined with the fine-grained lesion discovery power of CNNs. Experimental results on publicly available cotton leaf disease dataset show that the proposed model obtains 99.95% classification accuracy, which outperforms standard CNN-only methods. Moreover, our model does bring explainability with greater precision, recall, and F1-scores in identifying complex diseases like Fusarium Wilt. This will ensure better disease control of the cotton farmers by making available timely and spot interventions.

Keywords: Agricultural disease detection, convolutional neural networks, cotton leaf diseases, disease classification, explainable AI (XAI), hybrid deep learning, image processing, lesion detection, multiclass classification, vision transformer

Introduction

The agriculture sector is vital to the world economy, with cotton being among the most valued crops. Cotton cultivation sustains the livelihoods of millions of farmers, particularly in underdeveloped nations. However, cotton plants are vulnerable to various diseases, including bacterial blight, curl virus, Fusarium wilt, and powdery mildew, which can drastically reduce crop yield. Early detection and treatment are critical for minimizing economic losses.

To tackle these issues, we present an innovative hybrid approach that integrates the advantages of VIT and CNNs. Vision transformers (ViTs) have shown superior performance in image classification tasks due to their ability to process the entire image as a sequence of patches, capturing global contextual information. By integrating ViT with CNN, we aim to leverage the global feature extraction capability of ViTs and the local feature sensitivity of CNNs, resulting in a robust model that can accurately classify cotton leaf diseases.

The proposed technology has two primary components: the ViT for feature extraction and a CNN for fine-tuning and lesion detection. The ViT analyzes the cotton leaf image by segmenting it into patches and utilizing self-attention mechanisms to discern correlations among various sections of the image. This enables the model to discern global patterns, such as the comprehensive distribution of lesions, which are critical for diseases like curl virus. The CNN component, on the other hand, fine-tunes these features by focusing on specific areas of the image, enabling precise detection of small, localized lesions typical of bacterial blight or Fusarium wilt. Our model was trained and evaluated on a publicly available dataset of cotton leaf images, which contains both healthy and diseased samples. The dataset was expanded by approaches including rotation, flipping, and scaling to improve the model's robustness. The ViT underwent pretraining on the ImageNet dataset and subsequent fine-tuning on the cotton leaf dataset, whilst the CNN was developed from the ground up to guarantee precise lesion identification.

The experimental findings indicate that our hybrid model surpasses conventional CNN-based approaches in both binary and multiclass classification tasks. The ViT component strengthens the model's capacity to discern global illness patterns, whilst the CNN augments its precision in detecting small lesions. The model attained an overall accuracy of 99.95%, demonstrating notable enhancements in precision, recall, and F1-score across all disease categories. The amalgamation of the ViT and CNN enhances the model's explainability by facilitating the presentation of attention maps that underscore the picture regions most pertinent to the model's

[a]dhanalakshmi.metta@giet.edu, [b]bidushsahoo@giet.edu, [c]rajendragk@kluniversity.in, [d]dash.sachikanta@gmail.com

DOI: 10.1201/9781003658221-36

predictions. This is especially beneficial in agricultural contexts, where it is crucial for farmers and agronomists to comprehend the rationale underlying the model's diagnosis.

The hybrid Vit and CNN model serves as an effective instrument for the early identification of cotton leaf diseases. The model attains exceptional performance in accuracy, precision, and explainability by integrating the advantages of both architectures. This method has the capacity to markedly enhance disease management strategies in cotton agriculture,

resulting in elevated crop yields and diminished financial losses.

Contributions

a. Achieved accuracy, outperforming traditional CNN-based methods.
b. Provides better explainability using attention maps to visualize lesion detection.
c. Improved precision, recall, and F1-scores, particularly for challenging diseases like Fusarium Wilt.

Related Work

Author et al	Year	Proposed method	Merits	Demerits	Perfor-mance metrics	Numer-ical results
Amin et al.	2022 [1]	Explainable CNN for cotton disease	High accuracy	Limited generaliz-ability	Accuracy, F1-score	Accuracy:
Memon et al.	2024 [17]	Modified ResNet50 for crops	Good generali-zation	High complexity	Precision, F1-Score	Accuracy:
Nazeer et al.	2024 [19]	CNN for CLCuD detection	High accuracy	Limited scalability	Accuracy	Accuracy: 99%
Suleman et al.	2024 [28]	Hybrid ECNN for leaf diseases	Mobile integrat-ion	Environ-mental sensitivity	Accuracy, Precision	Accuracy: 98.17%

This comprehensive approach introduces a novel methodology using hybrid vision transformer and CNN for improved cotton leaf disease detection, addressing gaps in previous methodologies by increasing accuracy and explainability.

An Ensemble DL based Vision Transformer and CNN for Cotton Leaf Disease

This methodology combines the strengths of ViT for handling larger image patches and CNN for extracting fine-grained features. The ViT provides efficient handling of the entire image, while the CNN fine-tunes the specific lesion areas, improving the overall accuracy and explainability of disease detection.

Algorithm 1: Hybrid vision transformer for feature extraction

---

Input: Preprocessed cotton leaf images I
Output: Feature map (F_v)
1. Initialize parameters of the vision transformer (W v)
2. Patch division: Divide the input image I into (n \times n\ ) patches $\ (P = \ {P1, P2, \ dots, Pn \} \)$

3. Flatten each patch P_i and embed into a linear projection space
   $P_i' = W_p . P_i$, where (W_p) is the learnable patch embedding matrix
4. Add positional encodings to the embedded patches:
   $P_i'' = P_i' + PosEnc(P_i)$
5. Pass the patches through Transformer layers:
   $\ (Z = Transformer)$
6. Apply multi-head attention $\ (A = MultiHead (Z))$
7. Generate final feature map by concatenating patch outputs:
   $F_v = [Z_1, Z_2, ..., Z_n]$
   Return (F_v)

Patch Embedding:

$$P_i' = W_p \cdot P_i \qquad (1)$$

Positional Encoding:

$$P_i'' = P_i' + PosEnc(P_i) \qquad (2)$$

Multi-head attention:

$$A = \text{Concat} (\text{Head}_1, \text{Head}_2, ..., \text{Head}_h) \cdot W_o \qquad (3)$$

Transformer output:

$$Z = \text{Transformer} (P'') \qquad (4)$$

Final feature map:

$$F_v = [Z_1, Z_2, ..., Z_n] \qquad (5)$$

Algorithm 2: CNN fine-tuning on lesion regions

Input: Feature map $F - v$ from ViT
Output: Lesion classification output C
   1. Initialize CNN with pre-trained weights W_c
   2. Apply convolutional layers on feature map
      $(\overline{F\_v} \backslash)$
      $\backslash (F \subset c = \text{Conv}(Wc, F = v) \backslash)$
   3. Apply max-pooling to reduce dimensionality:
      $\backslash (F\_\{pool\} = \text{MaxPool}(F\_c) \backslash)$
   4. Pass through fully connected layers:
      $\backslash (F\_\{fc\} = FC(W\_\{fc\}, F\_\{pool\}) \backslash)$
   5. Apply soft max to get class probabilities:
      $\backslash (C = \text{Softmax}(F\_\{fc\}) \backslash)$
      Return $\backslash (C \backslash)$

CNN Convolution:

$$F_c = \text{Conv} (W_c, F_v)$$

Max-pooling:

$$F_{pool} = \text{MaxPool} (F_c)$$

Fully connected layer:

$$F_{fc} = W_{fc} \cdot F_{pool}$$

Softmax for classification:

$$C = \frac{e^{F_{fc}}}{\sum_j e^{F_{fc_j}}}$$

Experiments and Results

The Hybrid ViT and CNN model markedly surpasses current models in cotton leaf disease classification, offering enhanced accuracy and interpretability. The integration of extensive feature extraction via the ViT and subsequent fine-tuning with CNN improves the model's capacity to identify nuanced variations in lesion patterns. This is particularly evident in diseases like Fusarium Wilt, where traditional CNN models often struggled.

In terms of computational efficiency, while the Hybrid model requires more processing power, the trade-off is worthwhile given the substantial improvement in accuracy and precision. Moreover, the use of cross-validation indicates that the model is not overfitted, showing consistent performance across different subsets of the data.

Shows the comparison between traditional CNN, Vision Transformer, and the proposed Hybrid model. The Hybrid model achieves 99.95% accuracy, outperforming standalone methods.

Figures 36.1–36.5 give a complete visual representation of how well the model is performing in terms of important metrics. A heatmap visualization of the lesion regions detected by the Hybrid model. It highlights the ability of the model to focus on the diseased areas effectively. The experimental setup involved training the model on of the dataset, while

Table 36.1 Model comparison.

Component	Description
Hardware	NVIDIA RTX 3090 GPU, Intel i9 CPU, 64 GB RAM
Framework	PyTorch 1.9, TensorFlow 2.4
Preprocessing techniques	Data augmentation (rotation, flipping, scaling), color normalization
Model architecture	Vision transformer with CNN fine-tuning
Optimization	0.001
Loss	Cross-entropy loss
Evaluation metrics	Precision, accuracy, F1-score, Recall,

Source: Author

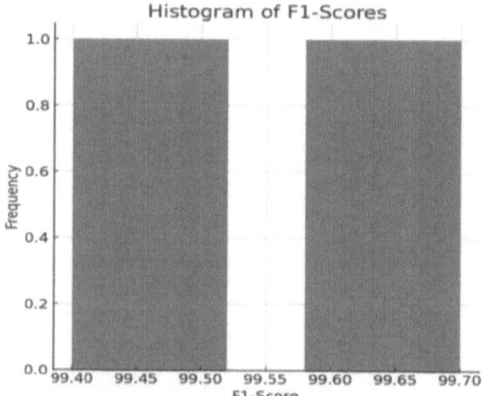

Figure 36.1 The precision, recall, and F1-Score for each disease class (Healthy, Bacterial Blight, Curl Virus, Fusarium Wilt)
Source: Author

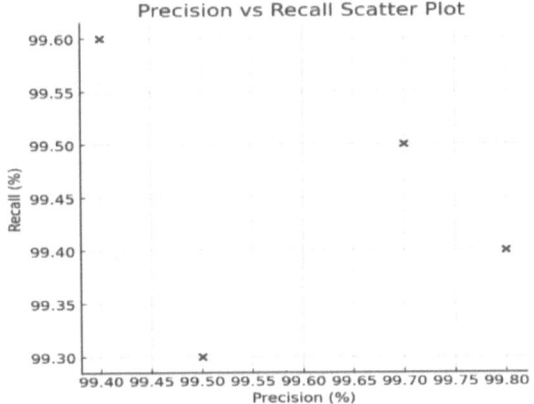

Figure 36.2 The relationship between Precision and Recall for different disease classes
Source: Author

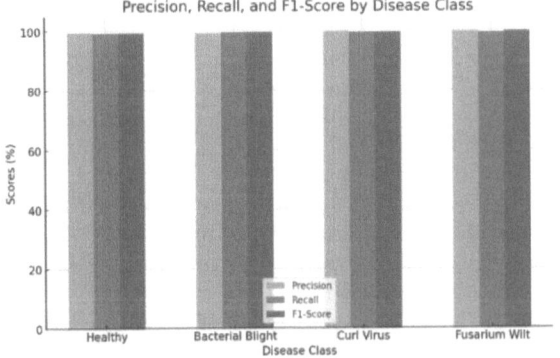

Figure 36.3 Display of the distribution of F1-scores across all disease classes
Source: Author

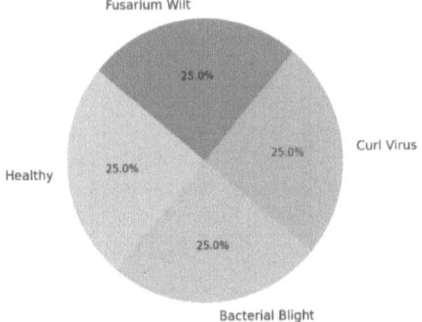

Figure 36.4 Representation of the accuracy distribution across different disease classes, all showing high accuracy
Source: Author

was reserved for testing. The ViT was pre-trained on ImageNet, and the CNN was fine-tuned on the lesion areas extracted from the ViT. The model was evaluated using accuracy, precision, recall, and F1score.

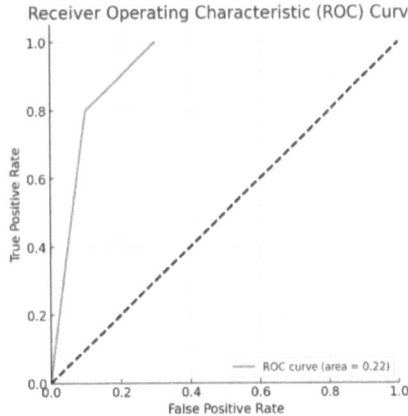

Figure 36.5 Depicts the receiver operating characteristic curve, with an area under the curve (AUC) close to 1, indicating excellent model performance
Source: Author

- Accuracy: The hybrid model achieved an accuracy of , outperforming both the standalone CNN (98.5%) and Vision Transformer (99.2%).
- Precision: For the Fusarium Wilt class, the model achieved a precision of, a significant improvement over the precision of the CNN-only model.
- Recall: The recall for Bacterial Blight improved from (CNN) to in the hybrid model.
- F1-score: The average F1-score for all classes was , indicating balanced precision and recall across the disease categories.

The confusion matrix showed that the model had minimal misclassifications, with the highest confusion observed between Curl Virus and Bacterial Blight due to similar lesion characteristics. The ROC curves demonstrated that the hybrid model consistently achieved a high AUC for all disease classes, with the lowest AUC being 0.995.

Conclusion

In this paper, we propose a new hybrid model combining vision transformers (ViT) and convolutional neural networks for classifying cotton leaf diseases. In this study, we combine the global feature extraction capabilities of ViTs with the fine-grained lesion detection skills of the CNN model. We demonstrate that this fusion technique can reach state-of-the-art performance across many measures. Traditional models: The new model, which of course is a small part of the new model achieves an accuracy rating as high as 99.5% compared to traditional models that have significantly lower ... The model also provides better interpretability using attention maps, which is extremely useful for agriculture professionals. In the future, it is hoped that this hybrid approach can be applied to additional crops and integrated with real-time disease detection into the farming routine.

References

[1] Amin, J., Anjum, M. A., Sharif, M., Kadry, S., & Kim, J. (2022). Explainable neural network for classification of cotton leaf diseases. *Agriculture*, 12(12), 2029.

[2] Mary, X. A., Raimond, K., Raj, A. P. W., Johnson, I., Popov, V., & Vijay, S. J. (2022). Comparative Analysis of Deep Learning Models for Cotton Leaf Disease Detection. In Disruptive Technologies for Big Data and Cloud Applications - Proceedings of ICBDCC 2021 (pp. 825-834). Springer

[3] Azad, S. M., Padhy, S., & Dash, S. (2023). A case study on the multi-hopping performance of IoT network used for farm monitoring. *Automatic Control and Computer Sciences*, 57, 70–80. https://doi.org/10.3103/S0146411623010029.

[4] Baig, M. D., Haq, H. B. U., Asif, M., & Tanvir, A. (2024). Leaf diseases detection empowered with transfer learning model. *Precision Agriculture Journal*, 2(1), 2358.

[5] Das, S. K., Pani, S. K., Padhy, S., Dash, S., & Acharya, A. K. (2023). Application of machine learning models for slope instabilities prediction in open cast mines. *International Journal of Intelligent Systems and Applications in Engineering*, 11(1), 111–121.

[6] Dash, S., Das, R. K., Guha, S., Bhagat, S. N., & Behera, G. K. (2021). An interactive machine learning approach for brain tumor MRI segmentation. In Das, S., & Mohanty, M. N. (Eds.), Advances in Intelligent Computing and Communication. Lecture Notes in Networks and Systems, (Vol. 202). Singapore: Springer.

[7] Dash, A. B., Dash, S., Padhy, S., Mishra, B., & Singh, A. N. (2022). Analysis of brain function effecting form the tumour disease using the image segmentation technique. In 2022 Second International Conference on Computer Science, Engineering and Applications (ICCSEA), (pp. 1–6). IEEE.

[8] Dash, S., Padhy, S., Azad, S. M. A. K., & Nayak, M. (2022). Intelligent IoT-based healthcare system using blockchain. In Ambient Intelligence in Health Care: Proceedings of ICAIHC 2022, (pp. 305–315). Singapore: Springer Nature Singapore.

[9] Dash, S., Padhy, S., Parija, B., Rojashree, T., & Patro, K. A. K. (2022). A simple and fast medical image encryption system using chaos-based shifting techniques. *International Journal of Information Security and Privacy (IJISP)*, 16(1), 1–24.

[10] Abouhawwash, M., & Mostafa, N. N. (2024). Deep Learning for Coffee Leaf Diseases Detection in Precision Agriculture. Optimization in Agriculture, 1, 129-136.

[11] Gülmez, B. (2023). A novel deep learning model with the grey wolf optimization algorithm for cotton disease detection. *Journal of Universal Computer Science*, 29(6), 595.

[12] Nazeer, R., Ali, S., Hu, Z., Ansari, G. J., Al-Razgan, M., Awwad, E. M., & Ghadi, Y. Y. (2024). Detection of cotton leaf curl disease's susceptibility scale level based on deep learning. Journal of Cloud Computing, 13(1), 50.

[13] Amin, J., Anjum, M. A., Sharif, M., Kadry, S., & Kim, J. (2022). Explainable neural network for classification of cotton leaf diseases. agriculture, 12(12), 2029.

[14] Iftikhar, M., Kandhro, I. A., Kausar, N., Kehar, A., Uddin, M., & Dandoush, A. (2024). Plant disease management: A fine-tuned enhanced CNN approach with mobile app integration for early detection and classification. Artificial Intelligence Review, 57(7), 167.

[15] Kumar, D., & Dash, S. (2022). Epileptic Seizures Detection Using Hybrid Deep Learning. *International Journal of Intelligent Systems*, 12, 560–575.

[16] Suganya, P. V. R., Sathyabama, A. R., Abirami, K., & Nalayini, C. M. (2023). Id+E1entification of Leaf Disease Using Machine Learning Algorithm—CNN. In Inventive Computation and Information Technologies (pp. 857–866). Springer, Singapore.

[17] Sarkar, M. S., Lifat, M. A. S., Hasan, R., & Hassan, M. M. (2024, April). Discovering the depths of cotton leaf disease detection: Integrating hypertuned residual networks with GradCAM XAI for enhanced understanding and diagnosis. In 2024 3rd International Conference on Advancement in Electrical and Electronic Engineering (ICAEEE) (pp. 1–6). IEEE.

[18] Bahadure, N. B., Dash, S., Padhy, S., Satpathy, A., & Routray, S. (2023). Rare diseases severity prediction system using a machine learning-based technique. In 2023 International Conference on Artificial Intelligence for Innovations in Healthcare Industries (ICAIIHI), Raipur, India, (pp. 1–6). doi:10.1109/ICAIIHI57871.2023.10489179.

[19] Nazeer, R., Ali, S., Hu, Z., Ansari, G. J., Al-Razgan, M., Awwad, E. M., & Ghadi, Y. Y. (2024). Detection of cotton leaf curl disease's susceptibility scale level based on deep learning. Journal of Cloud Computing, 13(1), 50.

[20] Naeem, A. B., Senapati, B., Chauhan, A. S., Kumar, S., Orosco Gavilan, J. C., & Abdel-Rehim, W. M. F. (2023). Deep Learning Models for Cotton Leaf Disease Detection with VGG-16. International Journal of Intelligent Systems and Applications in Engineering, 11(2), 550–556.

[21] Padhy, S., Dash, S., Shankar, T. N., Rachapudi, V., Kumar, S., & Nayyar, A. (2024). A hybrid crypto-compression model for secure brain mri image transmission. *Multimedia Tools and Applications*, 83(8), 24361–24381.

[22] Padhy, S., Dash, S., Shankar, T. N., Rachapudi, V., Kumar, S., & Nayyar, A. (2023). A hybrid crypto-compression model for secure brain mri image transmission. *Multimedia Tools Application*, 83(8), 24361–24381. https://doi.org/10.1007/s11042-023-16359-w.

[23] Padhy, S., Alowaidi, M., Dash, S., Alshehri, M., Malla, P. P., Routray, S., et al. (2023). AgriSecure: a fog computing-based security framework for agriculture 4.0 via blockchain. *Processes*, 11(3), 757.

[24] Patnaik, R., Rath, P. S., Padhy, S., & Dash, S. (2022). Histopathological colorectal cancer image classification by using inception V4 CNN model. In International Conference on Robotics, Control, Automation and Artificial Intelligence, (pp. 1003–1014). Singapore: Springer Nature Singapore.

[25] Parashar, N., & Johri, P. (2024). Enhancing apple leaf disease detection: A CNN-based model integrated with image segmentation techniques for precision agriculture. International Journal of Mathematical, Engineering and Management Sciences, 9(4), 943.

[26] Panigrahy, S., Dash, S., Padhy, S., Kumar, N., & Dash, Y. (2024). Predictive modelling of diabetes complications: insights from binary classifier on chronic diabetic mellitus. In 2024 International Conference on Communication, Computer Sciences and Engineering (IC3SE), Gautam Buddha Nagar, India, (pp. 1912–1920). doi: 10.1109/IC3SE62002.2024.10593308.

[27] Padhy, S., Dash, S., Routray, S., Ahmad, S., Nazeer, J., & Alam, A. (2022). IoT-based hybrid ensemble machine learning model for efficient diabetes mellitus prediction. *Computational Intelligence and Neuroscience*, 2022, 2389636. https://doi.org/10.1155/2022/2389636(SCIE-IF-3.633).

[28] Iftikhar, M., Kandhro, I. A., Kausar, N., Kehar, A., Uddin, M., & Dandoush, A. (2024). Plant disease management: A fine-tuned enhanced CNN approach with mobile app integration for early detection and classification. Artificial Intelligence Review, 57(7), 167.

[29] Chaitanya, V. S. K., Rakesh, D., Dash, S., Sahoo, B. K., Padhy, S., & Nayak, M. (2022). Tomato leaf disease detection using neural networks. In 2022 International Conference on Machine Learning, Computer Systems and Security (MLCSS), Bhubaneswar, India, (pp. 53–58). IEEE. doi: 10.1109/MLCSS57186.2022.00018.

[30] Bhagat, M., Kumar, D., & Kumar, S. (2024). Optimized transfer learning approach for leaf disease classification in smart agriculture. Multimedia Tools and Applications, 83(20), 58103-58123.

37 Sensitivity analysis based performance assessment of a MGWO based PID control approach for regulating frequency in an interconnected power system

Debayani Mishra[1,a], Anurekha Nayak[1,b], Manoj Kumar Kar[2,c] and Nayan Ranjan Samal[1,d]

[1]Department of EE, School of Engineering and Technology, DRIEMS University, Odisha, India

[2]Department of EEE, Tolani Maritime Institute, Talegaon, Maharashtra, India

Abstract

This study presents a novel approach to frequency management in a three-part thermal power system with varying capacities: the modified grey wolf optimization (MGWO)-designed PID control strategy. In this proposed strategy the main objective adapted here to reduce frequency deviations brought on by step load fluctuations. S The performance of the MGWO-PID control technique is compared to the optimization techniques for adjusting PID settings. Simulation outcomes demonstrate that the MGWO-based approach provides superior system stability and dynamic response compared to traditional optimization techniques. Additionally, sensitivity analysis is conducted by altering the system's inertia constant to evaluate the robustness of the controller. Findings reveal that even with significant changes in system parameters, the MGWO-PID controller performs effectively without requiring frequent retuning, demonstrating its versatility and resilience for controlling load frequency in networked thermal systems.

Keywords: Frequency regulation, modified grey wolf optimization, PID, sensitivity

Introduction

Modern power systems strive to deliver consistent and uninterrupted electricity by ensuring a balance between power generation and consumption, which is essential for maintaining frequency stability. Any imbalance can lead to frequency deviations, potentially compromising system stability. LFC has a crucial role in managing these fluctuations, that is required to govern the frequency of the system helping to regulate system frequency and also manages and regulates the tie-line power exchanges in interconnected networks. LFC adjusts the generator's mechanical input to balance power production and consumption and facilitates power transfer between areas to sustain system operation. Recent advancements in LFC include the exploration of multi-area systems with renewable energy integration and the adoption of hybrid controllers. Fractional-order controllers like FOPI, FOPID, and others have been widely studied for their robustness in LFC applications [3]. Additionally, other optimization techniques have been employed to enhance LFC performance [5]. Studies have also utilized strategies like the imperialist competitive algorithm for tuning fuzzy PI controllers [2] in multi-area systems, highlighting efforts to address diverse operational scenarios and ensure stable frequency and power flow in Latif et al. [6].

Power system performance has been analyzed using PID controllers in multi-area setups, including four-area systems and sliding mode PID controllers optimized via the Ant Lion Optimizer (ALO) [1]. Differential evolution techniques [7] have been employed to tune PID controllers for thermal systems with governor constraints and mixed generation sources like thermal, hydro, and diesel. Bio-inspired optimization methods have gained prominence in LFC research due to their effectiveness. Conducting sensitivity analysis is vital for assessing the impact of changes in system parameters on the dynamic performance of power systems. It evaluates how factors like generation capacity, governor response, damping coefficients, and load variations affect frequency and power exchange stability. In this paper discusses the proposed model, optimization strategy, findings, and concludes with insights on system stability and resilience.

Proposed Model

The case study examines a three-region interconnected system with thermal generators, each having unique parameters, and includes governor, turbine, generation, and load control.

[a]debayanim@gmail.com, [b]anurekha2611@gmail.com, [c]manojkar132@gmail.com, [d]dr.nsamal@driems.ac.in

DOI: 10.1201/9781003658221-37

The system follows the model proposed by Gupta et al. [4], where each area operates as an interconnected subsystem for load frequency control.

A novel MGWO-PID controller is applied in each area to improve system stability and dynamic performance, with the system's structure represented using transfer function-based models.

The PID controller consists of three components, each serving a unique purpose. Its parameters are optimized using the Modified Grey Wolf Optimization algorithm to achieve optimal performance, with its effectiveness assessed through a function

$$J = ITAE = \int_0^{t_{sim}} \left(\begin{array}{l} |\Delta f_1| + |\Delta f_2| + |\Delta f_3| + |\Delta P_{tie,12}| + \\ |\Delta P_{tie,23}| + |\Delta P_{tie,31}| \end{array} \right) t \, dt \quad (1)$$

where, Δf_1, Δf_2, Δf_3 are the frequency variation in areas 1, 2, and 3, correspondingly, $\Delta Ptie_{12}$, $\Delta Ptie_{23}$, $\Delta Ptie_{31}$ are deviation in tie-line powers, and t_{sim} is simulation stop time.

Suggested Approach

The GWO is a nature-inspired, population-based algorithm that emulates the social hierarchy and hunting behavior of grey wolves. It replicates the structured roles within a wolf pack, where the alpha governs decision-making, assisted by beta and delta wolves, while omega wolves hold the lowest position. The, hierarchy of a grey wolf pack is categorized into leadership level of four tiers: alpha (α), beta (β), delta (δ), and omega (ω). In optimization, the alpha signifies the optimal solution, while beta and delta represent the second and third best solutions, respectively, with omega encompassing the remaining candidates. The algorithm mathematically simulates the hunting process, where the alpha, beta, and delta wolves determine potential prey positions, and omega wolves adjust their locations accordingly. This collaborative approach iteratively refines solutions to minimize the objective function, simulating

the wolves' coordinated movements to efficiently locate prey.

$$\vec{P_\alpha} = |\vec{C_1}.\vec{Y_\alpha} - \vec{Y_i}(t)|,$$
$$\vec{P_\beta} = |\vec{C_2}.\vec{Y_\beta} - \vec{Y_i}(t)|, \quad (2)$$
$$\vec{P_\delta} = |\vec{C_3}.\vec{Y_\delta} - \vec{Y_i}(t)|$$

$$\vec{Y_1} = |\vec{Y_\alpha} - \vec{A_1}.\vec{P_\alpha}|,$$
$$\vec{Y_2} = |\vec{Y_\beta} - \vec{A_2}.\vec{P_\beta}|, \quad (3)$$
$$\vec{Y_3} = |\vec{Y_\delta} - \vec{A_3}.\vec{P_\delta}|$$

$$\vec{Y}(t+1) = \frac{\vec{Y_1}+\vec{Y_2}+\vec{Y_3}}{3} \quad (4)$$

$$\vec{M} = 2\vec{a}.\vec{s_1} - \vec{a} \quad (5)$$

$$\vec{N} = 2.\vec{s_2} \quad (6)$$

$$\vec{a} = 2\left(1 - \frac{t}{T_{max}}\right) \quad (7)$$

where, $\vec{Y_\alpha}$, $\vec{Y_\beta}$, $\vec{Y_\delta}$ and $\vec{P_\alpha}$, $\vec{P_\beta}$, $\vec{P_\delta}$ denotes the position vectors and coefficient vectors of the four membership tiers. The current and subsequent iterations, the solution is represented by. $\vec{Y_i}(t)$ and $\vec{Y_i}(t+1)$. The vectors for the coefficients are denoted by $\vec{M}$ and $\vec{N}$. During the iterations, the components of reduces from 2 to 0., while $\vec{s_1}$ and $\vec{s_2}$ are random vectors between [0, 1]. The maximum number of allowable iterations is T_{max}. In the suggested approach, assigned weightages of the α, β, and δ wolves are given by the values of 50%, 33.33%, and 16.66% weightage respectively. The revised equation is given by eq. 8,

$$\vec{Y_i}(t+1) = \frac{3\vec{Y_1}+2\vec{Y_2}+\vec{Y_3}}{6} \quad (8)$$

The optimal solution is determined by calculating the average of assigned values.

Results and Discussion

The block diagram for the three-area interconnected power system is depicted in Figure 37.1. On an interconnected system, the efficacy of the stated MGWO-PID controller is assessed at time t=0 seconds with a step load perturbation of 18%. With MATLAB software, the interconnected system is simulated. The Integral Time Absolute Error (ITAE) performance values are as follows: 0.331 for PSO-PID, 0.00868 for GWO-PID, and 0.01023 for MGWO-PID, highlighting the superior performance of the GWO-PID optimization approach.

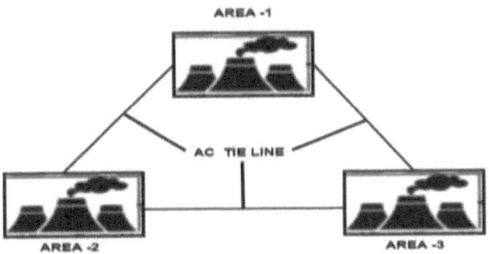

Figure 37.1 An integrated three-area thermal power plant's architecture
Source: Author

Analysis of controllers

This section primarily aims to demonstrate the benefits of the suggested MGWO approach over GWO and PSO performances. The PID controller is first put into use in the test system. Three sophisticated optimization methods PSO, GWO, and the suggested MGWO algorithm are then used to tune the system. The fluctuations in frequency and power transfer between areas are analyzed in Figure 37.2.

It is apparent from Figure 37.2 that the MGWO–PID controller performs better than the other controllers, exhibiting quicker convergence. It also exhibits a reduced settling time and less undershoot for tie-line power and frequency variation. The three area fitness values using different objective techniques with 18% SLD is shown in Table 37.1. The performance metric comparison is presented in Table 37.2. Due to the efficacy of the MGWO–PID controller on other controllers, it will be employed as the LFC controller for sensitivity analysis of the suggested model. and achieving the lowest objective function value.

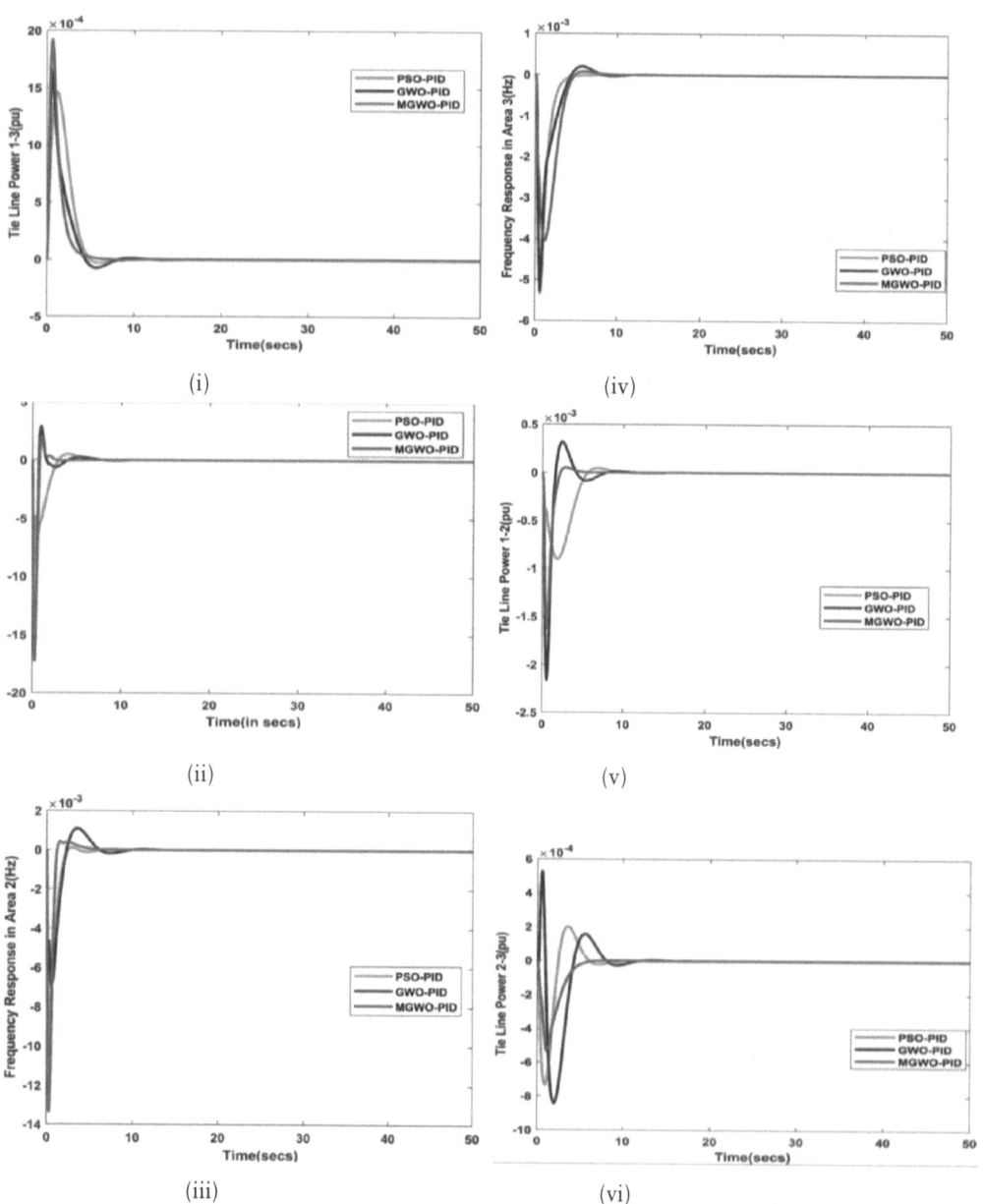

Figure 37.2 Three area system transient response variance in (i) System Area 1 (ii) System Area 2 (iii) System Area 3 (iv) Interaction between Area 1 and Area 2 (v) Interaction between Area 2 and Area 3 (vi) Interaction between Area 1 and Area 3

Source: Author

Table 37.1 Comparing the three-area system's fitness values using different optimization techniques with a 18% SLD.

ITAE Performance Values	PSO-PID	GWO-PID	MGWO-PID
	0.331	0.00868	0.01023

Source: Author

Table 37.2 Comparison of performance parameters.

Algorithm	ΔF_1	ΔF_2	ΔF_3	ΔP_{12}	ΔP_{23}	ΔP_{13}
PSO-PID Settling Time(secs)	10	14	13	8.6	11.33	9.11
GWO-PID Settling Time(secs)	8	12	12	9.34	11.27	8.69
MGWO-PID Settling Time(secs)	6	9	9	7.29	8.75	8.05

Source: Author

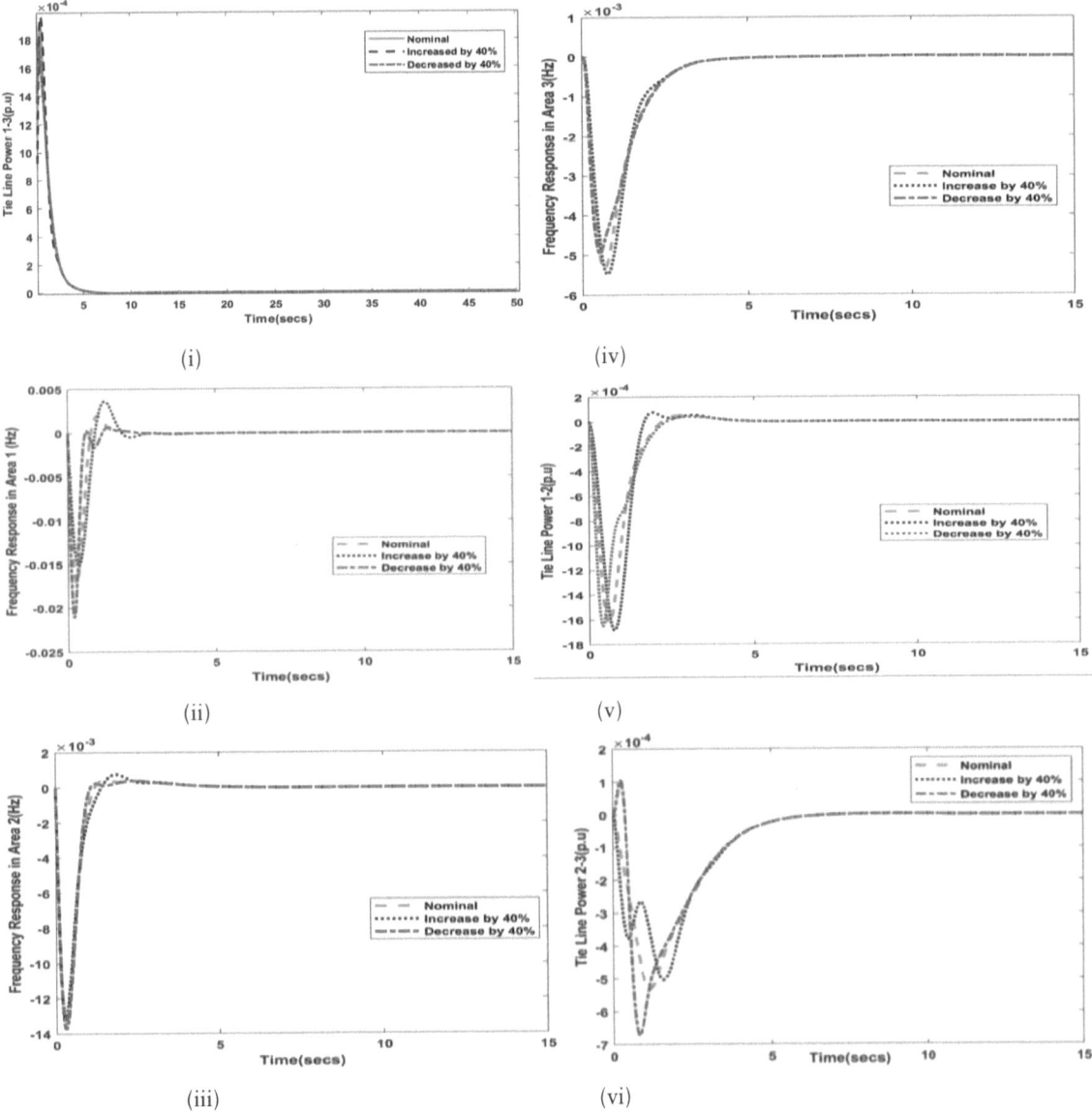

(i)   (iv)

(ii)   (v)

(iii)   (vi)

Figure 37.3 Dynamic response under parameter variation of M (i) Area 1 (ii) Area-2 (iii) Area 3 (iv) Area 1-2 (v) Area 2-3 (vi) Area 1-3

Source: Author

Table 37.3 Percentage change in performance index due to variation of M of Area-1.

Change in M	ITAE	% Change in Nominal Value
Increase by 40%	0.010562	3.24%
Decrease by 40%	0.00993	-2.93%

Source: Author

Due to the efficacy of the MGWO–PID controller on other controllers it will be employed as the LFC controller for sensitivity analysis of the suggested model and achieving the lowest objective function value.

Sensitivity analysis of controllers

Performing sensitivity analysis on the inertia constant 'M' and it is crucial for evaluating a power system's risk of frequency deviation. A lower value of 'M' further increases the system's exposure to rapid frequency changes and larger deviations, necessitating quicker and more sophisticated control strategies to ensure stability. Conversely, higher inertia enhances frequency stability but results in slower responses to disruptions. As power systems incorporate more renewable energy sources, effectively monitoring and managing inertia becomes vital for maintaining system reliability. In this paper the sensitivity study investigates how changes in inertia constant by 40% of area 1 affect the MGWO-PID controller ability to maintain frequency stability and regulate power flows between the interconnected areas. The system parameters are adjusted by decreasing and increasing them by 40% from their nominal value. Figure 37.3 illustrates the system's frequency deviation response.

The responses in Figure 37.3. display either overlap or minimal variations, even with significant changes to the system parameters. This indicates that large adjustments to system parameters have minimal influence on performance when the controller parameters remain unchanged. Even with significant system parameter changes, an in-depth investigation of all dynamic responses indicates that frequent controller adjustment is not required to maintain acceptable performance. Table 37.3 presents the objective function values, allowing for a 5% tolerance variation, with only a few minor exceptions. These outcomes were attained without altering the MGWO-PID controller's ideal parameters. The percentage change in performance Index due to variation of M in area-1has illustrated in Table 37.3.

Conclusion

This research suggests a dependable MGWO-PID controller for a thermal power system's frequency regulation in three different zones. The suggested controller is efficient to enhance dynamic performance and preserve system stability under various load scenarios. Comprehensive simulations are performed to illustrate the controller's efficacy, which reveals that it greatly enhances the frequency regulation and decreased settling times in every control area. A thorough sensitivity study was carried out, taking into account changes in system characteristics, to demonstrate the resilience of the recommended controller. The outcomes demonstrated the controller's flexibility and resistance to parameter uncertainties by confirming that it continued to operate steadily even in the face of notable variances. When compared to GWO-PID and PSO-PID controllers, the proposed MGWO-PID controller performs better. Future research might concentrate on expanding this strategy to include renewable energy sources and delving more into how various disturbances affect the system's functionality.

References

[1] Gomaa Haroun, A. H., & Yin-Ya, L. (2020). Ant lion optimized hybrid intelligent PID-based sliding mode controller for frequency regulation of interconnected multi-area power systems. *Transactions of the Institute of Measurement and Control*, 42(9), 1594–1617.

[2] Arya, Y. (2018). Automatic generation control of two-area electrical power systems via optimal fuzzy classical controller. *Journal of the Franklin Institute*, 355(5), 2662–2688.

[3] Das, S., Nayak, P. C., Prusty, R. C., & Panda, S. (2024). Design of fractional order multistage controller for frequency control improvement of a multi-microgrid system using equilibrium optimizer. *Multiscale and Multidisciplinary Modeling, Experiments and Design*, 7(2), 1357–1373.

[4] Gupta, N. K., Kar, M. K., & Singh, A. K. (2022). Design of a 2-DOF-PID controller using an improved sine-cosine algorithm for load frequency control of a three-area system with nonlinearities. *Protection and Control of Modern Power Systems*, 7(3), 1–18.

[5] Khokhar, B., Dahiya, S., & Parmar, K. P. S. (2021). Load frequency control of a microgrid employing a 2D sine logistic map based chaotic sine cosine algorithm. *Applied Soft Computing*, 109, 107564.

[6] Latif, A., Hussain, S. M. S., Das, D.C., Ustun, T. S., & Iqbal, A. (2021). A review on fractional order (FO) controllers' optimization for load frequency stabilization in power networks. *Energy Reports*, 7, 4009–4021.

[7] Mishra, D., Maharana, M. K., Kar, M. K., & Nayak, A. (2023). A modified differential evolution algorithm for frequency management of interconnected hybrid renewable system. *International Journal of Power Electronics and Drive Systems*, 14(3), 1711–1721.

38 Real time garbage detection using CNN and YOLO algorithms

Dhanushiya, S.[1,a], Navaneetha, K.[1,b] and Vijayalakshmi, V.[2,c]

[1]Department of Networking and Communications, School of Computing, SRM Institute of Science and Technology, Kattankulathur, Chennai, India

[2]Department of Networking and Communications, Faculty of Engineering and Technology, School of Computing, SRM Institute of Science and Technology, Kattankulathur, Chennai, India

Abstract

Modern metropolitan environments require real-time trash management since poor rubbish disposal can result in several environmental problems. In this paper, a novel garbage detection system based on convolutional neural network (CNN) and You Only Look Once version (YOLOv8) algorithms is proposed. By the use of live video feeds from webcams or CCTV, the system is intended to detect people handling rubbish and identify and categorize it. YOLOv8 is utilized to record video frames and identify things, and a CNN is used to categorize the objects into distinct garbage groups. A pre-trained dataset is used by the suggested method to attain high accuracy, with a mean average precision (mAP) of 0.94 and a 94% detection accuracy. According to experimental data, this method is effective at automating waste management operations, which may find use in smart cities.

Keywords: Automated classification, CCTV monitoring, CNN, deep learning, environmental sustainability, object detection, realtime garbage identification, YOLOv8

Introduction

In contemporary civilizations, waste management is a crucial issue since the amount of waste produced has increased significantly due to the fast rate of urbanization. Maintaining a clean environment and avoiding potential health risks requires efficient and timely rubbish disposal. Particularly in places with a high population density, manual trash monitoring is frequently ineffective, time-consuming, and prone to human error. The efficiency of waste management systems can be greatly increased by integrating cutting-edge technologies for garbage identification and classification.

Deep learning algorithms have become extremely effective instruments for automating activities related to object detection and categorization across a wide range of fields Pathak et al. [14]. Among these techniques, convolutional neural networks (CNN) are frequently employed for image classification tasks, while You Only Look Once (YOLOv8) has become well-known for its real-time object identification capabilities Vijayalakshmi et al. [19] A reliable method for identifying items and categorizing them into distinct groups is provided by the combination of YOLOv8 and CNN.

This study suggests a real-time garbage detection method that makes use of CNN for classification and YOLOv8 for object recognition. Technology analyzes live video feeds from webcams or CCTV cameras, recognizing people who handle rubbish and identifying different kinds of waste (such metal, glass, and plastic). The suggested system seeks to automate waste monitoring in residential and public locations by utilizing deep learning models and pre-trained datasets, offering a scalable solution for intelligent waste management.

In contemporary civilizations Khan et al. [8], waste management is a crucial issue since the amount of waste produced has increased significantly due to the fast rate of urbanization. Maintaining a clean environment and avoiding potential health risks requires efficient and timely rubbish disposal. Particularly in places with a high population density, manual trash monitoring is frequently ineffective, time-consuming, and prone to human error. The efficiency of waste management systems can be greatly increased by integrating cutting-edge technologies for garbage identification and classification.

Literature Survey

The detection and classification of waste using deep learning models has garnered substantial research interest, especially with the advancement of automated waste management systems. Various studies have employed different deep learning models to

[a]ds8543@srmist.edu.in, [b]nk8360@srmist.edu.in, [c]vijayalv@srmist.edu.in

DOI: 10.1201/9781003658221-38

tackle waste management challenges, predominantly using CNN and YOLO models.

Several studies have utilized CNN to classify waste images into predefined categories [13]. The CNN model was developed to classify waste images, achieving high classification accuracy across a large-scale dataset. However, this study did not address real-time performance, which is critical for automated waste management systems. Similarly, Kumar et al. [9] CNN is combined with transfer learning to improve classification accuracy on a small and imbalanced waste dataset. Although this model performed well in controlled conditions, its real-time capabilities require further optimization.

The work by Manivannan et al. [11] applies CNN-based saliency detection techniques, which enables the model to focus on the most informative parts of an image, leading to improved classification results. This approach was particularly effective in enhancing the precision of waste classification but introduced additional computational overhead. Another study by Ma et al. [12] uses image segmentation with CNN to detect and classify waste materials, demonstrating the potential of segmentation for distinguishing different types of waste. However, the computational demands of segmentation models are higher compared to object detection-based approaches, making them less suitable for real-time applications.

Setyawan et al. [18] introduces an ensemble approach combining CNN and Decision Trees for waste classification. This method improved classification accuracy but at the cost of increased computational complexity, making real-time deployment in edge devices challenging. Another interesting application of CNN in robotics was explored by Kumar [10], where a robotic arm was developed for waste sorting. This system used CNN for real-time visual feedback to guide the sorting process, achieving high accuracy. However, the reliance on a pre-defined dataset limited the system's adaptability to new types of waste, posing challenges for dynamic environments.

YOLO-based models have been widely adopted for waste detection due to their efficiency in real-time object detection tasks. Alvarico et al. [2] applied YOLOv3 to detect plastic waste in aquatic environments, highlighting the model's capability in processing frames in real time on low-power edge devices. Putra et al. [15] implemented YOLO for garbage detection in urban areas, focusing on common litter such as plastic bottles and cans. They noted YOLO's efficiency in detecting multiple objects per frame, making it well-suited for high-traffic areas.

The study by Xu et al. [20] explored the use of a hybrid approach that integrated YOLO for object detection with CNN for classifying detected objects. This combination improved the system's accuracy by reducing false positives and enhancing classification precision, proving effective for real-time applications. Similarly, Bhattacharya et al. [3] leveraged YOLOv5 in a garbage detection system, achieving higher accuracy than earlier models, and demonstrated its potential for deployment in smart city applications.

Babu et al. [4] implemented a MobileNet-based YOLO model for waste detection on mobile devices to enable deployment in resource-constrained environments. Although the accuracy of this model was lower compared to full-sized YOLO models, MobileNet's efficiency on edge devices made it ideal for mobile applications. In another work, Ren et al. [16] combined YOLO with cloud-based systems, where detection results were transmitted to a cloud server for further analysis and reporting. This approach enables scalable waste management solutions, although it introduces latency due to reliance on cloud processing.

Recent research has also investigated the integration of YOLO with Internet of Things (IoT) systems. In Chowdhury et al. [5] a smart bin was introduced that combines YOLOv4 and IoT sensors for real-time waste classification and segregation. This fully autonomous system demonstrated the potential of IoT-based waste management, although it involved higher implementation complexity. Kunieda and Suzuki [6] integrated saliency detection with YOLO to improve garbage classification accuracy by focusing on the most relevant parts of the image, thus addressing the challenges in classifying similar-looking waste items.

A novel YOLO-based approach was explored in Kaladevi et al. [7] for robotic waste management, here YOLO detected waste types, and the classification results guided a robotic arm to sort waste accordingly. This system demonstrated high sorting accuracy and proved effective in dynamic environments, though its reliance on specific datasets limited adaptability to unseen waste categories.

Methodology

Data gathering, model selection, system architecture design, and real-time implementation are some of the essential elements of the suggested system. Figure 38.1.

Gathering and preparing data
For deep learning models to be trained successfully, a large dataset is essential. The proposed work makes use of a pretrained dataset that includes 59 different waste material classes representing a variety of rubbish products [1].

(i) Dataset selection: The dataset was chosen due to its size, diversity, and applicability to the classification of rubbish in the actual world. There are enough pictures in each lesson to encourage generalization.

(ii) Data Augmentation: By adding variability to the training data, data augmentation techniques improve the resilience of the model. The following is a mathematical representation of the augmentation process:

$$I' = I \cdot T \tag{1}$$

Model selection

Two main models are used by the suggested system to identify and categorize trash:

(i) YOLOv8: YOLOv8 was selected due to its effectiveness in detecting objects in real-time Reis et al. [17]. YOLOv8 is trained using a loss function that combines classification loss, confidence loss, and localization loss:

$$L_{total} = L_{coord} + L_{conf} + L_{class} \tag{2}$$

(ii) CNN: The objects that YOLOv8 detects are classified using a CNN. A SoftMax output layer generates class probabilities after a series of convolutional, pooling, and fully connected layers make up the CNN architecture:

$$P(y = k|x; \theta) = \frac{e^{f_k^{(x;\theta)}}}{\sum_{j=1}^{K} e^{f_j^{(x;\theta)}}} \tag{3}$$

Implementation

Programming language and libraries

Python is used to implement the system, along with libraries like PyTorch for YOLOv8 integration, TensorFlow/Keras for CNN construction, and OpenCV for video processing.

Hardware configuration

A typical workstation with a webcam is used to test the system. It is intended to be deployed on edge devices in the future for wider uses. Training Process: To take advantage of transfer learning, the YOLOv8 model is pre-trained using the COCO dataset. The garbage dataset is used to train the CNN, and performance is tracked, and overfitting is prevented through training and validation:

$$Loss = \frac{1}{N} \sum_{i=1}^{N} L(y_i, \hat{y}_i) \tag{4}$$

Model performance is assessed through the use of metrics like Mean Average Precision (mAP), Precision, and Recall F1-score.

$$Precision = \frac{TP}{TP + FP}, Recall = \frac{TP}{TP + FN}, \tag{5}$$

$$F1 = 2 * \frac{Precision*Recall}{Precision+Recall}$$

where FP represents false positives, FN represents false negatives, and TP represents true positives. The mean average precision (mAP) is calculated:

$$mAP = \frac{1}{N} \sum_{C=1}^{N} AP_C \tag{6}$$

Experimental Setup and Result

A mean average precision (mAP) of 0.94 and a 94% detection accuracy were demonstrated by the experimental evaluation. According to these findings, the system does a good job of differentiating between different waste categories in real time as shown in Figure 38.2. High precision was found in the confusion matrix analysis for categories like "plastic bottle" and "glass bottle," although there were some misclassifications

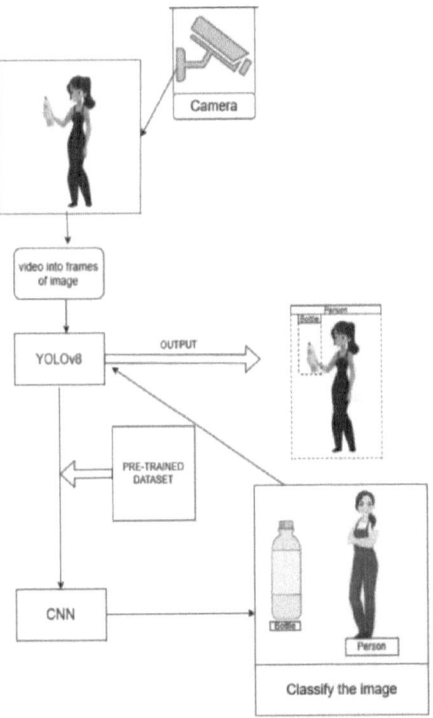

Figure 38.1 System architecture diagram
Source: Author

Table 38.1 Waste category with metric score.

Index	Waste category	Precision	Recall	F1 Score
1.	Drink can	95	93	94
2.	Clear plastic bottle	96	95	95
3.	Corrugated carton	94	93	93
4.	Food can	93	92	92
5.	Tissues	94	92	93
6.	Shoe	90	80	89
7.	Glass bottle	95	94	94
8.	Squeezable tube	93	91	92
9.	Tupperware	92	90	91
10.	Battery	91	89	90

Source: Author

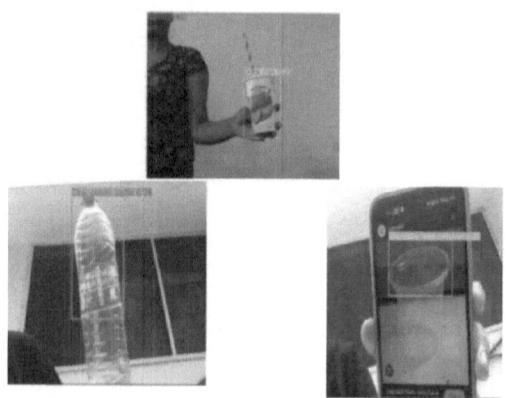

Figure 38.2 Object detection
Source: Author

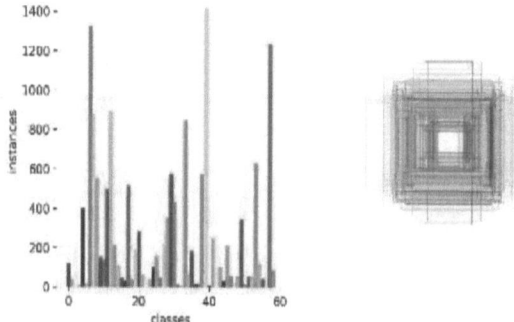

Figure 38.3 Trained dataset
Source: Author

between related categories like "plastic film" and "Plastic lid." Table 38.1.

Figure 38.2 shows the various objects detected using the proposed algorithms and the accuracy of the trained dataset is shown in Figures 38.3 and 38.4.

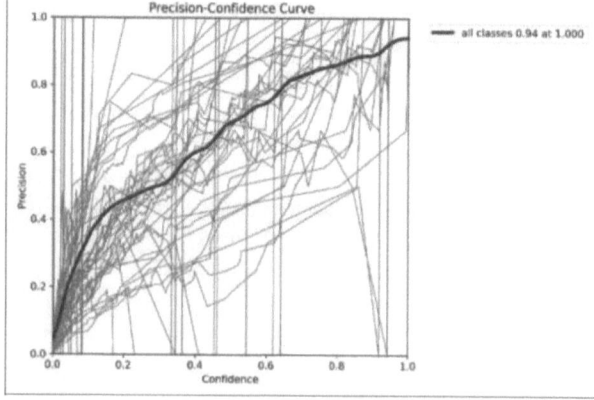

Figure 38.4 Overall trained accuracy
Source: Author

Conclusion and Future Scope

The proposed system utilizing CNN and YOLOv8 algorithms proves to be an effective solution for real-time detection and categorization of waste using live video feeds from webcams or CCTV cameras. With a mean average precision (mAP) of 0.94 and a detection accuracy of 94%, the system showcases its capability to automate waste management processes efficiently. This innovation holds significant potential to enhance urban waste management practices, contributing to cleaner environments and the development of sustainable smart cities. Future research will concentrate on enhancing detection accuracy in various environmental settings and incorporating more sensors for more comprehensive detection.

References

[1] Ahmed, M. I. B., Alotaibi, R. B., Al-Qahtani, R. A., Al-Qahtani, R. S., Al-Hetela, S. S., Al-Matar, K. A., et al. (2023). Deep learning approach to recyclable products classification: Towards sustainable waste management. *Sustainability*, 15(14), 11138.

[2] Alvarico, E. A. M., Arricivita, F. A. I., & Cruz, F. R. G. (2024). Real-time tracking and monitoring system with geofence notification for garbage collection vehicles. In 2024 6th International Conference on Electronics and Communication, Network and Computer Technology (ECNCT), (pp. 363–368). IEEE.

[3] Bhattacharya, S., Kumar, A., Krishav, K., Panda, S., Vidhyapathi, C. M., Sundar, S., et al. (2024). Self-adaptive waste management system: utilizing convolutional neural networks for real-time classification. *Engineering Proceedings*, 62(1), 5.

[4] Babu, M. M. S. K., Srevika, M. N., Sanjana, P. S., & Lavanya, M. (2024). Geo tracking of waste, triggering alerts and mapping areas with high waste index. *International Journal of Information Technology and Computer Engineering*, 12(3), 828–836.

[5] Chowdhury, S., Roy, S., Chowdhury, D. G., Chakraborty, S., Jain, K. P., & Bhattacharyya, B. (2024). SKIDS: an object classification and smart communication based waste bin. In 2024 3rd International Conference on Artificial Intelligence for Internet of Things (AIIoT), (pp. 1–6). IEEE.

[6] Kunieda, Y., & Suzuki, H. (2024). A detection method of garbage collection status from sound of garbage trucks. In 2024 IEEE International Conference on Consumer Electronics (ICCE), (pp. 1–6). IEEE.

[7] Kaladevi, R., Ganesh, K., Kishore, P., & Vijaya, R. V. (2024). Time littering detection—using pose estimation and object detection. In Disruptive Technologies for Sustainable Development, (pp. 156–160). CRC Press.

[8] Khan, S., Anjum, R., Raza, S. T., Bazai, N. A., & Ihtisham, M. (2022). Technologies for municipal solid waste management: current status, challenges, and future perspectives. *Chemosphere*, 288, 132403.

[9] Kumar, A., Ziya, Q., & Abid, K. (2024). IoT-based real-time waste management system with dust segregation & full bin notifications. In 2024 Sixth International Conference on Computational Intelligence and Communication Technologies (CCICT), (pp. 98–105). IEEE.

[10] Kumar, D. (2024). Smart garbage detection system for sustainable waste management using deep learning techniques. In 2024 IEEE International Conference for Women in Innovation, Technology & Entrepreneurship (ICWITE), (pp. 253–257). IEEE.

[11] Manivannan, C., Virgin, J., Suseendran, S., & Vani, K. (2024). Garbage monitoring and management using deep learning. *ISPRS Annals of the Photogrammetry, Remote Sensing and Spatial Information Sciences*, 10, 163–168.

[12] Ma, W., Chen, H., Zhang, W., Huang, H., Wu, J., Peng, X., et al. (2024). DSYOLO-trash: an attention mechanism-integrated and object tracking algorithm for solid waste detection. *Waste Management*, 178, 46–56.

[13] Mehadjbia, A., & Slaoui-Hasnaoui, F. (2024). Real-time waste detection based on YOLOv8. In 2024 4th Interdisciplinary Conference on Electrics and Computer (INTCEC), (pp. 1–6). IEEE.

[14] Pathak, A. R., Pandey, M., & Rautaray, S. (2018). Application of deep learning for object detection. *Procedia Computer Science*, 132, 1706–1717.

[15] Putra, M. T. D., Adiwilaga, A., Munggaran, J. P., Adhitama, M. A., As' Ad, R. A., Alhafidz, A. A., et al. (2024). Mini prototype of the futuristic bin with an automatic waste sortation system for managing the garbage problems in society. In 2024 10th International Conference on Wireless and Telematics (ICWT), (pp. 1–6). IEEE.

[16] Ren, Y., Li, Y., & Gao, X. (2024). An MRS-YOLO model for high-precision waste detection and classification. *Sensors*, 24(13), 4339.

[17] Reis, D., Kupec, J., Hong, J., & Daoudi, A. (2023). Real-time flying object detection with YOLOv8. arXiv preprint arXiv:2305.09972.

[18] Setyawan, E. B., Novitasari, N., & Zahira, A. D. (2024). Development of automatic object detection and IoT for garbage pickup assignment problem. *JOIV: International Journal on Informatics Visualization*, 8(2), 794–802.

[19] Vijayalakshmi, V., Gaur, R., & Singh, R. (2024). Real-time deep learning-based object recognition. In AIP Conference Proceedings, (Vol. 3075, No. 1). AIP Publishing.

[20] Xu, J., Liao, X., Zhang, N., Lin, W., & Huang, J. (2024). Research on garbage detection and classification based on YOLOv8. In 2024 5th International Conference on Computer Vision, Image and Deep Learning (CVIDL), (pp. 778–781). IEEE.

39 A novel modified SCA-PID based load frequency control

Prashanta Kumar Tripathy[1,a], Pratap Chandra Pradhan[2,b],
Manoj Kumar Kar[3,c] and Jeyagopi Raman[4,d]

[1]Department of EE, DRIEMS University, Cuttack, Odisha, India

[2]Department of EEE, DRIEMS University, Cuttack, Odisha, India

[3]Department of E&E, Tolani Maritime Institute, Pune, India

[4]Department of Engineering, INTI International University, Nilai, Malaysia

Abstract

This work presents a PID controller for load frequency control (LFC) that is tuned using the modified sine cosine algorithm (MSCA). Several modern techniques are applied to implement a stabilizing controller. An evaluation of the dynamic behavior of LFC utilizing optimization techniques has been conducted, and the issues has been addressed. In order to suppress oscillation, a MSCA-PID control method is implemented in this paper. Recently developed techniques are implemented to find out the optimal parameters. The controller performances are analyzed by using the MATLAB/SIMULINK tool.

Keywords: ACE, ITAE, LFC, MSCA, PID

Introduction

In modern power systems, maintaining stability [5] and reliability [11] is paramount to ensure uninterrupted and quality power supply. One of the key factors that influences the stability of a system is the regulation of frequency [2], which must be maintained within a narrow band to ensure the safe and efficient operation of electrical equipment. LFC is an essential mechanism that ensures a stable balance between power generation and load demand, thereby regulating the system frequency and ensuring power system stability.

The development of various control strategies has significantly enhanced the performance of LFC systems, especially in the context of modern power grids. PID controller is used [13] to provide a comprehensive performance analysis of LFC of a power network. A traditional PID controller is used to solve the LFC problem [7]. The dynamic characteristics of linked power systems using different sources of energy are compared [14]. A novel cascaded TID-FOPIDN controller was implemented [1] to improve the LFC with several sources. A cascaded 2DOF-PID controller is applied in a two-area networked system to control frequency [3, 10]. The microgrid system is subject to parametric fluctuations and load perturbations in order to show the controller's efficacy [8]. For the purpose of modelling virtual inertia in isolated microgrids under various test circumstances, a MDE tuned cascaded PIDFN controller is proposed in Mishra et al. [9]. In order to investigate practical power system analysis, a number of physical constraints have been considered [4]. DE optimized PI-PD method was proposed for LFC [12] under load variations.

It is evident from the literature that LFC is used to keep the system stable and reliable by making sure that power generation and consumption are equal. Ultimately, effective LFC enhances economic efficiency and prevents grid disturbances, safeguarding against potential blackouts. Although numerous optimization methods have been explored to enhance the performance of LFC systems, still there is a scope for improvement. In this work, a novel MSCA method is utilized for the LFC mechanism.

System Description

System configuration

It is a simplified model used in power system studies to represent a network divided into two distinct geographical or operational areas. These two areas are typically interconnected via a tie-line, allowing them to exchange power. The model is useful for understanding the dynamics of power exchange between different regions, frequency stability, and the effects of disturbances on interconnected power systems. The various sections are modelled using transfer functions, as shown in Figure 39.1.

PID controller

These controllers have been widely employed in LFC applications. A proportional, integral, and derivative term-based correction is applied by a PID controller, which continually computes an error value.

[a]prashanta.tripathy@gmail.com, [b]pratap_pin@yahoo.com, [c]manojkar132@gmail.com, [d]rjeyagopi@gmail.com

DOI: 10.1201/9781003658221-39

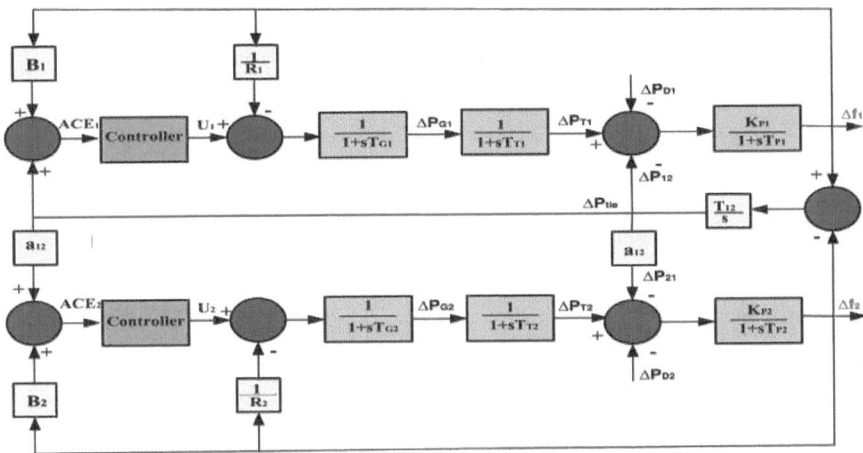

Figure 39.1 System configuration
Source: Author

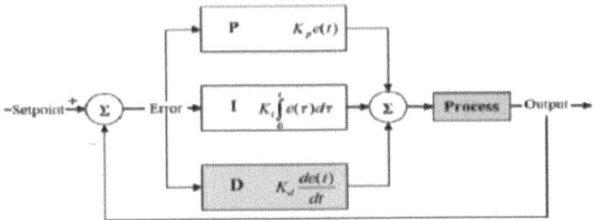

Figure 39.2 Controller structure
Source: Author

Figure 39.2 depicts the controller structure and is represented by eq. 1:

$$y(t) = K_p.q(t) + K_i.\int_0^t q(t)dt + K_d.\frac{dq(t)}{dt} \qquad (1)$$

where q = the control error and y = the control output. The three controller parameters are denoted by K_d, K_i, and K_p.

Problem Statement

Eq. 2 will provide the ITAE for the LFC as follows:

$$J = ITAE = \sum_{i=1}^{NA}\int_0^{t_{sim}}\left(|\Delta f_i| + \sum_{\substack{j=1\\j\neq i}}^{NA}|\Delta P_{tie_{12}}|\right).t.dt \qquad (2)$$

Proposed Method

SCA was developed by mimicking the periodic oscillatory behavior to update candidate solutions dynamically.

SCA ensures a balance between exploration (global search) and exploitation (local refinement) by adjusting its control parameters over iterations. A weighted combination of sine and cosine functions is used to

Table 39.1 Optimal values of PID controller.

Algorithm	K_p	K_I	K_D
MSCA	5	5	2.9215
	4.7476	5	2.6951
SCA	3.7126	5	2.1786
	4.6838	5	5
GWO	4.7813	5	2.7346
	4.0392	5	2.3404

Source: Author

update each candidate solution's position, ensuring diversification and intensification in the search process.

Equation 3 states that the optimization problem is solved by choosing the optimal solution from the earlier iterations.

$$X_i^{t+1} = \begin{cases} X_i^t + r_1 \times \sin(r_2) \times |r_3 \times P_i^t - X_i^t|, if\, r_4 < 0.5, \\ X_i^t + r_1 \times \cos(r_2) \times |r_3 \times P_i^t - X_i^t|, if\, r_4 \geq 0.5, \end{cases} \quad (3)$$

The parameters r_1, r_2 and r_3 can be obtained by using eq. 4-6.

$$r_1 = a * \left(1 - \frac{t}{T_{max}}\right) \qquad (4)$$

$$r_2 = 2 * pi * rand(0,1) \qquad (5)$$

$$r_3 = 2 * rand(0,1) \qquad (6)$$

The algorithm is improved as per [6].

Results and discussions

The MSCA technique is employed to obtain the optimized parameters of both the controllers. Table 39.1

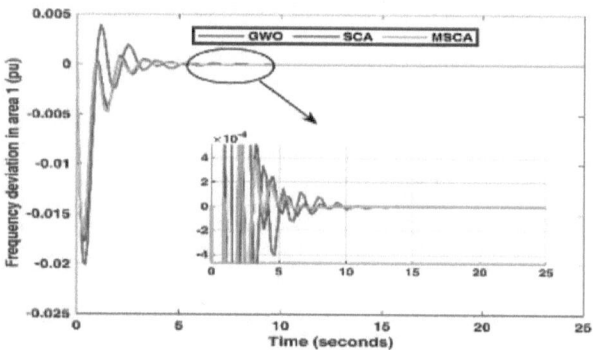

Figure 39.3 Δf_1 response
Source: Author

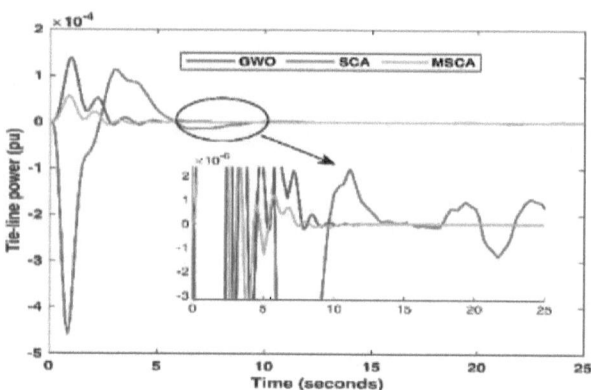

Figure 39.5 Variation in tie-line power
Source: Author

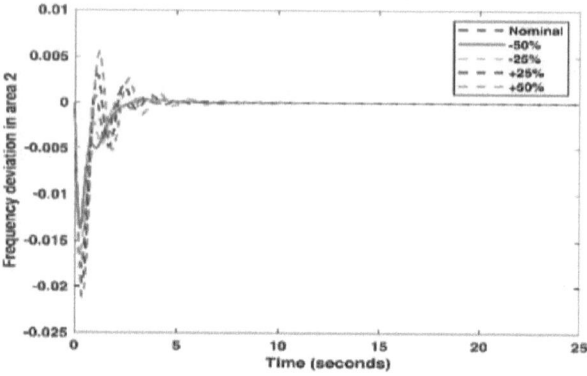

Figure 39.7 Δf_2 response under variation of T_G
Source: Author

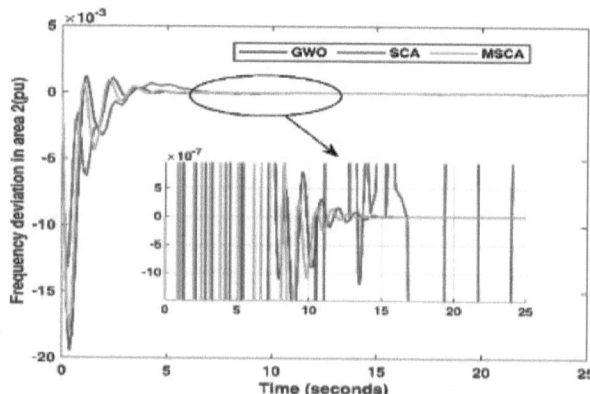

Figure 39.4 Δf_2 response
Source: Author

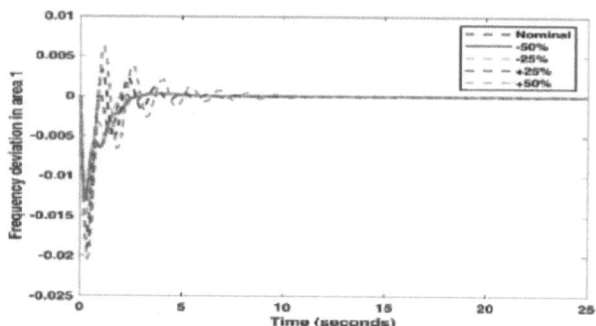

Figure 39.6 Δf_1 response under variation of T_G
Source: Author

displays the controller parameters for the various approaches.

The frequency variations for area 1, area 2, and tie-line power are shown in Figures 39.3–39.5, respectively. The MSCA-tuned controller outperforms the SCA and GWO results.

Figures 39.6 and 39.7 represent the frequency variation in both area under different variation of governor

time constant. It is observed that with the reduction of T_G, the response is improved.

Conclusion

By using the MSCA-PID controller, performance is improved. In order to determine the controller's optimal parameter, a modified SCA approach has been used. Three potential algorithms, including SCA and GWO, have been compared with the results produced. The findings highlight the efficacy of the suggested method to address complex challenges in the modern power systems. It helps in understanding the dynamics of load sharing, frequency control, and the impact of inter-area power transfers, making it an essential modern power grids.

References

[1] Ahmed, D., Ebeed, M., Kamel, S., Nasrat, L., Ali, A., Shaaban, M. F., et al. (2024). An enhanced jellyfish search optimizer for stochastic energy management of multi-microgrids with wind turbines, biomass and PV generation systems considering uncertainty. *Scientific*

Reports, 14, 15558. Nature Publishing Group UK. https://doi.org/10.1038/s41598-024-65867-8.

[2] Daraz, A., Khan, I. A., Basit, A., Malik, S. A., AlQahtani, S. A., & Zhang, G. (2024). Frequency regulation of interconnected hybrid power system with assimilation of electrical vehicles. *Heliyon*, 10(6), e28073. https://doi.org/10.1016/j.heliyon.2024.e28073.

[3] Gupta, N. K., Kar, M. K., & Singh, A. K. (2021). Load frequency control of two-area power system by using 2 degree of freedom PID controller designed with the help of firefly algorithm. In Control Applications in Modern Power System: Select Proceedings of EPREC 2020 (pp. 57–64). Springer Singapore.

[4] Gupta, N. K., Kar, M. K., & Singh, A. K. (2022). Design of a 2-DOF-PID controller using an improved sine–cosine algorithm for load frequency control of a three-area system with nonlinearities. *Protection and Control of Modern Power Systems*, 7(3), 1–18. https://doi.org/10.1186/s41601-022-00255-w.

[5] Kar, M. K. (2023). Stability analysis of multi - machine system using FACTS devices. *International Journal of System Assurance Engineering and Management*, 14(6), 2136–2145. (0123456789). https://doi.org/10.1007/s13198-023-02044-6.

[6] Kar, M. K., Kumar, S., Singh, A. K., & Panigrahi, S. (2021). A modified sine cosine algorithm with ensemble search agent updating schemes for small signal stability analysis. *International Transactions on Electrical Energy Systems*, 31(11), e13058.

[7] Memon, A. S., Laghari, J. A., Bhayo, M. A., Khokhar, S., Chandio, S., & Memon, M. S. (2023). Tunicate swarm algorithm based optimized PID controller for automatic generation control of two area hybrid power system. *Journal of Intelligent and Fuzzy Systems*, 45(2), 2565–2578.

[8] Mishra, D., Maharana, M. K., Kar, M. K., Nayak, A., & Cherukuri, M. (2023). Modified differential evolution algorithm for governing virtual inertia of an isolated microgrid integrating electric vehicles. *International Transactions on Electrical Energy Systems*, 2023(1), 8950650.

[9] Mishra, D., Maharana, M. K., Kar, M. K., Nayak, A., Islam, M. M., & Ustun, T. S. (2024). A metaheuristic algorithm for regulating virtual inertia of a standalone microgrid incorporating electric vehicles. *The Journal of Engineering*, 2024(5), 1–15.

[10] Mohanty, P., Sahu, R. K., Pradhan, P. C., & Panda, S. (2022). Design and analysis of the 2DOF-PIDN-FOID controller for frequency regulation of the electric power systems. *International Journal of Ambient Energy*, 43(1), 4463–4476. https://doi.org/10.1080/01430750.2021.1909134.

[11] Pani, S. R., Kar, M. K., & Bera, P. K. (2023). Reliability assessment of distribution power network. In 2022 2nd Odisha International Conference on Electrical Power Engineering, Communication and Computing Technology (ODICON), (pp. 1–5). IEEE.

[12] Pradhan, P. C., Sahu, R. K., Kar, M. K., Mehena, J., Sethy, S. K., & Samal, N. R. (2024). Analysis of differential evolution optimization-based cascade controller for frequency regulation of power system. In International Conference on Advanced Computing and Intelligent Engineering, (pp. 227–237). Singapore: Springer Nature Singapore. http://dx.doi.org/10.1007/978-981-99-5015-7_20.

[13] Qu, Z., Younis, W., Wang, Y., & Georgievitch, P. M. (2024). A multi-source power system's load frequency control utilizing particle swarm optimization. *Energies*, 17(2), 517.

[14] Sharma, A., & Singh, N. (2024). Load frequency control of connected multi-area multi-source power systems using energy storage and lyrebird optimization algorithm tuned PID controller. *Journal of Energy Storage*, 100(PB), 113609. https://doi.org/10.1016/j.est.2024.113609.1168.

40 Challenges and innovations in EV charging infrastructure: A focus on India's path to electrified transportation

Sushree Madhusmita Mishra[1,a], Anurekha Nayak[2,b], Nayan Ranjan Samal[2,c] and Debayani Mishra[2,d]

[1]Research Scholar, Department of EE, School of Engineering and Technology, DRIEMS University, Odisha, India

[2]Department of EE, School of Engineering and Technology, DRIEMS University, Odisha, India

Abstract

The rise of electric vehicles (EVs) represents a crucial shift toward sustainable transportation and a significant opportunity to reduce greenhouse gas emissions. This paper provides an in-depth examination of EV charging technologies, focusing on the evolving power supply methods, including alternating current (AC) and direct current (DC), as well as the functions of onboard and offboard chargers In addition to assessing the efficiency and practicality of various charging systems, this study explores the current status of EV charging infrastructure in India, identifying key initiatives, government programs, and private sector contributions to infrastructure expansion. The interoperability standards critical to ensuring compatibility across EVs and charging stations is also discussed in this paper. By addressing challenges such as grid resilience, charging demand forecasting, and energy distribution optimization, this research underscores the potential of EVs not only as transportation solutions but as pivotal elements of a sustainable energy framework. The findings and insights provided aim to guide future infrastructure planning and policy development, laying the groundwork for a resilient and efficient EV ecosystem.

Keywords: Battery swapping, charging demand forecasting electric vehicle supply equipment (EVSE), electric vehicles (EVs), sustainable transportation, vehicle-to-grid (V2G) technology EV charging infrastructure

Introduction

The advancement of electrified transportation presents a promising strategy to address the challenges posed by climate change. Numerous initiatives are currently in progress to mitigate emissions from the transportation sector. The primary objective is to create alternative fuels and implement advanced clean technology in vehicles, with the intention of decreasing greenhouse gas emissions while simultaneously enhancing vehicle performance [2]. The intermittency issues of non-conventional sources can be overcome with the implementation of vehicle-to-grid technology in power system. Concerns regarding EV charging's impact on the electrical grid are raised by the spread of EVs. Large EV fleets connecting to the grid for charging may have detrimental effects on it, including harmonics, system losses, voltage drop, phase imbalance, increased power consumption, equipment overloading, and stability problems [3]. The potential effects are further complicated by the dynamic behavior of EVs and the range of possible charging rates. However, controlling EV charging or using EVs as small distributed generators especially when using Vehicle to Grid mode would demonstrate the advantages of EV adoption [4]. Beyond only transportation, EVs are a symbol of the future of our world EVs are

further classified as battery electric vehicles (BEVs), hybrid electric vehicles (HEVs), and fuel cell electric vehicles (FCEVs). The grid-to-vehicle (G2V) system is the conventional approach for charging electric vehicles. Vehicle-to-grid (V2G) technology involves the capab0ility of a vehicle to act as a power source. In V2G mode, EVs can transfer their stored energy backup supply into the power grid. This technology serves various functions, including the stabilization of frequency and voltage, cost optimization, reduction of peak energy consumption, integration of renewable energy sources, and load balancing. Employing the V2G optimal logic control strategy is particularly beneficial for achieving a flattened load profile. V2G technology provides a viable approach to tackle the challenges associated with peak demand on the power system or distribution network [1]. The adoption of efficient strategies contributes in minimizing the various problems linked to distribution networks. The future EVs are expected to not only place demands on the electrical grid but also function as a form of decentralized power generation. Furthermore, they will be instrumental in load distribution across the system. The plug-in hybrid electric vehicle (PHEV) has the unique ability to both draw and supply power, allowing it to serve as both a load and a source for the grid [3]. Therefore, examining the effects of EV charging

[a]sushreemadhusmita03@gmail.com, [b]anurekha2611@gmail.com, [c]dr.nsamal@driems.ac.in, [d]debayanim@gmail.com

DOI: 10.1201/9781003658221-40

on the distribution network is of great theoretical significance and practical importance [5].

EV charging technologies can be assessed based on various criteria, including the method of battery charging, the direction of power flow, the type of chargers (onboard or offboard), and the power supply technique, which may vary according to specific requirements and locations. The fundamental components of an EV charging system are the electric vehicle supply equipment (EVSE), which facilitates the connection between the EV and the local electricity supply. Onboard and offboard chargers are utilized for integrating EVs with the grid, utilizing either AC or DC power [8].

Additionally, EV charging systems are designed to support either unidirectional or bidirectional power flow. The majority of commercially available onboard chargers feature unidirectional power flow, specifically grid-to-vehicle (G2V) capabilities, owing to their simplicity, reliability, cost-effectiveness, and straightforward control strategies. Conversely, bidirectional chargers possess the ability to transfer power back to the utility grid through vehicle-to-grid (V2G) functionality. As a result, bidirectional chargers are recognized as active distributed resources, equipped with specific control modes that facilitate load leveling, integration of renewable energy sources (RES), and the reduction of power losses within the utility grid.

In order to identify current issues and provide solutions, a thorough analysis of EV battery charging technologies, the power supply method, onboard or offboard chargers is necessary. EV battery charging technologies have evolved significantly to meet the growing demands for efficient, fast, and reliable energy delivery. The method of power supply, along with the type of charger, whether onboard or offboard is crucial in determining the speed, efficiency, and practicality of EV charging. EV battery charging technologies play a vital role in the EV ecosystem, affecting the efficiency, speed, and practicality of energy transfer to the vehicle's battery [7]. These technologies can be classified based on the power supply method—AC or DC and the type of charger, which can be either onboard or offboard. Onboard chargers are integrated into the vehicle and convert AC power from standard outlets into DC power for battery storage. They are typically utilized for home or workplace charging, where convenience and flexibility are key. However, their power capacity is limited, leading to longer charging times. On the other hand, offboard chargers are external devices that supply DC power directly to the battery, eliminating the need for onboard conversion. This enables much faster charging, especially at high-powered DC fast charging stations, which are crucial for minimizing downtime during extended journeys. Offboard chargers necessitate specialized infrastructure and are generally located in commercial or highway settings. As the adoption of EVs continues to grow, it is important to understand the interplay between onboard and offboard chargers and to optimize charging technologies to improve the overall EV experience and support the sustainability of charging infrastructure.

This research paper offers an extensive examination of EV charging technologies, emphasizing the various power supply methods, including AC and DC, as well as the functions of onboard and offboard chargers. It evaluates the efficiency, practicality, and infrastructure needs associated with each charging system, underscoring their influence on charging speed and adaptability.

Types of Charging System

Electric vehicle (EV) charging systems are categorized as: power levels, charging station, physical connection types, and the direction of energy flow. The following is an overview of these categories:

Types of power

In electric vehicle (EV) charging, power classification categorizes charging stations according to their charging speeds and the type of current utilized. This classification illustrated below helps clarify the different capabilities of EV charging stations, enabling better planning and implementation of EV infrastructure.

a) *AC Power:* AC charging delivers power to EVs through alternating current sourced from standard electrical outlets or dedicated chargers. Because EV batteries can only store DC, the vehicle's onboard charger is responsible for converting AC power into DC before it can be stored.

b) *DC Power:* DC charging delivers DC directly to an EV battery, bypassing the onboard charger, enabling much faster energy transfer compared to AC charging. This system uses high-voltage direct current provided by specialized DC fast chargers, allowing for rapid charging speeds. A concise representation of various charging levels is illustrated in Table 40.1.

Table 40.1 outlines the key characteristics of AC and DC charging levels for electric vehicles (EVs), offering a detailed look at their functional attributes and the contexts in which they are most appropriate for charging.

Table 40.1 Comparison level characteristics of EVs.

Attributes	Charging level (AC)		Charging level (DC)	
	Level 1	Level 2	DC fast	Ultra-Fast DC
Voltage/power	120V	240V	50-150kW	250-350kW
Charging speed	Slow	Moderate	Fast	Ultra -Fast
Range/Hr	3-5 miles	20-40 miles	100-150 miles	>200 miles
Connector types	Type 1 Type 2	Type 1 Type 2	CCS CHADeMO	Tesla, Ionity
Applications	Home for overnight charging	Residential & Public where daily recharge	Public charging networks minimizing downtime during trips	High-turnover locations recharge less than 20 mins

Source: Author

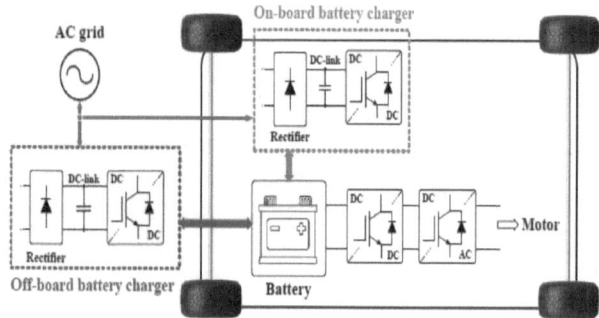

Figure 40.1 System for charging EVs
Source: Author

State of charging system

In the context of EV charging, onboard and offboard charging systems are two crucial elements that impact user experience, car design, and the overall efficacy of the development of electric vehicle infrastructure. To maximize charging options and promote the use of electric vehicles, it is essential to comprehend these categories.

a) *Onboard charging:* An onboard charger (OBC) is integrated directly within EV, enabling it to draw power from external AC sources, such as home outlets or public AC charging stations. It converts AC from the grid into DC to charge the vehicle's battery.

b) *Off board charging:* An offboard charger is located outside the vehicle and delivers DC power directly to the battery, eliminating the need for the vehicle's charger. EV battery chargers can be installed inside the car (on-board) or outside (off-board), as the following Figure 40.1 illustrates [5].

A comparative study between the above chargers is illustrated in Table 40.2.

Physical contact

Understanding the energy transmission from charging sources to automobiles in EV charging requires categorizing techniques according to physical contact. According to this framework, there are three main categories of EV charging: conductive, inductive, and battery swapping. These reflect various methods of providing energy to electric vehicles. Conductive charging involves a direct physical connection that enables the EV to charge from the power grid. This method can utilize two types of chargers: onboard and offboard chargers. In contrast, inductive charging, commonly referred to as wireless charging, does not require a physical connection between the EV and the power grid; instead, power is transmitted through an electromagnetic field. Battery swapping represents one of the quickest ways to obtain a fully charged battery for an EV.

In this approach, the owner replaces the depleted battery with a freshly charged one at a designated battery swapping station. Table 40.3 illustrates from various perspectives, the conductive charging, inductive charging, and battery switching techniques.

Direction of energy flow

Depending on how energy is used and transported inside their systems, electric vehicles can be divided into different categories. Understanding these divisions is crucial to developing effective energy management plans and enhancing EV performance in general. These vehicles' energy flow can be broadly categorized into the following groups: unidirectional and bidirectional chargers. EVs equipped with unidirectional chargers only draw power from the grid to charge the vehicle's battery, without feeding any electricity back into the grid. A bidirectional charger consists of two main power stages: the first is an active bidirectional AC-DC converter connected to the grid, which maintains a unity power factor (PF), and the second

Table 40.2 Comparative analysis of onboard and offboard charging sysyem.

Attributes	On-Board	Off-Board
Size	Small	Medium/large
Weight	Light	Heavy
Charging duration	Long	Short
Range of power	<50 kW	0-400 Kw
Merits	a. Provides flexibility for charging at various locations. b. Offers a more cost-effective solution for electric vehicle owners	a. Allows electric vehicles to charge at higher power levels so has fast charging capability b. Enhances overall efficiency and the performance of the EV.
Limitations	a. Charges at reduced power levels. b. Provides slow charging speeds. c. Increases the weight of the electric vehicle.	a. Potential battery heating problems. b. Limited flexibility for charging at different locations. c. More complicated and costlier to implement.

Source: Author

Table 40.3 Analysis of EV charging methods.

Features	Conductive charging	Inductive charging	Battery swapping
Charging Method	Direct electrical contact via cables	Wireless charging using magnetic fields	Replacing depleted batteries with charged ones
Infrastructure Ccst	Moderate; requires charging stations	Higher; requires specialized equipment	High; involves development Of swapping stations
Charging speed	Fast; depending on power output	Generally slower than conductive	Instant; swap process takes minutes
User experience	Simple; Plug and charge	Convenient; no need for physical connection	Quick and easy; less waiting time
Vehicle compatibility	Widely compatible with most EVs	Limited to vehicles designed for inductive	Requires vehicles designed for battery swapping
Maintenance	Low; occasional maintenance of stations	Higher; maintenance of coils and systems	Medium; requires managing battery inventory
Energy loss	Lower energy loss during transfer	Higher energy loss due to inefficiencies	Minimal during battery transfer
User adoption	High; established and familiar method	Growing; increasing in commercial settings	Still emerging; depends on infrastructure

Source: Author

is a bidirectional DC-DC converter that controls the charging current. The process of energy flow is illustrated in Figure 40.2.

The key differences between unidirectional and bidirectional energy flow in EVs are shown in the following comparison Table 40.4.

Current Status of EV Charging Infrastructure

The widespread use of electric vehicles and the development of a sustainable transportation system depends heavily on the infrastructure for charging them. Reducing driving range anxiety among prospective EV users requires the construction of a

Figure 40.2 Energy flow in EV
Source: Author

Table 40.4 Comparative analysis of energy flow in EV.

Features	Unidirectional	Bidirectional
Charging process	Energy flows from the charging station to the battery	Energy flows from the grid to the battery and vice versa
Discharging process	Energy flows from the battery to the electric motor	Energy can flow from the motor back to the battery
Regenerative braking	Not applicable (energy does not return to the battery)	Energy is recovered and sent back to the battery during braking
Efficiency	Generally efficient in single-direction scenarios	Can improve overall system efficiency through energy recovery
Applications	Common in standard battery electric vehicles (BEVs)	Used in vehicles designed for V2G and energy recovery systems

Source: Author

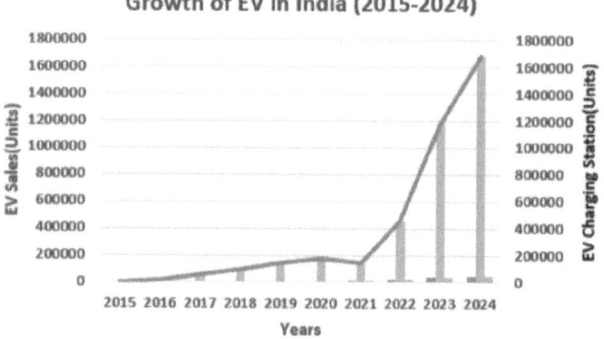

Figure 40.3 Rise of EVs in india
Source: Author

comprehensive and effective EV charging infrastructure. To promote a cleaner and greener future, governments, corporations, and energy suppliers are making significant investments in the development of this infrastructure. The faster adoption and manufacturing of hybrid & electric (FAME) Vehicles legislation, which was put into effect in 2015, marked the beginning of India's public charging station installations, with 520 charging stations installed nationwide. As of February 2024, 12,146 public charging stations had been installed in India, according to official data. Delhi, the nation's capital, is home to 1,886 of these charging stations. The growth of EV sales and charging infrastructure in India between 2015 and 2024 has seen remarkable increases, attributed to favorable policies, incentives, and a surge in consumer interest as represented in Figure 40.3. based on *NITI Aayog Reports*, 2024.

The red line indicates the increase in EV sales, while the orange bar line shows the rise in the number of charging stations. This growth highlights India's substantial strides in EV adoption and the expansion of its charging infrastructure, reflecting significant advancements in recent years. The rapid growth of

EVs in India has led to a considerable emphasis on building a complete infrastructure for charging EVs. This research outlines the key components of India's present EV charging infrastructure, including government initiatives, available infrastructure, challenges, and upcoming advancements.

a) *Government programs:* The FAME India scheme has approved the installation of thousands of charging stations across highways and cities to alleviate range anxiety. The Bureau of Energy Efficiency (BEE) has also revised guidelines to simplify tariffs and speed up electricity connections

b) *Private sector participation*: Companies like Tata Power, Ather Energy, and Statiq are leading the effort by building thousands of charging stations. Tata Power aims to establish 25,000 charging points, while Statiq plans to deploy 20,000 by 2023-24.

c) *Charging protocols and patterns:* While the CCS2 standard is becoming the norm for four-wheelers, standardization remains a challenge for two- and three-wheelers. Home charging remains common, though public infrastructure is more critical for long-distance travel and commercial fleet operations.

d) *Renewable energy integration*: The push to connect chargers with renewable energy aims to increase sustainability. By 2030, India plans to expand its renewable energy capacity to 500 GW, ensuring stable power for EV charging [6].

e) *Challenges and future plans*: Despite progress, the sector faces challenges such as interoperability, funding, and maintenance. Attracting investment and ensuring policy support will be essential to scaling the infrastructure as demand grows. Additionally, efforts to raise public awareness, streamline payment systems, and expand

Table 40.5 Standardization of EV charging and battery swapping in India.

Standard code & description	Charging type	Important attributes
IS 17017 (Indian Standard for EV Supply Equipment (EVSE)	AC	Provides performance and safety for AC charging.
IS 17017 (Part 2) (Specific requirements for AC EV chargers)	AC	Addresses safety procedures, testing, and design.
IS 14687 (Indian Standard for DC EV Charging Infrastructure)	DC	Focuses on performance, interoperability, and safety.
IS 17017 (Part 3)	DC	Emphasizes power levels and connectors.
BIS 20236 (Standard for battery swapping stations)	Battery Swapping	Highlights the compatibility and safety requirements for battery changing systems.
IS 16444 (Specification for battery swapping infrastructure)	Battery Swapping	Outlines specifications for battery design that must be standardized and modular.

Source: Author

rural infrastructure will be crucial for a smooth EV transition.

India's EV charging infrastructure is expected to grow due to cutting-edge technology, and cooperative efforts from government and commercial organizations. There is an emphasis on developing an extensive, easily accessible, and effective charging network that meets the various demands of the Indian population in order to promote a more seamless shift to electric vehicles by 2030.

EV Charging Demand Standards for Interopability

For users to have a seamless charging experience, India must adopt interoperability standards for EV charging. Standardized protocols facilitate successful communication between various EVs and charging stations as the demand for electric mobility increases. An outline of the main EV charging regulations that are presently in effect in India are discussed below in Table 40.5.

Assessing EV Charging Demand

The swift increase in electric EVs in India necessitates a detailed evaluation of charging demand together with the estimation of necessary infrastructure. This evaluation starts with comprehensive data collection and analysis, focusing on current trends in EV adoption, consumer behavior, and geographic factors. Findings indicate that urban regions possess a greater potential for charging stations due to higher population density and vehicle usage, emphasizing the need for infrastructure that caters to specific regional requirements.

After gathering data, demand forecasting utilizes utilization models and scenario analyses to anticipate future charging needs based on estimated EV sales. This method allows for the identification of various potential outcomes, aiding stakeholders in preparing for diverse market scenarios. As EV sales grow, the corresponding demand for charging stations is expected to rise, highlighting the critical need for infrastructure development. Estimating infrastructure requirements involves defining the necessary types and capacities of charging stations, including both fast chargers and level 2 chargers, based on anticipated EV usage patterns.

Geographic assessments help identify optimal locations for these charging stations, focusing on areas with high traffic and convenient access to services, thereby enhancing user convenience.

Conclusion

The study examined the evolution of EV charging technologies, assessing their impacts on infrastructure, efficiency, and practicality, with special emphasis on the distinctions between AC and DC systems, as well as onboard and offboard chargers. In India, government programs and private sector investment are fueling the expansion of EV charging infrastructure, though challenges like interoperability and standardization remain. The adoption of global and local standards for EV charging systems is vital to ensuring compatibility across vehicle types and charger designs, which will accelerate the transition to a sustainable transportation future. A comprehensive approach, integrating renewable energy and addressing regional EV charging demands, will be essential for scalable, efficient, and resilient EV infrastructure development, ensuring sustainability, grid stability, and widespread

EV adoption while minimizing environmental impact and operational costs.

References

[1]   Ahmad, F., Iqbal, A., Ashraf, I., & Marzband, M. (2022). Optimal location of electric vehicle charging station and its impact on distribution network: a review. *Energy Reports*, 8, 2314–2333.

[2]   Barman, P., & Dutta, L. (2024). Charging infrastructure planning for transportation electrification in India: a review. *Renewable and Sustainable Energy Reviews*, 192, 114265.

[3]   Dharmakeerthi, C. H., Mithulananthan, N., & Saha, T. K. (2011). Overview of the impacts of plug-in electric vehicles on the power grid. In 2011 IEEE PES Innovative Smart Grid Technologies, (pp. 1–8). IEEE.

[4]   Yong, J. Y., Ramachandaramurthy, V. K., Tan, K. M., & Mithulananthan, N. (2015). A review on the state-of-the-art technologies of electric vehicle, its impacts and prospects. *Renewable and sustainable energy reviews*, 49, 365–385.

[5]   Nasr Esfahani, F., Darwish, A., & Williams, B. W. (2022). Power converter topologies for grid-tied solar photovoltaic (PV) powered electric vehicles (EVs)—A comprehensive review. Energies, 15(13), 4648.

[6]   Sinha, P., Paul, K., Deb, S., & Sachan, S. (2023). Comprehensive review based on the impact of integrating electric vehicle and renewable energy sources to the grid. Energies, 16(6), 2924.

[7]   Pareek, S., Sujil, A., Ratra, S., & Kumar, R. (2020). Electric vehicle charging station challenges and opportunities: a future perspective. In 2020 International Conference on Emerging Trends in Communication, Control and Computing (ICONC3), (pp. 1–6). IEEE.

[8]   Saraswathi, V. N., and Ramachandran, V. P. (2024). A comprehensive review on charger technologies, types, and charging stations models for electric vehicles . *Heliyon*. 10,(20).

41 Utilization of areca fibre and bottom ash on geotechnical properties of expansive soil

Tapaswini Dash[a], Biplab Behera[b], Subham Priyadarshan[c] and Jayashree Bhuyan[d]

Department of CE, School of Engineering and Technology, DRIEMS University, Odisha, India

Abstract

The crucial requirements for every building that has to be made Important considerations in the field of any construction are the geotechnical and engineering characteristics of expansive or weak soil that show undesirable engineering characteristics, such as considerable swelling and shrinking. Stabilizing expansive soil requires the use of various admixtures made from industrial and agricultural waste to get around this undesired issue in building projects like road pavement or foundation work. The main aim and objective of this study is to find out using areca fiber and bottom ash to stabilize soil is practical. In coal-based thermal power plants, bottom ash is produced as a byproduct of burning coal, along with fly ash. In this investigation, bottom ash served as a stabilizing agent. As this natural areca nuts found in many parts of the India its application in geotechnical engineering is limited. The use of areca fiber and bottom ash is illustrated by a thorough experimental examination. Numerous tests, including the CBR test and the SPCT (standard proctor compaction test), were conducted. Different percentages of bottom ash (0%, 5%, 10%, 15%, and 20% by soil weight) and areca fiber (0%, 0.5%, 1%, and 1.5% by soil weight) were used. Cement was used in tiny amounts (3%) to increase the pozzolanic reaction. Finding the best way to use bottom ash by partially replacing fly ash is the primary goal of this study. According to the test results, the ideal proportions of areca fiber and bottom ash for subgrade soil are 1.5% and 15%, respectively. MDD rises and OMC values fall when bottom ash is added to the expanding soil compaction features. By adding bottom ash, the soil's CBR values improved, and the addition of areca fiber resulted in notable increases in the soil's engineering qualities. The presence of areca fiber also raised the soil's strength values. Therefore, using areca fiber and bottom ash together creates a safe and affordable material for stabilizing expansive soil.

Keywords: Areca fibre, bottom ash (BA), CBR, expansive soil, MDD, OMC, soil stabilization

Introduction

Several stabilizing approaches, including mechanical, thermal, and chemical treatments, can improve the characteristics of black cotton soil. Given that this soil contains montmorillonite, it is essential to comprehend its geotechnical and engineering properties. Significant volume variations between the wet and dry seasons are caused by this mineral, which can make building difficult. In conclusion, although there are significant difficulties in building with black cotton soil, creative stabilizing methods provide encouraging ways to enhance its engineering qualities [1]. The best result was found to be 1.2% material fiber and the percentage of fly ash is 15% by weight of given soil sample, which demonstrated enhanced in`strength metrics. When 30% bottom ash is added, the maximum dry density value exceeds the traditional [2]. The CBR value likewise rises as the amount of areca fiber increases. Brescia-Norambuena et al. [3] Experimental studies revealed that when BA was added to the soil up to 30%, OMC decreased and MDD rose; however, when 40% BA was added, this tendency reversed.

Additionally, it was shown that the CBR of the soil-BA mix increased up to 30% BA concentration and decreased with 40% BA. Chaduvula et al. [4] with 40% BA. The inclusion of fiber considerably increased the soil-BA-cement mix's strength, according to the results of CBR tests conducted on the mixture with different percentages of areca fibers. Areca fibers, which are currently discarded in significant quantities, have been discovered to be a viable resource for soil reinforcement. Chauhan et al. [5]. Because bottom ash had a lower cohesiveness than soil particles and a higher internal friction value, it displayed a mixed character. Chethan and Ravi Shankar [6]. CBR values have been inconsistent. The percentage of the ideal combination may vary based on the source. Choudhary et al. [7]. Bottom ash has been demonstrated to be a viable alternative to traditional granular material for a number of geotechnical applications. Chauhan et al. [5]. As more clay and cement are added, the maximum dry density rises. The cement's ability to bind particles together is what causes this rise in cohesiveness Cheriaf et al. [8]. Some researchers presented the effect of areca fiber

[a]tapaswini@driems.ac.in, [b]dr.biplab@driems.ac.in, [c]spriyadarshan@driems.ac.in, [d]jayashree@driems.ac.in

DOI: 10.1201/9781003658221-41

and 3% cement on lateritic soil of low volume pavement. When Areca fiber was added to lateritic soil, the characteristics of the soil improved somewhat, with the ideal content being 0.6% by soil weight. Das et al. [9]. The CBR values were increased when 3% cement by soil weight was added, along with Areca nut coir.

Findings from Literature Review

Civil engineering material, i.e. ash produced by various power plants is one of the wastes that need safe disposal techniques. The combustion products which are considered as output in thermal power plants for following the burning of coal are two different forms of ash: FA and BA. Bottom ash, which stays in the bottom of the combustion chamber, and fly ash, which are tiny particles that leave the chamber and either settle or are eliminated by air pollution control equipment. With the help of superior reactivity and pozzolanic reaction, it has been effectively used in cement and concrete, subgrade, and several other civil engineering applications.

Materials

Expansive soil
This is the foundational material utilized in the research. The weight of the dry soil partially replaces it with areca fiber for strengthening and bottom ash for stabilization.

Bottom ash
In thermal power plants, bottom ash is extracted from the dry boilers bottom part and generated as one type of granular substance. It is also produced during the burning of coal and is coarser than fly ash.

Areca fiber
Many sections of Asia, the tropical Pacific, and East Africa are home to the areca palm. In northern China, Bangladesh, India, Taiwan, Maldives, Ceylon, Laos, Cambodia, as well as in the West Indies, areca cultivation is quite common.

Experimental Program

Particle size distribution of soil: (Is – 2720 Part 4)
There is soil everywhere, varying in size and shape. In this case, the soil was the focus of dry sieve analysis. Plotting the analysis's findings on a semi-log graph yields a particle size distribution curve, with particle size serving as the abscissa and % finer N as the ordinate. If all of the particle sizes are adequately represented in nature, the soil is well graded; if there are too many or too few distinct sizes, the soil is badly or consistently graded. 10%, 60%, and 30% of the particles are finer than the diameters D10, D60, and D30, which are shown in millimeters. CU (coefficient of uniformity) = D60/D10 CC (coefficient of curvature) =D2 30/D60xD10 Again If the CU is > 4 for gravel and 6 for sand, and the CC is between 1 and 3, the soil is considered well graded; if not, it is considered badly graded.

Proctor compaction test (PCT)
The process of densifying soil by minimizing air spaces is known as compaction. A rammer is used to compress the sample in a standard proctor mold after expansive soil and water have been combined as shown in figure.

$$\gamma d = \frac{m/v}{1+w}$$

Where total mass of the soil = 'M', volume of soil='V', water content='w'. Similarly, different percentages of areca fiber (0.5%, 1%, and 1.5%) and bottom ash (5%, 10%, 15%, and 20%) by weight of soil were used in this test to measure OMC and MDD.

California bearing ratio test (Is 2720-16-1987)
A penetration rate of approximately 1.25 mm per minute was attained by loading the penetration plunger. The load readings were obtained at penetrations of 0, 0.5, 1.0, 1.5, 2.0, 2.5, 4.0, 5.0, 7.5, 10.0, and 12.5 mm as shown in Figure 41.3.

Results and Discussions

Both expansive soil with and without admixtures were subjected to a variety of tests. It has been investigated how bottom ash and areca fiber together affect the soil's MDD and OMC, shear parameters, CBR strength, and unconfined compressive strength. Table 1 represents the virgin properties of the soil obtained from the test.

Effect on dry density and moisture content
Figure 41.3 shows the relationship between percent of BA on MDD whereas Figure 41.4 shows the relationship between optimum moisture content and bottom ash percentage. Figure 41.5 illustrates the effect of fibre on MDD. The MDD rises up to 15% as the percentage of BA rises and then falls with additional additions. Figure 41.6 illustrates how OMC and MDD fluctuate with the percentage.

Figure 41.1 Proctor mould
Source: Author

Figure 41.2 CBR with mould
Source: Author

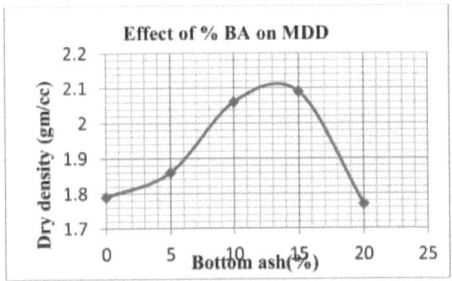

Figure 41.3 Comparison graph between dry densities with varying BA content
Source: Author

Table 41.1 Expansive soil virgin properties obtained.

Sl No	Conducted tests	Test value
1	FSI (Free swell index)	50%
2	LL (Liquid limit)	52.2%
3	PL (Plastic limit)	20.92%
4	PI (Plasticity index)	23.47%
5	'g' (Specific gravity)	2.64
6	OMC	20.2%
7	MDD	1.79gm/cc
8	CBR unsoaked	5.83%
9	CBR soaked	2.05%
10	cohesion	0.511kg/cm2
11	Angle of internal friction	10°

Source: Author

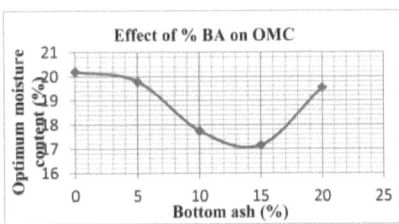

Figure 41.4 Comparison graph between OMC with varying BA content
Source: Author

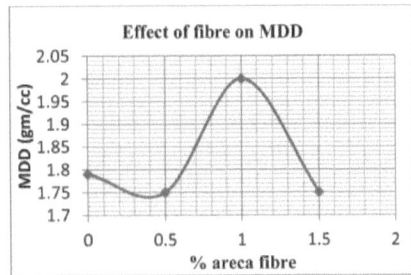

Figure 41.5 Comparison graph between MDD and varying AF content with 15% BA
Source: Author

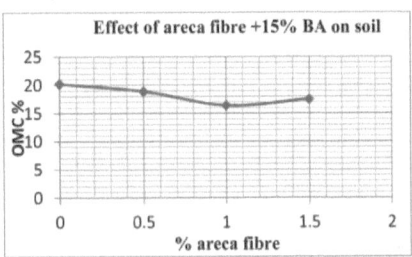

Figure 41.6 Comparison graph between OMC and varying of AF content with 15% BA
Source: Author

Effect on CBR strength

CBR test is used to determine the bearing capacity of soil and used in pavement design. CBR values show an increment till the % of BA up to 15% and then decreases beyond 15% of BA mix. It was observed that the un-soaked CBR values greater than soaked CBR values at 2.5mm penetration. The effect of BA and areca fibre mix effect on soil was given in figure below. Figure 41.7 illustrates the effect of BA on unsoaked CBR.

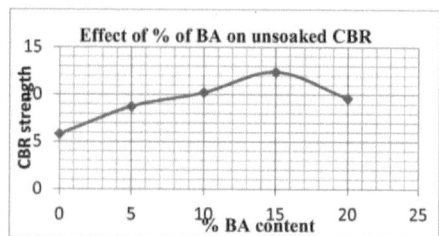

Figure 41.7 Effect of BA content on un-soaked CBR strength
Source: Author

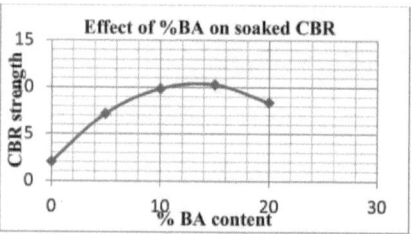

Figure 41.8 Effect of BA content on soaked CBR strength
Source: Author

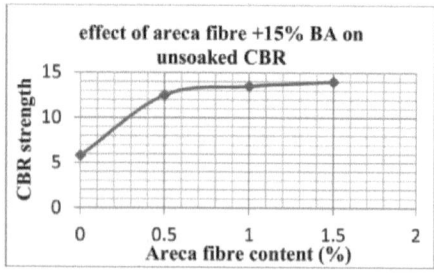

Figure 41.9 Effect of areca fiber content on un-soaked CBR strength
Source: Author

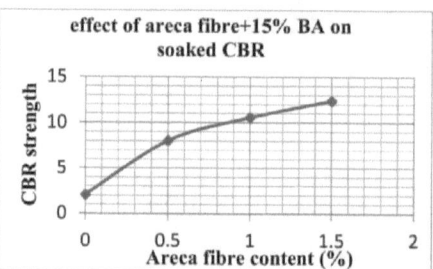

Figure 41.10 Effect of areca fiber content on soaked CBR strength
Source: Author

Conclusions

The soil is well-graded sand, with coefficients of uniformity and curvature of 9.37 and 1.043, respectively. The BA values of Cu and Cc are 2.60 and 1.03, indicating uniform grading. Virgin soil and bottom ash have specific gravities of 2.64 and 2.46, respectively. In Figure 41.8 BA has a lower specific gravity than dirt. When BA was added to the soil up to 15%, it was shown that OMC decreased and MDD increased; however, this was reversed when 20% BA was added as shown in Figure 41.9. Maximum CBR value for unsoaked condition is obtained at 15% of BA (12.40%) and soaked value is obtained at 15% of BA (10.29%) as shown in Figure 10.

The above outcomes are based on experimental investigations with limited range of materials. It is concluded that the agricultural waste areca fiber and industrial waste bottom ash in combination can be used in constructions of pavement safely. Bottom ash and areca fiber lead to cost effective and eco-friendly construction with satisfying strength requirements.

Acknowledgment

The author is very great thankful to Dr. Biplab Behera H.O.D, Department of Civil Engineering, for his guidance, which helped to improve the present paper; and Mr. Subham Priyadarshan, Assistant Professor , Department of Civil Engineering, for reading and preparing the research paper and giving valuable suggestions to improve the quality of the paper.

References

[1] Abdel-Gawwad, H. A., Rashad, A. M., Mohammed, M. S., & Tawfik, T. A. (2021). The potential application of cement kiln dust-red clay brick waste-silica fume composites as unfired building bricks with outstanding properties and high ability to CO2-capture. *Journal of Building Engineering*, 1(42), 102479.

[2] Abou-Taleb, A. N., Musaiger, A. O., & Abdelmoneim, R. B. (1995). Health status of cement workers in the United Arab Emirates. *Journal of the Royal Society of Health*, 115(6), 378–81.

[3] Brescia-Norambuena, L., Gonzalez, M., Avudaiappan, S., Flores, E. I., & Grasley, Z.(2021). Improving concrete underground mining pavements performance through the synergic effect of silica fume, nanosilica, and polypropylene fibers. *Construction and Building Materials*, 285, 122895.

[4] Chaduvula, U., Viswanadham, B. V., & Kodikara, J. (2016). Desiccation cracking behavior of geofiber-reinforced expansive clay. In Geo-Chicag, (pp. 368–377).

[5] Chauhan, M. S., Mittal, S., & Mohanty, B. (2008). Performance evaluation of silty sand subgrade reinforced with fly ash and fibre. Geotextiles and Geomembranes, 26(5), 429–435. https://doi.org/10.1016/j.geotex-mem.2008.02.001.

[6] Chethan, B. A., & Ravi Shankar, A. U. (2022). Areca fiber reinforced alkali-activated black cotton soil using class f fly ash and limestone powder for pavements. In

Road and Airfield Pavement Technology: Proceedings of 12th International Conference on Road and Airfield Pavement Technology, (pp. 331–346). Cham: Springer International Publishing.

[7]   Choudhary, R., Gupta, R., Alomayri, T., Jain, A., & Nagar, R. (2021). Permeation, corrosion, and drying shrinkage assessment of self-compacting high strength concrete comprising waste marble slurry and fly ash, with silica fume. In Structures, (Vol. 33, pp. 971–985). Elsevier.

[8]   Cheriaf, M., Rocha J. C., & Pera, J. (1999). Pozzolanic properties of pulverized coal combustion bottom ash. *Cement and Concrete Research*, 29(9), 1387–1391. https://doi.org/10.1016/S0008-8846 (99)00098-8.

[9]   Das, B., Yen, S., & Das, R. (1995). Brazilian tensile strength test of lightly cemented sand. *Canadian Geotechnical Journal*, 32(1), 166–171. https://doi.org/10.1139/t95-013.

42 Nature-inspired metaheuristic algorithm optimized TIDN controller for frequency management in networked power system with storage device and HVDC link

Sushanta Kumar Sethy[1,a], Pratap Chandra Pradhan[2,b], Manoj Kumar Kar[3,c] and Binod Kumar Sahu[4,d]

[1]Department of Electrical Engineering, SOET, DRIEMS University, Cuttack, Odisha, India

[2]Department of Electrical and Electronics Engineering, SOET, DRIEMS University, Cuttack, Odisha, India

[3]Electrical and Electronics Department, Tolani Maritime Institute, Pune, Maharashtra, India

[4]Department of Electrical Engineering, ITER, SOA Deemed to be University, Bhubaneswar, Odisha, India

Abstract

This research paper presents the TIDN controller for LFC in a multi-area thermal power system (PS), employing an AGTO. Additionally, the integration of an HVDC link and CES has been suggested to further enhance the transient performance of the system. The controller performance has been validated using a two-area PS model with governor deadband (GDB) characteristics. The proposed controller's performance was evaluated against established control methods. The results show that the AGTO-TIDN controller outperforms the AGTO-PI and AGTO-PID controllers, with a 19.91% and 16.58% reduction in ITAE value, respectively. Simulation results demonstrate that the AGTO-based system, incorporating the HVDC link and capacitive energy source, exhibits better transient.

Keywords: Artificial gorilla troops optimizer, capacitive energy storage, governor dead band, HVDC link, load frequency control, tilt–integral–derivative with filter controller

Introduction

The importance of load frequency control (LFC) is particularly evident in large-scale interconnected PS, where multiple areas are connected through tie-lines. In such complex systems, maintaining the frequency of each individual region and the planned power transfer via tie-lines poses a significant challenge [11].

Different soft computing algorithms and control techniques such as modified volleyball premier league (MVPL)-optimized 3-DOF (FOPI)–FOPD [11], hFA-PS tuned fuzzy PID [12], SSA tuned PIDF (1+FOD) [7], hDE-PS tuned FOPID [8], DE tuned 2DOF-PIDN-FOID [4], DE based hPIDN-FOPD [5], DE tuned hFPID [9], ICA based integral [10], Improved ACO tuned FPID Chen et al. [2], HBA optimized FOTID [6] Controllers are taken in the literature for effective LFC of multi area single/multi sources PS.

The literature review indicates that novel control structures and optimization techniques have the potential to enhance PS control. Motivated by the success of these developments, the authors propose an AGTO-optimized TIDN controller for the current study.

Configuration of the System Model

The model of a 2-area LFC system with GDB is depicted in Figure 42.1. To ensure a more realistic analysis, the GDB must be incorporated, which introduces nonlinearity to the PS. Sahu et al. [12], Chen et al. [2] presents the typical values of the LFC model parameters.

In this investigation, Integral of Time multiplied Absolute Error (ITAE) is employed as the objective function (OF) for the LFC analysis.

Expression for ITAE is provided in (1).

$$J = ITAE = \int_0^{t_{sim}}(|\Delta P_{tie}| + |\Delta F_1| + |\Delta F_2|).t.dt \tag{1}$$

where, t_{sim} is simulation period. $\Delta F_1, \Delta F_2, \Delta P_{tie}$ are frequency change and tie-line power change respectively.

CES unit comprises a supercapacitor and a Power Conversion System (PCS) [10].

Controller Structure

The TIDN controller structure, depicted in Figure 42.2, resembles the traditional PID controller [3, 6].

[a]sushantasethy@driems.ac.in, [b]pratap_pin@yahoo.com, [c]manojkar132@gmail.com, [d]binoditer@gmail.com

DOI: 10.1201/9781003658221-42

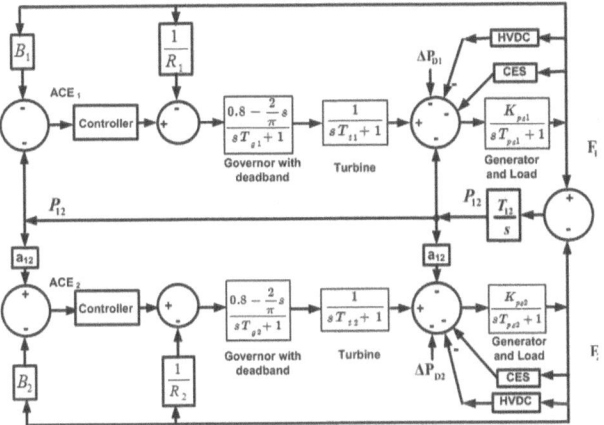

Figure 42.1 Transfer function (TF) Model of 2-area test system with HVDC & CES
Source: Author

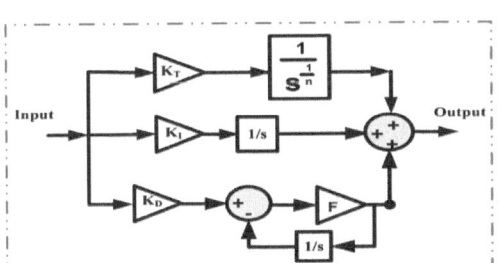

Figure 42.2 TIDN controller [6]
Source: Author

The mathematical model of a TIDN controller's transfer function is given below in (2).

$$T(s) = \frac{K_p}{S^{\frac{1}{n}}} + \frac{K_i}{S} + \frac{K_d\, NS}{S+N} \qquad (2)$$

Artificial Gorilla Troops Optimizer (AGTO)

This study employs the AGTO technique. Gorillas are highly social animals that live in troops led by a dominant silverback male.

The AGTO algorithm uses five operators based on gorilla behaviors. It employs a distinct method for the exploration and exploitation phase change process.

To enhance comprehension, the AGTO phase change procedure is illustrated in Figure 42.3, and formulation algorithm's steps and mathematical equations are thoroughly explained in Abdollahzadeh et al. [1].

Methodology

This research investigates a 2-area thermal-thermal PS with HVDC link and storage device, as depicted in Figure 42.1. Each PS area contains a generator, GDB,

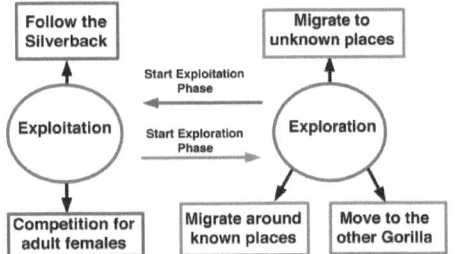

Figure 42.3 Different phases of AGTO [1]
Source: Author

Table 42.1 AGTO-tuned TIDN /PI/PID controller parameters with HVDC link and CES.

Parameters	AGTO-PI	AGTO-PID	AGTO- TIDN (Proposed)
K_{T1}	-	-	5.0000
K_{P1}	1.3438	1.9112	-
K_{I1}	5.0000	5.0000	5.0000
K_{D1}	-	0.4227	1.6712
F_1	-	-	100.0000
n_1	-	-	1.4444
K_{T2}	-	-	1.8081
K_{P2}	-1.5796	-2.0073	-
K_{I2}	2.5213	2.2981	2.7929
K_{D2}	-	-0.3509	5.0000
F_2	-	-	100.0000
n_2	-	-	1.0000

Source: Author

turbine, HVDC link, and CES unit. TIDN controllers are employed in the respective areas. The ITAE criterion is employed as the OF, and an AGTO is used to tune the controller gains. The efficacy of the recommended AGTO-tuned TIDN controller is tested and compared to that of PID and PI controllers. The proposed approach successfully stabilizes the frequency and enhances the overall PS responses.

Results and Discussion

The researchers used the MATLAB/SIMULINK platform to model a two-area PS incorporating an HVDC link and a storage device. In the proposed model TIDN controller is used to meet the LFC efficiently. The AGTO algorithms were coded in MATLAB script files. The parameter of the controllers was tuned using the AGTO algorithm, and the tuned gain of the TIDN/PID/PI controllers is given in Table 42.1. A comparative assessment of the PI, PID, and suggested TIDF controllers was carried out for each case study.

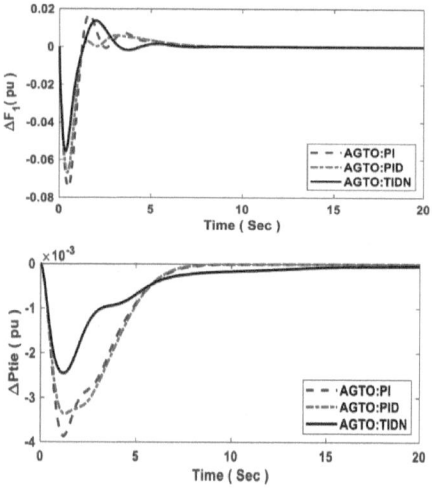

Figure 42.4 Dynamic response of PS in terms of deviation (ΔF_1, ΔP_{tie}) for 10% SLP in area 1
Source: Author

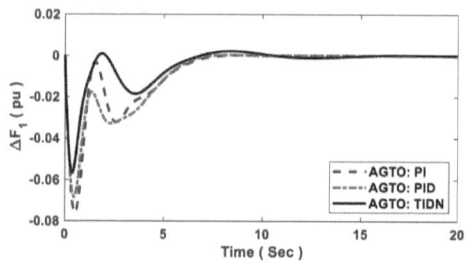

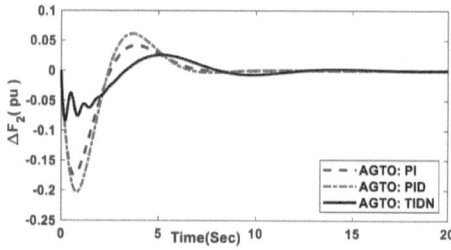

Figure 42.5 Dynamic response of PS in terms of deviation (ΔF_1, ΔF_2) with 10% SLP (in area-1) and 20% SLP (in area-2)
Source: Author

Table 42.2 Comparative performance of TIDN controller with PI/PID controllers.

	Controllers			
		PI	PID	TIDN
ΔF_1	U.S.	-0.076	-0.07	-0.058
	S.T.	15	13	12
ΔF_2	U.S.	-0.18	-0.2	-0.08
	S.T.	17	15	13
ΔP_{tie}	O.S.	$15*10^{-3}$	$12*10^{-3}$	$6*10^{-3}$
	S.T.	18	16	14
J	ITAE	0.2706	0.2598	0.2167

U.S. = undershoot, S.T. = settling time, O.S. =overshoot
Source: Author

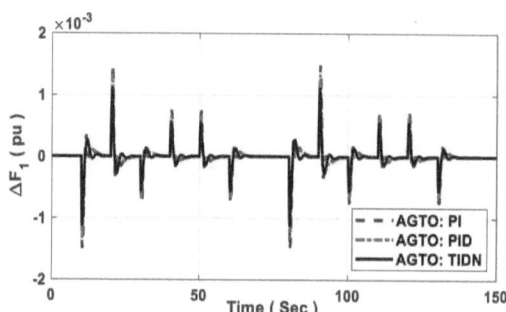

Figure 42.6 Variation of frequency in area-1 due to random load variation
Source: Author

Table 42.2 clearly shows that proposed AGTO optimized TIDN controllers produce the minimum errors (ITAE) as compared to PI and PID controller. Comparing AGTO-TIDN to AGTO-PI and AGTO-PID controllers, the ITAE (J) value decreased by 19.91% and 16.58%, respectively.

The results presented in Table 42.2 and Figures 42.4 and 42.5 indicate that the suggested technique is effective in mitigating overshoots, undershoots, and reducing settling time.

Scenario1: Single area disturbance

The behavior of the PS is assessed by applying a step load perturbation (SLP) of 10% (area-1). To demonstrate the performance of the PS, the response of the TIDN controller is compared to that of PID and PI controllers. The dynamic deviations (ΔF_1, ΔP_{tie}) are shown in Figure 42.4.

Scenario 2: Both area disturbance

The behavior of the PS is assessed by applying a SLP of 10% (area-1) & 20% (area-2). The variations (ΔF_1, ΔF_2) are presented in Figure 42.5.

Scenario 3: Random load variation

The growing industrial activities have led to increased electricity demands from nearby industries, households, and other loads. As a result, these electricity requirements are treated as random load variations impacting each area of the PS.

The analysis of Figures 42.4–42.6 indicates that the power system equipped with an HVDC link and CES exhibits superior performance. The results presented indicate that the suggested AGTO technique and TIDN controllers demonstrate superior performance compared to other controllers.

Conclusion

This paper validates the superior performance of a TIDN controller in LFC within a two-area PS consisting of an HVDC link and CES. The ITAE is chosen as the OF, and AGTO is used as the optimization algorithm to optimize controller gains. The investigation evaluates the controller's capability in maintaining PS frequency and minimizing undershoot, overshoot, and settling time. Compared to PID and PI controllers, the proposed TIDN controller demonstrates improved dynamic performance in terms of ST and undershoots, as illustrated in Figures 42.4 and 42.5. When comparing AGTO-TIDN to AGTO-PI and AGTO-PID controllers, the ITAE value is reduced by 19.91% and 16.58%, respectively. The ITAE value presented in Table 42.2 suggests that the suggested control technique is effective in mitigating the damping in the system. Lastly, effectiveness of the proposed method is validated under random load patterns.

References

[1] Abdollahzadeh, B., Soleimanian Gharehchopogh, F., & Mirjalili, S. (2021). Artificial gorilla troops optimizer: a new nature-inspired metaheuristic algorithm for global optimization problems. *International Journal of Intelligent Systems*, 36(10), 5887–5958.

[2] Chen, G., Li, Z., Zhang, Z., & Li, S. (2019). An improved ACO algorithm optimized fuzzy PID controller for load frequency control in multi area interconnected power systems. *IEEE Access*, 8, 6429–6447.

[3] Lurie, B. J. (1994). Three-parameter tunable tilt- integral-derivative (TID) controller. US Patent 5.

[4] Mohanty, P., Sahu, R. K., Pradhan, P. C., & Panda, S. (2022). Design and analysis of the 2DOF-PIDN-FOID controller for frequency regulation of the electric power systems. *International Journal of Ambient Energy*, 43(1), 4463–4476.

[5] Mohanty, P., Sahu, R. K., Pradhan, P. C., Samal, N. R., & Panda, S. (2022). Simulation and hardware-in-the-loop real time testing of different controllers for frequency regulation of electrical power systems. *International Journal of Ambient Energy*, 43(1), 7560–7575.

[6] Naik, A. K., Jena, N. K., Sahoo, S., & Sahu, B. K. (2024). Optimal design of fractional order tilt-integral derivative controller for automatic generation of power system integrated with photovoltaic system. *Electrica*, 24(1), 140–153.

[7] Prakash, A., Kumar, K., & Parida, S. K. (2020). PIDF (1+FOD) controller for load frequency control with SSSC and AC–DC tie-line in deregulated environment. *IET Generation, Transmission and Distribution*, 14(14), 2751–2762.

[8] Pradhan, P. C., Sahu, R. K., & Panda, S. (2021). Comparative performance analysis of hDE-PS technique for frequency control of the electric power system. *Journal of Electrical Systems and Information Technology*, 8, 1–21.

[9] Pradhan, P. C., & Sahoo, J. R. (2019). Design and analysis of hybrid fuzzy PID controller for diverse source of power system using DE optimization technique. *ARPN Journal of Engineering and Applied Sciences*, 14(7), 1309–1316.

[10] Ponnusamy, M., Banakara, B., Dash, S. S., & Veerasamy, M. (2015). Design of integral controller for LFC of SSSC and CES based multi area system consisting of diverse sources of generation employing imperialistic competition algorithm. *Electrical Power and Energy Systems*, 73, 863–871.

[11] Sariki, M., Priyesh, S., Kumar, A., Ravi, S., & Parida, S. K. (2024). Improved LSTM-Based load forecasting embedded 3DOF (FOPI)-FOPD controller for proactive frequency regulation in power system. *IEEE Transactions on Industry Applications*. doi -10.1109/ TIA:3443243.

[12] Sahu, R. K., Panda, S., & Pradhan, P. C. (2015). Design and analysis of hFA-PS based fuzzy PID controller for LFC of multi area power systems. *International Journal of Electrical Power & Energy Systems*, 69, 200–212.

43 Effect of compression on floating ice-floe with an irregular porous seabed

Dumesh Meher[1,a], Sagarika Khuntia[2,b] and Manas Ranjan Sarangi[1,c]

[1]Department of Mathematics, Gandhi Institute of Engineering and Technology University, Odisha, Gunupur, India

[2]Department of Mathematics, DRIEMS University, Odisha, India

Abstract

The study provides a solution to the water wave scattering problem, whereby the upper surface is obscured by ice floe under a compressive force, treated as a lean flexible plate, where the down surface is regarded as permeable with small distortion. Perturbation technique is applied to reduce the given problem into simpler form and then Fourier transform method is applied to get the first-order correction of reflected as well as transmitted energies. A special kind of base distortion topography is considered to calculate the aforesaid energies. The results obtained here closely align with the energy-balance relation, which represents a significant advantage of this study.

Keywords: Base distortion, compressive force, energy identity, Fourier transform technique, ice-floe, porous bed, reflected energy, transmitted energy

Introduction

The scattering of water waves by floating or immersed structures, including versatile floating plates on the seabed, is crucial for coastal regions and marine construction.

Porter and Porter [5] analyzed the water wave scattering problem by an ice-sheet with the undulating bed of the fluid. Jeng [1] investigated the transcendental equation of water waves with porous seabed by utilizing the complex number in pro-elastic model. Considering permeable seabed, Mohapatra and Sarangi [4] analyzed the water wave diffraction problem and shown that the results are satisfying the energy balance relation. Mohapatra [3] and Khuntia and Mohapatra [2] obtained the interaction problem of surface water waves by a permeable base with base distortion of ice-covered ocean.

This study examines the propagation of incident waves by a porous surface with little base distortion, where the upper surface is coated with ice floes under the influence of a compressive force, designed as a lean flexible plate. The problem is reduced to simpler form by perturbation technique and then Fourier transform method is applied to get the first order correction of reflected as well as transmitted energies. The reflected and transmitted energies depend on the shape of the base distortion. This study employs a particular type of base distortion topography to compute the aforementioned energies. The "energy balance relation" is computationally verified to guarantee the accuracy of the analytical and numerical conclusions involving reflected and transmitted energies.

Problem Statement

Consider an inviscid, incompressible fluid with a porous, distorted base surface and an ice-floe in presence of compressive force covering the top surface, which is supposed to be a thin, flexible plate. The base surface of the fluid is described by $y = h_b + \varsigma f(x)$, where $f(x)$ specifies the shape function and ς examines the smallness of the base distortion and $f(x)$ tends to 0 as $|x|$ tends to ∞. It is believed that the fluid flows in an irrotational, time-harmonic manner. With the above assumptions, the velocity potential in a fluid with density ρ can be expressed as $Re[\Psi(x, y) e^{-i\omega t}]$, where Re represents the real component and $\Psi(x, y)$ fulfils

$$\frac{\partial^2 \Psi}{\partial x^2} + \frac{\partial^2 \Psi}{\partial y^2} = 0 \ , |x| < \infty \ , \tag{2.1}$$

$$y \in [0, h_b + \varsigma f(x)]$$

$$\left(D \frac{\partial^4}{\partial x^4} + N \frac{\partial^2}{\partial x^2} + 1 \right) \frac{\partial \Psi}{\partial y} + K\Psi = 0 \ , \tag{2.2}$$

$$|x| < \infty, \ y = 0$$

$$\Psi_n - P\Psi = 0 \ , y = h_b + \varsigma f(x) \tag{2.3}$$

$$\Psi = \begin{cases} (e^{is_0 x} + R e^{-is_0 x}) \chi(s_0, y), & x \to -\infty \\ T e^{is_0 x} \chi(s_0, y), & x \to \infty \end{cases} \tag{2.4}$$

[a]dumeshmeher@giet.edu, [b]sagarikakhuntia113@gmail.com, [c]manasranjan@giet.edu

DOI: 10.1201/9781003658221-43

where $K = \frac{\omega^2}{g}$, is the flexural rigidity of the flexible plate, $N = \frac{Q}{\rho g}$ represents the compressive force acting on the ice-floe, $\frac{\partial}{\partial n}$ indicates the normal derivative to the sea bed at a point (x, y), P is the porosity parameter of the porous bed. The linearized boundary condition for the bottom surface (2.3) may be expressed as follows, taking ς as the parameter of the little bottom deformation of the porous bed and ignoring the second and higher order terms:

$$\Psi_y - \varsigma \left\{ \frac{d}{dx} [f(x)\Psi_x(x, h_b)] \right\} - P[\Psi + \varsigma f(x)\Psi_y] + O(\varsigma^2) = 0 \quad (2.5)$$

Method of Solution

Using Perturbation technique, we can express Ψ, R and T as

$$\left. \begin{array}{c} \Psi(x, y) = \Psi_0(x, y) + \varsigma \Psi_1(x, y) + O(\varsigma^2), \\ R = \varsigma R_1 + O(\varsigma^2), \\ T = 1 + \varsigma T_1 + O(\varsigma^2), \end{array} \right\} \quad (3.1)$$

where $\Psi_0(x, y) = e^{i s_0 x} \chi(s_0, y)$ with $\chi(s_0, y)$

$$= \left[\frac{\cosh s_0 (h_b - y) - \left(\frac{P}{s_0}\right) \sinh s_0 (h_b - y)}{\cosh s_0 h_b - \left(\frac{P}{s_0}\right) \sinh s_0 h_b} \right], \quad (3.2)$$

$$|x| < \infty, \, y \in [0, h_b],$$

with $\cosh s_0 h_b - \left(\frac{P}{s_0}\right) \sinh s_0 h_b \neq 0$ and s_0 satisfies the following dispersion equation

$$\begin{array}{c} H(s) \equiv \left[(Ds^4 - Ns^2 + 1)s + \\ \left(\frac{P}{s}\right) K \right] + \tanh s h_b - \\ [(Ds^4 - Ns^2 + 1)P + K] = 0 \end{array} \quad (3.3)$$

Using Equation (3.1) in equations (2.1), (2.2), (2.4) and (2.5) and comparing $O(\varsigma^1)$ terms, we will obtain the first-order boundary value problem. Solving the first-order boundary value problem (BVP) using Fourier transform technique, we obtain R_1 and T_1 as

$$R_1 = \frac{i(Ds_0^4 - Ns_0^2 + 1)(P^2 + s_0^2)}{H'(s_0) \left[\cosh s_0 h_b - \left(\frac{P}{s_0}\right) \sinh s_0 h_b \right]} \times$$

$$\int_{-\infty}^{\infty} f(x) e^{2 i s_0 x} \, dx, \quad (3.4)$$

$$T_1 = \frac{i(Ds_0^4 - Ns_0^2 + 1)(P^2 - s_0^2)}{H'(s_0) \left[\cosh s_0 h_b - \left(\frac{P}{s_0}\right) \sinh s_0 h_b \right]} \times$$

$$\int_{-\infty}^{\infty} f(x) \, dx \quad (3.5)$$

Hence, the reflected as well as transmitted energies can be calculated, once $f(x)$ is known.

Particular form of Bottom Profile

Let us consider the shape function $f(x)$ which can be stated as

$$f(x) = \begin{cases} a \sin \lambda x, & \frac{-m\pi}{\lambda} \leq x \leq \frac{m\pi}{\lambda} \\ 0 & otherwise \end{cases} \quad (4.1)$$

where a is amplitude and λ is wavenumber, of the sinusoidal undulation and $m \in z^+$. For m number of sinusoidal ripples, R_1 and T_1 can be determined.

It should be highlighted that the energy balance relation is crucial for verifying the outcome of mixed BVPs. The energy identity equation concerning reflected and transmitted energy is known to satisfy the following equation when series of incident water waves passes through a permeable sea-bed of variable depth beneath an infinitely enlarged floating flexible plate under compressive force:

$$|R|^2 + |T|^2 = 1. \quad (4.2)$$

Numerical Outcomes and Discussion

For the above shape function, the first order transmitted energy T_1 vanishes. The numerical estimates for R_1 and T_1 associated with particular type of base profile are shown in the preceding section. The reflected energy R_1, as specified in Eq. (3.4), is calculated analytically and is plotted versus $K h_b$. The value of $|R_1|$ is shown in Figure 43.1, for various flexural rigidity parameters which are considered as $a/h_b = 0.1$, $\lambda/h_b = 1$, $m = 4$, $P h_b = 0.02$, $N/h_b^2 = 0.1$. This figure shows that decreasing the flexural rigidity parameter leads to an increase in the absolute value of R_1. Moreover, it is noted that the maximum values of R_1 are reached as the base distortion wavenumber on the porous seabed approaches nearly double that of the drifting ice plate. Figure 43.2 shows R_1 for various porosity parameters. Here, we set $D/h_b^4 = 0.8$, $m = 4$, $N/h_b^2 = 0.1$, $\lambda h_b = 1$. It is evident from this figure that as the porosity parameter rises, so does the absolute value of R_1. This indicates that the change in the porosity parameter of the porous seafloor affects the reflected energy. Figure 43.3 show

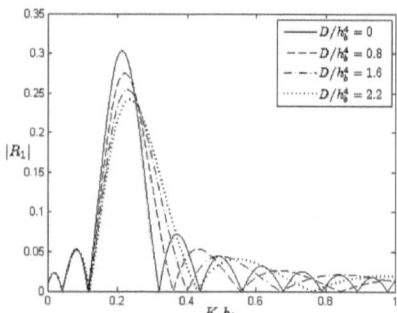

Figure 43.1 $|R_1|$ vs Kh_b for $Ph_b = 0.02$ and $N/h_b^2 = 0.1$
Source: Author

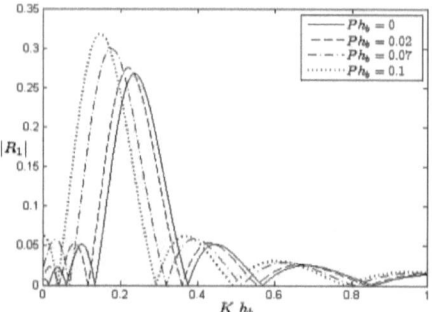

Figure 43.2 $|R_1|$ vs Kh_b for $N/h_b^2 = 0.1$ and $D/h_b^4 = 0.8$
Source: Author

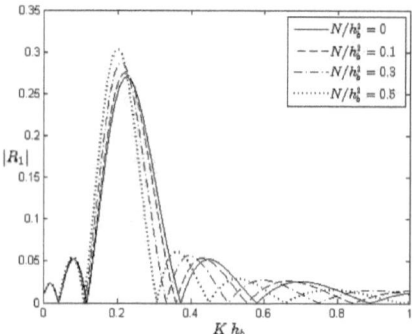

Figure 43.3 $|R_1|$ vs Kh_b for $D/h_b^4 = 0.8$ and $Ph_b = 0.02$
Source: Author

R_1 is depicted for different compressive force acting on the ice-floe with $D/h_b^4 = 0.8$, $m = 4$, $Ph_b = 0.02$, $\lambda h_b = 1$. This figure indicates that when the compressive force exerted on the ice floe increases, the global maximum value of R_1 also increases. Additionally, the surface wave exhibits minimal sensitivity to variations in the compressive force parameter, indicating that $|R|$ remains relatively unaffected.

Within the framework of an infinitely enlarged floating adaptable plate under the influence of a compressive force, the interaction of an occurring surface water wave and a porous bed characterized by undulation leads to fluctuations in $|R|$, $|T|$, and the energy

Table 43.1 Numerical estimations of $|R|$, $|T|$ and $|R|^2 + |T|^2$ for different Kh_b values.

Kh_b	$|R|$	$|T|$	$|R|^2 + |T|^2$
0.03	0.00017	1	1.0000
0.23	0.00272	1	1.0000
0.43	0.00054	1	1.0000
0.63	0.00024	1	1.0000
0.83	0.00002	1	1.0000
1.03	0.00015	1	1.0000

Source: Author

identity equation, as defined in Table 43.1 for a range of wave numbers. This table presents the numerical estimations for s_0 h_b, $|T|$, $|R|$ and $|R|^2 + |T|^2$ corresponding to a section of sinusoidal ripples on the permeable base. With the current scenario, we established $D/h_b^4 = 1$, $N/h_b^2 = 0.1$, $Ph_b = 0.02$, $m = 4$ and $\lambda h_b = 1$. In this instance, we note that we achieved a high level of accuracy in the fulfilment of the energy identity equation.

Conclusion

A one-layer fluid flow is analyzed, in which the bottom surface is bounded by a permeable base exhibiting distortion, and the upper surface is covered by an ice sheet subjected to a compressive force, which is assumed to behave as a thin elastic plate. Applying linear water wave theory, the governing BVP is reduced to simpler form by using perturbation technique and then solved by Fourier transform approach. The energies that are reflected and transmitted are evidently contingent upon the configuration of the bed deformation. Here, a particular form of base distortion, referred to as a sinusoidal bed, is employed to compute the aforementioned energies. It is observed a decline in the maximum estimation of the reflected energy as the ice parameter rises. Also, it is found that the reflected energy is raising with raise in the value of the permeability parameter and compressive force parameter. The primary benefit of this work is that the numerical values of the reflected and transmitted energies derived here closely adhere to the energy-identity equation. The findings are expected to be quantitatively beneficial in resolving base distortions and water wave scattering problems on a porous bed in arctic regions.

References

[1]   Jeng, D. S. (2001). Wave dispersion equation in a porous seabed. *Ocean Engineering*, 28(12), 1585–1599.

[2] Khuntia, S., & Mohapatra S. (2021). Interaction of oblique waves by base distortion on a permeable bed in an ice-covered sea. In Advances in Fluid Dynamics. Lecture Notes in Mechanical Engineering, (pp. 315–326). Singapore: Springer.

[3] Mohapatra, S. (2014). Scattering of surface waves by the edge of a small undulation on a porous bed in an ocean with ice-cover. *Journal of Marine Science and Application*, 13(2), 167–172.

[4] Mohapatra, S., & Sarangi, M. R. (2017). A note on the solution of water wave scattering problem involving small deformation on a porous channel-bed. *Journal of Marine Science and Application*, 16, 10–19.

[5] Porter, D., & Porter, R. (2004). Approximations to wave scattering by an ice sheet of variable thickness over undulating topography. *Journal of Fluid Mechanics*, 509, 145–179.

44 Smart hotel automation system

Aman Goel[a], Jayanth Nair[b], Rishaan Yadav[c], Dhruv Veragiwala[d] and Jeyasekar, A.[e]

Department of CSE, SRM Institute of Science and Technology, Chennai, India

Abstract

As technology continues to advance, automation in hotels and restaurants is gradually increasing. Smart hotels and restaurants improve comfort and security by implementing automatic controlling of equipment and appliances using sensors. It helps to save energy, monitor and control the hotels remotely and avoid accidents. It improves the overall operation of the hotels and restaurants. In this paper, we propose a smart hotel automation system using cisco packet tracer in which various sensors are deployed to automate the operation of the hotel and prevent the accidents happening in the hotels due to gas leakages etc. Since the cisco packet tracer is a suitable platform to simulate the real IoT devices and visualization, we simulate the smart hotel automation system using cisco packet tracer. The smartphone, gateway and IoT server are used to remotely control the IoT devices. The simulation provides higher security because all the IoT devices are registered and authenticated by the administrator.

Keywords: Automation, internet of things, packet tracer, sensors, smart hotel

Introduction

Internet of things became popular last decades and provides high impacts in local and remote automations. It is utilized in many fields like agriculture, healthcare, industries, transport, smart cities, smart homes, smart car etc. It provides comfort and security to the people [5]. Hotel automation is an initiative where guests can easily access and control various appliances remotely using smartphones or computers without the need for human intervention. The sensors are deployed to operate the devices which react to the changes in the room's temperature, motion, and air quality. Motion-activated lights and fans, window opening and alarm activation when a carbon monoxide sensor detects a leak, and humidity and air conditioning regulation for maximum comfort are examples of automated features [4, 7].

Therefore, in this paper, we propose a smart hotel automation system using Cisco Packet Tracer, enabling IoT devices to be controlled via smartphones and laptops. The main contributions of this work include the installation of temperature-based fans and air conditioning systems for temperature control, carbon monoxide alarms to prevent fatal incidents, and a smart temperature controller that enhances guest comfort and safety. By adding IoT-based controls, the system improves guest satisfaction by creating more responsive and personalized experience, makes better use of resources, and simplifies facility management. This innovative approach demonstrates how IoT can transform traditional hotel operations into intelligent, efficient, and guest-focused experiences.

Literature Review

The smart hotel automation system enables automatic appliance control and remote monitoring. Various smart systems using Cisco Packet Tracer have been proposed [1–10].

Azhari highlighted IoT in office networks with VLAN, SSL VPN, and ASA Firewall to optimize management and security [1]. Smart home automation studies demonstrated IoT-enabled device control, improving convenience, energy efficiency, comfort, and security using motion and temperature sensors for practical applications in homes and offices [2, 4, 6].

Smart classroom designs emphasized automated devices to improve safety, discipline, and efficiency, advocating for network redundancy and transparency [3, 9].

Roy et al. introduced a three-phase smart home automation design enhancing convenience, safety, and security through IoT integration and network security [5]. IoT integration simulated smart hospitals automating lighting, HVAC, security, water management, and parking to enhance patient care and efficiency [7]. Shemsi I. showcases the implementation of a smart home system using Cisco Packet Tracer that integrates IoT devices for home automation, emphasizing security and environmental management through sensors and microcontrollers [10].

[a]ag2351@srmist.edu.in, [b]jn0075@srmist.edu.in, [c]ry1178@srmist.edu.in, [d]dv4794@srmist.edu.in, [e]ajeyasekar@yahoo.com

DOI: 10.1201/9781003658221-44

Smart Hotel Automation System

Because of the tremendous growth in the tourism and hospitality industry around the world, we propose a smart hotel automation system which controls appliances like CCTV, air conditioners, fans, doors, lamps, humidifiers, and coffee makers. The proposed system uses motion detectors that activated CCTV recording when the door opened and triggered lamps and the coffee maker upon guest detection. Temperature monitors alternate fan and AC usage based on room temperature, while a humiture monitor controlled the humidifier.

The proposed smart hotel automation system consists of five modules which are further explained in Section 4. All the connected appliances shown in Figure 44.3 are controlled and monitored through a tablet or PC as illustrated in Figure 44.2.

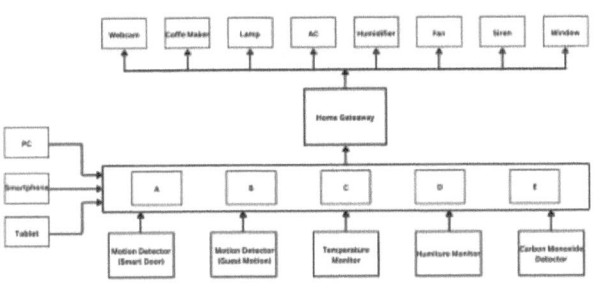

Figure 44.1 System architecture
Source: Author

Figure 44.2 Operational condition for IoT devices
Source: Author

Experimental Setup and Result Analysis

We use Cisco packet tracer to simulate a smart environment using various sensors, appliances, gateway, laptops, smartphones and IoT server. The network architecture of "smart hotel automation system" is shown in Figure 44.4.

All appliances are controlled via a tablet connected to the home gateway as shown in Figure 44.4. Motion sensors and a temperature monitor are included to monitor room temperature and alternately regulate the operation of the air conditioner and fan. A humiture monitor is integrated to activate the humidifier when the temperature decreases. Additionally, a motion detector is implemented to activate lights and appliances when guest motion is detected. A carbon monoxide sensor detects CO leaks, triggering a siren and opening a window for ventilation as seen in Figure 44.4.

We described various smart devices in the system, listing their IP addresses and the conditions that control their actions in Figure 44.2. A temperature monitor (192.168.25.118) keeps track of the temperature, while the fan (192.168.25.126) turns on

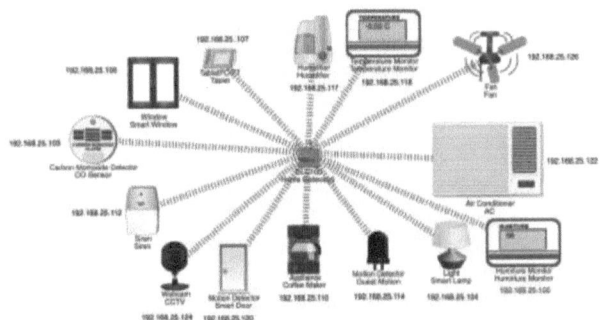

Figure 44.3 Available devices
Source: Author

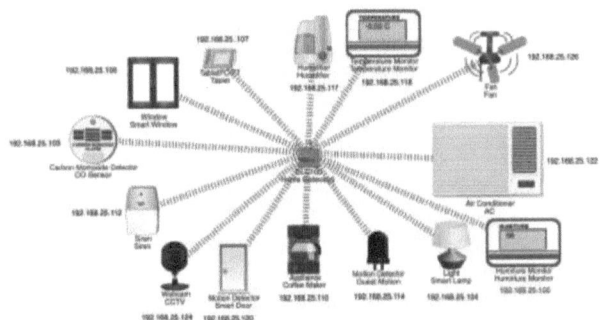

Figure 44.4 Topology
Source: Author

when the temperature lies between 0.0°C and 20.0 °C, and the humidifier (192.168.25.117) activates when the temperature is between -5.0°C and 10.0°C. The air conditioner (192.168.25.122) runs when the temperature is 20.0°C or higher and switches off below 25.0°C. A humiture monitor (192.168.25.100) also tracks both temperature and humidity. The smart lamp (192.168.25.104) and coffee maker (192.168.25.110) are triggered by guest movement, activating when motion is detected. The smart door (192.168.25.120) and CCTV (192.168.25.124) also respond to motion. For safety, a carbon monoxide sensor (192.168.25.103) detects CO levels, activating both the siren (192.168.25.112) and smart window (192.168.25.106) if CO levels go above 1. A tablet (192.168.25.107) acts as a control panel, helping to manage or monitor these devices. Altogether, these devices create a smart environment that reacts to changes in temperature, humidity, and motion.

The Smart Hotel Automation system demonstrates how automation with IoT can improve the safety, comfort, and productivity of guests in a hospitality environment. The tested core functionalities include motion sensors, carbon monoxide detectors, thermal sensors, and webcams, which were all successfully tested in a simulated setting using Cisco Packet Tracer. Motion-activated webcams enhance security by providing authentication and real-time authorization, restricting building access to authorized individuals and ensuring guest safety. Motion and temperature sensors improve comfort by automating lights, appliances, and climate control. Carbon monoxide detectors ensure safety by triggering alarms and ventilation during gas leaks.

Discussions

The Smart Hotel Automation demonstrates how IoT can enhance guest convenience and safety in hospitality environments. By incorporating sensors that automatically respond to environmental changes this system minimizes the need for human intervention, creating a responsive and secure environment for hotel guests. Its capabilities effectively reduce risks associated with hazardous situations like elevated carbon monoxide levels, which remain significant threats. Integrating a cloud server for remote management and real-time data monitoring would be a significant advancement for this project. Additionally, cloud integration would extend the technology beyond hotels, making it applicable in both residential and workplace settings. Also implementing encryption at the home gateway level would increase data transmission security, thereby protecting guest information and preventing unauthorized access.

Conclusion

The design of the proposed smart hotel automation system employs Cisco Packet Tracer to enhance understanding of automating IoT devices in real-world applications. This prototype serves as a foundational design for implementing appliance automation within the smart hotel. The functionalities of the IoT devices align with those in Cisco Packet Tracer. Mainly, the design includes AC automation for temperature adjustment, fan automation to regulate airflow, carbon monoxide alarm automation for safety, and door automation for secure and seamless entry. Additionally, other appliances are incorporated into the smart hotel automation system to create a comfortable and efficient environment for guests.

References

[1] Azhari, A. D., Sulaiman, N. A., & Kassim, M. (2021). Secured internet office network with the internet of things using packet tracer analysis. In 2021 IEEE 11th International Conference on System Engineering and Technology (ICSET), (pp. 200–205). Shah Alam, Malaysia.

[2] Alfarsi, G., Jabbar, J., Tawafak, R. M., Malik, S. I., Alsidiri, A., & Alsinani, M. (2019). Using cisco packet tracer to simulate smart home. International *Journal of Engineering Research and Technology (IJERT)*, 8(12), 670–674.

[3] Ayesha, B. B., & Murthy, S. (2022). Implementation of network design for universities with IoT.

[4] Ratnala, B., Anuradha, T., Maddali, V., & Chintalapudi, H. V. (2023). Designing smart room using cisco packet tracer simulator. In 2023 5th International Conference on Smart Systems and Inventive Technology (ICSSIT), (pp. 183–188). Tirunelveli, India.

[5] Roy, B., Nivethika, S. D., Manju, G., & Pandian, M. S. (2023). Comprehensive smart home automation: network setup, fire prevention, and network security using cisco packet tracer. In 2023 Intelligent Computing and Control for Engineering and Business Systems (ICCEBS), (pp. 1–4). Chennai, India.

[6] Dogman, A., & Jewiley, M. (2020). Design and implement IoT smart home via Cisco Packet Tracer: Applications and simulations. Faculty of Information Technology, University of AL-Zintan, AL-Zintan, Libya.

[7] Madhav, G. S., Kommana, C., Chandana, B. S., & Khanna, M. (2023). Design and simulation of a healthcare unit. In 2023 IEEE 3rd International Conference on Technology, Engineering, Management for Societal impact using Marketing, Entrepreneurship and Talent (TEMSMET), (pp. 1–6). Mysuru, India.

[8] Gururani, H., Kumar, A., Waghamode, D., Jain, E., & Garg, A., (2022). Smart city using IOT simulation design in cisco packet tracer. *International Journal for Research in Applied Science and Engineering Technology*, 10, 2544–2551.

[9] Duvvuri, K., Kanisettypalli, H., Kunisetty, J., & Khanna, M. (2023). Design and implementation of smart classroom using cisco packet tracer. In 2023 International Conference on Advances in Electronics, Communication, Computing and Intelligent Information Systems (ICAECIS), (pp. 116–120). Bangalore, India.

[10] Shemsi, I. (2018). Implementing smart home using cisco packet tracer simulator. *International Journal of Engineering Science Invention Research and Development*, 4(VII).

45 A novel hybridized method for moving object detection based on optical flow and edge detection technique in a dynamic scene environment

Maheswar Mishra[1,a], Tusharkanta Samal[1,b], Manas Ranjan Mishra[2,c], Jnanaranjan Mohanty[3,d] and Kamalakanta Muduli[4,e]

[1]Department of CSE, DRIEMS University, Tangi, Cuttack, Odisha, India

[2]Department of CSE, C.V. Raman Global University, Bhubaneswar, Odisha, India

[3]Department of Humanities and Social Sciences, Parala Maharaja Engineering College, Berhampur, India

[4]Department of Mechanical Engineering, Papua New Guinea University of Technology, Lae, Morobe Province, Papua New Guinea

Abstract

Moving objects has played a significant role in various applications like surveillance, recognition of different activities, monitoring road condition, protection of marine border, etc. However, one of the biggest challenges facing researchers is accurately tracking and detecting moving objects. Because the current traditional approaches have not produced an accurate result due to a lack of information about the surrounding environment, this paper focusses on discovering an effective strategy to detect and track a moving item with regard to the camera. The main objective is real-time detection and tracking of moving objects in real life, along with providing accurate information about the surrounding environment. Therefore, an attempt has been made to improve algorithms that can detect moving objects with respect to the camera along with to some extent eradicating the limitations of other algorithms. The proposed strategy has been used to employ the sparse optical flow method on picture frames, which approximate interesting features, i.e. pixels displaying just the corners and edges of moving objects in the scene. In addition to all, the proposed method has claimed higher accuracy and a consistent value for different videos in the dataset that is around 97%. This provides a better solution for moving object detection in different environments than its existing counterparts in terms of object speed, recall and f1score.

Keywords: Computer vision, HOG, moving object detection, optical flow, sobel edge detection

Introduction

Recently computer vision is the emerging technology and being densely empowered by the outbreak of artificial intelligence (AI) and the advanced deep learning techniques. The majority of academics in the late 19th century were constantly working to create an intelligent vision system that could be applied to space exploration, surveillance, mobile robot path planning, and many other fields. However, object detection is crucial computer vision task that involves detecting visual objects instances. The purpose of object detection is just to create models and approaches for computation that would provide one of the first most fundamental data a computer needs. Environmental conditions like occlusion, low illumination or sudden changes in intensities, waiving trees, cameras positioning are the most sensitive issues to be addressed here for better detection of an object in a dynamic scene.

This paper's primary contribution is explained as follows:

- An approach towards linear feature space combination and better representation of images by means of a hybrid optical flow and edged Sobel edge detection algorithm has been proposed.
- To find the sparse flow features which reduce computation time and experimentation towards fast moving object detection which is directly proportional to view space window size.
- The suggested method's performance has been compared with existing counterparts.

A brief review of related works done so far regarding object detection is presented in Section 2. The detailed description of suggested moving object detection is described in Section 3 and the simulation results of our suggested framework along with its counterparts

[a]maheswar.mishra@driems.ac.in, [b]dr.tusharkanta.samal@driems.ac.in, [c]manascvrce@gmail.com, [d]jmohanty2304@gmail.com, [e]kamalakanta.muduli@pnguot.ac.pg

DOI: 10.1201/9781003658221-45

are discussed in Section 4 and Lastly, a summary of the conclusion is provided in Section 5.

Related Work

Choi et al. [1] proposed a sparse optical flow network that fully utilizes the rich information that exists between adjacent frames. The method proposed by Fan et al. [2] fully utilizes the rich information between adjacent frames and is primarily based on a light-weight optical flow network. By studying the GMM_BLOB method that is described by Luo et al. [3] it uses the Homograph method that does not require camera parameters. Similarly, the method projected by Chen et al. [4] uses computer vision technology to do moving object detection. The primary benefits of this are the longest running time, lower time consumption, and less mistake caused by the tracking box's deviation from the target's areal position. Above all, the highest tracking accuracy. By combining visual similarity and spatiotemporal consistency, the UnOVOST-Unsupervised algorithm tracks and segments a vast variety of objects in complicated senses [5]. However, to enhance the CCTV camera surveillance system, there are some methods that are proposed by Swalaganata et al. [6] integration method between the cam shift method and the Kalman filter to track objects. To enhance the image there is another method proposed by Sharma et al. [7] that assists in eliminating dynamic and static elements of the background scene and identifying significant moving regions. A robust, accurate, and high- performance approach for object recognition using video analytics. Threshold-based filtering technique proposed by Vijayakumar et al. [8] in which the PSNR rate was reduced by 23% and object detection accuracy improved by 8%. The main disadvantage of the method is that Algorithm may fail in multiple planes structure, video data limitation, and resolution limitation. To build an effective method for developing an efficient video processing system. All the above methods have lower accuracy in either detecting a moving objects in the fast, medium or slow detection screen.

Methodology

This method presents a hybrid technique for object reorganization that combines Sobel edge detection with optical flow to increase robustness and object detecting accuracy. The function of Sobel method is to identify picture edges. The edges may then be used to pick the feature points that the optical flow approach will track. This method offers various benefits over the optical flow method on its own. Secondly, by just tracking a subset of feature points, it decreases the computational cost of the optical flow technique. Second, it enhances feature point tracking accuracy by picking points that are more likely to be stable over time. Finally, by combining both edge and motion information, it delivers a more robust object detection technique.

Mathematical model

Video images are taken as input from the video frame set. It is then split into two parts. The input two-dimensional image (I) which have the values x, y, t, here x gives the x axis of the image and y gives the y axis of the image and t is for the time change in position of the image with relevance to the frame. The primary part is that the frames have collected and therefore the background is extracted. Then the second part calculates the merchandise sums. There the image is further formed to Wx, Wxy, Wy, Wxt, Wyt. After that its further segregated to dynamic object and static object. The term Wx, Wxy, Wy represents the static image because the time factor t is not present and also the term Wxt, Wyt represent the dynamic image as it has the time factor t which is able to change with regard to the frame. After that, the static and dynamic image part will be distributed to the updated part added to their respective frames. Figure 45.1.

Based on the mathematical representation the intensity of the image is I $(x + \Delta x, y + \Delta y, t + \Delta t)$. Thus, an object's pixel intensities are constant frames. So:

$$I(x, y, t) = I(x + \Delta x, y + \Delta y, t + \Delta t), \qquad (1)$$

$$I(x + \Delta x, y + \Delta y, t + \Delta t) = I(x, y, t) + I(x, y, t) + I(Wx + Wxy + Wy + Wxt + Wyt), \qquad (2)$$

Now have to calculate the optical flow from the sparse region

$$I(x + \Delta x, y + \Delta y, t + \Delta t) = Vx(t) + Vy(t), \qquad (3)$$

Using (2) and (3) We get

$$I(x, y, t) + I(Wx + Wxy + Wy + Wxt + Wyt) = Vx(t) + Vy(t), \qquad (4)$$

It simplifies the detection of the object in the moving dynamic scene.

Proposed method

1: Video sequence as input for each group N $Im(t)$, $t = 1, 2, ..., \infty$, and the number of frames
2: for each group of N frames in $\{Im(t)\}$ do

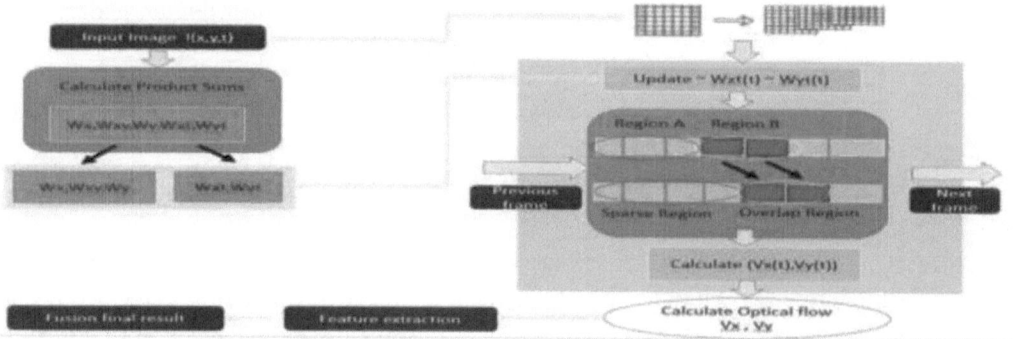

Figure 45.1 Proposed hybrid optical flow moving object detection mathematical model
Source: Author

3: $Im0 = Im(t–1)$ initialization

4: for $t = 2$ to $N = 1$ do

5: $f_{sobel(t)} = F\_sobel(Im(t–1), Im(t)$ Sobel edge detection feature extraction of adjacent frames

6: f_lucas_kanade(t) = F_lucas_kanade(Im(t–1), Im(t)) Lucas-Kanade optical flow feature extraction of adjacent frames

7: end for

8: $f_{sobel(t)} = \Sigma i = t$ to $t + N–2$ $wif\_sobel(i)$ Sobel edge detection feature maps combined in a group.

9: $^f\_lucas\_kanade(t) = \Sigma i = t$ to $t + N–2$ wif_lucas_kanade(i) Lucas-Kanade optical flow feature maps combined in a group

10: $\Sigma 1$ to $N–1$ $wi = 1$ 11: for $t = 1$ to N do

11: $^Im(t) = (Im(t), ^f\_sobel(t), ^f\_lucas\_kanade(t))$ mask for the original image

12: $z(t) = Ndet[Nfeat(I(t))]$ object detection using feature maps

13: end for

14: $Z = \{z(t)\ t = 1, 2, ..., N$

15: end for

16: Output: Video object detected Z

Simulation Result and Analysis

Experimental setup

Optical flow in python and PyCharm have been implemented by taking the real-world image. This experiment has been successfully evaluated on the laptop with specifications as follows: 11th generation Intel(R) Core (TM) i3-1115G4 processor @3.00GHz 2.90GHz, RAM:8GB, System Type:64-bit operating system, x64-based processor. In addition to this, HOG has been implemented with different edge detection algorithm in Google Colab having the same dataset and got the result accordingly.

Dataset

Two large datasets namely CAVIAR and PETS2009 Choi et al. [1] have been implemented in the proposed

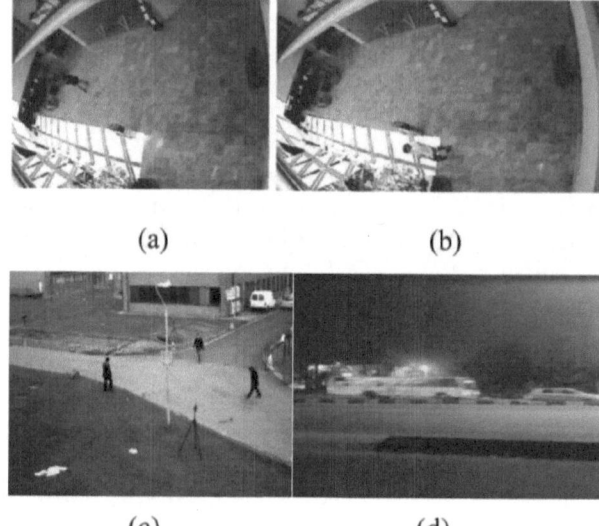

(a) (b)

(c) (d)

Figure 45.2 Examples of dataset (CAVIAR, PETS2009 and actual camera) (a) person walking in a straight path (b) Person walking back and forth. (c) Multiple pedestrian (d) Real time traffic scene
Source: Author

method and performed a comparative analysis with the methods used in the Choi et al. [1]. We basically took two videos namely 1,2 where video 1 represents travelling of a person in a straight path in a hallway, video 2 represent flow statistics for movement back and forth. The resolution of the dataset is 384 × 288. The third video is of the second dataset that demonstrates that there are more people in the surrounds than there were in the first set. The resolution is 768 × 576. The dataset has been presented in Figure 45.2.

Experimental analysis

The input video list Im(t) and the frames per group is N. Then that image is initialized as well as the method exploit and group N-1 optical flow feature map. This process basically masks the image by making it into a

 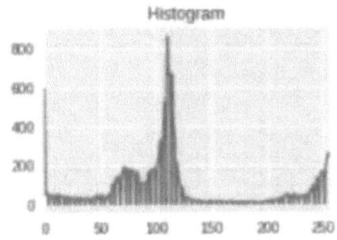

Figure 45.3 Input image histogram computation
Source: Author

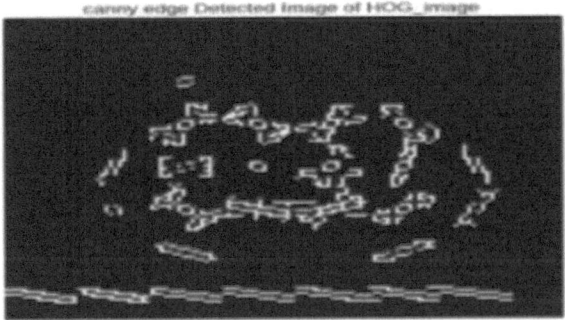

Figure 45.4 Computation of canny and HOG hybrid features for object detection
Source: Author

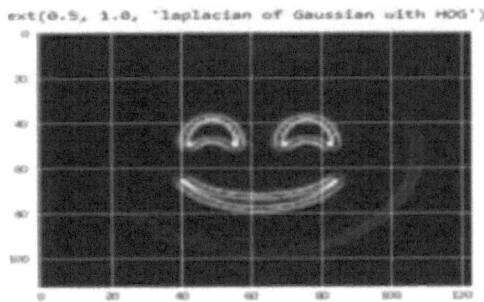

Figure 45.5 Experimental result Laplacian of Gaussian (LOG) with HoG feature combination
Source: Author

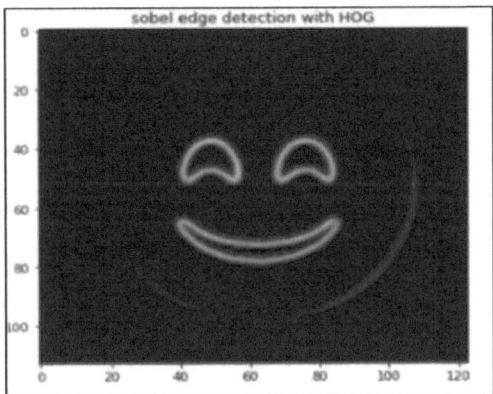

Figure 45.6 Obtained hybrid features: Sobel edges and histogram oriented gradient of input object
Source: Author

fused image. Here Wi is basically the fused image that masks the current video frame and extracts the feature. After masking the image, it determines the object and derives its feature. This loop continues until the number of frames becomes null. Finally, Y(t) is determined, and the video object is detected. The histogram of oriented gradients (HOG) is a feature description technique in the field of computer vision that counts the occurrence of direction of greatest change in the intensity in the locality of an image for the purpose of object detection.

Further to improve the algorithm some modifications have been done in the HOG algorithm by applying it with different edge detection algorithm like Laplacian of Gaussian, Canny, and Sobel. Then merged HOG with Canny have been merged to get a better desired result. Figure 45.3.

In the Figure 45.4 the result for Canny edge detection algorithm combined with HOG algorithm in which most of the noise present in the background was removed and can be able to get accurate result. Furthermore, Laplacian of Gaussian combined with HOG algorithm have been implemented.

From the Figure 45.5 we get that the image after conversion gets more distinctive and as a result can be able to detect the external features of the image. This helps us to get a better idea about the actual image after its conversion, but it fails to show the exact

outline of the image. Considering the similar experimental setup we have also implemented Sobel edge detection combined with HOG and we get the following result:

In Figure 45.6 it has shown that all the background outliers were removed, and the focused image was clearly visible. Out of all the three images in the algorithm that combines Sobel with HOG most of the erroneous threshold value were removed. Optical flow is a method to detect moving objects. For an object in an image, it identifies some points in the object and then uses the speed of those points to determine if the object in the image is static or in motion. Figure 45.7.

Moving object detection using our proposed hybrid method

The proposed method helps used in detecting as well as tracking objects in a real time environment. The algorithm takes the video input and converts it into different frames. In each of the frames it marks the moving objects as red dots which signifies that it has already tracked down the objects and based on that it calculates true positive, true negative, false positive, false

Figure 45.7 Moving object detection in real time national highway camera capture video
Source: Author

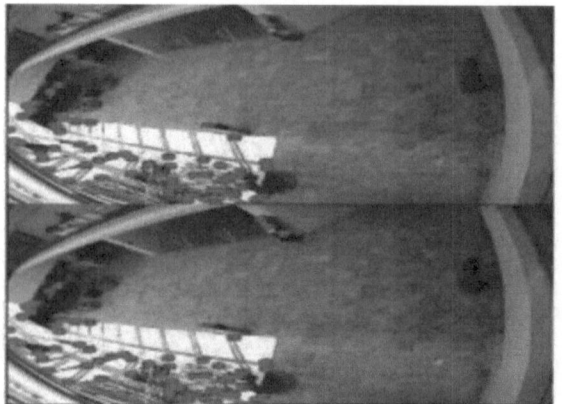

Figure 45.8 Slow moving object detection on PETS2009 and CAVIAR dataset images
Source: Author

Figure 45.9 Moderately moving object detection on PETS2009 and CAVIAR dataset images
Source: Author

Figure 45.10 Fast moving object detection on PETS2009 and CAVIAR dataset images
Source: Author

negative. Using these parameters, it calculates accuracy, precision, recall, f1score. The output obtained in Figure 45.8 is from the real-life video environment and the result obtained is marked with red dots.

The proposed method is again tested on dataset which was used in Choi et al. [1] which uses dataset that is CAVIAR and PETS2009 and from that dataset it was first tested on videos which contained slow moving objects. The result obtained from that video frame is shown in Figure 45.8. After identifying the moving object, it calculates true negative, true positive, false negative, false positive. Using these parameters, it calculates accuracy, precision, recall, f1score.

Our proposed method is again tested on dataset which was used in Choi et al. [1] which uses dataset that is CAVIAR and PETS2009 and from that dataset it was tested on videos which contained medium moving objects. The result obtained from that video frame is shown in Figure 45.9. It computes true negative, true positive, false negative, and false positive after determining the moving object. Using these parameters, it calculates accuracy, precision, recall, F1-score.

Our proposed method is again tested on dataset which was used Choi et al. [1] which uses a dataset that is CAVIAR and PETS2009 and from that dataset,

Table 45.1 Presentation of output [1].

Video	Recall	F-score
1	0.77	0.80
2	0.70	0.77
3	0.86	0.87

Source: Author

it was tested on videos which contained fast moving objects. The result obtained from that video frame is shown in Figure 45.10. After identifying the moving object it calculates true negative, true positive, false negative, and false positive. Using these parameters, it calculates accuracy, precision, recall, f1score.

Table 45.1 represents output Choi et al. [1] and Table 45.2 represents the output which we are getting from the proposed method. Here video1,2,3 represents slow moving, medium moving and fast-moving video objects from the dataset CAVIAR and PETS2009. We have also additionally tested our proposed algorithm

Table 45.2 Representation of using proposed method.

Video	Recall	F-score
1	0.94	0.97
2	0.95	0.97
3	0.94	0.97
4	0.95	0.97

Source: Author

with video 4 that is live feed from real world environment. By comparing the result from the table it has been noted that the suggested approach provides a higher F1-score.

Since we found that Sobel algorithm gives the best result as compared to other edge detection algorithm therefore, we have used this with optical flow which constitutes our proposed method. This combination of optical flow with Sobel gives a high value algorithm that is our proposed method and on comparing the proposed method with the method used in the base research paper we found that our method gives higher accuracy.

Conclusion and Future Scope

The proposed algorithm combines Sobel with an optical flow method. The proposed hybrid method has obtained a higher accuracy result than its existing counterpart. Moreover, this work provides a better end-to-end method framework for detecting an object by focusing on improved performance and the relationship between the frames concerning their surroundings. The augmentation method will help us to look at objects from a different perspective and help us to determine various features of the object. Translation, Scaling, Shearing, Rotation and Reflection can be determined by augmentation methods in an accurate

way as a result image feature extraction can be implemented with high accuracy.

References

[1] Choi, H., Kang, B., & Kim, D. (2022). Moving object tracking based on sparse optical flow with moving window and target estimator. *Sensors*, 22(8), 2878.

[2] Fan, L., Zhang, T., & Du, W. (2021). Optical-flow-based framework to boost video object detection performance with object enhancement. *Expert Systems with Applications*, 170, 114544.

[3] Luo, X., Wang, Y., Cai, B., & Li, Z. (2021). Moving object detection in traffic surveillance video: new MOD-AT method based on adaptive threshold. *ISPRS International Journal of Geo-Information*, 10(11), 742.

[4] Chen, C., & Li, D. (2021). [Retracted] Research on the detection and tracking algorithm of moving object in image based on computer vision technology. *Wireless Communications and Mobile Computing*, 2021(1), 1127017.

[5] Luiten, J., Zulfikar, I. E., & Leibe, B. (2020). Unovost: unsupervised offline video object segmentation and tracking. In Proceedings of the IEEE/CVF Winter Conference on Applications of Computer Vision, (pp. 2000–2009).

[6] Swalaganata, G., & Affriyenni, Y. (2018). Moving object tracking using hybrid method. In 2018 International Conference on Information and Communications Technology (ICOIACT), (pp. 607–611). IEEE.

[7] Sharma, L., & Yadav, D. K. (2017). Histogram-based adaptive learning for background modelling: moving object detection in video surveillance. *International Journal of Telemedicine and Clinical Practices*, 2(1), 74–92.

[8] Vijayakumar, P., & Senthilkumar, A. V. (2016). Threshold based filtering technique for efficient moving object detection and tracking in video surveillance. *International Journal of Research in Engineering and Technology*, 5(2), 303–310.

46 YOLO & ML based crack detection and strength prediction for structural health monitoring of bridges

Surajit Mohanthy[a], Niva Tripathy[b], Shekharesh Barik[c], Rojalin Dash[d], Rajeev Agrawal[e] and Prabhu Prasad Nanda[f]

Department of CSE, Driems University, Cuttack, India

Abstract

Machine learning (ML) and You Only Look Once (YOLO)-based approaches have shown significant promise in crack detection and strength prediction for structural health monitoring (SHM) of bridges. These techniques offer improved accuracy, real-time monitoring capabilities, and efficient data processing for bridge maintenance and safety. YOLO-based algorithms have been successfully applied to bridge crack detection, demonstrating high accuracy and real-time performance.

Keywords: Crack detection, Internet of Things, machine learning, structural health monitoring, You Only Look Once

Introduction

The growth of information technology has made the Internet of Things (IoT) widely available, widely used, and has many advantages. The area of civil engineering has been especially impacted by these developments, which have encouraged the creation of sophisticated, intricate, and networked structures. Given the potential for disastrous outcomes from undiscovered structural flaws, it is crucial to detect cracks in vital infrastructure, particularly in buildings like bridges. Cracks are signs of structural flaws that, if ignored, could jeopardize the integrity of the entire system. Our focus on assessing these models—Yolov8 demonstrates outstanding performance—highlights the vital necessity of utilizing cutting-edge technology to precisely and quickly detect structural flaws. Based on research and application viewpoints, the current methods for detecting structural damage may be loosely divided into two groups: dynamic-based methods and static force-based approaches. The traits of the structures themselves serve as the foundation for both categories. However, because to algorithmic limitations and variations in the loads that structures experience while in operation, these approaches often encounter challenges in practical implementations. Thanks to developments in intelligent identification terminal technology, the use of IoT) technology for structural security monitoring has grown in popularity. Real-time construction site monitoring is made feasible by IoT technology through intelligent terminals. This facilitates the processing and analysis of data to produce early warning indicators for security. IoT enables automatic identification and tracking, data transmission across networks, computerized alarms, digitization, information construction, and full real-time monitoring of the safety state of structures.

Our work constitutes a major advance in infrastructure monitoring and maintenance. Predictive modeling of concrete strength, IoT-based monitoring, and YOLO models for crack identification work together to meet the need for a detail descriptive approach to structural integrity. This integrated approach establishes a precedent for the ongoing development of technology-driven solutions in civil engineering and has immediate consequences for the durability and safety of essential infrastructure. Our study serves as a groundbreaking endeavor that closes the gap between conventional inspection techniques and state-of-the-art technology, guaranteeing the structural integrity of vital infrastructure as we work to construct safer and more robust structures.

Related Work

For the quick and precise identification of bridge surface cracks, a convolution block attention module (CBAM) in conjunction with MobileNets and an enhanced You Only Look Once (YOLO) v3 algorithm has been presented [3]. Using depth wise separable convolution and inverted residual blocks, this method lowers network parameters without sacrificing accuracy.

---

[a]surajit.mohanty@driems.ac.in, [b]nivatripathy@driems.ac.in, [c]shekharesh.barik@driems.ac.in, [d]rojalin.dash@driems.ac.in, [e]rajev.agarwal@driems.ac, [f]prabhupnanda@gmail.com

DOI: 10.1201/9781003658221-46

Similarly, a lightweight YOLO v4-based approach has been designed that requires just 23.4 MB of storage space and processes 140.2 frames per second, while attaining good recall (90.12%), and F1-score (92%) precision (93.96%), [4].

Interestingly, while YOLO-based methods show promise, other ML techniques have also been explored for bridge SHM. For instance, the K-Star algorithm has outperformed locally weighted learning (LWL) in predicting vertical deflection of composite bridges, offering an efficient tool for rapid prediction and facilitating more effective bridge health monitoring and management [1]. Additionally, a novel ML-based framework using strain gauge sensors and Bluetooth communication has been developed for non-destructive crack detection and localization in composite structures, which could potentially be applied to bridge SHM [2].

Zhang et al. [3] suggest the CrackUnet using deep learning model. 1200 sub images are produced by manually labeling, enhancing, and resizing a heterogeneous dataset to $256 \times 256 \times 3$ dimensions. To identify pixel-level accurately the CrackUnet model and a Unet-based technique, is used. Superior performance is demonstrated by evaluation on test and Crack Forest datasets, with precision, recall, and F1-score reaching up to 91.45%, 88.67%, and 90.04%, respectively. Comparative results show CrackUnet's super set- in terms of dataset properties, training time and accuracy. Testing on noisy images confirms robustness and generalization, confirming its effectiveness in evaluating concrete structures. A novel beginning to end break finding approach employing a convolutional neural network (CNN) is presented by Xu H. et al. [4]. Without pre-training, the model uses profundity-wise different convolution, atrous convolution, and the atrous spatial pyramid pooling (ASPP) module to reach a noteworthy 96.37% detection accuracy. In order to preserve traffic safety, it has the potential to identify concrete bridge fractures accurately and automatically. It also gives flexible integration into other convolutional networks for effective feature extraction, outperforming traditional classification models.

This study concludes the review of literature by outlining its distinct contributions to the field and distinctly setting it apart from earlier approaches.

(i) The YOLO method for crack identification is combined with a machine learning approach as a prediction model in the suggested method.

(ii) The suggested model is contrasted with YOLO V3 and V8, two other competing algorithms.

Integrated YOLO-ML Model

IoT architecture

With the ability to link commonplace devices to the internet and make data collecting and sharing easier, the IoT has become a seamless part of our everyday life. Understanding the IoT architecture, a tiered framework that coordinates the seamless flow of data between devices, networks, and applications, is crucial to understanding how this complex network operates. Four interconnected levels, each of which is essential to the system, can be used to broadly classify the IoT architecture:

- Bottom layer:
  It contains sensor, mobile devices. These will generate huge amount of data which will be further forwarded to the upper layer for processing.
- Communication layer:
  This layer creates a link between the bottom layer and data processing layer. This layer contains protocols like WiFi, Bluetooth, or cellular networks. This is responsible for data transmission between two consecutive layers.
- Processing layer:
  This layer is responsible for collection and analysis of the data received from bottom layer. This may be edge or cloud computing devices.
- User interface layer:
  The last layer uses applications developed for mobile or web platforms to communicate with users.

YOLO model

There are similarities between GoogleNet and the YOLO (You Only Look Once) architecture. It makes use of 4 max-pooling layers, 24 convolutional layers, and 2 fully linked layers. All input images are arranged in a sequential fashion to evaluate and extract relevant

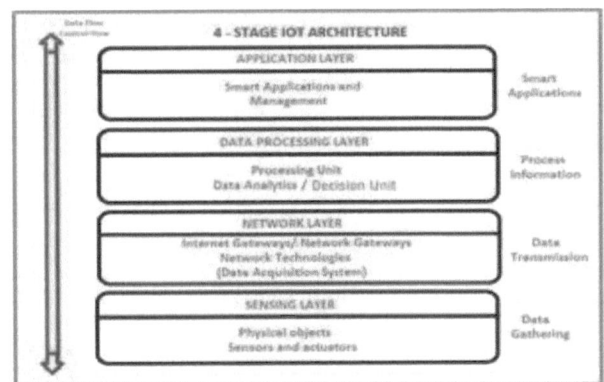

Figure 46.1 IoT architecture stages
Source: Author

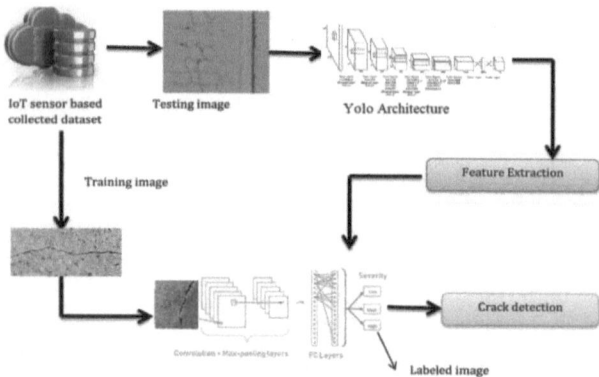

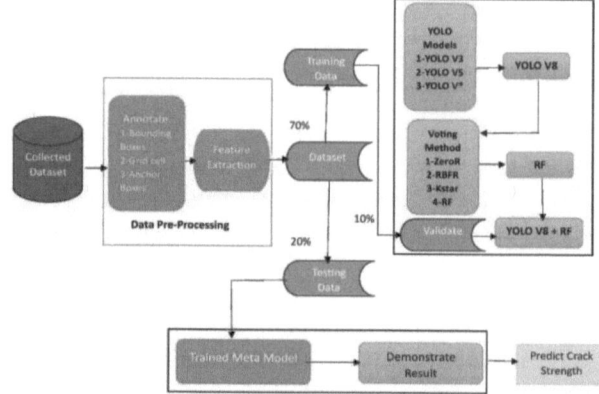

Figure 46.2 Flow of crack detection
Source: Author

Figure 46.3 Proposed model flow chart
Source: Author

structures. However, it is important to recognize that depending on the form and variant in question, YOLO's organization and structure may vary.

ML model

The ML models comprise of a wide range of tactics, from a basic baseline (ZeroR) to more advanced techniques for instance-based learning and non-linear connections, such as RBFR and Kstar. Random forest is a flexible option that improves accuracy and generalization across a variety of datasets because to its ensemble learning methodology.

Figure 46.1 describes the IoT architecture stages and the working of all stages.

Figure 46.2 describe the detail flow of crack detection process.

Figure 46.3 describe the flow chart of the detail process.

Figure 46.4 shows the vector analysis of the crack detection.

Figure 46.5 show the confusion matrix.

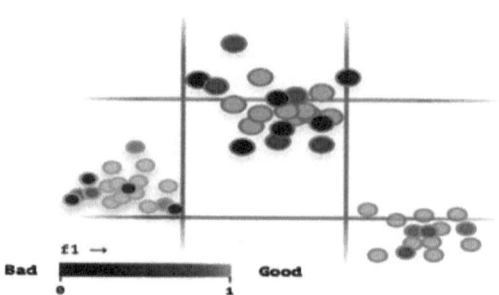

Figure 46.4 Vector analysis
Source: Author

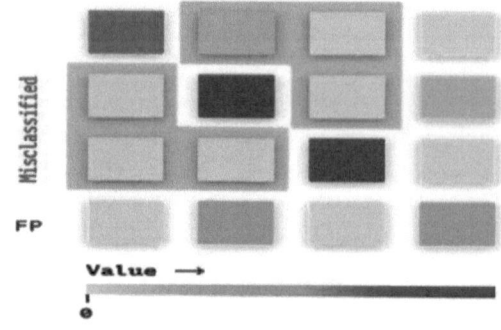

Figure 46.5 Confusion matrix
Source: Author

Methodology

Data collection phase

Both manually and through IoT devices, we gather the dataset for our study. Below is a discussion of some of the gadgets we utilize for this. Finding structural flaws is crucial to ensure stability and security of mining infrastructure which includes shafts, tunnels, and support systems. A variety of approaches and technologies are used in traditional structural deformation detection methods to find alterations in the integrity, alignment, and shape of mining structures.

Annotation and data pre-processing phase

The YOLO algorithm aims to predict the class of an object as well as the hopping box, which shows the object's location in the input image. Each bounding box is identified by four digits: the height (bh), width (bw), and center (bx, by) of the box.

Moreover, YOLO predicts not only the consistent c number for the expected class, but also the probability of the prediction Pc.

Extraction of feature phase

To extract features, the image is run through a CNN. YOLO extracts hierarchical characteristics from the image using a variety of convolution, pooling, and activation function layers.

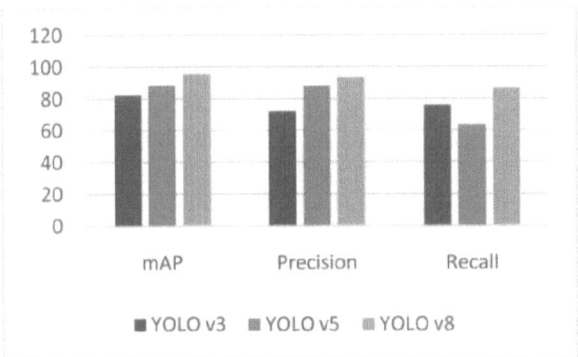

Figure 46.6 Comparison of YOLO model
Source: Author

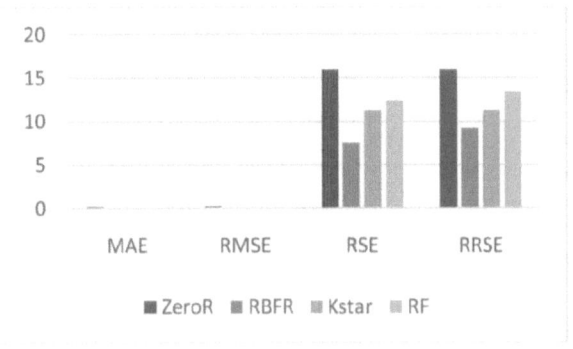

Figure 46.9 Comparison of error value in thrust and displacement datasets
Source: Author

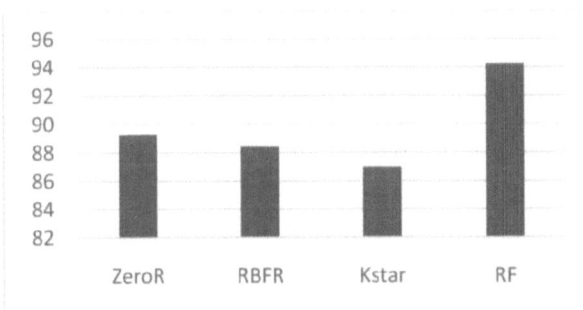

Figure 46.7 Accuracy comparison
Source: Author

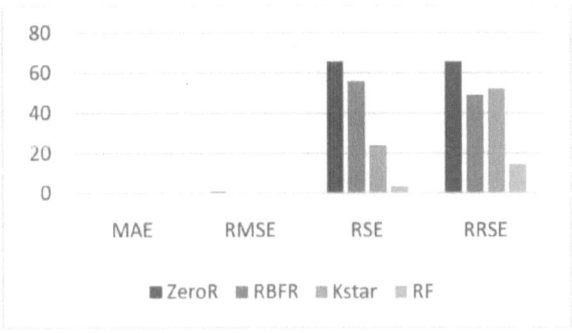

Figure 46.10 Comparison of error value in response time data set
Source: Author

Figure 46.8 Comparison of error value in pressure data set
Source: Author

Figure 46.7 shows the Accuracy comparison of different algorithm.

Figure 46.8 shows the error value comparison in pressure data set.

Figure 46.9 shows the error value comparison in thrust and displacement datasets.

Figure 46.10 shows the error value comparison in response time data set.

Simulation Result

YOLO V3, V5, and V8 are three variants of the YOLO approaches used to finish the recognition task. The models listed above have received nominations due to their notable advancements in crack detection.

It was discovered that the suggested model has the highest accuracy for identifying bridge cracks. Using numerical data gathered by IoT devices and trained using machine learning models (i.e., RBFR, RF, ZeroR, and, Kstar) for prediction of accuracy over pressure, thrust, displacement, and response time, we were able to assess the strength of that construction. We discovered that Random Forest provides the lowest error numbers and the best accuracy. RF yields 94.23%,

Division phase
The figure is separated into a cell grid. The task of identifying items inside its borders falls to each cell.

Evaluation metrics phase
A few evaluation metrics, such MAP (mean average precision), are employed to compare various models. For object identification methods, performance metrics like recall and precision are employed.

Figure 46.6 shows the precision and recall comparison of YOLO model.

Kstar gives 86.99%, RBFR gives 88.44%, and ZeroR delivers 89.28%.

Because the 3 YOLO approach considered various parameters randomly initialized, the results of the simulation change from one run to the next. Two popular statistical tests on different procedures pairs, the Sign test and the Wilcoxon Signed rank test, were employed in our study to conduct an equitable evaluation of performance. The dominance of the suggested YOLOv8 algorithm is assessed using the Sign test and the Wilcoxon Signed rank test. The three YOLO algorithms have been executed 25 times each, and performance measurements have been recorded at each iteration. In order to achieve significant levels of α = 0.05 and α = 0.01; several methods are used 5–25 times.

Conclusion

The suggested model combines DCGAN and YOLOv8 to detect connection crashes more quickly and accurately. We have established the parameters and organizational structures of the DCGAN and YOLOv8 neural networks, as well as the training environment and associated theories. The trained DCGAN can quickly produce an endless number of fictitious bridge crack images by learning the properties of the cracks. This is used to increase the size of the real dataset. This study analyzes the model's performance and contrasts it with the results of the larger dataset. The results show that YOLOv8 created on the larger dataset was just as accurate as when created on the original dataset, providing a whole new perspective for large-scale concrete structure monitoring and support cost control. The results of the simulation show that the suggested method produces better results than the others. Furthermore, the statistical analysis demonstrates the suggested model's superiority over the others.

References

[1] Ha, H., Manh, L. V., Pham, B. T., Hiep, L. V., & Nguyen, D. D. (2024). Using machine learning algorithms for predicting bridge vertical deflection. Proceedings of the Institution of Civil Engineers - Engineering and Computational Mechanics, 177.1, 15–21. https://doi.org/10.1680/jencm.23.00016.

[2] Shah Mansouri, T., Lubarsky, G., Finlay, D., & Mclaughlin, J. (2024). Machine learning-based structural health monitoring technique for crack detection and localization using Bluetooth strain gauge sensor network. Journal of Sensor and Actuator Networks, 13(6), 79. https://doi.org/10.20944/preprints202409.1639.v1.

[3] Zhang, Y., Huang, J., & Cai, F. (2020). On bridge surface crack detection based on an improved YOLO v3 algorithm. IFAC-Papers OnLine, 53(2), 8205–8210. https://doi.org/10.1016/j.ifacol.2020.12.1994.

[4] Zhang, J., Tan, C., & Qian, S. (2022). Automated bridge crack detection method based on lightweight vision models. Complex and Intelligent Systems, 9(2), 1639–1652. https://doi.org/10.1007/s40747-022-00876-6.

47 Real-time vehicle accident detection using computer vision and deep learning

Rasmita Kumari Mohanty[1,a], Satyabrat Sahoo[2,b], Manas Ranjan Kabat[3,c], Satya Prakash Sahoo[3,d], Piyush Mohapatra[4,e] and Basim Alhadidi[5,f]

[1]Department of CSE-(CyS,DS) and AI&DS, VNR Vignana Jyothi Institute of Engineering and Technology, Hyderabad, India

[2]Department of CSE, SOET, DRIEMS University, Cuttack, Odisha, India

[3]Department of CSE, Veer Surendra Sai University of Technology, Burla, Odisha, India

[4]Modeling and Data Science Corporate Technologies, Donaldson, India

[5]Department of Computer Information System, Prince Abullah Bin Ghazi Faculty of Information and Communication Technology, AI-Balqa Applied University, Salt, Jordan

Abstract

Vehicle accidents are a significant cause of injuries and fatalities worldwide. Prompt detection and response to accidents can greatly mitigate the consequences and improve road safety. In recent years, advancements in computer vision and machine learning (ML) have paved the way for innovative solutions to address this issue. This abstract presents a comprehensive overview of vehicle accident detection using video, highlighting key techniques, challenges, and future directions. The primary goal of vehicle accident detection using video is to automatically identify and classify accidents from video footage captured by surveillance cameras, dashcams, or other sources. By leveraging computer vision algorithms and machine learning models, accidents can be detected in real-time or post-event analysis, enabling timely emergency response, and aiding in accident investigation. This paper explores various approaches for vehicle accident detection, including object detection, motion analysis, and deep learning techniques. The object detection algorithms YOLO and faster region based convolutional neural networks (R-CNN) assist in identifying and detecting vehicles together with other objects in video Frames to analyze vehicle movements for potential accident detection. Challenges and future directions in accident detection are also discussed.

Keywords: Computer vision, deep learning techniques, machine learning techniques, object detection, vehicle accident detection

Introduction

This paper addresses the critical concern of road safety through the utilization of advanced technology. By employing real-time video analysis, computer vision, and artificial intelligence, the project aims to detect potential vehicle accidents promptly and implement preventive measures. The system's objectives include real-time accident identification, rapid alert dissemination to relevant parties, and reduction of response time to minimize accident severity.

Challenges involve adapting to diverse scenarios, ensuring high-quality training data, and addressing privacy issues associated with video data. The project intends to achieve accuracy and reliability in accident detection by training the system on a varied dataset and fine-tuning its algorithms. The successful implementation of this project could lead to safer roads, decreased fatalities, and improved traffic management, ushering in a new era of accident prevention through cutting-edge technology [1, 2].

Vehicle accident detection using video holds significant potential for improving road safety. Through the fusion of machine learning and deep learning techniques and computer vision, accidents can be detected and responded to promptly, minimizing the impact on human lives [3]. Despite the challenges, ongoing research and advancements in the field continue to push the boundaries of what is achievable in accident detection, offering hope for safer roads. Potential benefits of the proposed vehicle accident detection system using YoloV7 are Reduced response time for first responders, Improved survival rates for accident victims, Reduced traffic congestion, Reduced insurance costs and improved public safety.

The objective of accident detection using video analysis is to develop a system that can automatically

[a]rasmita.atri@gmail.com, [b]satyabratsahoo@driems.ac.in, [c]manas_kabat@yahoo.com, [d]sahoo.satyaprakash@gmail.com, [e]piyush.mohapatra@donaldson.com, [f]hadidi72@hotmail.com

DOI: 10.1201/9781003658221-47

detect vehicle accidents in real-time using video footage from traffic cameras. Such a system could be used to reduce the time it takes for first responders to arrive at the scene of an accident, thereby improving the chances of survival for those involved.

Video analysis systems for accident detection typically work by first identifying vehicles in the video footage. Once vehicles have been identified, the system can track their movement and identify any unusual behavior, such as sudden braking or swerving. The system can also use other factors, such as the time of day and weather conditions, to assess the likelihood of an accident.

Literature Review

In recent years, various technologies have been explored to improve vehicle accident detection systems. The study by [4] focuses on the use of easily accessible cellphone sensors, such as accelerometers and gyroscopes, to detect accidents. The research emphasizes the need to broaden detection beyond high-speed crashes to include low-speed impacts as well. By analyzing changes in the motion and orientation of a phone, the proposed solution can trigger emergency notifications. While cost-effective and promising, the system faces challenges, including false positives from normal phone movements and rapid battery depletion due to continuous sensor use.

In another approach, [5] explores the use of Vehicular Ad-Hoc Network VANETs () for vehicle-to-vehicle communication and cloud computing for real-time data processing. VANETs enable the direct exchange of real-time accident information between vehicles, while the cloud processes this data to facilitate faster emergency response protocols. This system offers improved situational awareness and response times compared to traditional, infrastructure-based systems. However, the deployment of VANET infrastructure is costly, and delays in VANET communication may cause latencies in transmitting data to the cloud for analysis. CAPDATA data is used in this study to simulate and evaluate the system.

Similarly, [6], provides a comprehensive analysis of Internet of Things (IoT) technology for vehicle accident detection. This study explores the integration of various sensors—such as accelerometers, GPS, and gyroscopes—with cloud platforms for real-time data analysis and emergency notifications. The combined use of IoT sensors and cloud-based processing offers a robust, data-driven solution for accident detection. However, the study identifies significant challenges, including the cost of sensor system implementation,

energy consumption, and scalability issues as the number of connected vehicles increases.

Moreover [7], highlights the development of a real-time accident detection system using IoT devices embedded in vehicles. The system relies on a variety of sensors to detect impacts, and it uses cloud connectivity to transmit and analyze the data. Upon detecting an accident, the system triggers alarms and sends emergency notifications to designated responders. Although this system can improve response times and potentially save lives, concerns over false positives and the high bandwidth requirements for real-time data transmission remain unresolved.

In contrast [8], investigates the application of deep learning methodologies such as CCNN, for automatic accident detection in video footages. CNNs are effective in analyzing video data to identify patterns that indicate accidents, such as abrupt movements, vehicle deformation, or smoke. This method is particularly promising for areas where traditional detection methods are constrained by limited infrastructure, and traffic management systems equipped with cameras are already operational. However, the requirement for high-resolution video, along with the complexity of collecting and labeling accident footage for training deep learning models, presents significant obstacles.

In [9], the authors proposed a fault tolerant methodology to provide optimal solutions for execution and response times in case of faulty nodes. The model enables us to get accurate information on vehicle routing so that any catastrophe can be avoided.

System Architecture

You Only Look Once (YOLO) V7 brings enhancements in actual-time object detection by adapting the system for advanced accuracy and speed. The YOLO V7 model structure makes use of a deep CNN in particular designed to recognize gadgets in pix with a single forward skip. This new release builds on the achievement of previous variations of YOLO by including enhancements for feature extraction and object localization.

The foremost features are featuring extraction backbone: YOLO V7 uses a deep CNN backbone, along with Darknet or ResNet, to efficaciously extract features from input pix. These functions setup a vital functionality in the fairly accurate detection of objects of the hobby shown in Figure 47.1.

The head network of YoloV7 would predict several bounding boxes for the objects in the image. When processing a bounding box both the head network produced a class prediction score. The class probability measures the likelihood of finding a car within the bounding box area.

YOLO V7

The real-time vehicle detection system uses YoloV7 as its current top-tier object detection algorithm. The input video is preprocessed to improve the performance of YoloV7. This includes tasks such as frame resizing, color normalization, and noise reduction. YoloV7 is used to detect vehicles in the preprocessed video frames. YoloV7 outputs bounding boxes and confidence scores for each detected vehicle. Once an accident is detected, the system can notify the appropriate authorities for immediate response.

The YoloV7 algorithm consists of three main components: Backbone: The backbone is a CNN that extracts features from the input image or video. Neck: The neck is a network that aggregates the feature maps from the backbone and produces a set of feature pyramids. Head: The head is a network that predicts the class probabilities and bounding boxes for the objects in the feature pyramids.

Modules implemented
Video data collection and preprocessing

Collecting a diverse dataset of real-world driving scenarios and accident situations. - Preprocessing the video data to ensure consistent format, resolution, and quality. - Annotation of video frames to label objects of interest, such as vehicles, pedestrians and road elements.

Object detection using YOLO v7

Implementing YOLO v7, a cutting-edge object detection model. YOLO's ability to detect objects in real time with high accuracy and efficiency is crucial for timely accident detection. - Fine-tuning the YOLO v7 model on our annotated dataset to adapt it specifically for accident-related object detection.

Feature engineering

Extracting relevant features from the detected objects, including object size, speed, trajectory, and relative distances. - These features provide the foundation for identifying potentially hazardous situations and distinguishing normal traffic patterns from accidents.

Anomaly detection using machine learning

Utilizing machine learning algorithms to build an anomaly detection system. - Training the model on both normal and accident-related scenarios to enable it to differentiate between regular and critical events. - Incorporating techniques such as clustering, outlier detection, and time-series analysis to enhance accuracy.

Real-time analysis and decision making

Developing a real-time analysis pipeline to process incoming video streams. - Integrating the YOLO v7 model and the anomaly detection system to assess ongoing traffic situations. - Generating immediate alerts or notifications to appropriate authorities or stakeholders when potential accidents are detected.

Data visualization and reporting

Creating intuitive visualizations to represent traffic patterns, potential accidents, and system alerts. - Generating comprehensive reports that highlight accident trends, locations, and contributing factors. These reports provide valuable insights for further analysis and policymaking.

Result Analysis

The application receives training by detecting images showing vehicle collisions leading to accidents while also analyzing test videos of vehicle collisions to identify actual accidents. The training process utilizes TensorFlow together with CNN algorithms as depicted in Figure 47.1.

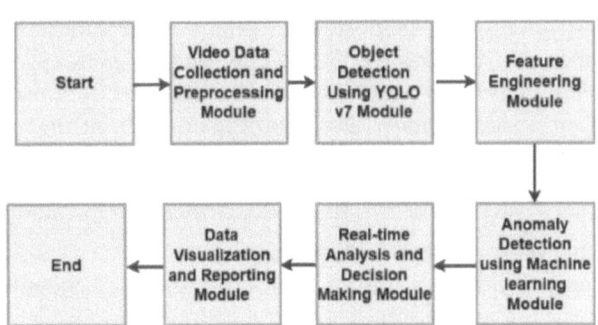

Figure 47.1 System architecture
Source: Author

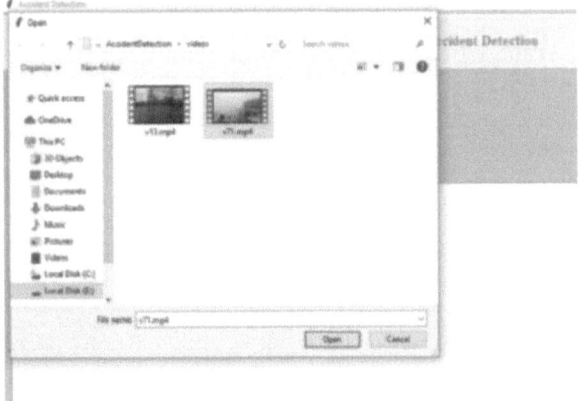

Figure 47.2 Video uploading
Source: Author

Figure 47.3 Accident detector
Source: Author

As shown in the screen of Figure 47.2, after loading the TensorFlow model users should select the 'Browse System Video' button to upload their video.

As depicted in the screen of Figure 47.3, video starts playing and upon accident detection will get below the screen.

Conclusion

In summation, the "real-time vehicle accident detection using computer vision and deep learning" initiative emerges as a highly promising endeavor poised to significantly enhance road safety. Leveraging cutting-edge technology, it not only endeavors to proactively prevent accidents but also seeks to minimize their repercussions. Through the incorporation of advanced real-time video analysis, the project aspires to establish a sophisticated system adept at swiftly identifying potential accidents and facilitating timely interventions. The realization of its objectives holds the potential for multifaceted contributions, encompassing the creation of safer road conditions, a reduction in the loss of life, and the augmentation of traffic control mechanisms. In essence, this pioneering initiative holds the prospect of fostering a safer and more secure road environment for all stakeholders involved. As it progresses, the project stands as a testament to the synergy between technological innovation and the imperative to prioritize public safety on our roadways. With advancements in technology and the increasing availability of high-resolution cameras, there are several exciting possibilities to explore. One potential direction is the integration of artificial intelligence and machine learning approaches to improve the accuracy and efficiency of accident detection. This could involve training models on large datasets of accident scenarios to enhance the system's ability to recognize and classify different types of accidents accurately. Additionally, incorporating real-time communication with emergency services and nearby vehicles could

further enhance response times and facilitate immediate assistance. This could enable the system to not only detect accidents but also provide valuable insights for accident reconstruction and prevention. Overall, the future holds immense potential for advancing the capabilities of vehicle accident detection using video analysis, making our roads safer for everyone.

References

[1] Jegan, J., Suguna, M. R., Shobana, M., Azath, H., Murugan, S., & Rajmohan, M. (2024). IoT-Enabled black box for driver behavior analysis using cloud computing. In 2024 International Conference on Advances in Data Engineering and Intelligent Computing Systems (ADICS), (pp. 1–6). IEEE.

[2] Langa, R. M., Moeti, M. N., & Kgoete, S. F. (2024). An automated smartphone-capable road traffic accident notification system. *Journal of Advanced Computational Intelligence and Intelligent Informatics*, 28(4), 939–952.

[3] Mohanty, R. K., Sahoo, S. P., & Kabat, M. R. (2024). Thermal-energy-efficient-secured-link reliable and delay-aware routing protocol (TESLDAR) for wireless body area networks. *International Journal of Computers and Applications*, 46(9), 715–727.

[4] Mohanty, R. K., Sahoo, S. P., & Kabat, M. R. (2023). Sustainable remote patient monitoring in wireless body area network with Multi-hop routing and scheduling: a four-fold objective based optimization approach. *Wireless Networks*, 29(5), 2337–2351.

[5] Mohanty, R. K., Sahoo, S. P., & Kabat, M. R. (2023). A network reliability based secure routing protocol (NRSRP) for secure transmission in wireless body area network. In 2023 8th International Conference on Communication and Electronics Systems (ICCES), (pp. 663–668). IEEE.

[6] Mohanty, R. K., Kumar, A. P., Padmaja, R., & Prashanthi, V. (2024). Deep learning for analyzing user and entity behaviors: techniques and applications. In Consumer and Organizational Behavior in the Age of AI, (pp. 219–250). IGI Global.

[7] Sahoo, S., Sahoo, S. P., Barik, R. C., & Kabat, M. R. (2023). A fault-tolerant technique for future Social IoV using evolutionary computation -based whale optimization model. In 2023 OITS International Conference on Information Technology (OCIT), (pp. 334–339). IEEE.

[8] Sahoo, S., Sahoo, S. P., & Kabat, M. R. (2024). A pragmatic review of QoS optimization in IoT driven networks. *Wireless Personal Communications*, 137(1), 325–366.

[9] Verma, A., & Khari, M. (2024). Vision-based accident anticipation and detection using deep learning. *IEEE Instrumentation and Measurement Magazine*, 27(3), 22–29.

48 Heuristic traffic management and load balancing techniques for wireless body area networks

Aishwarya Nayak[1,a], Tusharkanta Samal[1,b], Sunayana Das[1,c] and Bhabendu Kumar Mohanta[2,d]

[1]Department of Computer Science Engineering, DRIEMS University, Tangi, Cuttack, India

[2]Department of Information and System Security, College of information Technology, United Arab Emirates University, AI Ain, UAE

Abstract

The progress of wireless body area networks (WBANs) have transformed humanoid existence through its use in sports, entertainment, fitness, and healthcare, among other fields. However, ensuring QoS and energy economy are two of the key design problems for WBAN. When designing an energy-efficient and dependable system, the load balance of the various packet queues in a WBAN is crucial. Based on the IEEE 802.15.6 architecture, this research presents heuristic traffic management and load balancing techniques for WBANs. The main objective of this work is to boost WBAN throughput by decreasing packet drop in the queues. To ensure that no packet has to wait a long time in the designated queue before being transferred to the base station, the suggested work has taken into account both the packet's priority and the location from whence it was received. The results of the simulation demonstrate that the suggested approach outperforms the current techniques.

Keywords: Delay, JAYA, load, packet queue, WBAN

Introduction

The world population is growing at an astonishing rate, but there are three main problems with it: the baby boomer generation, the aging population, and the costly healthcare system. Thus, to promote proactive wellness and illness prevention, an intelligent healthcare system is needed. The Wireless Sensor network which is used particular healthcare application is called as wireless body area networks (WBANs) [11]. The information is transmitted to a isolated specialist via the Internet for a medical diagnostic Kurunathan [6], in WBANS. For WBANs, a variety of wireless communication protocols have been developed, including IEEE 802.15.4–2006 [8] and IEEE 802.15.6 [7]. Due to its many advantages, such as several frequency bands, access modes, low power consumption, short range, and high data rates, IEEE 802.15.6 is far more widely used than its predecessor.

Different bio medical sensor nodes make up the WBAN architecture. The purpose of these nodes is to measure various physical parameters coming from human beings. WBAN biosensors are typically small and have a small energy supply. In addition, several WBANs collaborate in a certain region or area to offer a dynamic healthcare service. However, each WBAN's QOS requirement may be impacted by inadequate buffering and unforeseen traffic surges. In a standard WBAN design, the PD (personalized device) picks the subsequent PD or AP to transfer the data once it has been received from sensors. For a QOS-aware packet delivery with increased energy efficiency, an effective algorithm is needed to buffer and schedule the data packets according to their user priorities at various PDs.

Related Work

In a network of several WBANs, scheduling is a crucial problem that is necessary to ensure dependable and energy-efficient real-time data packet transmission. This section demonstrates some of developed works for scheduling of packets.

Cheng and Huang [14] proposed an inter-WBAN scheduling based on graph coloring to minimize interference among the WBANs. They used spatial-reuse coloring to implement a dense sensor architecture and The body sensors were planned for various time windows. Xie et al. [15] developed a novel clique-based WBAN scheduling (CBWS) algorithm that separated the WBANs into several groups and assigned each group a specific time slot in order to lessen the effect of interference. Yan et al. [16] suggested a QoS-driven scheduling strategy that maintains a threshold value to adjust the transmission order of WBANs and distribute the best slots in accordance with the WBANs' QoS requirements. In order to improve the packet delay ratio, Torabi et al. [12] proposed an economical

[a]aishwaryanayak22@gmail.com, [b]samaltushar1@gmail.com, [c]Sunayana.das1722@gmail.com, [d]bhabendu@uaeu.ac.ae

DOI: 10.1201/9781003658221-48

dynamic scheduling technique for WBANs with relay node placement. To equitably distribute resources among several WBANs. An approach to resource allocation that considers link quality was put forth by Samanta et al. [1]. Additionally, in Samanta et al. [2], the effectiveness of WBANs was examined in crucial scenarios with varying traffic circumstances. Zhou et al. [9] investigate how execution time affects adaptive video scheduling. Yi et al. [5] developed a price-based capacity sharing scheme for WBANs that takes priorities into account. Furthermore, Yi et al. [4] and Yi et al. [3] suggested a precedence aware protocol for delay-sensitive data transmission in WBANs and an incentive mechanism for communication scheduling, respectively.

M. Ambigavati et al. (2012) [10] suggested a load-balanced, energy-efficient priority queue method for WBAN. In order to buffer and schedule data packets according to their priorities, the authors take into account four distinct kinds of queues. Even though this technique efficiently transfers important data with the least amount of latency, the load across the four queues is not appropriately balanced, increasing energy consumption and delaying the transmission of important data. Based on the IEEE 802.15.6 architecture, Samal et al. [13] introduced an algorithm for load balancing in various priority queues in WBANs. The proposed protocol has improved Quality of Service (QoS) parameters, including throughput, latency, and packet delivery ratio, ensuring more efficient data transmission and better overall network performance in WBANs.

Proposed Model

We proposed heuristic traffic management and load balancing techniques for WBANs by using JAYA algorithm for dependable packet transfer in IEEE 802.15.6-based WBAN. The objective is to scheduling packets based on their priority along with the load on personal assistance device (PD) which is balanced by using a population-based meta heuristic optimization approach named as JAYA approach. The degree of imbalance issue among the WBAN system's PDs is addressed by this approach. The biosensors provide the essential signals they have collected from the patient's body to the sink node either directly or via the depend node. A dynamic queue is kept at each PD to hold incoming data that are received from its bio-sensor nodes. Four subqueues, each with a variable length determined dynamically, are conceptually separated from this queue. Additionally, each PD's load is dynamically determined, and an overloaded PD uses JAYA technology to move the newly created packet

to an underloaded PD. The packets from these queues are then sent, using the time slot allocation mechanism, to the closest AP.

The proposed model works in the following steps. In the first step, the incoming packets received from the biosensors are coherently buffered in the dynamic queue maintained at each PD. In the second step, in order to make a QoS-aware packet transmission JAYA approach which is an effective load balancer mechanism is implementing to balances the loads of each PD by distributing the loads fairly among them. In the third step, an effective time-slot allocation scheduling mechanism is composed over IEEE 802.15.6 standard to schedule the data from PDs to AP according to remaining time left and criticality (node location) of the packets.

The WBAN system model and packet prioritization
The proposed network consists of N TDMA-based WBANs that are exist in a specific area like a hall in a big hospital. Each WBAN is represented as WS_i, $1 \leq i \leq N$. Each WBAN consists of a single personal assistance device (PD/HUB) represented as PD_i, and up to K bio-sensors that are represented as BS_{ij}, $1 \leq j \leq K$. Additionally, let there are M no. of access points represented as APS, $1 \leq S \leq M$. The Figure 48.1 shows the multiple WBAN system that consists of N no multiple WBANs/PDs, each with K biosensor nodes and M number of APs.

In WBAN, the bio-sensors capture various types of packets that need to be scheduled based on their criticality. Priority plays a major role in achieving QoS-based reliable packet transmission. Priority assignment to the packets may be done in many ways but as here the real-time packets are required to be scheduled so, in WBAN system a dynamic priority assignment strategy has been taken on where the personal gadget

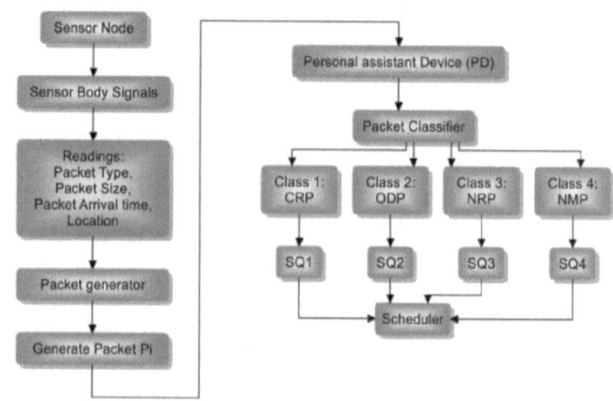

Figure 48.1 Procedure of packet classification and assignment of subqueues to packets
Source: Author

Table 48.1 Priority of different data packets.

Priority	Service category
CRP (The highest)	Information on emergency
ODP	Regular information
NRP	Aperiodic medical data (body temp, hemoglobin, ESR, etc)
NMP (lowest)	Audio/Video packets

Source: Author

device (PD) infers the priority levels for the packets by examining the essential factors such as location of the bio-sensor node generating the packet (distance from PD to bio-sensor node), kind of data pattern and frequency of packet generation. Depending on its usage, the WBAN can handle both medical and non-medical traffic types. In medical applications the traffic includes various vital monitoring signals such as ECG (heart), EEG (brain), EMG (brain), body temperature, BP rate, and sugar level and so on. In nonmedical applications the traffic includes audio and video signals.

As indicated in Table 48.1, all sensor-generated packets in the proposed work are categorized into one of four classes:

Dynamic packet buffering
Before delivering data packets to the access point (AP), each personalized assistant device (PD) buffers them in a physical queue. The PD received the data packet from its biosensor nodes and added it to its physical queue. There are four virtual subqueues within the PD physical queue, known as priority subqueues (SQ1, SQ2, SQ3, and SQ4), whose lengths are dynamically altered. They are primarily designed to keep the packets of the same priority index together, and each subqueue is given the same priority index value, which correlates to the packet priority. For example, SQ1 assigns a priority of 4 to keep the highest priority packet, which is the CRP type; SQ2 assigns a priority of 3 to keep the moderately important packet, which is the ODP type; SQ3 assigns a priority of 2 to keep the NRP type; and SQ4 assigns a priority of 1 to keep a data packet of NMP type. Each subqueue's length is dynamically modified based on how many data packets of that priority level are received. The process of classifying packets and allocating subqueues to them is summarized in Figure 48.1.

At the PD, when a packet is received from a bio-sensor node of its own WBAN, it checks the space availability at the corresponding subqueue. It assigns the respective subqueue to packet based on packet priority level i.e. SQ1 is allocated to packet type CRP, SQ2

is allocated to ODP type, SQ3 and SQ4 is allocated to NRP and NMP respectively.

Each pre-defined subqueue can be dynamically changed its length and calculated by the PD. When a new data packet comes, the PD calculates the length of each subqueue dynamically using the given eq (1)

$$L_i = \frac{N_i}{\sum_{i=1}^{4} N_i} * L_t, \tag{1}$$

Where, L_i indicates the length of the subqueue (i = 1, 2, 3,4), L_t indicates the total length of the physical queue at PD, N_i is the no. of data packets received with priority level *Pi*.

Performing load balancing using JAYA approach
Each WBAN connects several bio-sensor nodes to collect the vital physiological signals from the patient's body and forward this information to PD (or Hub). Each PD has its own physical queue i.e. splitted into four virtual subqueues to buffer these packets then transmit these to the AP. Due to the heterogeneous characteristic of the received data packet load at each PD is different means some PDs are underloaded while some PDs are overloaded. Hence a dynamic optimized load balancing approach is required to balance the load among PDs. For an efficient load balancing here we adopted a population-based optimization approach named as JAYA approach.

Effective load computation at PD
Each WBAN composed of several heterogeneous bio-sensor nodes and each node has different loads. The load at each bio-sensor BSij is the combination of both computational and communicational load, termed as Intra-WBAN communication load (£_BSij). Mathematically, it is represented as

$$£_{BSij} = \sum_{t=1}^{T} (CM_t^{(BSij \to PDi)} + CP_t^{(BSij \to PDi)}), \tag{2}$$

Where, $CM_t^{(BSij \to PDi)}$ denotes the communicational cost calculated based on packet length to a bio-sensor node BS_{ij} at time t, while $CP_t^{(BSij \to PDi)}$ denotes computational processing cost and is expressed mathematically as

$$CP_t^{(BSij \to PDi)} = \frac{\rho_{rec}^{PDi}}{\rho_{max}^{PDi}}, \tag{3}$$

Where, ρ_{rec}^{PDi} and ρ_{max}^{PDi} represents the received and maximum signal power from the bio-sensor node BSij to PDi.

Similarly, the load of each WBAN ($£_{PDi}$), termed as Inter-WBAN communication is calculated as the combination of computational and communicational load of PD to AP. It is mathematically expressed as

$$£_{PDi} = \sum_{t=1}^{T}(CM_t^{(PDi \to APm)} + CP_t^{(PDi \to APm)}),\quad (4)$$

Where, $CM_t^{(PDi \to APm)}$ is the communicational cost on the basis of packet length at PD_i to communicate with AP_m and $CP_t^{(PDi \to APm)}$ is the computational processing cost of a particular PD_i, which is mathematically expressed as

$$CP_t^{(PDi \to APm)} = \frac{\rho_{rec}^{APm}}{\rho_{max}^{APm}},\quad (5)$$

Where, ρ_{rec}^{APm} and ρ_{max}^{APm} denotes the received and maximum signal power from PDi to APm, respectively.

Now, the total load of each PDi or WBAN i.e $£_t^{PDi}$ is the combination of all connected sensors' load and computational processing load of that PDi to communicate with an APm, means the combination of Intra-WBAN and Inter-WBAN communication load at a particular time period t and is computed as in eq. (8)

$$£_t^{PDi} = \sum_{j=1}^{k} £_{BSij} + \sum_{i=1}^{m} £_{PDi},\quad (6)$$

Where, m is the total no of APs available at a particular time period t.

Grouping of PDs
The state of each PDi is identified by comparing its load with the average WBAN system load i.e. $£\_ASL$,which is mathematically represented as

$$£_{ASL} = \frac{1}{N} \sum_{i=1}^{N} £_t^{PDi},\quad (7)$$

$£\_ASL$, divides each PDi into three different categories based on their respective current load such as overloaded PD (OPD), underloaded PD (UPD) and balanaced PD (BPD). This classification is recognized as follows

If $£_t^{PDi} < £_{ASL}$, then it indicates PD_i is underloaded
Else If $£_t^{PDi} > £_{ASL}$, then it indicates PD_i is overloaded
Else PD_i is balanced.

After receiving a packet from the bio-sensor node, load at each PD_i is calculated and if that PD_i is overloaded, then the receiving packet is migrated to other PD_i i.e. underloaded PD_i.

Load balancing decision
In this part load balancing and scheduling of packets for transmission are decided. Depending on the load at each PDi and standard deviation (SD) calculated using eq 8 and eq 10 respectively, the system decides whether the load balancing is to be applied or not. This decision will depends on two states.

Simulation vs delay has been presented in Figure 48.2.

Simulation vs packet delivery ratio has been presented in Figure 48.3.

   i.   Whether the system is balanced or not
   ii. Whether the whole system is overloaded or not

I.   Finding state of the system
In order to take the decision for load balancing in the system, the standard deviation (σ) can be computed using the following eq. (10)

$$\sigma = \sqrt{\frac{1}{N} \sum_{i=1}^{N}(PT_i - PT)^2},\quad (8)$$

Table 48.2 Simulation parameters.

Parameters	Value
Time spent simulating (second)	200
WBANs	50
Aps Count	5/10
Tx-circuit consumption of Energy	16.7 nJ
Rx-circuit consumption of Energy	36.1 nJ
The quantity of customized gadgets	50
Bio sensors	300 (6/WBAN)
Rate of packet	4 packets/sec
Size of packet	512 Bytes

Source: Author

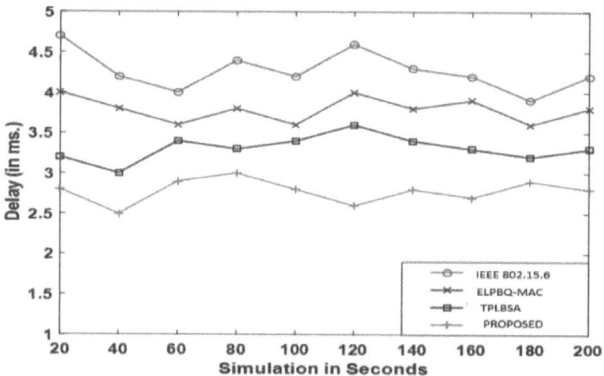

Figure 48.2 Simulation vs delay
Source: Author

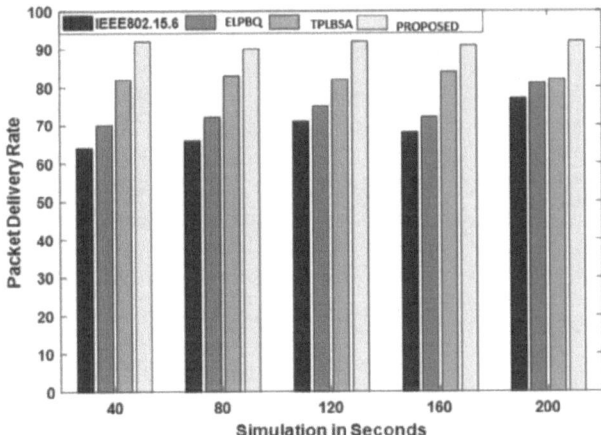

Figure 48.3 Simulation vs packet delivery ratio
Source: Author

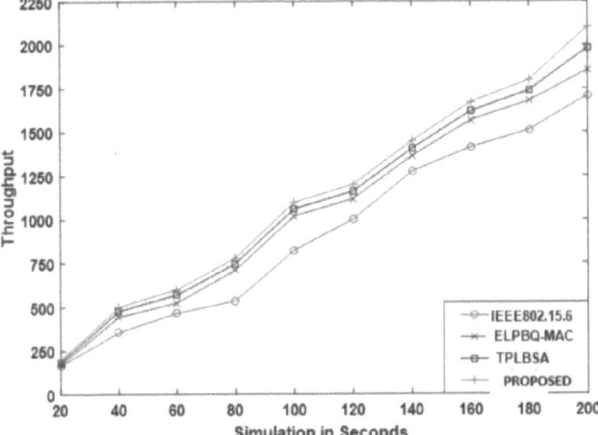

Figure 48.4 Simulation vs throughput
Source: Author

Where N is the total no. of PDs, PT_i and PT are the processing time of PDi and whole system correspondingly. The PT_i and PT are mathematically expressed as in the following Eqs. (11) and (12).

$$PT_i = \frac{\pounds_t^{PDi}}{\pounds_{ASL}}, \tag{9}$$

$$PT = \sum_{i=1}^{N} PT_i, \tag{10}$$

Simulation Parameters and Result Analysis

The performance analysis of our suggested method is shown in this section along with comparisons to the current TPLBSA Ambigavathi and Sridharan, [10], and ELBPQ-MAC (IEEE Std 802.15.6-2012) algorithm. Castalia, a recently created tool by Omnet ++ Simulator, is used to implement the suggested

algorithm and its alternatives in order to examine their performance. The suggested algorithm's simulation settings are shown in Table 48.2.

Simulation vs throughput has been presented in Figure 48.4. It is observed from the figures that the proposed protocol has minimum delay and maximum packet delivery ratio and throughput as compared to its counterparts.

Conclusion and Future Scope

The packets are scheduled according to priority in the proposed work schedule along with load on personal assistance device (PD) which is balanced by using a population-based meta heuristic optimization JAYA approach. It works in four steps .In the first step, the incoming packets received from the biosensors are coherently buffered in the dynamic queue maintained at each PD. In the second step, in order to make a QoS-aware packet transmission JAYA approach which is an effective load balancer mechanism is implementing to balances the loads of each PD by distributing the loads fairly among them. In the third step, an effective time-slot allocation scheduling mechanism is composed over IEEE 802.15.6 standard to schedule the packets from PDs to AP based on the remaining time left and criticality (node location) of the packets. Through simulation results, the suggested model was contrasted with the most advanced algorithms available today, and it has discovered to perform better than its predecessors.

The performance of WBANs with transitory link quality between Personal Devices (PDs) and Access Points (APs) may be the subject of future extensions of this article. In the proposed algorithm, we have considered WBANs, PDs, and APs as static. Therefore, future research could explore the impact of mobility on WBAN performance, addressing challenges such as dynamic link variations, energy efficiency, and real-time data transmission to enhance the overall reliability and efficiency of WBAN communication systems.

References

[1] Samanta, A., Bera, S., & Misra, S. (2015). Link-quality-aware resource allocation with load balance in wireless body area networks. *IEEE Systems Journal*, PP(99), 1–8.

[2] Samanta, A., Misra, S., & Obaidat, M. S. (2015). Wireless body area networks with varying traffic in epidemic medical emergency situation. In Proceedings of IEEE International Conference on Communications, (pp. 6929–6934).

[3] Yi, C., & Cai, J. (2017). A priority-aware truthful mechanism for supporting multi-class delay-sensitive medical packet transmissions in e-health networks. *IEEE Transactions on Mobile Computing*, 16(9), 2422–2435.

[4] Yi, C., Alfa, A. S., & Cai, J. (2016). An incentive-compatible mechanism for transmission scheduling of delay-sensitive medical packets in e-health networks. *IEEE Transactions on Mobile Computing*, 15(10), 2424–2436.

[5] Yi, C., Zhao, Z., Cai, J., de Faria, R. L., & Zhang, G. M. (2016). Priority-aware pricing-based capacity sharing scheme for beyond-wireless body area networks. *Computer Networks*, 98, 29–43.

[6] Kurunathan, J. H. (2015). Study and overview on WBAN under IEEE 802.15.6. *U. Porto Journal of Engineering*, 1, 11–21.

[7] Fourati, H., Idoudi, H., & Saidane, L. A. (2016). Overview of four emerging mechanisms for e-health communications. *International Journal of Sysems Control and Communication*, 7, 337–359.

[8] Samal, T., & Kabat, M. R. (2019, July). Energy efficient real time reliable mac protocol for wireless body area network. In 2019 10th International Conference on Computing, Communication and Networking Technologies (ICCCNT) (pp. 1-6). IEEE.

[9] Zhou, L., Yang, Z., Wang, H., & Guizani, M. (2014). Impact of execution time on adaptive wireless video scheduling. *IEEE Journal on Selected Areas in Communications*, 32(4), 760–772.

[10] Ambigavathi, M., & Sridharan, D. (2018). Energy efficient and load balanced priority queue algorithm for wireless body area network. *Future Generation Computer Systems*, 88, 586–593.

[11] Chen, M., Gonzalez, S., Vasilakos, A., Cao, H., & Leung, V. C. (2011). Body area networks: a survey. *Mobile Network Applications*, 16, 171–193.

[12] Torabi, N., Kaur, C., & Leung, V. (2012). Distributed dynamic scheduling for body area networks. In Proceedings of IEEE Wireless Communications and Networking Conference, (pp. 3177–3182).

[13] Samal, T., & Kabat, M. R. (2022). A prioritized traffic scheduling with load balancing in wireless body area networks. *Journal of King Saud University-Computer and Information Sciences*, 34(8), 5448–5455.

[14] Cheng, S. H., & Huang, C. Y. (2013). Coloring-based inter-WBAN scheduling for mobile wireless body area networks. *IEEE Transactions on Parallel and Distributed Systems*, 24(2), 250–259.

[15] Xie, Z., Huang, G., He, J., & Zhang, Y. (2014). A clique-based WBAN scheduling for mobile wireless body area networks. *Procedia Computer Science*, 31(0), 1092–1101.

[16] Yan, Z., Liu, B., & Chen, C. W. (2012). QoS-driven scheduling approach using optimal slot allocation for wireless body area networks. In Proceedings of IEEE International Conference on e-Health Networking, Applications and Services, (pp. 267–272).

49 Deep learning based optimized diagnostic model for diabetic retinopathy detection

Shekharesh Barik[a], Ramesh Kumar Mohapatra[b] and Ratnakar Dash[c]

Department of CSE, National Institute of Technology, Rourkela, Odisha, India

Abstract

Diabetic retinopathy constitutes a severe complication of diabetes with a pessimistic perspective. It is one of the main leading causes of vision loss due to its late discovery, high prevalence rate, and poor prognosis. The diseased retina varies widely in location, size, and shape. As a result, non-invasive and repeatable treatments are desired. This research proposes a system for detecting diabetic retinopathy in retinal images that combines deep learning and optimization approaches. Combining these two strategies will improve detection on clinical datasets (images). Based on clinical imaging, an optimization technique and multilevel thresholding are used to detect diabetic retinopathy in its early stages and lower prevalence rates. The optimization approach provides the optimal parameters for maximizing prediction rate with a deep learning classifier such as CNN. Finally, in this work 95.92% of accuracy was obtained.

Keywords: Deep learning, diabetic retinopathy, optimization technique, retinal images, segmentation

Introduction

One of the most devastating complications of diabetes is diabetic retinopathy, characterized by late identification, significant prevalence, and a dismal prognosis [1]. As a result, non-invasive approaches for predicting the progression of diabetic retinopathy are necessary. Medical imaging has lately been used to examine retinal changes in a non-invasive manner [2]. This includes blurred vision, floaters, dark areas in vision, and difficulty perceiving colors. In most cases, detailed imaging techniques like fundus photography (FP) and optical coherence tomography (OCT) are employed to precisely grade the severity of the disease. It is critical for doctors to be able to develop successful treatment strategies [3]. Assessment has become common in diabetic retinopathy imaging research. The assessment entails extracting highly dimensional information derived from images and translating this information into relevant characteristics for objectively determining retinal morphology and variability. Medical imaging is used to assess retinal changes in a non-invasive manner. The diseased retina varies widely in location, size, and shape. The existence of such variables has made segmentation of diabetic retinopathy lesions problematic [4, 5]. According to the American Diabetes Association (ADA) in 2022, there are anticipated to be millions of new instances of diabetic retinopathy, deep learning and optimization approaches must be used to perceive the disease in its premature phases. Most imaging techniques, like fundus photography and OCT, can assist in the identification of retinal abnormalities [6]. The most significant screening modality and mathematical technique for screening and detection of diabetic retinopathy are optical coherence tomography and optimization, respectively [7]. The suggested method iteratively trains the model through optimization, yielding an estimate of the maximum and minimum function. It represents one of the greatest deep learning phenomena that can improve performance [8]. The purpose of optimization is to develop the best design possible in regard to a set of priority criteria or constraints. Deep learning-based CNNs were used to extract diabetic retinopathy features [9]. It was also used to detect and categorize essential qualities [10].

Literature Review

The field of diabetic retinopathy (DR) detection and classification has seen significant progress with the integration of DL and optimization techniques to improve accuracy and the speed of diagnoses. This literature survey reviews recent studies focused on improving Diabetic retinopathy detection, which is critical due to its high potential for causing severe vision impairment or blindness, if not detected early [11].

Deep learning models for DR detection

Thanikachalam et al. [11] developed an optimized deep CNN model specifically targeting both DR and diabetic macular edema. Similarly, Mishra et al. [7] Bilal et al. [3] explored a hybrid model that combines

[a]919cs5107@nitrkl.ac.in, [b]mohapatrark@nitrkl.ac.in, [c]ratnakar@nitrkl.ac.in

DOI: 10.1201/9781003658221-49

CNN with singular value decomposition (SVD) optimized SVM. Das and Pumrin [4] have used the lightweight CNN architectures pre-trained on large, imbalanced datasets, achieving efficient classification.

Specialized architectures and hybrid techniques
Upreti et al. [12] employed a segmentation-based approach, integrating CNN with blood vessel segmentation improves classification accuracy. Alam et al. [2] used a segmentation-assisted fully convolutional network (FCN) of DR. Gadekallu et al. [6] demonstrated that optimization layers combined with CNNs for predicting DR.

Addressing dataset imbalances and real-world challenges
Singh et al. [10] used a fine-tuned deep learning model applied to fundus images, balancing datasets to improve accuracy. Saha [8] emphasized early detection and applied extensive data augmentation techniques.

Hybrid optimization and enhanced accuracy metrics
Thanikachalam et al. [11] integrated a CNN with multilevel thresholding techniques to increase prediction accuracy. Additionally, studies by Ebrahimi et al. [5] on OCTA layer fusion for DR detection using multiple imaging layers.

Research Gap

Some limitations may be seen after reviewing numerous research articles based on diverse applications of human posture detection and its uses in various fields:

- Many systems detect individual body components like the head, trunk, or limbs but fail to capture full-body posture, requiring improved error detection.

- Challenges exist in computing stress data, targeting specific regions, optimizing sensor placement, and selecting appropriate materials for accurate measurements.

Proposed Methodology

The proposed methodology leverages a deep learning-based approach, specifically utilizing the DenseNet121 architecture, to detect diabetic retinopathy from retinal fundus images. Key features, including blood vessels and lesions, are extracted through segmentation techniques to highlight critical regions. Different models, fine-tuned with a transfer learning strategy, is trained on this processed dataset, while a hybrid optimization technique dynamically adjusts hyperparameters like learning rates and dropout values to enhance prediction accuracy. Evaluation metrices ensure robust performance assessment, achieving the highest detection accuracy of 97.92%.

Figure 49.1 shows the System flow diagram of DR detection.

Methodology followed during this research work shown below
Data set used
Indian Diabetic Retinopathy Image Dataset (IDRiD) dataset is used during this paper research work.

Preprocessing
The preprocessing Steps includes noise reduction, contrast enhancement, and normalization to ensure data consistency. The dataset was split into training (80%) and testing (20%) subsets to ensure robust evaluation. Contrast limited adaptive histogram equalization (CLAHE) was employed to enhance the visibility of lesions and blood vessels.

Figure 49.2 shows the preprocessed image for a sample image.

Empirical Results

The proposed DenseNet121 model demonstrated robust performance on the IDRiD dataset, achieving

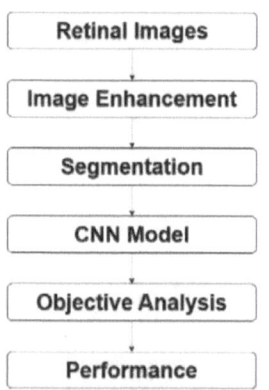

Figure 49.1 System flow diagram
Source: Author

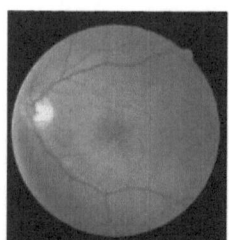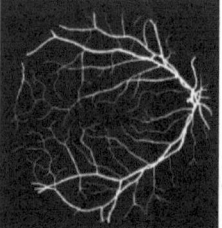

Figure 49.2 Pre-processed image
Source: Author

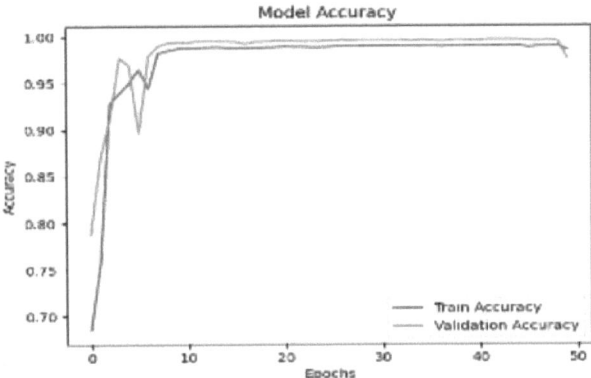

Figure 49.3 Model accuracy
Source: Author

Table 49.1 Performance metrics.

Attributes	Image-i	Image-ii
Sensitivity(%)	96.3174	98.5605
Specificity(%)	73.4516	75.1622
Accuracy(%)	95.4896	97.9218
MSE	0.001	0.0071
Entropy	0.3489	0.3378
Correlation	0.7817	0.6012

Source: Author

97.92% accuracy and a Cohen's Kappa score of 0.89, outperforming the baseline InceptionNetV3. The training progression, shown in Figure 49.3, highlights a consistent reduction in cross-entropy loss and a steady improvement in Kappa score, indicating efficient learning and robust generalization. Preprocessing techniques like noise reduction, contrast enhancement, and data augmentation significantly contributed to handling imbalanced datasets and enhancing classification precision. Table 49.1 confirms DenseNet121's superior accuracy and lower loss, establishing its potential as a reliable tool for diabetic retinopathy detection.

Conclusion

This research proposed a deep learning-based diagnostic model for diabetic retinopathy detection, integrating DenseNet121 architecture with preprocessing and optimization techniques. The methodology addressed key challenges such as dataset imbalance and lesion variability, achieving a high accuracy of 97.92% on the IDRiD dataset. Comparative analysis with InceptionNetV3 reinforced DenseNet121's superiority. Future work will focus on real-time application and multi-dataset generalization.

References

[1] Al-ahmadi, R., Al-ghamdi, H., & Hsairi, L. (2024). Classification of diabetic retinopathy by deep learning. *International Journal of Online and Biomedical Engineering*, 20(1), 74.

[2] Alam, M., Zhao, E. J., Lam, C. K., & Rubin, D. L. (2023). Segmentation-assisted fully convolutional neural network enhances deep learning performance to identify proliferative diabetic retinopathy. *Journal of Clinical Medicine*, 12(1), 385.

[3] Bilal, A., Imran, A., Baig, T. I., Liu, X., Long, H., Al-zahrani, A., et al. (2024). Improved Support Vector Machine based on CNN-SVD for vision-threatening diabetic retinopathy detection and classification. *Plos One*, 19(1), e0295951.

[4] Das, P. K., & Pumrin, S. (2024). Diabetic retinopathy classification: performance evaluation of pre-trained lightweight CNN using imbalance dataset. *Engineering Journal*, 28(7), 13–25.

[5] Ebrahimi, B., Le, D., Abtahi, M., Dadzie, A. K., Lim, J. I., Chan, R. P., et al. (2023). Optimizing the OCTA layer fusion option for deep learning classification of diabetic retinopathy. *Biomedical Optics Express*, 14(9), 4713–4724.

[6] Gadekallu, T. R., Khare, N., Bhattacharya, S., Singh, S., Maddikunta, P. K. R., & Srivastava, G. (2023). Deep neural networks to predict diabetic retinopathy. *Journal of Ambient Intelligence and Humanized Computing*, 14, 1–14.

[7] Mishra, R., Satpathy, R., Pati, B., Mohapatra, D., Mishra, P., & Palai, G. (2024). Diabetic retinopathy image classification and blind-ness detection using deep learning techniques. *Machine Intelligence Research*, 18(1), 1056–1077.

[8] Saha, A. (2024). Deep learning for the early detection of diabetic retinopathy. *Journal of Advances in Computational Intelligence Theory*, 6(2), 50–67.

[9] Shekhar, S., & Thakur, N. (2024). Deep learning framework for forecasting diabetic retinopathy: an innovative approach. *International Journal of Innovative Research in Computer Science and Technology*, 12(1), 17–20.

[10] Singh, S. P., Gupta, P., & Dung, R. (2024). Diabetic retinopathy detection by fundus images using fine tuned deep learning model. *Multimedia Tools and Applications*, 83, 1–23.

[11] Thanikachalam, V., Kabilan, K., & Erramchetty, S. K. (2024). Optimized deep CNN for detection and classification of diabetic retinopathy and diabetic macular edema. *BMC Medical Imaging*, 24(1), 227.

[12] Upreti, K., Kapoor, A., Hundekari, S., Upreti, S., Kaul, K., Kapoor, S., et al. (2024). Deep dive into diabetic retinopathy identification: a deep learning approach with blood vessel segmentation and lesion detection. *Journal of Mobile Multimedia*, 20(2), 495–524.

50 Analyzing PM2.5 levels across diverse zones in Cuttack, Odisha

Jayashree Bhuyan[a], Aditya Prasad Das[b], Tapaswini Das[c] and Subham Priyadarshan[d]

Department of CE, School of Engineering and Technology, DRIEMS University, Odisha, India

Abstract

High exposure to PM2.5 significantly endangers public health. This research presents a comprehensive assessment of PM 2.5 concentration in seven different places (Lunahar, Balisahi, SCB Medical college, CMC Dumpyard, Netaji Bus terminal, Jagatpur Industrial Estate and Tangi) of Cuttack, Odisha starting from the month November 2023 to March 2024. Digital air quality monitoring equipment was utilized for data collection purposes. The above study also focuses on establishing a relationship between the pollutant concentration, temperature fluctuation and humidity. Data analysis revels significance disparities in pollutant concentration among locations and months, with industrial areas exhibiting consistently higher level compared to residential or commercial areas. From November to March, air quality data revealed fluctuating PM2.5 levels across locations. In November, SCB Medical College recorded the highest morning PM2.5 levels, peaking at 203.00 μg/m³. January showed significant increases at Lunahar, alongside consistently high PM2.5 levels at SCB Medical College and CMC Dump yard. By February and March, PM2.5 levels decreased, yet SCB Medical College and Jagatpur Industrial Estate still exhibited elevated readings.

Keywords: Air quality monitoring, humidity, industrial areas, PM2.5 exposure, pollutant concentration, residential areas, temperature fluctuation

Introduction

Air is crucial in regulating Earth's climate, but human activities have raised greenhouse gas levels, driving global warming and climate change [1]. Our health and well-being are directly impacted by the quality of the air we breathe. Particulate matter, O_2, NO_2, H_2SO_4, and other pollutants can have a high negatively impact on the cardiovascular including respiratory systems [2], resulting in conditions like bronchitis, asthma, and cardiovascular disorders. Public health requires that air quality be maintained at a high level. The respiratory system can be deeply penetrated by particulate matter (PM), PM2.5, which can cause respiratory infections, asthma, bronchitis, and cardiovascular disorders. and perhaps early demise. These particles can evade the body's natural defenses, reaching deeper into the lungs and bloodstream. PM also impairs visibility by scattering and absorbing sunlight, affecting aviation, transportation, and scenic views, and contributing to haze and smog in urban areas, creating unpleasant living conditions. Originating from natural sources such as wildfires, volcanic eruptions, and dust storms, as well as human activities, such as vehicle emissions, industrial processes, construction activities, and agricultural practices. Combustion of fossil fuels, biomass burning, and diesel-powered vehicles are significant contributors to urban air pollution. According to the National Ambient Air Quality Standards (NAAQS) and the Central Pollution Control Board (CPCB) in India. Areas closer to industrial zones exhibit moderate air pollution levels, ranging from 51 to 75 on the AQI scale. The region surrounding steel industries shows signs of transitioning from moderate to heavy pollution in the near future. The study conducted in Rohtak city, Haryana, monitored ambient air quality using High Volume Sampler [3]. NO_2 exceeded standards at four sites in winter. Ozone levels were below standards but higher in summer. SPM concentrations surpassed safety limits across all sites and seasons. Transport emissions and increasing construction activities contribute to pollution, exacerbated by rising vehicle volume and traffic patterns. Results revealed that criteria pollutants SPM, CO, SO_2, and NO_2 either exceeded or approached limits [4], highlighting the need for continuous monitoring and control mechanisms. Prolonged exposure to PM2.5 increases the likelihood of developing type 2 diabetes mellitus, although further research is needed to definitively establish the link between PM2.5 and GDM (gestational diabetes mellitus). Implementing effective measures to reduce PM2.5 exposure in vulnerable populations, particularly pregnant women, would be prudent. The various surveys show that

[a]jayashree@driems.ac.in, [b]adityadas@driems.ac.in, [c]tapaswini@driems.ac.in, [d]spriyadarshan@driems.ac.in

DOI: 10.1201/9781003658221-50

there is a positive relationship between air pollution levels and sick days, suggesting that reducing pollution may decrease illness. In Sambalpur, Odisha, air pollutant monitoring was conducted at four stations for a year, following CPCB guidelines. Gaseous pollutants remained within standards, but particulate matters exceeded limits. Meteorology and anthropogenic activities influenced pollutant dispersion. Continuous monitoring and source reduction for particulate matters are crucial for improving air quality. Meteorological analysis showed a significant negative correlation between relative humidity and pollutants. The study conducted, focusing on Odisha's remote opencast coal mining region, confirms that proximity to mining increases respiratory illness (RI) likelihood [4], potentially influenced by under-reporting bias. Regression analyses highlight the significance of variables like distance from mine, treatment, and per capita income, indicating higher RI likelihood in closer proximity to mines and in treatment villages.

Literature Review

A comprehensive summary is presented of the seasonal variation of PM2.5 concentration in seven different places (Lunahar, Balisahi, SCB Medical College, CMC Dump yard, Netaji Bus terminal, Jagatpur Industrial Estate, and Tangi) of Cuttack, Odisha, starting from November 2023 to March 2024. Based on ground observations taken by using digital air quality monitor. This analysis provides useful information for comparative research by identifying patterns of human activity and local temporal emission characteristics [5] based on observation. To improve air quality and meet Indian NAAQS regulations, city-level mitigation programs must be designed and implemented [6] using the findings as a foundation.

Instrument Used and Methodology

Study area
Cuttack is situated at latitude 20°30' north and longitude 85°50' east, located on a fertile delta at the confluence of the Mahanadi and Kathajodi rivers. The city is approximately 35 kilometres from the Bay of Bengal and experiences significant tidal influence from the rivers. With a population of around 1.5 million, Cuttack serves as a major urban and industrial center in Odisha. Given its proximity to industrial areas and the heavy vehicle traffic in the city, measuring the ambient concentration of PM2.5 is critical. Regular monitoring helps assess exposure levels and identify pollution sources [7], facilitating effective management of air quality and safeguarding the health of residents.

The GPS locations of the study area are shown in Figure 50.1.

Table 50.1 lists the GPS coordinates of the selected study locations.

GPS positions of sampling locations
Instrument specifications
By combining many air sensors and incorporating a built-in fan, this scientific air quality detection system allows for real-time monitoring of formaldehyde (HCHO), total volatile organic compounds (TVOC), PM2.5/10, AQI, temperature, and humidity [2], all of which are shown on its digital LCD screen.

The digital air quality monitor used in the study is depicted in Figure 50.2.

Data Collection and Analysis Technique

The entire research endeavor spanned from November 2023 to March 2024, encompassing duration of five

Table 50.1 GPS locations of study area.

Category	Location	Longitude	Latitude
Residential	Balisahi	86.110002	20.483637
Residential	Lunahar	86.108904	20.482
Construction site and Health care	SCB medical college	85.892237	20.477769
Waste disposal site	CMC Dumpyard	85.849514	20.474956
Traffic crossing/bus stand	Netaji Bus terminal	85.903008	20.447488
Industrial	Jagatpur Industrial Estate	85.920215	20.495293
Traffic crossing	Tangi	85.997602	20.555828

Source: Author

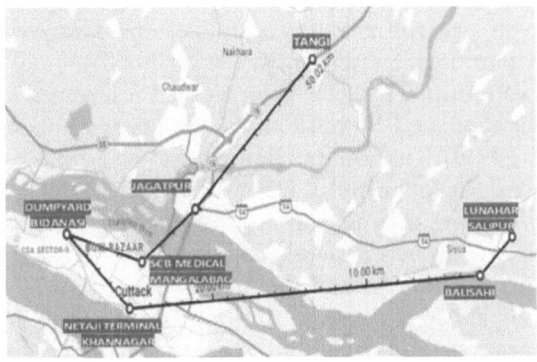

Figure 50.1 GPS locations of study area
Source: Author

Figure 50.2 Digital air quality monitor
Source: Author

Figure 50.3 Field image
Source: Author

months [6]. Data collection occurred at seven designated locations on the 14th day of each month, encompassing three time slots: morning (6:00 am), mid-day (1:00 pm), and evening (7:00 pm). Digital air quality monitoring equipment was utilized for data collection purposes. Subsequently, the air quality index (AQI) values for PM2.5 were computed utilizing the formula stipulated by the Central Pollution Control Board (CPCB). The computation process was entirely conducted using formulas within the Excel application. Following the calculation of AQI values for pm 2.5 for each month, comprehensive analysis ensued. The principal objective of the analysis was to ascertain compliance with permissible limits for pollutant concentrations and to establish correlations between meteorological parameters (temperature and humidity) and pollutant (PM2.5) concentrations.

Table 50.2 Standard AQI values for different pollutants (as per CPCB standards).

AQI Category (Range)	PM$_{10}$ 24-hr	PM$_{2.5}$ 24-hr	NO$_2$ 24-hr	O$_3$ 8-hr	CO 8-hr (mg/m^3)	SO$_2$ 24-hr	NH$_3$ 24-hr	Pb 24-hr
Good (0-50)	0-50	0-30	0-40	0-50	0-1.0	0-40	0-200	0-0.5
Satisfactory (51-100)	51-100	31-60	41-80	51-100	1.1-2.0	41-80	201-400	0.6-1.0
Moderate (101-200)	101-250	61-90	81-180	101-168	2.1-10	81-380	401-800	1.1-2.0
Poor (201-300)	251-350	91-120	181-280	169-208	10.1-17	381-800	801-1200	2.1-3.0
Very poor (301-400)	351-430	121-250	281-400	209-748*	17.1-34	801-1600	1201-1800	3.1-3.5
Severe (401-500)	430+	250+	400+	748+*	34+	1600+	1800+	3.5+

One hourly monitoring (for mathematical calculation only)

Image credit: National Air Quality Index Report by Central Pollution Control Board

Source: Author

A field image of the monitoring setup is presented in Figure 50.3.

Table 50.2 presents the standard AQI values for different pollutants as per CPCB standards.

Table 50.3 shows the AQI values of the study area during different time periods.

Formula used for AQI calculation

$$Ip = \left(\frac{IHi - ILow}{BPHi - BpLow} \right)(Cp - BpLow) + ILow$$

Ip : The air quality index

Cp : The pollutant concentration

BpLo : The concentration break point that is ≤ Bp

BpHi : The concentration break point that is ≥ Bp

ILow : The index break point corresponding to Bplow

IHigh : The index breakpoint corresponding to Bphigh

Results and Discussion

Table 50.3 AQI values of study area during different time period.

Sl no.	Month: November Place	Morning	Timing Mid-day	Evening
1.	Lunahar	561.86	104.41	53.21
2.	Balisahi	145.38	114.66	83.93
3.	SCB Medical College	121.48	112.24	22.48
4.	CMC Dumpyard	22.48	112.00	295.59
5.	Netaji Bus Terminal	87.34	80.83	115.66
6.	Jagatpur Industrial Estate	200.00	159.03	60.03
7.	Tangi	60.03	112.24	165.86
	December			
1.	Lunahar	330.16	261.86	244.98

2.	Balisahi	352.42	269.53	254.19
3.	SCB Medical College	304.84	254.19	312.51
4.	CMC Dumpyard	296.40	269.53	363.16
5.	Netaji Bus Terminal	309.44	250.35	297.16
6.	Jagatpur Industrial Estate	317.88	268	286.42
7.	Tangi	265.70	261.09	268.77
January				
1.	Lunahar	286.34	166.86	825.72
2.	Balisahi	221.48	91.76	777.93
3.	SCB Medical College	979.34	276.10	207.83
4.	CMC Dumpyard	1027.14	265.86	344.38
5.	Netaji Bus Terminal	610.66	163.45	388.76
6.	Jagatpur Industrial Estate	125.90	433.14	170.28
7.	Tangi	78.10	474.10	170.28
February				
1.	Lunahar	327.09	284.12	296.40
2.	Balisahi	330.93	254.19	281.81
3.	SCB Medical College	289.49	226.56	234.23
4.	CMC Dumpyard	244.98	228.09	223.49
5.	Netaji Bus Terminal	223.49	229.63	225.02
6.	Jagatpur Industrial Estate	250.35	286.42	282.58
7.	Tangi	259.56	317.88	301.00
March				
1.	Lunahar	108.45	89.86	138.86
2.	Balisahi	100.00	110.14	130.41
3.	SCB Medical College	52.69	34.10	32.41
4.	CMC Dumpyard	51.00	35.79	29.03
5.	Netaji Bus Terminal	54.38	29.03	23.97
6.	Jagatpur Industrial Estate	115.21	267.28	91.55
7.	Tangi	91.55	240.24	203.07

Source: Author

Conclusion

From the above study and analysis, it was found out those Industrial areas like Jagatpur Industrial Estate show higher pollutant levels compared to residential or commercial areas [5] like Netaji Bus Terminal. There are fluctuations in pollutant concentrations across months. For example, PM2.5 levels at SCB Medical College seem to decrease from November to January but increase again in February.

As the temperature and humidity increases, the pollutant concentration is also getting increased. The analysis also aimed to ascertain compliance with permissible limits for pollutant concentrations. By comparing the measured concentrations with established standards, it was possible to identify areas of concern and prioritize mitigation efforts.

Continued monitoring of air quality parameters beyond the study period will provide valuable insights into seasonal trends, long-term variations, and the effectiveness of mitigation measures implemented over time. Further research focusing on the health impacts of air pollution in Cuttack, particularly among vulnerable populations, can inform public health policies and interventions aimed at reducing health risks associated with poor air quality.

Integrating meteorological data with air quality measurements will enhance our understanding of the factors influencing pollutant dispersion and accumulation, enabling more accurate air quality forecasting and management.

Acknowledgment

The author is very great thankful to Dr. Biplab Behera H.O.D, Department of Civil Engineering, for his guidance, which helped to improve the present paper; and Mr. Subham Priyadarshan, Mr. Aditya Prasad Das, and Ms. Tapaswini Dash, Assistant Professor, Department of Civil Engineering, for reading and preparing the research paper and giving valuable suggestions to improve the quality of the paper.

Graphical Abstract

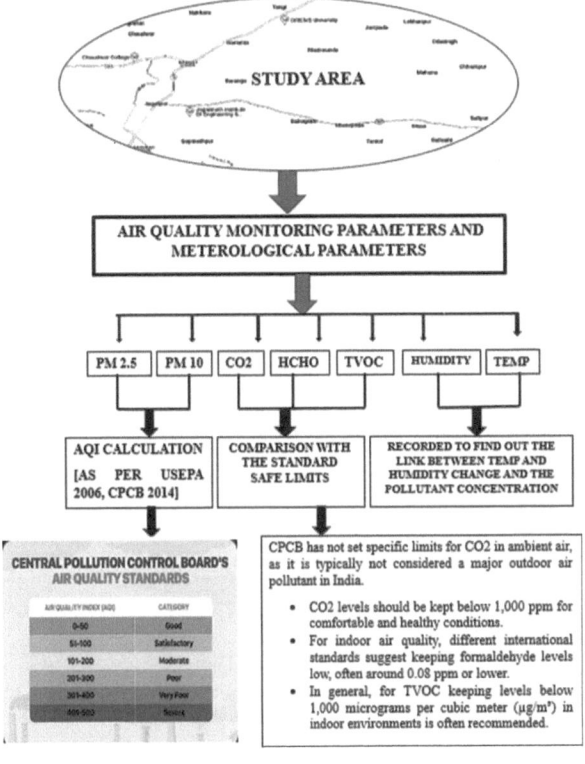

References

[1] Balashanmugam, P., Ramanathan, A. R., & Kumar, V. N. (2012). Ambient air quality monitoring in Puduch-

erry. *International Journal of Engineering Research and Applications (IJERA)*, 2(2), 300–307. ISSN: 2248-9622.

[2]   Bhuyan, P. K., Samantray, P. & Rout, S.P. (2021). Ambient air quality status in choudwar area of Cuttack District. *International Journal of Environmental Sciences*, 1(3). ISSN 0976–4402.

[3]   Dash, S. K., & Dash, A. K. (2015). Determination of air quality index status near bileipada, joda area of Keonjhar, Odisha, India. *Indian Journal of Science and Technology*, 8(35), 1–7. DOI: 10.17485/ijst/2015/v8i35/81468.

[4]   Mahapatra, P. S., Ray, S., Das, N. & Mohanty. (2012). Urban air-quality assessment and source apportionment studies for Bhubaneshwar, Odisha. *Theoretical and Applied Climatology*, 112, 243–251.

[5]   Panigrahi T., Das, K. K., Dey, B. S., Mishra, M., & Panda, R. B. (2012). Air pollution tolerance index of various plants species found in F.M. university campus, Balasore, Odisha, India. *Journal of Applicable Chemistry*, 1(4), 519–523.

[6]   Sarella, G., & Khambete, A. K. (2015). Ambient air quality analysis using air quality index – a case study of Vapi. *JIRST - International Journal for Innovative Research in Science and Technology*, 1(10), 1–4.

[7]   Shukla, V., Dalal, P., & Chaudhry, D. (2010). Impact of vehicular exhaust an ambient air quality of Rohtak city, India. *Journal of Environmental Biology*, 31, 929–932.

51 Smart fitness tracker for exercise monitoring integrated with voice assistant

Karmajit Patra[1,a], Shekharesh Barik[1,b], Maheswar Mishra[1,c], Surajit Mohanty[1,d], Niva Tripathy[1,e] and Deepankar Rout[2,f]

[1]Department of CSE, School of Engineering and Technology, DRMU, Odisha, India

[2]Department of R&D, Silicon Tech. Lab Pvt. Ltd, Bhubaneswar, Odisha, India

Abstract

Exercise can have many benefits, including physical, mental and spiritual benefits. There are many exercise videos and popular exercise tracking programs that are freely available on the internet. The goal is to help people to do these exercises independently. In this paper, the human body is represented as a collection of joint parts and the angle between the joints are calculated by the fitness tracker. The system provides suggestions for the user's exercise by using pose estimation techniques and then coordinate joint from the estimated pose is extracted. In this paper, the fitness tracker detects and locates defects very successfully according to the threshold deviation between the angles of the body parts.

Keywords: Coordinate, exercise, fitness tracker, pose estimation

Introduction

Now a days, fitness trackers is widely accepted among employee as they can monitor and improve their health and fitness. This paper presents a comprehensive approach for detecting human posture during home workouts and providing real-time guidance through voice-based as well as textual feedback.

The approach has the potential to improve the safety and effectiveness of home-based exercise routines and can benefit individuals who do not have access to professional guidance or fitness trainers. This paper provides an overview of the system, including its design, implementation, and evaluation, as well as a discussion.

Related Works

Xi and Shi [8] proposed several elements to refine the whole material body, divide the subject into parts to evaluate the head, torso and limbs, and isolate and process individual training part classifiers for advanced human anatomical subjects.

Subasi et al. [7] developed a HAR system based on smart phone sensor data using Bagging and Adaboost ensemble classifiers.

Gu et al. [3] present an intelligent PC vision framework sufficiently particular to work with different wellsprings of human posture assessment.

Sacchetti et al. [6] describe a method for detecting and classifying human postures in personal situations, particularly in classrooms. Curone et al. [1] performs a human activity and posture assessment with triaxial accelerometers which provides useful information about functional ability.

Estrada et al. [2] Used gyroscope readings at various points in the human spine(thoracic, thoracolumbar and lumbar spine) via a mobile device attached to these points.

Panini and Cucchiara [5] proposed a method of classifying human posture for indoor observation at home.

Drawbacks of the Existing Systems

The following limitations are found after reviewing different research articles based on diverse applications of Posture detection and its uses in various fields:

Some systems are capable of detecting different body components, such as head, trunk, limbs, but not posture. Therefore, error detection should be covered in the training process. In other systems, computing the stress data and getting information about each point on the targeted region is a challenging task. Few systems do not use any criteria or indications to classify the acquired data to obtain the accurate categorization. Some systems did not provide a record of the user's current status.

[a]patra.karmajit@gmail.com, [b]shekharesh@gmail.com, [c]maheswar.mishra@driems.ac.in, [d]mohanty.surajit@gmail.com, [e]nivatripathy@driems.ac.in, [f]deepankarrout99@gmail.com

DOI: 10.1201/9781003658221-51

Proposed Methodology

Problem statement

The fitness tracker will ensure that the relevant exercises are done correctly with a proper posture. If the exercise is performed incorrectly, the voice assistant will provide feedback and the counting meter will not advance the repetition count; letting the user know that he or she made a mistake in the current repetition and may remedy the error in subsequent repetitions.

Architecture of the system

The system works by greeting the user; the user is then prompted to select the kind of exercise they want to do. After the user chooses the workout, the system displays a video for the user's reference. The user will get a general idea of how to perform the exercise from the video displayed on the screen [4]. The user then continues to perform the corresponding exercise when the movie is finished. The user's motions are captured by the camera and the corresponding joint coordinates are shown. To determine the angle between the various spots, these points are plotted. These criteria help to determine whether or not the user is performing the exercise appropriately. After each repetition, the counter gets incremented. But if the repetition was done incorrectly, the counter does not increase. After the completion of desired repetitions the voice assistant gives relevant feedback and the system close.

Figures 51.1 shows the overall architecture of the working model.

Figure 51.2 shows us the DFD Level 0 diagram or the Conceptual Data Model. The figure represents the entire system as a single process, showing the high-level view of inputs, processes, and outputs. Figure 51.3 shows the DFD Level 1 diagram or the logical data model.

Figure 51.3 represents a more detailed view of the system by breaking down the Level 0 diagram into smaller sub-processes or modules. It shows the flow of data between these sub-processes and external entities. This model shows the processes on a deeper level and how the user interacts with OpenCV. It also shows the contributions of joint coordinates.

To give the user the ability to correct errors. The human body is modeled as a set of limbs and the angle between the limb pair is measured. Based on the critical deviation between the limb angles, we can successfully identify and locate faults in the activity of the user. The system starts by acquiring a real-time image of the user. The user then selects the exercise they wish to perform. Users will then watch an instructional video that teaches them how to perform the exercise in order to familiarize them with the program.

Results and Discussion

The first step is to check how MediaPipe collects photographs and videos in real time system testing. The library's viability and compatibility with the hardware device were examined. After checking the visibility of media pipe and OpenCV, the pose Landmarks function was tested to see how accurately the joint points of the user are being captured, which is shown in Figure 51.4 and Figure 51.5.

A function called calculate_angle was defined and tested. It takes 3 parameters that are the joint coordinates of the user and then displays the angle.

The user does the movements for the exercises as per the reference videos and accordingly the system recognizes whether the repetition was done successfully or not and the counter for repetitions gets incremented. We have implemented various exercises like

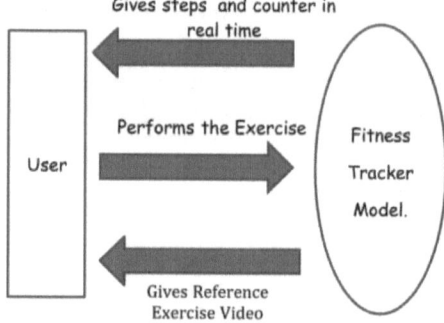

Figure 51.2 Conceptual data model
Source: Author

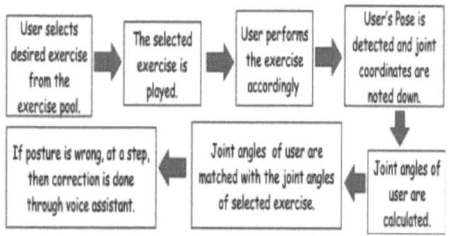

Figure 51.1 Architecture of the working model
Source: Author

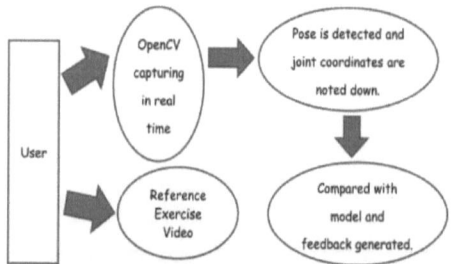

Figure 51.3 Logical data model
Source: Author

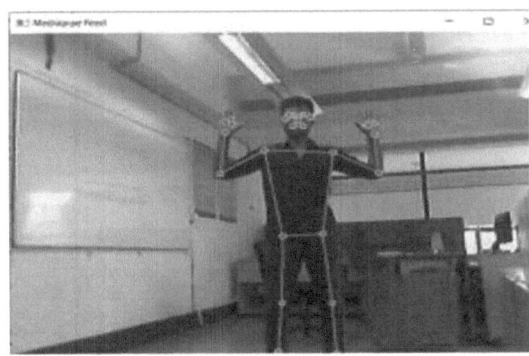

Figure 51.4 Pose landmark function plots points over the user's body
Source: Author

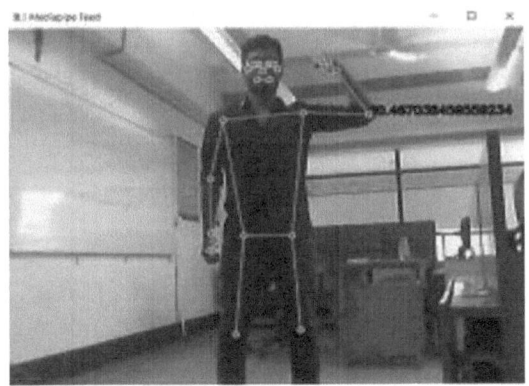

Figure 51.5 Calculating angles according to user's movement
Source: Author

Bicep curls, shoulder press, squats, lunges, frontal raises.

Conclusion and Future Scope

This paper presents a novel fitness tracker application that is capable of detecting and correcting user posture during workouts by providing voice and text-based feedback. Our experiments demonstrated that the proposed approach is effective in identifying and correcting incorrect postures, and can significantly improve the overall workout experience of users. Additionally, the system's ability to adapt to different users and workout routines makes it a versatile and useful tool for fitness enthusiasts of all levels. Overall, the results presented in this paper demonstrate the potential of this approach to help users achieve better fitness outcomes and prevent workout-related injuries. The user benefits greatly from the description of the exercise's stage of execution. Our model is quite user-friendly for beginners and doesn't require a lot of technological expertise.

One potential direction for future work is to expand the capabilities of the system to include additional types of workouts and exercises. For example, the system could be adapted to detect and correct posture during Yoga, Pilates or Physiotherapy exercises, which require different types of movements and postures than traditional workouts. Another potential area for improvement is the integration of artificial intelligence and machine learning algorithms into the system. By leveraging these technologies, the system could be trained to adapt to the individual needs and preferences of users, providing more personalized feedback and coaching.

References

[1] Curone, D., Bertolotti, G. M., Cristiani, A., Secco, E. L., & Magenes, G. (2010). A real-time and self-calibrating algorithm based on triaxial accelerometer signals for the detection of human posture and activity. *IEEE Transactions on Information Technology in Biomedicine*, 14(4), 1098–1105.

[2] Estrada, J. E., & Vea, L. A. (2016). Real-time human sitting posture detection using mobile devices. In 2016 IEEE Region 10 Symposium (TENSYMP), (pp. 140–144). IEEE.

[3] Gu, Y., Pandit, S., Saraee, E., Nordahl, T., Ellis, T., & Betke, M. (2019). Home-based physical therapy with an interactive computer vision system. In Proceedings of the IEEE/CVF International Conference on Computer Vision Workshops, (pp. 0–0).

[4] Nagarkoti, A., Teotia, R., Mahale, A. K., & Das, P. K. (2019). Realtime indoor workout analysis using machine learning & computer vision. In 2019 41st Annual International Conference of the IEEE Engineering in Medicine and Biology Society (EMBC), (pp. 1440–1443). IEEE.

[5] Panini, L., & Cucchiara, R. (2003). A machine learning approach for human posture detection in domotics applications. In 12th International Conference on Image Analysis and Processing, 2003. Proceedings, (pp. 103–108). IEEE.

[6] Sacchetti, R., Teixeira, T., Barbosa, B., Neves, A. J., Soares, S. C., & Dimas, I. D. (2018). Human body posture detection in context: the case of teaching and learning environments. In SIGNAL 2018: The Third International Conference on Advances in Signal, Image and Video Processing, (Vol. 87, pp. 79–84).

[7] Subasi, A., Fllatah, A., Alzobidi, K., Brahimi, T., & Sarirete, A. (2019). Smartphone-based human activity recognition using bagging and boosting. *Procedia Computer Science*, 163, 54–61.

[8] Xi, C., & Shi, D. (2019). Analysis of moving human body detection and recovery aided training in the background of multimedia technology. *Multimedia Tools and Applications*, 82, 1–21.

52 A systematic literature review of public sentiments on health initiatives: A case study of COVID-19

Subhasis Mohapatra[a], Sudhir Kumar Mohapatra[b] and Sweta Samantaray[c]

Faculty of Engineering and Technology, Sri Sri University, Cuttack, Odisha, India

Abstract

SARS-CoV-2 is the pathogen that causes coronavirus disease 2019 (COVID-19). In Wuhan, China, the first incidence was recorded in December 2019. Without specific treatment, many people infected with viruses will recover from mild to moderate respiratory infections. Some people, nevertheless, will get sick and need medical care. The COVID-19 virus continues to spread and poses health risk to people, particularly to the elderly, those with chronic diseases, people with compromised immune systems, and expecting moms. Effective and safe immunizations help reduce the chance of severe disease and fatalities from COVID-19. Globally, COVID-19 immunizations will prevent 14.4 million deaths in 2021 alone. Vaccine hesitancy, however, is one of the biggest dangers to world health. Social media misinformation about vaccines is the root cause of vaccination reluctance. With the help of a thorough literature analysis, this study analyses many quantitative and qualitative data sources on COVID-19 vaccine reluctance. Concerning how people view the COVID-19 vaccination, this systematic literature review examines over seventy-four papers published between 2020 and 2024 that satisfy our research questions (RQs) and related topics. This study identifies research gaps, evaluates several techniques and resources for obtaining quantitative findings, and suggests future research avenues.

Keywords: COVID-19, literature exploration, SARS-CoV-2, vaccine reluctance

Introduction

In December 2019, Wuhan, China, reported discovering the first recorded case of COVID-19. Following phylogenetic analysis, the International Committee on Taxonomy of Viruses designated the coronavirus as severe acute respiratory syndrome coronavirus 2 (SARS-CoV-2) [11]. Because of its quick worldwide spread, the World Health Organization classified it as an epidemic in 2020 [11]. Over 2.6 million people died from this disease contagion in the first half of March 2021, out of around 120 million cases reported globally. COVID-19 vaccinations are extremely successful at avoiding the most serious effects of a COVID-19 infection. On December 21, 2020, BioNTech/Pfizer received approval from the European Union for the first vaccination, BNT162b2 [10]. The efficacy of these vaccinations in lowering the incidence of symptomatic COVID-19 contagions and the seriousness of the disease was shown in the third phase of clinical trials [10]. The problem of vaccine reluctance is not new; it existed before the epidemic. Before the official launch of the SARS-CoV-2 vaccine, social networking websites such as Facebook, LinkedIn, and Twitter saw a surge of remarks that discouraged COVID-19 immunization during the pandemic. Even after the vaccination campaign began, there was still a negative reaction on social media, even though millions of individuals had received the vaccine, proving its effectiveness and safety. Hence a systematic literature review is necessary for summarizing and analyzing the people's perception about the COVID-19 vaccine.

Review Methods

The systematic literature review uses a methodical and rigorous approach to identify, evaluate, and synthesize study data on public discrimination of the COVID-19 vaccination. By adhering to SLR principles, the review will ensure the transparency and dependability of the outcomes.

The strategies specified in the PRISMA guidelines will be employed in the current study [7].

This research synthesis aims to discuss how individuals perceive their desire to get vaccinated against COVID-19 and outline some of the review study findings. These are-:

- SLR identifies and integrates the intention of the COVID-19 vaccine.
- SLR outlines many techniques for analyzing reluctance to get the COVID-19 antidote.
- It describes the eligibility and exclusion standards for primary studies, which helps us find pertinent materials for our evaluations.

[a]subhasis22@gmail.com, [b]sudhir.mohapatra@srisriuniversity.edu.in, [c]swetasamantaray21@gmail.com

DOI: 10.1201/9781003658221-52

Research questions

Research questions (RQs) preparation is a crucial step in conducting a systematic literature survey Page et al. [7]. This is significant because the way research questions are formulated can have a direct impact on how relevant primary studies (PSs) are found and how data extraction (DE) is carried out. Effectively gathering and organizing the data through a comprehensive data analysis process is crucial to providing answers to the RQs.

For this review, three RQs have been formulated as follows:

RQ1: What methodologies have been utilized in developing opinion assessment measures to evaluate public views of COVID-19 vaccines, and what are the key findings from these analyses?

RQ2: What data sources are employed to observe people's attitudes towards COVID-19 immunization, and how do they contribute to understanding public sentiment?

RQ3: Which machine learning techniques are used to analyze public opinion on COVID-19 vaccines, and what are their respective strengths and weaknesses?

Inclusion and exclusion criteria

A set of standards, referred to as inclusion and exclusion, is developed to estimate the importance of primary research. The inclusion criteria define the standards that must be met for a primary study to be included, while the exclusion criteria help to eliminate studies that do not meet certain requirements to be considered primary research. Table 52.1 displays the inclusion and exclusion criteria used in the current research.

Table 52.1 Criteria for inclusion and exclusion [7].

Inclusion criteria	Exclusion criteria
IC1: studies should be written in English.	EC1: Studies that are written before 2020.
IC2: Studies are required to be published between 2020 and 2024.	EC2: There was no addition of gray literature, such as speeches, reports, or newsletters.
IC3: Studies must be published in journal issues.	EC3: Studies that do not use English as the language of writing.
IC4: studies that examined the factors influencing the adoption of the COVID-19 vaccination based on quantitative global surveys included.	EC4: Studies that are not fully accessible.
IC5: Studies that discuss perception towards COVID-19 vaccination.	EC5: Duplicate studies.
IC6: studies that address emotions and readiness to receive the COVID-19 vaccine.	

Source: Author

Exploration approach

To classify pertinent research, a thorough exploration approach was employed across multiple digital databases, which included PubMed, Wiley, Scopus, Springer, IEEE Xplore, and Google Scholar. The exploration strategy comprised a combination of specific keywords related to RQs proposed in Section 2.1; the main keywords are public perception, COVID-19 vaccination, and machine learning. To guarantee that all pertinent studies are found, the review will adhere to the PRISMA guidelines framework [7].

Search string formulation

The RQs that are created have a significant impact on how the search string is formed. The search strings for this review are stated as follows:

RQ1: What methodologies have been utilized in developing opinion-mining tools to evaluate public perception of COVID-19 vaccines, and what are the key findings from these analyses?

Search String: ("methodologies" OR "approaches" OR "techniques" OR "methods") AND ("sentiment analysis" OR "opinion mining" OR "sentiment evaluation") AND ("public perception" OR "public opinion" OR "public sentiment") AND ("COVID-19 vaccines" OR "COVID-19 vaccination") AND ("key findings" OR "results" OR "insights")

RQ2: What data sources are employed to monitor public attitudes towards the COVID-19 vaccine, and how do they contribute to understanding public sentiment?

Search String: "COVID-19 vaccine public attitudes data sources" OR "monitoring public sentiment COVID-19 vaccine" OR "understanding public attitudes COVID-19 vaccine sources"

RQ3: Which machine learning techniques are used to analyze public opinion on COVID-19 vaccines, and what are their respective strengths and weaknesses?

Search String: ("machine learning methodologies" OR "machine learning approaches" OR" machine learning techniques") AND ("public perception" OR "public opinion" OR "public sentiment") AND ("COVID-19 vaccines" OR "COVID-19 vaccination") AND ("pros and cons" OR "advantages and disadvantages" OR "strengths and shortcomings "OR "merits and demerits ")

Study selection process

The goal of this research is to evaluate and contrast the different methods, machine learning strategies, and data sources that are employed to analyze public discrimination of the COVID-19 vaccination, together with the benefits and drawbacks of each. The review complies with the PRISMA guidelines, which involve

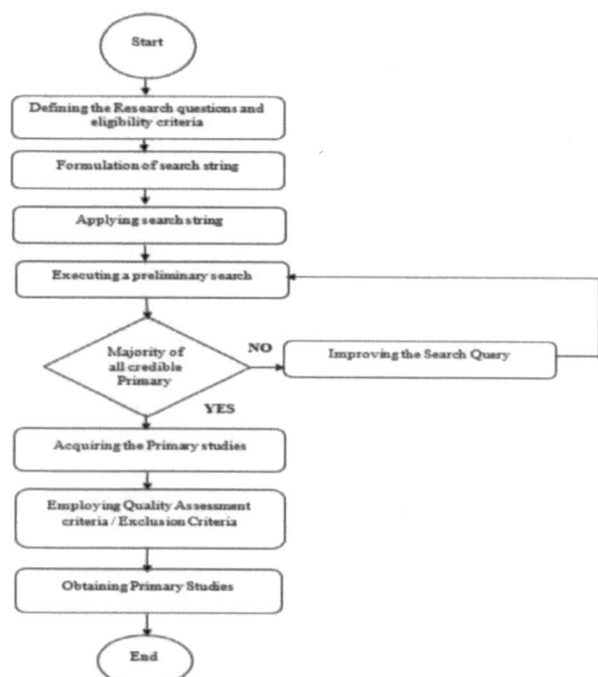

Figure 52.1 Overall primary studies selection process
Source: Author

Table 52.2 Distribution of papers based on the databases [7].

No.	Database	Before exclusion	After exclusion
1	IEEE Xplore	28	9
2	Wiley	9	1
3	Springer	39	18
4	Elsevier	81	20
5	Taylor & Francis	38	3
6	Others	214	23
Total	___	409	74

Source: Author

four main phases: identification, screening, eligibility, and inclusion [7].

In the identification step, we applied a systematic search strategy to find relevant studies from various electronic databases. The total number of papers in each research database and the number of papers before and after being included as Primary research are presented in Table 52.2.

The SLR searched for papers relevant to RQs using three main concepts. This resulted in 409 papers. After that, the review excluded 165 papers that were not directly related to this research. The study also removed 67 duplicate papers. The researcher applied inclusion and exclusion criteria based on the title, abstract, keywords, and publication type to the remaining 177 papers. The research assessed their quality and excluded 75 unpublished materials and 35 studies that were not published between 2020 and 2024. The researcher added 7 more research from other sources. Then the study used 74 studies for further evaluation. For a better understanding of the general PS selection process, please refer to Figure 52.1.

Results for RQ3 revealed the greatest number of papers, with RQ2 and RQ1 following suit.

A pie chart depicting the total number of primary studies for each RQ is demonstrated in Figure 52.2.

Quality assessment standards
Quality assessment standards ensure that the included studies are valid, dependable, and pertinent to the

research issue. These criteria include the study's methodology, the data's quality, the depth of the research, and the reporting's transparency. The Quality Assessment Criteria questions are stated below:

- Does the study specifically discuss how the general population views the COVID-19 vaccination using the Machine Learning technique?
- Are the data utilized in the study of good quality and pertinent to the research question?
- Were adequate statistical techniques applied to the analysis of quantitative data?

Primary studies selection
The primary study selection is presented in Table 52.3.

Related Work

Here, we attempted to examine relevant literature on public discrimination on COVID-19 immunization. The outcomes of the linked work are revealed as follows:

Alvarado-Socarras et al. [1] aimed to evaluate Colombian physicians' opinions about COVID-19 vaccination by presenting two COVID-19 vaccine frameworks. Using the acceptance of free vaccination as the dependent variable and optimizing for age and sex, a binomial regression analysis was carried out with efficacy rates of 60 and 80%.

El-Deeb et al. [4] employed machine learning to uncover public opinions on the COVID-19 immunization procedure on Twitter. The tweet clustering is accomplished via the K-means clustering algorithm so that sentiment analysis can be performed using the Amazon Comprehend module [12].

Cotfas et al. [3] surveyed the views regarding COVID-19 immunization by concentrating on the period that elapsed between the initial broadcast of the vaccine and the UK's first immunization, during that time civil society showed a greater level of interest

Table 52.3 Selected primary studies.

Study No.	Author name	Publication year	Objective of the study	RQ1	RQ2	RQ3
PS1	Cotfas et al.	2021 [3]	The goal of this research is to examine the dynamics of attitudes regarding the COVID-19 serum in the month following the vaccine's initial broadcast.	X	X	X
PS2	Baj-Rogowska	2021 [2]	This research used the 6As taxonomy framework to directly identify the qualities related to vaccination from people's thoughts on Twitter.	X		X
PS3	Elhadad et al.	2020 [5]	This paper presents a technique that uses the COVID-19 pandemic to identify false information in the English language.	X	X	X
PS4	Yu et al.	2021 [13]	This study introduces Sentimenti-COVID19, and a collective graphical analytics tool for measuring, interpreting, and identifying social media sentiment fluctuations as well as public sentiment.	X		X
PS5	Tiwari et al.	2022[12]	This paper examined the applications and techniques of machine learning as they are related to the COVID-19 investigation and other domains.	X		X
PS6	Naseem et al.	2021[8]	The primary goal of this research was to classify the topics and patterns of community opinion regarding COVID-19 as they were communicated on Twitter.	X	X	X
PS7	El-Deeb et al.	2021[4]	The goal of this study was to apply machine learning techniques to uncover public observations on the COVID-19 immunization procedure on Twitter.	X	X	X
PS8	Nyawa et al.	2024[9]	This research intends to assess the effectiveness of several deep learning methods and machine learning models in identifying tweets from people who are unwilling to be immunized during the COVID-19 infection.	X	X	X
PS9	Mahdin et al.	2023 [6]	This study examines the emotion expressed in tweets regarding the COVID-19 immunization.	X	X	X

Source: Author

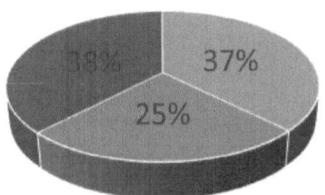

■ RQ1   ■ RQ2   ■ RQ3

Figure 52.2 Primary studies count for RQs
Source: Author

in the immunization procedure. A total of 2349659 tweets have been collected, investigated, and associated with the circumstances covered in the news broadcasts.

It finds an accuracy of 78.94%, the suggested method has BERT to classify the tweet messages into three primary categories: those who are in favor of, against, and neutral of COVID-19 immunization.

Elhadad et al. [5] developed a model to identify misleading COVID-19 findings collected from various certified digital sources from February 4, 2020, to March 10, 2020 [8]. Then, they constructed detection models with ten popular classification machine-learning algorithms.

Nyawa et al. [9] aimed to highlight vaccination hesitancy behaviour by assessing the usefulness of machine and deep learning models in recognizing vaccine-reluctant tweet messages during the worldwide COVID-19 immunization drives. They found that, with 86% and 83% accuracy, LSTM and RNN models identified vaccine-hesitant statements on social media better than classic machine learning models.

Mahdin et al. [6] gathered tweets as part of a dataset from Kaggle and used five machine learning algorithms—LR, Naive Bayes, SVM, and others—to survey people's convictions on the COVID-19 serum. Among these, the SVM scored the maximum correctness of 88.7989%, according to the performance criteria used to measure the models' performance.

Baj-Rogowska [2] aimed to ascertain if all the factors related to COVID-19 serum uptake found in tweets could be accurately covered and identified the characteristics associated with vaccination straight from people's opinions on Twitter using the 6As taxonomy framework.

Tiwari et al. [12] defined its main purpose as to: 1) Examine the impact of the type and nature of the data as well as any processing obstacles specific to COVID-19 data. 2) Gain a deeper understanding of the significance of clever strategies like machine learning for the COVID-19 epidemic. 3) The creation of enhanced machine learning algorithms and ML kinds for COVID-19 prognosis. 4) Analyzing the impact and efficacy of different tactics during the COVID-19 outbreak.

Conclusion

With emphasis on machine learning approaches, this literature review searched and categorized a collection of existing work on sentiment investigation of COVID-19 Twitter data using the PRISMA principles. This study examined the effectiveness of the machine learning classification methods used in the included studies and identified areas for further research as well as their inadequacies. In this study, we discovered that the majority of the research used English datasets, which may restrict the findings' applicability. Therefore, in the future, for sentiment analysis, we must utilize datasets in a different language. We need to foster the creation and use of practical sentiment analysis apps in the framework of COVID-19.

References

[1] Alvarado-Socarras, J. L., Vesga-Varela, A. L., Quintero-Lesmes, D. C., Fama-Pereira, M. M., Serrano-Diaz, N. C., Vasco, M., et al. (2021). Perception of COVID-19 vaccination amongst physicians in Colombia. *Vaccines*, 9(3), 287.

[2] Baj-Rogowska, A. (2021). Mapping of the COVID-19 vaccine uptake determinants from mining Twitter data. *IEEE Access*, 9, 134929–134944.

[3] Cotfas, L. A., Delcea, C., Roxin, I., Ioanăş, C., Gherai, D. S., & Tajariol, F. (2021). The longest month: analyzing COVID-19 vaccination opinions dynamics from tweets in the month following the first vaccine announcement. *IEEE Access*, 9, 33203–33223.

[4] El-Deeb, R., El-Gamal, F. E. Z., Sakr, N., Elhishi, S., & El-Metwally, S. (2021). Unlocking the public perception of covid-19 vaccination process on social media. In 2021 Tenth International Conference on Intelligent Computing and Information Systems (ICICIS), (pp. 327–334). IEEE.

[5] Elhadad, M. K., Li, K. F., & Gebali, F. (2020). Detecting misleading information on COVID-19. *IEEE Access*, 8, 165201–165215.

[6] Mahdin, H., Ahmad, M., Darman, R., Haw, S. C., Shaharudin, S. M., & Arshad, M. S. (2023). Sentiment analysis on COVID-19 vaccine tweets using machine learning and deep learning algorithms. *International Journal of Advanced Computer Science and Applications*, 14(5), 32–41.

[7] Page, M. J., McKenzie, J. E., Bossuyt, P. M., Boutron, I., Hoffmann, T. C., Mulrow, C. D., et al. (2021). The prisma 2020 statement: an updated guideline for reporting systematic reviews. *International Journal of Surgery*, 88, 105906.

[8] Naseem, U., Razzak, I., Khushi, M., Eklund, P. W., & Kim, J. (2021). COVIDSenti: a large-scale benchmark Twitter data set for COVID-19 sentiment analysis. *IEEE Transactions on Computational Social Systems*, 8(4), 1003–1015.

[9] Nyawa, S., Tchuente, D., & Fosso-Wamba, S. (2024). COVID-19 vaccine hesitancy: a social media analysis using deep learning. *Annals of Operations Research*, 339(1), 477–515.

[10] Rzymski, P., Zeyland, J., Poniedziałek, B., Małecka, I., & Wysocki, J. (2021). The perception and attitudes toward COVID-19 vaccines: a cross-sectional study in Poland. *Vaccines*, 9(4), 382.

[11] Sohrabi, C., Alsafi, Z., O'neill, N., Khan, M., Kerwan, A., Al-Jabir, A., et al. (2020). World health organization declares global emergency: a review of the 2019 novel coronavirus (COVID-19). *International Journal of Surgery*, 76, 71–76.

[12] Tiwari, S., Chanak, P., & Singh, S. K. (2022). A review of the machine learning algorithms for COVID-19 case analysis. *IEEE Transactions on Artificial Intelligence*, 4(1), 44–59.

[13] Yu, X., Ferreira, M. D., & Paulovich, F. V. (2021). Senti-COVID19: an interactive visual analytics system for detecting public sentiment and insights regarding COVID-19 from social media. *IEEE Access*, 9, 126684–126697.

53 High-quality image generation from text descriptions using stable diffusion: A machine learning approach

Rasmita Kumari Mohanty[1,a], Maheswar Mishra[2,b], Satyabrat Sahoo[2,c], Satya Prakash Sahoo[3,d] and Manas Ranjan Kabat[3,e]

[1]Department of CSE-(CYS, DS) and AI&DS, VNR Vignana Jyothi Institute of Engineering and Technology, Hyderabad, Telengana, India

[2]Department of CSE, School of Engineering and Technology, DRIEMS University Tangi, Cuttack, Odisha, India

[3]Department of CSE, Veer Surendra Sai University of Technology, Burla, Sambalpur, Odisha, India

Abstract

It presents a cutting edge tool that turns textual descriptions into high-quality photographs through the use of machine learning algorithms. The application uses a state-of-the-art method known as stable diffusion to enhance the coherence and caliber of the images that are generated. Textual descriptions can be converted into aesthetically pleasing and contextually relevant visuals by the application by combining machine learning algorithms and natural language processing techniques. Because stable diffusion reduces artifacts and increases overall coherence, it is essential for generating images of higher quality. The project includes preprocessing, training deep learning models, and data collecting as part of its full workflow. The model is able to acquire meaningful associations and produce realistic images that correspond to supplied descriptions since it has been trained on a huge dataset of text-image pairs. The machine learning text to image application demonstrates its capabilities through various examples of text-to-image generation, showcasing its accurate interpretation and translation of textual descriptions into visually coherent images. The application yields impressive results, indicating the effectiveness of the stable diffusion technique in producing high-quality and contextually relevant images. The project's potential applications span creative expression, storytelling, and aiding individuals with visual impairments. The machine learning text to image application presents a promising future for machine learning based image generation and its impact across different domains.

Keywords: Deep learning models, machine learning, natural language processing, stable diffusion, text-to-image generation

Introduction

The ability to generate synthetic images from text descriptions is a powerful tool for a variety of applications, such as creating illustrations for stories, developing new video game assets, or generating realistic medical images for training purposes. A new text-to-image diffusion model called Stable Diffusion has demonstrated remarkable performance in producing high-quality images from text descriptions. This project uses machine learning and Stable Diffusion to turn text into high-quality images, enhancing their coherence and reducing artifacts. It has a workflow that includes data collection, preprocessing, and model training. It generates realistic graphics from word descriptions by utilizing a sizable text-image collection. This technique is aptly demonstrated by the ML Text to Image Application, which use the Stable Diffusion approach to produce visually striking and suited for the context. The project's potential applications span creative expression, storytelling, and aiding individuals with visual impairments. The ML text to image application shows that machine learning-based image production and its effects on several domains have a bright future.

A Markov chain Monte Carlo (MCMC) technique is used in the text-to-image diffusion model known as "Stable Diffusion" to produce images from text descriptions. MCMC is a sampling method that is used to approximate the distribution of a probability function. In the context of image generation, the probability function is the distribution of images that match a given text description. Stable Diffusion works by starting with a random image and gradually refining it over many steps. To estimate the likelihood that a particular image will match the written description, the model makes use of a neural network. The model then uses this probability to guide the refinement process.

The model can be used to generate a wide variety of images, from realistic photographs to abstract paintings. With further development, we can expect Stable

---

[a]rasmita.atri@gmail.com, [b]maheswar.mishra@driems.ac.in, [c]satyabratsahoo@driems.ac.in, [d]sahoo.satyaprakash@gmail.com, [e]manas_kabat@yahoo.com

DOI: 10.1201/9781003658221-53

Diffusion to become an even more powerful tool for synthetic image generation. Video analysis systems for accident detection typically work by first identifying vehicles in the video footage.

Literature Survey

Text-to-image generation has gained significant traction with advancements in deep learning and generative models, focusing on transforming textual descriptions into high-quality images. Several approaches have been developed, each addressing specific challenges in this domain.

Guo et al. [1] suggested one notable model Controllable Text-to-Image Generation model, which employs a pre-trained bidirectional LSTM as a text encoder, encoding text descriptions into sentence and word features. Further advancements in text-conditional image generation are observed by Ma et al. [3] Hierarchical Text-Conditional Image Generation with CLIP Latents approach. The AttnGAN model proposed by Rao et al. [8] takes fine-grained text-to-image generation a step further with attention-driven, multi-stage refining. MirrorGAN proposed by Mohanty et al. (2023) further explores the challenge of semantic congruence between text and images. The NViSII tool proposed by Mohanty et al. (2023) focuses on photorealistic image generation by enabling users to script complex 3D scenarios, encompassing object meshes, textures, and lighting. The Learn, Imagine and Create: Text-to-Image Generation from Prior Knowledge (LeicaGAN) proposed by Sharma et al. [9] introduces a three-phase approach to text-to-image generation. The authors Mohanty et al. [7] have applied long term and brief term reminiscence and convolutional Long-Short Term Memory (LSTM) to investigate temporal, spatial, and temporal dependence.

The authors Jadhav et al. [2] applied cross-attention layers to enable convolutional high-resolution synthesis. In order to achieve both the ability to alter semantics and high fidelity in identity preservation, the authors Li et al. [4] match semantically relevant human face latent space with text-to-image diffusion models. To match the model selection process with human preferences, the authors Qin et al. [10] present Advantage Databases, which supplement the Tree-of-Thought with human feedback.

System Architecture

It is a conceptual model as depicted in Figure 53.1 System architecture that explains a system's composition, perspectives, and actions. An architecture description is a formal description and representation of a system that is set up to make it easier to reason about the behaviors and structures of the system. A system architecture is made up of built-in subsystems and system components that cooperate to implement the system as a whole. The suggested System design begins with recognizing the species of fish, pulling out the necessary fish attributes, and finding features within the dataset. The UI calls the API after the system has scanned the fish image, and the API uses the model to forecast the species of fish and present the findings.

Proposed System

These are the steps to be followed for the Proposed system.

Dataset collection and preprocessing
The project will begin with gathering of relevant data of text and images. The data will cover a wide range of topics and include rich textual descriptions in addition to the photos to ensure a comprehensive learning process. Formatting consistency is maintained through preprocessing which is the essential step that entails image preprocessing to standardize image size and structure, tokenization that converts sentences into machine-readable tokens, and text normalization. We'll leverage the state-of-the-art augmentation techniques like flipping, scaling and color modifications to make the model more robust to overfitting and to replicate a more diverse training environment.

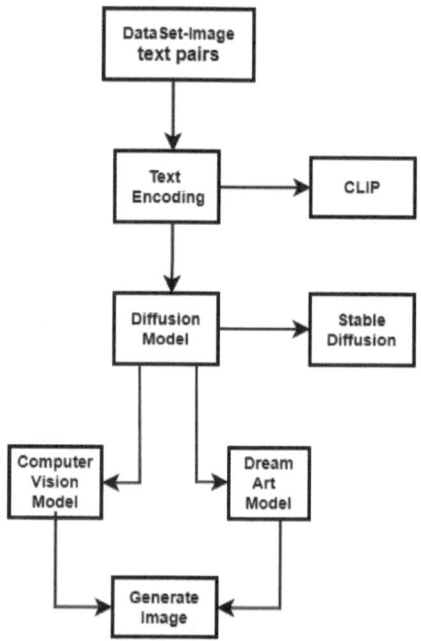

Figure 53.1 System architecture
Source: Author

Model architecture

Text encoder: Text encoding component of the model involves using the abilities of the CLIP model to generate detailed semantic text embeddings from Open AI. CLIP is a multimodal model pretrained on a broad spectrum of images and text pairs which makes it to generate and represent textual and visual complex concepts. The model is created to encode both visual and text into the same embedding space where they can be directly compared. This makes it highly suitable for tasks that requires understanding the relationship between text and images.

Stable diffusion architecture

Stable Diffusion is a state-of-the-art deep learning architecture designed for generating high-quality images from textual descriptions. At its core, the model is built upon a latent diffusion process, which begins with a random noise pattern and iteratively refines it into a coherent image that aligns with the input text prompt. The architecture leverages a denoising autoencoder, which operates within the latent space — a compressed representation of the image data — to facilitate efficient and effective image synthesis. This approach allows the model to handle high-resolution images with significantly reduced computational overhead compared to pixel-space models. Stable Diffusion also incorporates a transformer-based language model, which parses and understands the textual prompts, guiding the image generation process with semantic precision.

Diffusion model

Diffusion models belong to the category of generative models, which means that their purpose is to produce new data, frequently visuals. These models train by adding Gaussian noise to the training images at a series of time intervals and then learning to reverse this process of noising. Diffusion model involves forward and reverse diffusion process. A training image undergoes a forward diffusion procedure to add noise, gradually transforming it into an unusual noise image whereas reverse diffusion generates recovers image (could be any class) from the noise or meaningless image. We incorporated a Stable diffusion model which is a latent diffusion model because the intuitive diffusion process is computationally very slow and inefficient. Latent diffusion model first compresses the image into a latent space instead of operating with high dimensional images. This is achieved using variational autoencoder (VAE) which not only reduces the image size but also preserves most of the important information and the decoder of VAE is also responsible for recovering fine details in the final image. This model is also a lot faster in comparison to standard diffusion process.

Conditioning

Reversing of diffusion will make an image from the noise but the created image tumbles over. The aim of conditioning is to manage the impulse of the noise predictor so that the noise sign we are expecting is the one we want after noticed. First, every word in the text prompt is transformed into a token during the tokenization process. After that, each token is transformed into a vectorized Text to Image Generation 3 embedding (768 dimensions), for which the CLIP model is used. The embeddings are ready for use by a noise predictor following their processing by a text transformer. Throughout the U-Net, the noise predictor makes repeated use of the text-transformer's output.

Lora fine tuning

Low Rank Adaption (LORA) is an enhanced fine-tuning technique which preserves the knowledge base of pretrained model and allows model personalization with faster training on specific data. Full fine-tuning would involve optimizing and training all layers in neural network which is resource-intensive and time consuming. Lora is an improved fine-tuning technique in which two smaller matrices that roughly correspond to the bigger matrix are fine-tuned rather than all the weights that make up the weight matrix of the pre-trained large language model.

Training and evaluation

Lora Model is generated in kohya_ss library with the parameters: 1 repeat for each image, trained for 5 epochs, training batch size equals to one and with learning rate equals to 1E-5. Due to computational limitation, we trained only 3 classes of images and 20 handpicked images for each class along with the captioning (which took around 4hrs for each epoch). Then we merged the custom lora fine-tuned model with realistic vision model in Automatic1111 stable diffusion web UI to achieve best results. Both the training of Lora model and Generation of Synthetic data is performed on a PC with 16GB Ram, AMD Ryzen 7 5800H, NVDIA RTX 3050 GPU with 4gb VRAM and Windows 10 OS.

Result Analysis

After the training phase ended, we thoroughly tested our Stable Diffusion model to see how well it could produce images from text descriptions. The model performed remarkably well, as measured by the metrics attained and the qualitative studies performed.

The inception score (IS), which assesses the clarity and diversity of created images, also showed promising results. The high IS values obtained by our model confirm that the photos are not only unique from one another, but also contain well-defined objects and scenarios that correspond to the text descriptions.

Human evaluators offered further insight regarding the model's performance. Participants in the qualitative assessment consistently assessed the visuals as very relevant to the text prompts. The generated graphics were complimented for their visual attractiveness, as well as the model's ability to grasp the complex concepts and minute details given in the descriptions. This excellent feedback from human evaluators emphasizes the model's suitability for practical applications where image relevance and quality are critical.

Ablation studies were carried out to assess how different components of our methodology affected the model's performance. The integration of attention mechanisms in the U-Net architecture was found to considerably improve the relevance of generated images to text prompts. Fine-tuning on our curated dataset was discovered to be crucial for capturing textual semantics, since the model adjusted to the intricacies of the image-text combinations.

The web-based user interface developed for interacting with the latent diffusion model has proven to be both intuitive and efficient. Leveraging Chakra UI, the frontend provides a clean and modern aesthetic, with a straightforward navigation system that allows users of all levels of technical expertise to easily generate and view synthetic fish images. The 'Generate' interface simplifies the process of submitting text

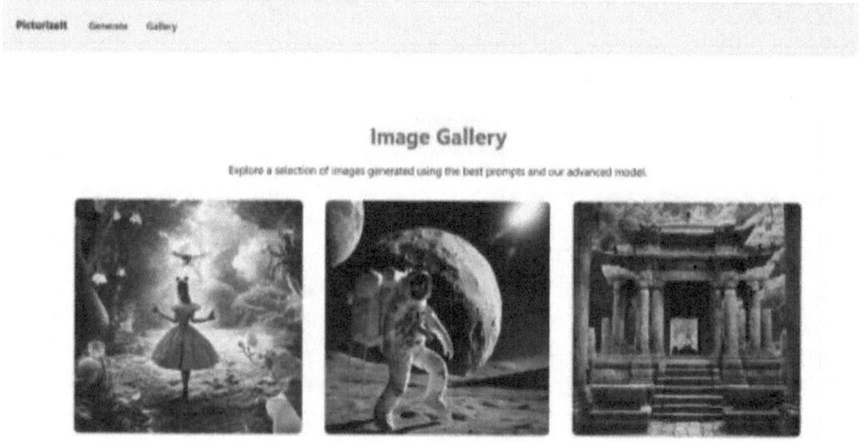

Figure 53.2 Gallery page of the web application
Source: Author

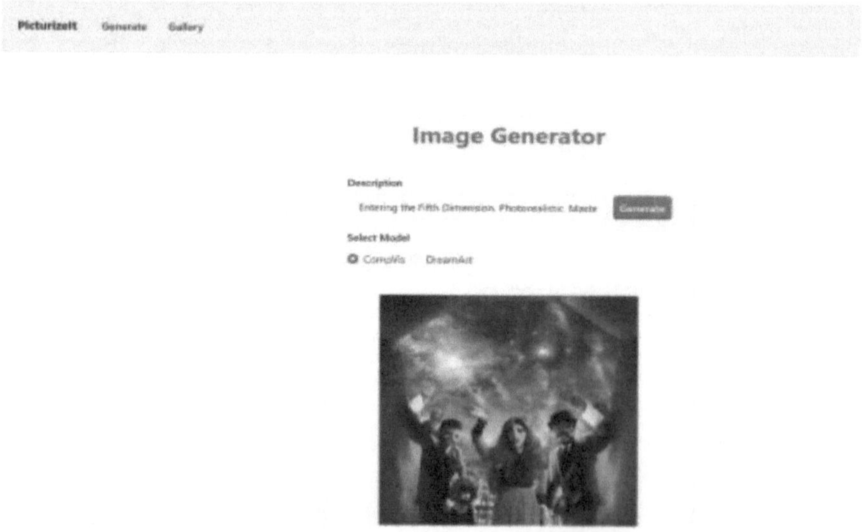

Figure 53.3 Generated an image by giving a text prompt and selecting appropriate model
Source: Author

prompts, while real-time feedback and a seamless user experience encourage iterative exploration of different descriptions as depicted in Figure 53.3. Additionally, the 'Gallery' component shown in Figure 53.2, presents the generated images in an organized manner, enabling users to quickly assess the quality and diversity of the synthetic dataset. User testing has indicated high levels of satisfaction with the interface's responsiveness and the speed at which images are rendered and displayed, emphasizing the practicality of the system in a research setting. Overall, the web UI's design and functionality have significantly contributed to the accessibility and usability of the latent diffusion model for synthetic image generation.

Conclusion

The goal of this project was to provide a formal framework for understanding text-to-image prompts using recognized art ideas and subject extraction methodologies. The goal was to bridge the gap between the computer process of creating images from verbal descriptions and the long history of artistic expression. We hoped to achieve a synergy between technology breakthroughs in artificial intelligence and pedagogical methodologies of visual arts instruction.

By establishing a connection between text-to-image generation technology, such as Stable Diffusion, and traditional art principles, we aim to enhance the educational experience. With this relationship, users can create more purposeful and effective suggestions that align with the fundamental ideas of different genres and artistic movements. As a result, images reflecting a deeper comprehension of creative purpose may be produced by more advanced image generation models that are tuned to the subtleties of well-established art concepts.

We recognize that Stable Diffusion is a novel technique in the field of visual arts, but it also carries certain risks and limitations, especially when it comes to the validity of artistic intention and expression. Questions concerning authorship and creativity may arise as a result of the technology's capacity to change how art is produced and appreciated. Nonetheless, we propose that educators might employ stable diffusion as a useful teaching tool by providing it with rigorous oversight and curation. A wide range of genres, movements, and aesthetics characterize cultural products, and it is through which experience and technical knowledge can be easily communicated.

Additionally, instructors can use Stable Diffusion to teach more comprehensive ideas of scene building, storytelling, and thematic inquiry by incorporating concepts of visual arts and media studies. This method can enhance the educational process by letting students explore and comprehend how different visual components interact in a particular setting.

In conclusion, our research contributes to the evolving dialogue on the convergence of artificial intelligence and art. The knowledge gained from this research has the potential to completely transform art practice and education by providing fresh avenues for creative expression. It is crucial that we keep a critical eye on things as we work through the.

References

[1] Guo, R., Wei, J., Sun, L., Yu, B., Chang, G., Liu, D., et al. (2023). A survey on image-text multimodal models. arXiv preprint arXiv:2309.15857.

[2] Jadhav, B., Jain, M., Jajoo, A., Kadam, D., Kadam, H., & Kakkad, T. (2024). Imagination made real: stable diffusion for high-fidelity text-to-image tasks. In 2024 2nd International Conference on Sustainable Computing and Smart Systems (ICSCSS), (pp. 773–779). IEEE.

[3] Ma, Y., Yang, H., Wang, W., Fu, J., & Liu, J. (2023). Unified multi-modal latent diffusion for joint subject and text conditional image generation. arXiv preprint arXiv:2303.09319.

[4] Li, X., Hou, X., & Loy, C. C. (2024). When stylegan meets stable diffusion: a w+ adapter for personalized image generation. In Proceedings of the IEEE/CVF Conference on Computer Vision and Pattern Recognition, (pp. 2187–2196).

[5] Mohanty, R. K., Sahoo, S. P., & Kabat, M. R. (2023). A network reliability based secure routing protocol (NRSRP) for secure transmission in wireless body area network. In 2023 8th International Conference on Communication and Electronics Systems (ICCES), (pp. 663–668). IEEE.

[6] Mohanty, R. K., Sahoo, S. P., & Kabat, M. R. (2023). Sustainable remote patient monitoring in wireless body area network with multi-hop routing and scheduling: a four-fold objective based optimization approach. *Wireless Networks*, 29(5), 2337–2351.

[7] Mohanty, R. K., Kumar, A. P., Padmaja, R., & Prashanthi, V. (2024). Deep learning for analyzing user and entity behaviors: techniques and applications. In Consumer and Organizational Behavior in the Age of AI, (pp. 219–250). IGI Global.

[8] Rao, S. B., Shetty, S. S., Singh, C., & Bekal, A. (2023). Text to image generation using attentional generative adversarial network. In International Conference on Advanced Computing, Machine Learning, Robotics and Internet Technologies, (pp. 70–82). Cham: Springer Nature Switzerland.

[9] Sharma, A., Sharma, G., Asiri, F. A., Bhutto, J. K., & Barnawi, A. B. (2024). Optimized mirror generative adversarial network with BERT Neural architecture for text caption to image conversion. *SN Computer Science*, 5(4), 334.

[10] Qin, J., Wu, J., Chen, W., Ren, Y., Li, H., Wu, H., et al. (2024). Diffusiongpt: LLM-driven text-to-image generation system. arXiv preprint arXiv:2401.10061.

54 Test scenario generation and optimization from UML behavioural diagrams

Kulshan Pattanaik[a] and Prateeva Mahali[b]

Department of Computer Science and Engineering, DRIEMS University, Odisha, India

Abstract

Optimization of test scenarios is one of the techniques that effectively manage the exponential extension in the testing period and test expenses. But the researchers often negotiate the coverage of the code while optimizing it. In this paper, we introduce a proposal to produce an optimized test scenario using a cuckoo search algorithm. The test suite is also reduced because of optimization while retaining the code coverage percentage unchanged. First, we implement this approach using a single UML model then we extend our work using a combined UML diagram for the same proposal, in short, we have to draw a system modelled using a sequence diagram (SD). After that, it transformed into a sequence diagram graph (SDG). Using the DFS algorithm and the cuckoo search we have generated test scenarios and also optimized the process. Test scenarios are generated by traversing the graph. Test scenario optimization is leading to a more reliable test suite in software testing concerning time, expense, and path coverage. it also helps during software under maintenance or fixing a bug that time we have to retest all the system to verify it will work fine and not affect other system functionality.

Keywords: Cuckoo search algorithm, sequence diagram, software testing, test scenarios, UML diagram

Introduction

Now a days, software Testing Ammann and Offutt [7] contribute a lot in software development life cycle (SDLC) as its main goal is to find errors, bugs as early as possible in development phase, and producing highly reliable systems while keeping the quality intact. Maintaining the quality of the software is a monotonous task due to experimental growth in the complexity of the software. Quality can be improved using some good testing technique, testing can be done in two ways one way is traditional manual testing and the other is very trendy now days is automated. Furthermore, due to an increase in the size of system program size also increasing as well as the complexity of the program, it is becoming much more difficult to produce effective test cases, The software testing process can be loosely divided into two sections. For example, manual testing and automated testing. Throughout the manual evaluation process, various testers attempt to manually check each of the application's features and functions process. Manual testing activities of different types follow test plans, test cases to explore and identify bugs in it. So that's why nowadays everybody moves on automated testing. Automation in testing means running one or more scripts that can adequately cover all test cases and test scenarios in a short period to do the job of the manual testers. We can also re-execute the test suite multiple times. In this light, automated test case creation, optimization, and prioritization of test suite play an impactful role in reducing manual effort and making the system.

Related Work

UML diagrams are quite common and are most extensively utilized to shape the system in the design aspect. In the current paper, we present former existed relevant work of test cases generation, their optimization and prioritization [8, 9]. Sarma et al. [3] presented a method for using UML sequence diagrams to construct test cases. Sequence diagrams were transformed into Sequence Diagram Graphs (SDG) for this process.

Test scenarios are formed by traveling or traversing the SDG, depending on coverage criteria and fault model.

Khurana et al. [6] suggested a method for creating test cases and optimizing them. By combining state chart and sequence diagrams, the system is transformed into a graph known as the system testing graph (SYTG). Considering the fault model and coverage they have applied Genetic Algorithm to create test samples automatically.

Jena et al. [1] introduces a proposal to produce test cases utilizing the UML activity diagram. The diagram was transformed into activity flow graph (AFG). The designer worked upon the activity coverage criterion and traveling the activity flow graph to produce

[a]kulshan@driems.ac.in, [b]dr.prateeva@driems.ac.in

DOI: 10.1201/9781003658221-54

test cases from graph. Then, genetic algorithms were applied for optimization process.

Proposed Approach

An approach to generate test scenarios using sequence diagram and optimization using a cuckoo search algorithm has been proposed. First, create a sequence diagram (SD) then generate all the scenarios from the sequence diagram and convert it into an intermediate graph called as sequence diagram graph (SDG) and test scenarios are generated using the DFS algorithm from the graph. Cuckoo search is used to optimize test scenarios. The proposed approach (given in Figure 54.1) is applied to an online voting system (OVS) case study for validation. The introduced method is also implemented in different case studies and the results are recorded.

Conversion of SD into SDG
In SDG the set of all nodes is denoted by state and proceedings among states are denoted as edge. The first node represents the beginning state and final node is the last. The 'message' sent is the collection of all activities occurring in the operation scenario. An event is said to be a subtext indicated via a tuple, event where message, from, to represents the label of message along with its signature, the one sending the message is denoted by from, the recipient of the subtext is represented by to, and the guard condition is non-compulsory part upon which the happening of an event depends. An event with * illustrates the repetitive event.

Concepts behind cuckoo search algorithm
Another nature-inspired optimization algorithm is the cuckoo search. This algorithm's primary concept is the use of the Levy flight to pick solutions technique which

we will discuss in the below sections in detail. This algorithm (given in Algorithm 1) is heavily inspired by the way cuckoos Mohamad et al. [2], Khari et al. [4] lay their eggs. In nature, the cuckoos lay their eggs in some other host bird's nests. Sometimes the host birds realize that the egg in their nest is not their own and they simply throw those away and leave the nest to construct a new one in a different location. It is also worth noticing that few female cuckoos have the capability of mimicking the color and patterns of some host birds [5].

Algorithm 1 cuckoo search algorithm
Output: Optimal solution
Objective function: $f(x) = (x_1, x_2, x_3, ..., x_d)$ initial population generation for n hosts

1. while (t < MaxGeneration) or (stop criterion) do
2. Select cuckoo randomly (say, i) and levy flights are performed to change its solution
3. Compute cuckoos fitness F_i

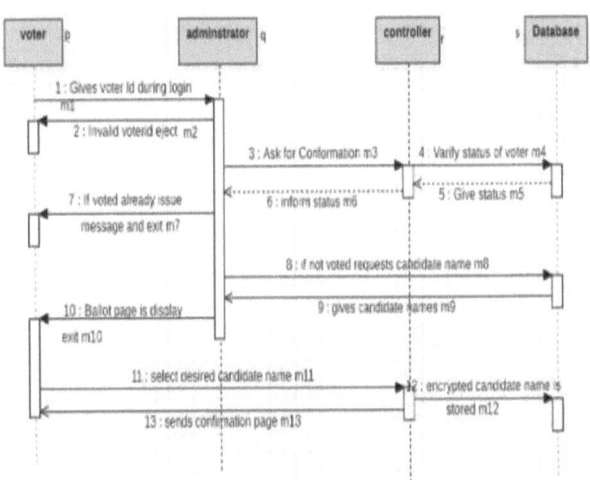

Figure 54.2 Sequence diagram for online voting system
Source: Author

scenario1	scenario2	scenario3	scenario4
			<scn4
			state a
		<scn3	s1:(m1,p,q)
	<scn2	state a	s3:(m3,q,r)
	state a	s1:(m1,p,q)	s4:(m4,r,s)
<scn1	s1:(m1,p,q)	s3:(m3,q,r)	s6:(m6,q,s)
state a	s3:(m3,q,r)	s4:(m4,r,s)	s7:(m7,s,q)
s1:(m1,p,q)	s4:(m4,r,s)	s6:(m6,q,s)	s8:(m8,q,p)|cond3
s2:(m2,q,p)|cond1	s5:(m5,q,p)|cond2	s7:(m7,s,q)	s9:(m9,p,r)
state b>	state b>	s8:(m8,q,p)|cond3	s10:(m10,r,s)
		state b>	s11:(m11,r,p)|cond4
			state b>

Figure 54.3 Scenarios obtained from the sequence diagram
Source: Author

Input

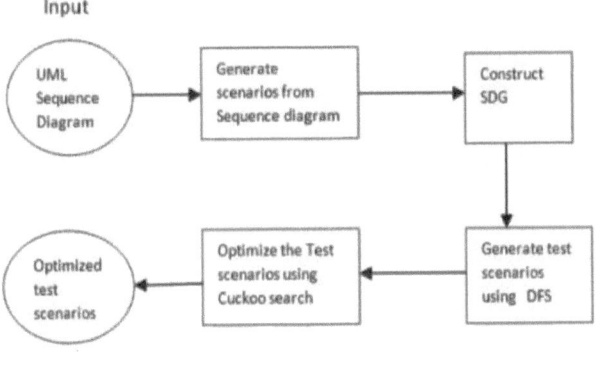

Output

Figure 54.1 Framework for test scenario optimization
Source: Author

4. Randomly select a nest among n different nest (say; j)
5. if $f_i > f_j$ then
6. Set j as new solution.
7. else
8. Continue
9. end if
10. Leave fraction Pa of the worst nests and build new ones
11. Best solutions are kept and ranked
12. Store current best solution in the next generation
13. end while

Cuckoo search algorithm to optimize test scenarios
The generated test scenarios are used as input for test scenarios optimization process. We have implemented the following variation of cuckoo search algorithm below:

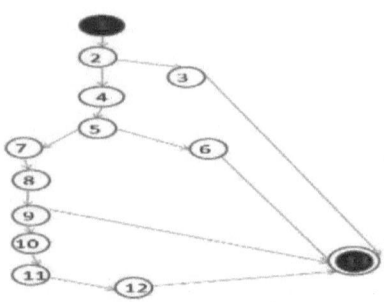

Figure 54.4 SDG for the sequence diagram
Source: Author

- Calculate fitness score of each test scenario: For each test scenario, the fitness score using an objective function is calculated. Mathematically, the fitness score of test scenario is represented as **Fitness-score = Wp*Np+ Wc*Nc**
- Find redundant and optimal test scenario
- Terminate when all nodes are covered

Implementation and Results

The proposed approach is discussed here taking a case study of the online voting system. The sequence diagram of online voting system is given in Figure 54.2. The generated scenarios are given in Figure 54.3 and the dependency graph is shown in Figure 54.4. After

Table 54.1 Test scenario generation and optimized result.

Sl. No.	Total Test Scenario	Optimized Test scenario
1	1→2→3→ 13	1→2→4→5 →7→8→9 →10→11→ 12 →13
2	1→2→4→5→ 6→13	1→2→4→5 → 6→13
3	1→2→4→5→7 →8→9→10→ 11 →12 →13	1→2→3→ 13
4	1→2→4→5→ 7→8→9→13	

Source: Author

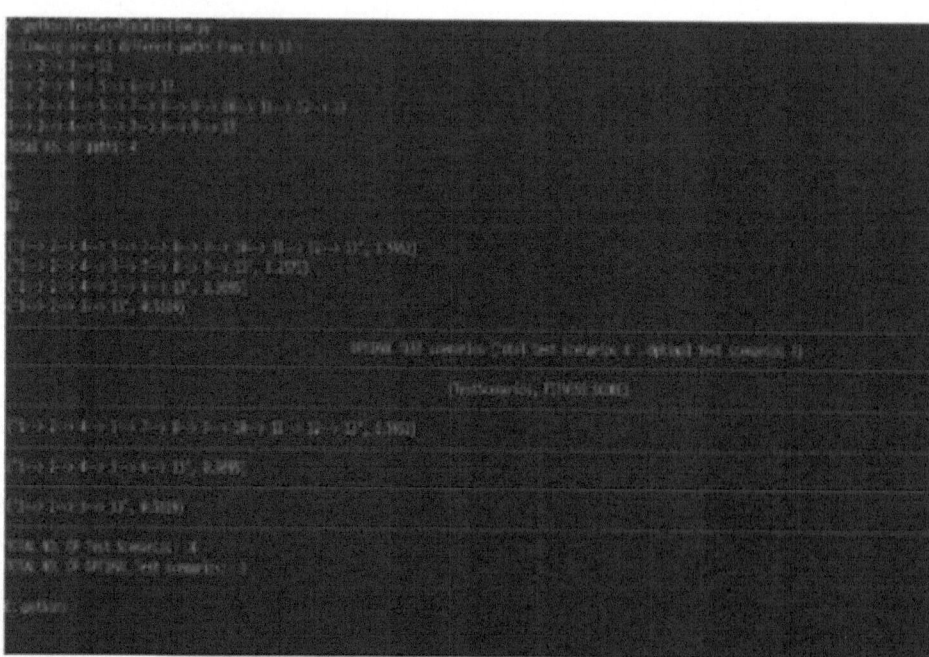

Figure 54.5 Optimized test scenarios for (Online voting system)
Source: Author

Table 54.2 Implementation of different case study.

Sl. No	Case study	Total no of test scenario	Optimized test scenario	% of reduction
1	Online voting system	4	3	25%
2	Atm pin verification	5	4	20%
3	Movie ticket system	6	4	33%
4	Atm System	8	7	12.5%
5	Hospital Management System	7	6	14.2%
6	Phone Communication System	7	5	28%

Source: Author

that, the test scenarios are generated using DFS algorithm and given in Figure 54.5. The generated test scenarios are optimized using cuckoo search algorithm. The implementation result of the generation and optimization of test scenarios are given in Table 54.1.

Here, we implement proposed approach on different case study like ATM pin verification, movie ticket system, ATM system, hospital management system, phone communication system.

The implementation of proposed approach with diffferent case study is given in Table 54.2.

Conclusion

The proposed approach generates test scenario using sequence diagram and optimizes them using cuckoo search. First, we create a sequence diagram then using that sequence diagram we generate all the scenarios and convert them into an intermediate graph and generate test scenarios using the DFS algorithm. Then we optimize these test scenarios using the cuckoo search algorithm. On validating our proposed approach on six different case study we have seen an optimization of 22.16% of test scenarios.

References

[1] Jena, A., Swain, S., & Mohapatra, D. (2014). A novel approach for test case generation from UML activity diagram. In 2014 International Conference on Issues and Challenges in Intelligent Computing Techniques (ICICT), (pp. 621–629). IEEE.

[2] Mohamad, A. B., Zain, A. M., & Bazin, N. E. N. (2014). Cuckoo search algorithm for optimization problems a literature review and its applications. *Applied Artificial Intelligence*, 28(5), 419–448.

[3] Sarma, M., Kundu, D., & Mall, R. (2007). Automatic test case generation from UML sequence diagram. In 15th International Conference on Advanced Computing and Communications (ADCOM 2007), (pp. 60–67). IEEE Computer Society.

[4] Khari, M., Kumar, P., Burgos, D. & Crespo, R. G. (2018). Optimized test suites for automated testing using different optimization techniques. *Soft Computing*, 22(24), 8341–8352.

[5] Mareli, M., & Twala, B. (2017). An adaptive cuckoo search algorithm for optimisation. *Applied Computing and Informatics*, 14(2), 107–115.

[6] Khurana, N., & Chillar, R. (2015). Test case generation and optimization using UML models and genetic algorithm. *Procedia Computer Science*, 57, 996–1004.

[7] Ammann, P., & Offutt, J. (2017). Introduction to Software Testing. Cambridge University Press.

[8] Jain, P., & Soni, D. (2020). A survey on generation of test cases using UML diagrams. In 2020 International Conference on Emerging Trends in Information Technology and Engineering (ic-ETITE), (pp. 1–6).

[9] Shunkun Yang, T. M., & Xu, J. (2014). Improved ant algorithms for software testing cases generation. *The Scientific World Journal, Hindawi Publishing Corporation*, 2014, 1–9.

55 Hybrid CNN-LSTM approach for sentiment analysis on IMDB movie reviews

Rahul Kumar Gupta[a], Binayak Ojha[b], Surya Prasad Yadav[c], Abhinav Kumar Singh[d], Prasant Kumar Dash[e] and Aadarsh Kumar Singh[f]

Department of CSE, C. V. Raman Global University, Bhubaneswar, Odisha, India

Abstract

Paper advances a hybrid model developed through an approach based on convolutional neural networks and LSTM for sentiment analysis tasks that would focus on the IMDB movie review data. Aim of this paper will be to distinguish the opinion into positive and negative emotions based on the strengths of both CNNs for the feature extraction and LSTMs for sequence modeling. We have conducted a vast number of experiments compared CNN-LSTM hybrids with basic models (LSTM model, CNN model, and Logistic Regression with term frequency-Inverse document frequency). Additionally, we carried out an ablation study to note the importance of various modules while building up the model, and done Hyperparameter optimization for the optimization of model accuracy, and Conducting adversarial attacks to test robustness on the models. The LIME and SHAP frameworks improved model interpretability, adding much-needed transparency to the classifications. The Proposed model CNN-LSTM has enhanced accuracy, recall, F1-score and precision comparison to baselines.

Keywords: CNN, deep learning, LIME, LSTM, NLP, sentiment analysis, SHAP

Introduction

Sentiment analysis, a key tool for opinion mining and customer satisfaction, has evolved from traditional techniques like Naive Bayes and logistic regression to advanced neural network-based models [14, 2]. Among these, hybrid CNN-LSTM architectures excel by combining CNNs' ability to capture local features such as n- grams Kovács and Tajti [8] with LSTMs' strength in modeling long-term dependencies [3]. This study uses IMDB movie reviews dataset to classify sentiment, comparing the proposed CNN-LSTM hybrid with baseline models including logistic regression with TF-IDF, CNN, and LSTM. Modern NLP models such as XLNet [17], RoBERTa [9], and BERT [4] offer exceptional performance, their computational cost and lack of interpretability limit practical applications. Our CNN-LSTM hybrid seeks a balance between performance and efficiency, incorporating LIME and SHAP for improved model transparency. Key contributions of this research include Developing a CNN-LSTM hybrid for sentiment analysis and demonstrating its superiority over baseline models. Enhancing interpretability using LIME and SHAP. Assessing model robustness against adversarial attacks to evaluate its practical applicability [1, 5, 6, 15, 16].

Literature Review

Approaches to sentiment analysis depended on machine learning techniques like SVMs, Naive Bayes and logistic regression, often paired with TF-IDF features or bag-of-words.

Advances in distributed word embeddings like Word2Vec and GloVe enabled models to incorporate contextual awareness, significantly improving performance. LSTM networks, part of the RNN family, demonstrated exceptional ability in modeling sequential dependencies and capturing complex relationships in text. CNNs, on the other hand, excelled at handling n-gram features and local patterns, with Kim's work showcasing their potential in NLP tasks. Combining CNNs and LSTMs further enhanced performance by leveraging both local and global dependencies in sentiment analysis. Transformer based models like BERT [4], RoBERTa [9], and XLNet [17] represented a massive advancement in NLP with Pre-training on large datasets and fine-tuning for target applications. However, although they lead to outstanding performance, they carry a high computational cost and low interpretability, which makes it hard to work with in practice. Methods like SHAP [10] and LIME [12] have been used to address transparency concerns in transformer forecasts. Hybrid models like CNNLSTM

---

[a]raahulgupta32@gmail.com, [b]binu7x@gmail.com, [c]suryadav903@gmail.com, [d]abhinav98444@gmail.com, [e]prasantdash@cgu-odisha.ac.in, [f]aadarshcngh9869@gmail.com

DOI: 10.1201/9781003658221-55

architecture find a compromise between performance and computational complexity and are good for the application of sentiment analysis. For instance, a CNN-LSTM model with GloVe embeddings achieved 85.46% accuracy in sentiment classification of football-related tweets [16]. Other studies combining CNN, LSTM, and MLP architectures reported successful improvements in multi-domain sentiment classification [11]. A graph-based Bi-LSTM CNN model achieved remarkable 98.61% accuracy in analyzing Industry 4.0 sentiment data [15]. Furthermore, SHAP and LIME have been effectively applied to hybrid models for explainability, as demonstrated in customer review analysis for food delivery services [1].

Proposed Methodology

Data preprocessing

The 50,000 reviews in the IMDB Movie Reviews dataset are evenly distributed between positive and negative sentiments, and was used for the study [2]. Preprocessing steps included: Lowercasing: To ensure uniformity. HTML tags and removal of special characters: To clean the text. Stopwords removal: To focus on meaningful words. Tokenization and padding: The text was tokenized and padded to a fixed sequence length for model ingestion. These steps prepared the data for feature extraction and sequence modeling.

Model architectures

The CNN-LSTM hybrid architecture combines CNN's ability to capture local patterns with LSTM's capability to model long-term dependencies. The architecture includes:

Embedding layer: Converts word indices into dense semantic vectors.

Convolutional layer: Extracts local patterns using ReLU activation.

Max pooling layer: Reduces dimensionality while preserving critical features.

Bidirectional LSTM layer: Captures context by learning from both past and future sequences.

Fully connected layers: Performs final sentiment classification.

$$f_t = \sigma(W_f[h_{t-1}, x_t] + b_f), \quad C_t = f_t C_{t-1} + i_t \tilde{C}_t \quad (1)$$

$$c_i = f(w \cdot X_{i:i+k-1} + b) \quad (2)$$

$$\text{TF-IDF}(t, d) = \text{TF}(t, d) \cdot \log(\frac{N}{DF(t)}) \quad (3)$$

Above Equations 1,2,3 [13, 7, 14] provides LSTM, CNN, logistic regression with TF-IDF models respectively.

$$\text{Accuracy} = \frac{TP + TN}{TP + FP + FN + TN} \quad (4)$$

$$\text{F1-Score} = \frac{2 \cdot \text{Precision} \times \text{Recall}}{\text{Precision} + \text{Recall}} \quad (5)$$

Equations 4, 5 is used for calculating accuracy and F1-scores.

The accuracy of the CNN-LSTM model was 88.76%, outperforming LSTM (84.1%) (Zou et al., 2021), CNN (82.5%) Kim, [7], and logistic regression (78.3%) [14]. This demonstrates the hybrid model's effectiveness in combining convolutional feature extraction and sequence learning.

Experiments and Results

Comparative Analysis with Baseline Models

The effectiveness of the hybrid CNN-LSTM model was evaluated in comparison to three baseline models: Logistic Regression with TF-IDF features, CNN, and LSTM. The IMDB dataset was split into

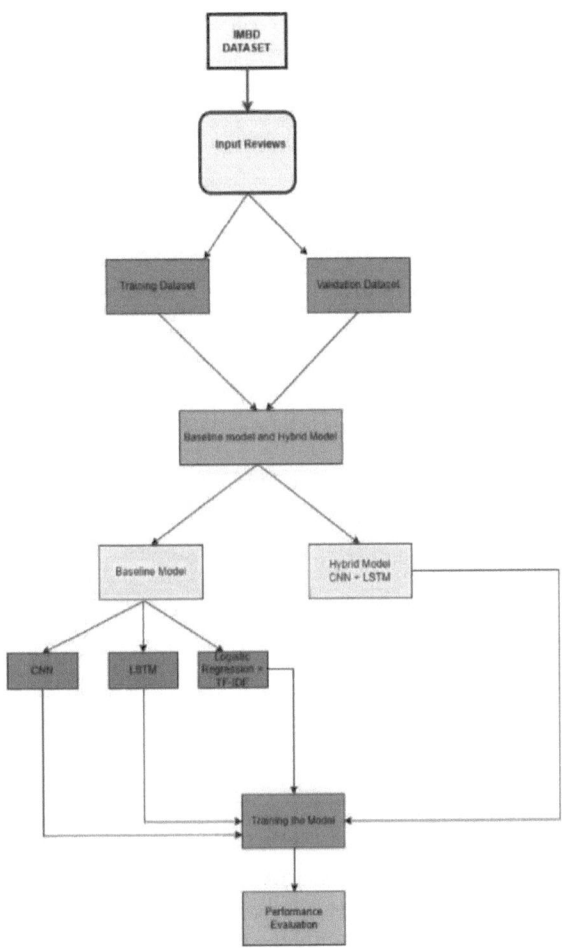

Figure 55.1 Model architecture of the proposed system
Source: Author

Table 55.1 Evaluation against various models.

Model	Accuracy (%)	Precision (%)	Recall (%)	F1-Score (%)
Logistic regression (TF-IDF)	87.0	88.0	87.0	89.0
CNN	87.69	88.0	88.0	88.0
LSTM	87.72	87.0	89.0	88.0
CNN-LSTM hybrid (Proposed)	88.76	90.0	87.0	89.0

Source: Author

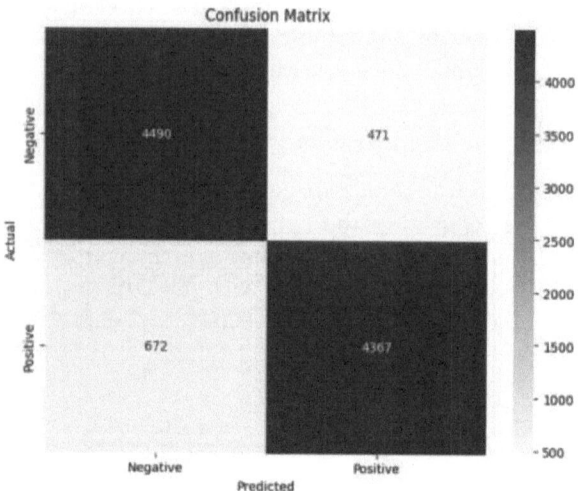

Figure 55.2 Confusion matrix
Source: Author

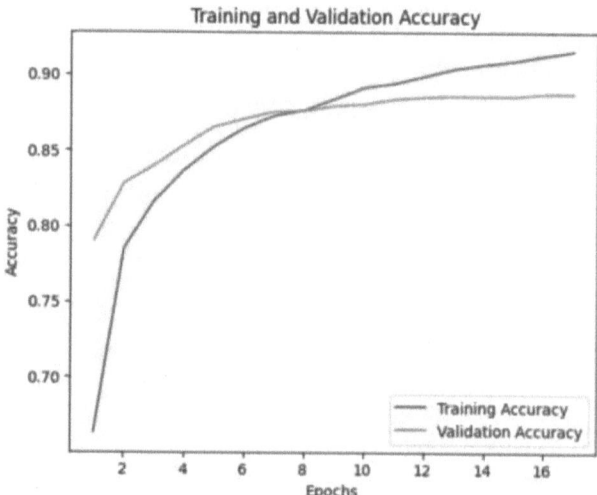

Figure 55.4 Training and validation accuracy
Source: Author

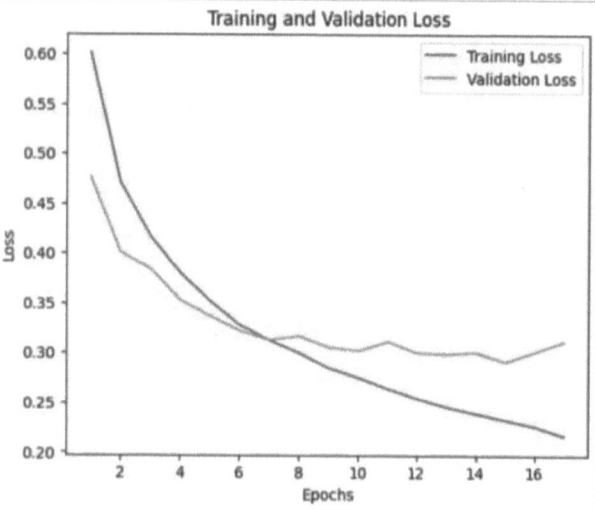

Figure 55.3 Training and validation loss
Source: Author

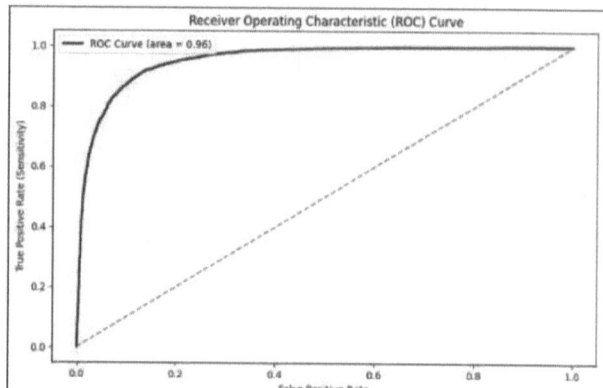

Figure 55.5 ROC curve of proposed model
Source: Author

eighty percent training and twenty percent testing. Table 55.1 highlights the evaluation metrics: The CNN-LSTM hybrid model achieved 88.76% accuracy, utilising the sequential modelling capabilities of LSTM and the feature extraction capabilities of CNN. Training for 100 epochs with early stopping and dynamic learning rate adjustments ensured robust learning and regularization. Validation accuracy peaked at the 17th epoch, with minimal overfitting.

Key results:

Training and validation: Stable accuracy and loss trends.

Confusion matrix: Balanced classification for positive and negative sentiments.

ROC curve: High sensitivity and specificity.

Precision-recall curve: Strong precision and recall across thresholds.

Label distribution: Consistent actual vs. predicted performance.

Conclusion

The proposed hybrid CNN-LSTM model effectively combines CNNs for feature extraction and LSTMs for sequence learning, achieving high accuracy (88.76%) on the IMDB movie reviews dataset. It outperforms standalone models and transformer-based methods with lower computational costs and improved interpretability. Techniques like LIME and SHAP enhance transparency, making it suitable for noisy environments and sensitive applications like healthcare. Future work will focus on multi-class sentiment analysis and fairness in real-world scenarios.

References

[1] Adak, A., Pradhan, B., Shukla, N., & Alamri, A. (2022). Unboxing deep learning model of food delivery service reviews using explainable artificial intelligence (XAI) technique. *Foods*, 11(14), 2019.

[2] Bosques Palomo, B., Velarde, F., Cantu- Ortiz, F., & Ceballos, H. (2024). Sentiment analysis of IMDb movie reviews using deep learning techniques. In Proceedings of the 2024 International Conference, (pp. 421–434). DOI: 10.1007/978-981-99-3236-8_33.

[3] Charitha, N. S. L. S., Yasaswi, K., Rakesh, V., Varun, M., Yeswanth, M., & Kiran, J. S. (2023). Comparative study of algorithms for sentiment analysis on IMDb movie reviews. In 2023 9th International Conference on Advanced Computing and Communication Systems (ICACCS), (pp. 824–828). DOI: 10.1109/ICACCS57279.2023.10113113.

[4] Devlin, J., Chang, M.-W., Lee, K., & Toutanova, K. (2018). BERT: pre-training of deep bidirectional transformers for language understanding. arXiv preprint arXiv:1810.04805.

[5] Hossain, S., Sumon, J., Alam, M. I., Kamal, K. M. A., Sen, A., & Sarker, I. (2022). Classifying sentiments from movie reviews using deep neural networks. In Proceedings of the 2022 International Conference, (pp. 399–409). DOI: 10.1007/978-3-031-19958-5_37.

[6] Jadia, H. (2023). Comparative analysis of sentiment analysis techniques: SVM, logistic regression, and TF-IDF feature extraction. *International Research Journal of Modernization in Engineering Technology and Science*.

[7] Kim, Y. (2014). Convolutional neural networks for sentence classification. arXiv preprint arXiv:1408.5882.

[8] Kovács, Á., & Tajti, T. (2023). Enhancing sentiment analysis accuracy on IMDb reviews through ensemble machine learning techniques. In 2023 IEEE 21st Jubilee International Symposium on Intelligent Systems and Informatics (SISY), (pp. 289–294). DOI: 10.1109/SISY60376.2023.10417873.

[9] Liu, Y., Ott, M., Goyal, N., Du, J., Joshi, M., Chen, D., et al. (2019). RoBERTa: a robustly optimized BERT pretraining approach. arXiv preprint arXiv:1907.11692.

[10] Lundberg, S. M., & Lee, S.-I. (2017). A unified approach to interpreting model predictions. *Advances in Neural Information Processing Systems (NeurIPS)*, 30, 4765–4774.

[11] Munandar, D., Rozie, A., & Arisal, A. (2021). A multi-domain short message sentiment classification using hybrid neural network architecture. *Indonesian Journal of Electrical Engineering and Informatics*. 9(3), 621–629.

[12] Ribeiro, M. T., Singh, S., & Guestrin, C. (2016). 'Why should I trust you?' explaining the predictions of any classifier. In Proceedings of the ACM SIGKDD International Conference on Knowledge Discovery and Data Mining, (pp. 1135–1144).

[13] Saeed, M. Q. (2020). Sentiment analysis of IMDb movie reviews using long short-term memory. In 2020 2nd International Conference on Computer and Information Sciences (ICCIS), (pp. 1–4). IEEE. DOI: 10.1109/ICCIS49240.2020.9257657.

[14] Talibzade, R. (2023). Sentiment analysis of IMDb movie reviews using traditional machine learning techniques and transformers. In Proceeding 2nd International Conference Computer and Information Science (ICCIS), (pp. 1–4). DOI: 10.13140/RG.2.2.29464.16644.

[15] Venkatesan, D., Kannan, S., Arif, M., Atif, M., & Ganeshan, A. (2022). Sentimental analysis of industry 4.0 perspectives using a graph-based Bi-LSTM CNN model. *Mobile Information Systems*, 2022(1), 5430569.

[16] Venkatesh, S., Hegde, S. U., Zaiba, A., S., & Nagaraju, Y. (2021). Hybrid CNN-LSTM model with GloVe word vector for sentiment analysis on football specific tweets. In Proceedings of the 2021 International Conference on Advances in Electrical, Computing, Communication and Sustainable Technologies (ICAECT).

[17] Yang, Z., Dai, Z., Yang, Y., Carbonell, J., Salakhutdinov, R. R., & Le, Q. V. (2019). XLNet: generalized autoregressive pretraining for language understanding. *Advances in Neural Information Processing Systems (NeurIPS)*, 32, 5754–5764.

56 An Internet of Things system application for monitoring and control of underground mine environment

Raghunath Rout[1,a], Prateeva Mahali[2,b] and Subhendu Kumar Pani[3,c]

[1]Department of Computer Science and Engineering, Biju Patnaik University of Technology, Rourkela, Odisha

[2]Department of Computer Science and Engineering, DRIEMS University, Odisha, India

[3]Department of Computer Science and Engineering, Krupajal Computer Academy, Odisha, India

Abstract

Alongside agriculture, mining is generally considered to be one of the earliest forms of economic activity in human history. It is imperative that improvements in safety measures be made if one is to achieve both the goal of maximizing the efficiency of the mining process and the goal of ensuring the miners' personal safety. Characteristics of the work environment in underground coal mines become unsafe with some precursors and warnings of adverse situations due to the presence of poisonous and flammable mine gases, improper condition of the mining equipment, and other factors in some typical underground mine conditions before accidents or disasters leading to loss of life and equipment's. Monitoring the environment is not only an essential use of wireless sensor networks (WSNs), but it is also important for the safety of underground coal mine activity. The purpose of this paper is to offer an attempt at conducting experimental research on the environment of a mine; it also makes suggestions for the implementation of an early warning system, as well as works to be carried out to promote safe working conditions for mine workers.

Keywords: Mine equipment, mine gases, real-time, sensors, WSN, zig bee

Introduction

Underground mining is one of the most dangerous locations to work since it is surrounded by hazardous gases such as methane, carbon monoxide, etc. The majority of these flammable gases are also very toxic in their unburned state. In spite of the presence of toxic gases, there is still some work that may be done below with the assistance of sufficient quantities of oxygen. It is of the utmost importance that continuous monitoring of the presence of these dangerous gases within a safe limit as well as the presence of levels of oxygen be carried out in order to ensure the safety of any children who may be there. The use of wireless sensor networks can make the job of unmanned monitoring of the parameters remotely easier. Despite this, the electromagnetic characteristics or behavior in an underground mine can be highly variable. The purpose of this effort is to provide a scalable infrastructure that will facilitate the monitoring of these important gaseous characteristics in an efficient manner. As a result, the deployment and experimentation of wireless sensor networks in real time is required for the purpose of monitoring harmful gases in the restricted areas of the mines.

The monitoring of the en2vironment of underground coal mines is typically carried out with a variety of monitoring devices. Online monitoring has been implemented in developed nations in order to keep track of the environmental status of underground coal mines.

Review of Literature

An investigation into many environmental monitoring platforms has been provided by Kim et al. [7] which found that these platforms do not give a high level of accuracy and sensitivity.

The research conducted by Somov et al. [1] on how to extend the life of sensor nodes beyond 187 days of operation in the gas wireless sensor network (GWSN) domain includes looking for the most efficient way to extract energy from waste while making use of the appropriate technology.

The method of forming the service line of a mine network for the purpose of remote monitoring was proposed by Lin et al. [4]. In this scenario, the sensor network collects data on the aboveground and underground networks of motes by connecting through gateway nodes and making use of optical fiber.

Chaulya et al. [5] provides a detailed description of my management to take appropriate action by continuous online monitoring studying the environmental parameters in normal conditions methane, carbon monoxide, air velocity, and temperature. The system is designed to provide on-line visualizations trend of

[a]raghunathrout78@gmail.com, [b]dr.prateeva@driems.ac.in, [c]pani.subhendu@gmail.com

DOI: 10.1201/9781003658221-56

all monitored parameters and give audio visual warning signal in the case that a particular value crosses a threshold limit, which is responsible for the early action reduction in emissions.

In order to monitor the ambient methane gas distribution and dynamics at all major areas of the coal mines with high precision, speed, and reliability, the author Liu et al. [6] employs a fiber Optic Raman distributed sensor.

The authors Qin et al. [3] employ a computational fluid dynamic (CFD) modelling approach to optimize the gas drainage design for underground long wall coal mining. This is done to have maximum methane capture and minimize fugitive emissions in gassy and multiple seam circumstances.

Table 56.1 Underground gas sample analysis report.

Gas name	Concentration	Units	Normalized (in %)
Helium	12.35	ppm	0.0012
Hydrogen	4153.42	ppm	0.4121
Oxygen	14.8989	%	14.7839
Nitrogen	80.4590	%	79.8382
Methane	0.1138	%	0.1130
Carbon monoxide	3238.66	ppm	0.3214
Carbon dioxide	3.6122	%	3.5843
Ethylene	76.10	ppm	0.0076
Ethane	184.83	ppm	0.0183
Acetylene	0.00	ppm	0.0000
Argon	0.9272	%	0.9200

Source: Author

Characteristics of Ambient Gases in Underground Mines

Methane (CH_4), nitrous gases, carbon dioxide (CO_2), carbon monoxide (CO), hydrogen sulfide (H_2S), Sulfur dioxide (SO_2), hydrogen (H_2), nitrogen (N_2), and acetylene (C_2H_2) are some of the most dangerous gases that can be found inside underground coal mines Walsh et al. [2].

The underground gas sample ananlysis report is given in Table 56.1.

1) Methane is a gas that is lighter than air, doorless, colorless, and highly flammable. In the presence of oxygen, it produces carbon dioxide and water vapor when it burns.

2) Nitrogen dioxide is a gas that has the appearance of a rusty orange-brown, has a smell that is described as being sharp and biting, and is toxic. Burning fossil fuels results in the production of this byproduct.

3) Carbon dioxide is an odorless, colorless, tasteless gas that cannot be burned under typical environmental conditions. At sufficiently high concentrations, it is poisonous to animals. The increase in carbon dioxide can be attributed to either human or animal respiration.

4) Carbon monoxide is an odorless, tasteless, and colorless gas that is extremely hazardous to the health of humans and other animals. It is poisonous because it prevents the hemoglobin in the blood from transporting oxygen from the lungs to the muscles and other tissues in human body.

5) Hydrogen sulfide is a gas that is odorless, very poisonous, and extremely flammable. Because it is heavier, it has a tendency to collect towards

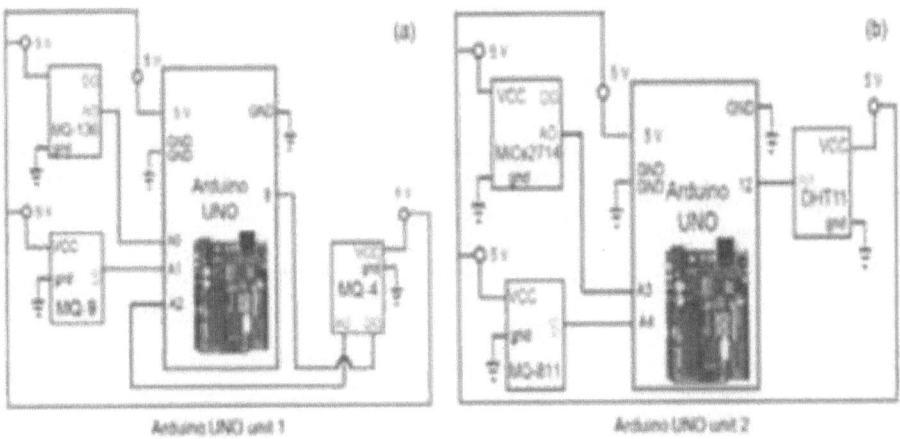

Figure 56.1 Circuit diagram of Arduino UNO (SN) (a) attached MQ4, MQ136 and MQ9 (b) attached MiCs2714, MQ811, and DTH11

Source: Author

Table 56.2 Sensors and their characteristics.

Characteristics	DTH-11	Sensor module		
		MQ4	MQ9	MQ811
Sensor type	Negative temperature coefficient (NTC)	MOS	MOS	MOS
Target gas	Temperature and humidity	CH4	CO	CO2
Typical detection range	D20-90% RH, 0-50 C	200-1000ppm	10-1000 ppm	350-10, 000 ppm

Source: Author

Table 56.3 Sample data collected from prototype sensor node.

Time in Mins	Temperature in °C	Humidity	CO2 level %
0	27	92	0.035
5	27	93	0.035
10	27	95	0.035
15	28	92	0.035
20	28	92	0.035
25	26	92	0.035
30	26	92	0.035
35	27	94	0.036
40	27	93	0.036
45	26	93	0.036
50	27	93	0.036
55	27	93	0.035
60	27	93	0.035
65	27	93	0.035
70	27	92	0.035

Source: Author

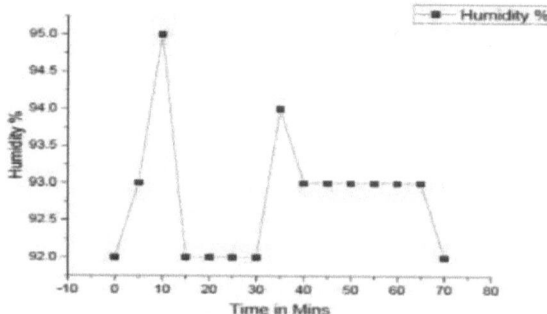

Figure 56.2 Statistics of humidity parameters
Source: Author

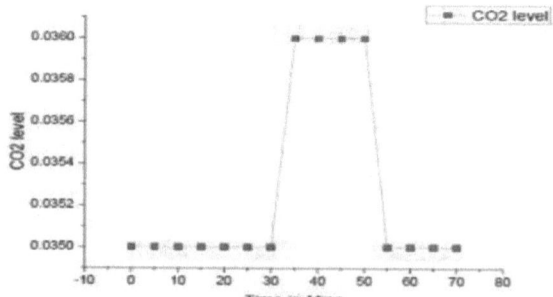

Figure 56.3 Statistics of CO2 parameters
Source: Author

the bottom in areas with inadequate ventilation.

System Architecture

Using sensor modules that are connected to an Arduino UNO provides the foundation for the sensing units utilized in this system. The primary purpose of an SN is to monitor and take readings of the many air quality indicators that are present within my environment. Sensor nodes (SNs) typically include a micro-controller, wireless transmitters, and other electronic components like a sensor module. In this particular investigation, SN is implemented using a micro-controller made by Arduino called a UNO.

In this particular investigation, SN is implemented using a micro-controller made by Arduino called a UNO as shown in Figure 56.1

The statistics of humidity and CO2 arameters is shown in Figure 56.2 and 56.3.

Arduino is an open-source platform that is also low-cost and uses a minimal amount of power i.e. 5 V DC and with capacity of 2 kB as well as 32 kB of RAM memory. The DTH-11 has the capacity to measure humidity in the range of 20-80% and has an operational temperature range that spans from -40 degrees Celsius to 120 degrees Celsius with an accuracy of +/-2 degrees Celsius. Table 56.2 shows the characteristics of different sensors.

Result Analysis

An effort was undertaken, utilizing the prototype model, to evaluate the various sensors' sensitivity,

Table 56.4 Mean and standard deviations of nodes.

Parameters	Sensor Node 1		Sensor Node 2		Commercial Equivalent	
	Mean	STDEV	Mean	STDEV	Mean	STDEV
Temperature in	26.42	1.09	26.4	1.13	26.54	1.06
Humidity %	92.33	2.11	92.43	2.08	93.24	2.13
CO2 (ppm)	22.24	1.49	22.27	1.45	22.25	1.5

Source: Author

stability, and environmental immunity. The data that was obtained at the sensor node was collected from gateway node, and a single file which contained all data had been recorded during a test. Table 56.3 contains a fragment of a smaller sample block of the data that was obtained at a particular node.

With the assistance of a hermetically sealed gas test box, SR3 (235 millimeters on a side, 180 millimeters on the other, and 210 millimeters on the top), each gas sensor was calibrated on a laboratory scale at a temperature of 25 degrees Celsius. The Arduino-based SNs that were constructed were validated using a commercial instrument called the Aeroqual-900 (Aeroqual, located in Avondale, Auckland, New Zealand), which also included other sensors for measuring temperature and humidity. Table 56.4 provides a summary of the means and standard deviations of the SNs as well as their respective commercial equivalents.

Conclusion

Monitoring the surrounding environment is essential to ensuring the health and safety of those working in coal mines. It is difficult to deploy the conventional cable monitoring system (CMS) in complex locations such as the mining and mined-out portions of mines. As a result, wireless sensor networks (WSN) are now being researched as a potential method for monitoring the environment in underground coal mines. By including a number of different sensing nodes into the Internet of Things based environmental monitoring system that has been developed, it is possible to obtain an immediate map of the many ambient characteristics that are present throughout a large area or in a specific location.

References

[1] Somov, A., Baranov, A., Savkin, A., Ivanov, M., Calliari, L., Passerone, R., et al. (2012). Energy-aware gas sensing using wireless sensor networks. In Wireless Sensor Networks: 9th European Conference, EWSN 2012, Trento, Italy, February 15-17, 2012. Proceedings 9, (pp. 245–260). Springer.

[2] Walsh, P., Evans, P., Lewis, S., Old, B., Greenham, L., Gorce, J.-P., et al. (2016). Technical guide on direct-reading devices for airborne and surface chemical contaminants. OSHwiki networking Knowledge.

[3] Qin, J., Qingdong, Q., & Guo, H. (2017). CFD simulations for longwall gas drainage design optimisation. *International Journal of Mining Science and Technology*, 27(5), 777–782.

[4] Lin, R., Wang, Z., & Sun, Y. (2004). Wireless sensor networks solutions for real time monitoring of nuclear power plant. In Fifth World Congress on Intelligent Control and Automation (IEEE Cat. No. 04EX788), (Vol. 4, pp. 3663–3667). IEEE.

[5] Chaulya, S., Bandyopadhyay, L., & Mishra, P. (2008). Modernization of Indian coal mining industry: vision 2025, CSIR,(pp. 28–35).

[6] Liu, T., Wei, Y., Song, G., Hu, B., Li, L., G. Jin, et al. (2018). Fibre optic sensors for coal mine hazard detection. *Measurement*, 124, 211–223.

[7] Kim, Y. W., Lee, S. J., Kim, G. H., & Jeon, G. J. (2009). Wireless electronic nose network for real-time gas monitoring system. In 2009 IEEE International Workshop on Robotic and Sensors Environments, (pp. 169–172). IEEE.

57 A novel U-Net with fine tuned VGG16 backbone model for ulcer detection and classification using 2-D endoscopy images

Debayush Behera[a], Satyaprakash Gharei[b], Vatsal Raj[c], Satyajit Singh[d] and Ram Chandra Barik[e]

Department of CSE, CV Raman Global University, Odisha, India

Abstract

Automated detection of ulcers in endoscopic images is essential for early diagnosis and treatment of gastrointestinal disorders. This paper presents a novel approach for 2D endoscopic ulcer detection using a customized deep neural network, designed to accurately segment ulcerous regions from endoscopic images. The proposed model employs a U-Net architecture with a VGG16 backbone, leveraging pre-trained feature extraction capabilities for enhanced segmentation performance. Our model is trained on publicly available datasets that include Kvasir-SEG, which contain annotated ulcer images and corresponding masks. The endoscopic images are resized to 224 × 224 for compatibility with the VGG16 architecture, and the model is trained using binary cross-entropy loss. In order to further improve interpretability, we incorporate bounding box data to generate heatmaps, providing clinicians with a visual overlay for better ulcer localization.

Keywords: CNN, endoscopy, ulcer classification, U-Net, VGG16

Introduction

Ulcers in the gastrointestinal tract, particularly in the stomach and intestines, are a significant medical concern, leading to severe complications if left undetected or untreated. Endoscopy is a standard procedure for diagnosing ulcers, where clinicians visually inspect the gastrointestinal tract using endoscopic images. However, the process is highly dependent on the clinician's expertise, and manual interpretation can be laborious and subject to human mistake. To overcome these obstacles, automated deep learning systems have emerged as powerful tools for assisting in the early diagnosis and detection of ulcers from endoscopic images. In this research, we present a novel approach for 2D endoscopic image-based ulcer detection using a customized deep neural network. Leveraging the power of CNNs, our system is designed to automatically identify and segment ulcerous regions in endoscopic images with high accuracy. We build upon a U-Net architecture, incorporating a VGG16 backbone to enhance feature extraction, while the custom decoder is optimized for precise segmentation. This approach enables the model to effectively learn both high-level and fine-grained features, which are essential for accurate detection of ulcers. Our model is trained and tested on the CVC-Clinic DB and Kvasir-SEG datasets, which consist of 2D endoscopic images of the gastrointestinal tract, providing both ground truth masks for ulcers and additional bounding box annotations for better localization. The images are pre-processed to a certain resolution of 224 × 224 pixels to ensure compatibility with the network architecture, and data augmentation techniques are applied to enhance model generalization. Furthermore, we integrate bounding box data to generate heatmaps, offering an interpretable visualization of the model's detection and segmentation capabilities.

The goal of this research is to develop an efficient and interpretable model that can assist clinicians in detecting ulcers from endoscopic images, thereby reducing diagnostic errors, and improving patient outcomes. Using a customized neural network, we aim to push the boundaries of automated ulcer detection, offering a novel solution to a critical problem in medical imaging and diagnosis.

Tools and Methodology

This section outlines techniques used in this study, the dataset description, preprocessing steps, model selection, training, and technical metrics used to assess the performance of our proposed ulcer detection model.

The dataset used for this study is the Kvasir-SEG dataset, which contains the collection of labeled

[a]debayush.2003@gmail.com, [b]satyaprakashgharei@gmail.com, [c]vatsalraj481@gmail.com, [d]satyajitsingh878@gmail.com, [e]ram.chandra@cgu-odisha.ac.in

DOI: 10.1201/9781003658221-57

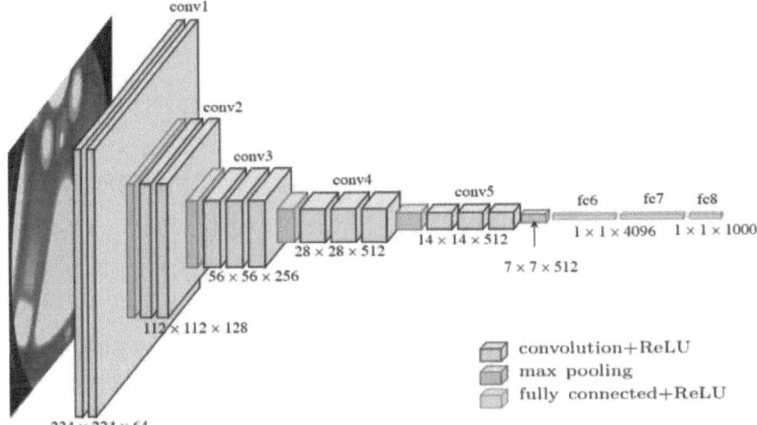

Figure 57.1 The VGG16 model architecture is depicted in three dimensions in this diagram
Source: Author

endoscopic images used for medical segmentation tasks. The dataset consists of 1,000 polyp images with corresponding ground truth masks, which were manually labeled by medical experts. Each image is resized to 224 × 224 pixels to ensure compatibility with the VGG-16 model used as the backbone for feature extraction.

Data preprocessing

To ensure optimal performance of the neural network and uniformity across all input samples, several pre-processing steps were applied:

Resizing: All images and masks were resized to 224 × 224 pixels to match the input requirements of the VGG-16 model.

Normalization: The pixel values of both images and masks were normalized to a range of 0 to 1 by dividing the pixel values by 255. This step facilitates faster convergence during training.

Splitting: To guarantee reproducibility, Using a random split with a fixed seed, the dataset has been split into training (80%) and testing (20%) sets.

Model selection

The core of our model architecture is based on a U-Net architecture, deep learning model for medical image segmentation tasks. To enhance performance and benefit from pre-trained knowledge, the encoder section of the U-Net model is replaced by a VGG-16 model pre-trained on the ImageNet dataset.

VGG-16 Backbone: Acts as the encoder, capturing image features through its convolutional layers.

Custom U- hierarchical Net Decoder: Consists of up-sampling layers and skip connections that help recover the spatial details lost during down-sampling in the encoder. A 3D diagram of the model architecture shown in Figure 57.1, which also shows the size

of feature maps at each convolution, pooling, and fully connected layer.

Training and testing

I.   The model was trained 50 times using the Adam optimizer, with an initial learning rate of 0.0001, which is appropriate for segmentation tasks with binary predictions (0 for background gradient and 1 for ulcer). The training batch size is 16, and the training and validation data ratio is 80/20. The purpose of early stopping is to avoid over-fitting, and sample checking is used to save the highest-performing model based on performance.

Data augmentation

To improve the overall ability of the model and prevent overfitting, more reliable information for the image is used:

Proposed Model Architecture

The architecture of the U-Net model with a VGG16 backbone can be described in detail with a focus on the encoder, decoder, and the connections between them also shown in Figure 57.4.

Input layer

The input to the model is an RGB image of size H × W × 3, in which case H = W = 224 and 3 represents the RGB channel. Input:

$$X \in R^{H \times W \times 3}$$

VGG16 encoder (downsampling path)

The encoder is a pre-trained VGG16 model, excluding the fully connected layers. As the number of channels (depth) increases, the input's spatial dimensions are gradually reduced. The encoder comprises the following blocks:

Block 1: Two 3 × 3 convolutional layers with ReLU activations.

Max-pooling operation with a 2 × 2 window and stride 2, reducing the spatial dimensions by half.

$$C_1 = \text{MaxPool}(\text{ReLU}(W_1 \cdot X + b_1))$$
$$\text{Output Dimensions: } C_1 \in R^{112 \times 112 \times 64} \tag{1}$$

Block 2: Two 3 × 3 convolutional layers with ReLU activations.

Max-pooling operation with a 2 × 2 window and stride 2.

$$C_2 = \text{MaxPool}(\text{ReLU}(W_2 \cdot C_1 + b_2))$$
$$\text{Output dimensions} - C_2 \in R^{56 \times 56 \times 128} \tag{2}$$

Block 3: Three 3 × 3 convolutional layers with ReLU activations.

Max-pooling operation with a 2 × 2 window and stride 2.

$$C_3 = \text{MaxPool}(\text{ReLU}(W_3 \cdot C_2 + b_3))$$
$$\text{Output dimensions} - C_2 \in R^{28 \times 28 \times 256} \tag{3}$$

Block 4: Three 3 × 3 convolutional layers with ReLU activations.

Max-pooling operation with a 2 × 2 window and stride 2.

$$C_4 = \text{MaxPool}(\text{ReLU}(W_4 \cdot C_3 + b_4))$$
$$\text{Output dimensions} - C_4 \in R^{14 \times 14 \times 512} \tag{4}$$

Block 5: Three 3 × 3 convolutional layers with ReLU activations.

Max-pooling operation with a 2 × 2 window and stride 2.

$$C_5 = \text{MaxPool}(\text{ReLU}(W_5 \cdot C_4 + b_5))$$
$$\text{Output dimensions} - C_5 \in R^{7 \times 7 \times 512} \tag{5}$$

Decoder (up sampling path)

The decoder dynamically up samples the feature maps and concatenates them with the comparing encoder layer yields (skip connection), permitting for exact localization.

Layer 6 (Up sample and concatenate with Block 4): Transposed convolution layer (Conv2DTranspose) to up sample the feature maps.

Concatenation with the output from Block 4. There will be less channels due to the convolutional layer

$$U_6 = \text{Conv2D}(\text{Concat}(\text{Conv2DTranspose}(C_5), C_4))$$
$$\text{Output dimensions} - U_6 \in R^{14 \times 14 \times 512} \tag{6}$$

Layer 7 (Up sample and concatenate with Block 3): Transposed convolution layer (Conv2DTranspose). Concatenation with Block 3. Convolutional layer.

$$U_7 = \text{Conv2D}(\text{Concat}(\text{Conv2DTranspose}(U_6), C_3)) \tag{7}$$

Output dimensions – $U_7 \in R^{28 \times 28 \times 256}$

Layer 8 (Upsample and concatenate with Block 2): Transposed convolution layer (Conv2DTranspose). Concatenation with Block 2. Convolutional layer.

$$U_8 = \text{Conv2D}(\text{Concat}(\text{Conv2DTranspose}(U_7), C_2)) \tag{8}$$

Output dimensions – $U_8 \in R^{56 \times 56 \times 128}$

Layer 9 (Upsample and concatenate with Block 1): Transposed convolution layer (Conv2DTranspose). Concatenation with Block 1. Convolutional layer.

$$U_9 = \text{Conv2D}(\text{Concat}(\text{Conv2DTranspose}(U_8), C_1)) \tag{9}$$

Output dimensions – $U_9 \in R^{112 \times 112 \times 64}$

Final output layer

After reducing the feature maps to a single channel that represents the pixel-wise probabilities for segmentation (binary classification), the final output is a 1 × 1 convolutional layer.

$$Y = \text{Conv2D}(U_9)$$
$$\text{Output dimensions} - Y \in R^{224 \times 224 \times 1} \tag{10}$$

Output dimensions – $Y \in R^{224 \times 224 \times 1}$

The sigmoid making sure that the output values are between 0 and 1. Can be interpreted as probabilities.

Loss function and optimization

Loss function: binary cross entropy

For binary segmentation tasks, where each pixel is classified into one of two classes (foreground or background), Binary Cross Entropy loss is employed.

$$L_{\text{binary cross-entropy}} = -\frac{1}{N} \sum_{i=1}^{N} [y_i \log(\hat{y}_i) + (1 - y_i) \log(1 - \hat{y}_i)] \tag{11}$$

where y_i is the ground truth and $\hat{y}_i$ is the predicted probability for pixel i.

Adam optimizer with 0.0001 training value. Adam optimizer uses gradients calculated by backpropagation to change the weights, and the adjustment value is based on the first and second term

$$\theta_t = \theta_{t-1} - \eta \frac{m_t}{\sqrt{v_t} + \epsilon} \qquad (12)$$

Empirical Simulation and Result Analysis

Several common metrics for image segmentation and classification tasks were used to assess the suggested model. The model's performance is thoroughly evaluated by these metrics, especially regarding segmentation accuracy, precision, and recall as shown in Table 57.1.

Table 57.1 Our proposed model evaluation chart.

Accuracy	F1 score	Precision	Recall
95%	84.8%	84.2%	85.4%

Source: Author

ROC *curve*

The ROC curve was plotted by evaluating the rates of false positives and true positives at various thresholds. With an AUC score near 1, the model demonstrated

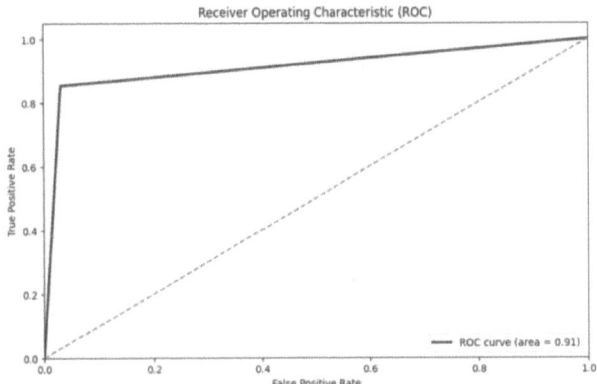

Figure 57.2 True positive rate is displayed against the false positive rate in this ROC curve, which has an AUC of 0.9

Source: Author

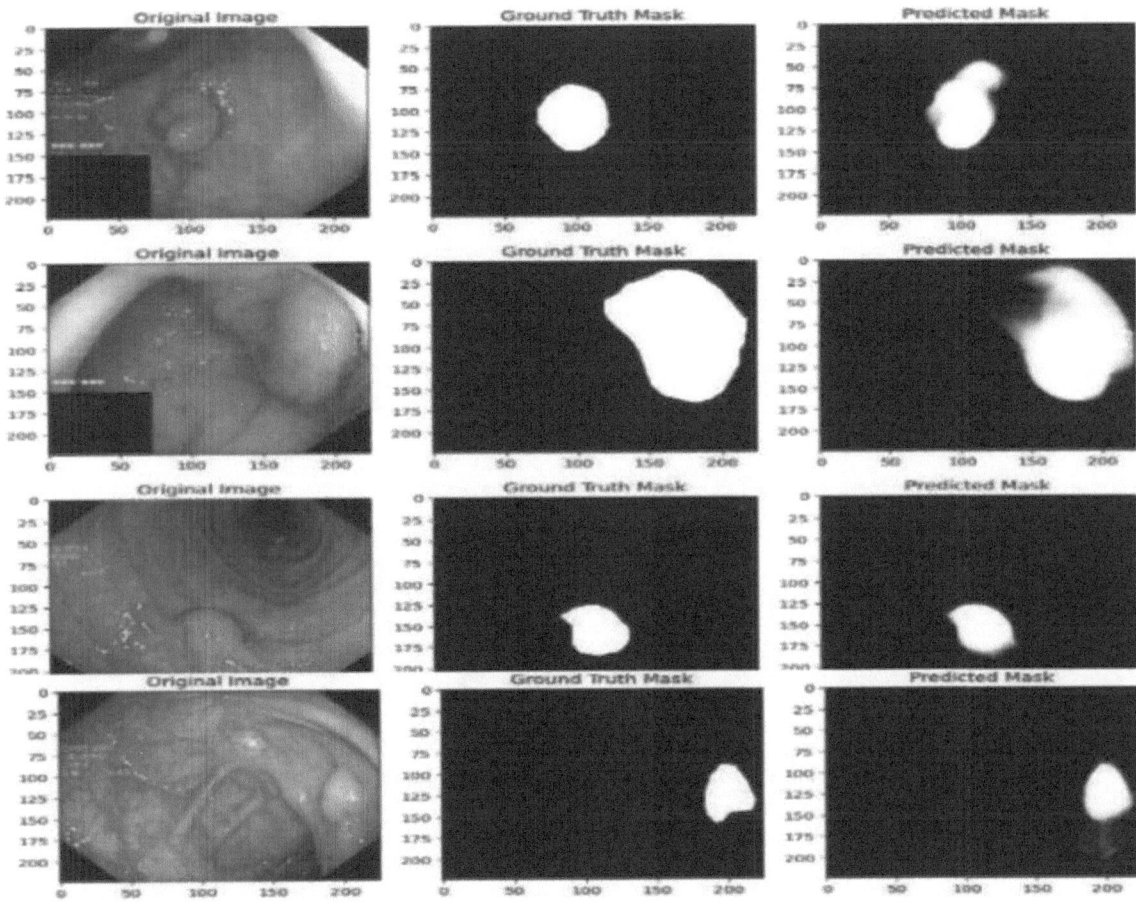

Figure 57.3 Sample comparison of ground truth with our model prediction, achieving high overlap with ground truth labels

Source: Author

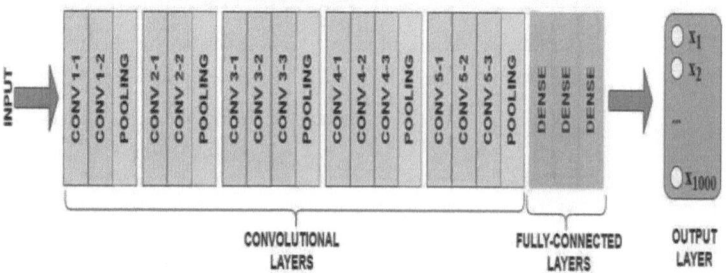

Figure 57.4. Diagram illustrating the architecture of the VGG16 model, detailing its convolutional and fully connected layers along with the input and output layers
Source: Author

Table 57.2 Performance metric comparison with existing models.

References	Models	Operation	Accuracy (Performance metric)
[1]	Single Shot Multibox Detector (CNN)	Detection	88.2%
[2]	VGGNet-based CNN	Classification	86.6%
[3]	(BEEMD) Lacunarity analysis	Ulcer detection classification	90%
[4]	U-Net	Biomedical image segmentation	90%
[5]	BIR (MobileNet + custom CNN)	Classification of bleeding images	94%
[6]	Recurrent attention neural network	Classification	90.85%
Proposed model	U-Net with VGG-16 backbone	Segmentation	95%

Source: Author

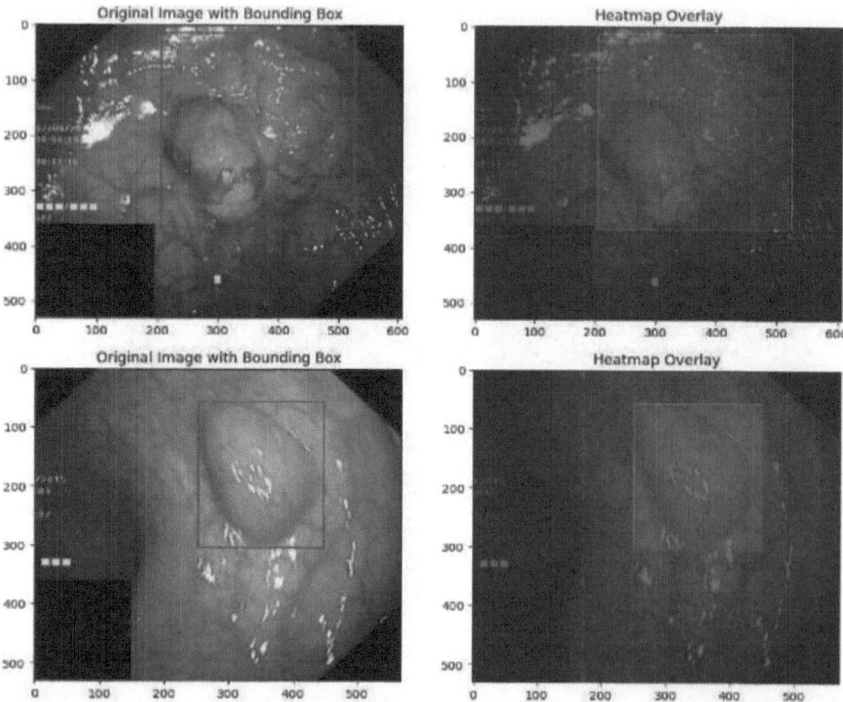

Figure 57.5 Model achieves significant improvements with bounding box and heatmap integration
Source: Author

a high level of efficacy in differentiating between ulcer and non-ulcer pixels. Our study computed AUC shows excellent performance on this task as shown in Figure 57.2.

This methodology demonstrates a clear, step-by-step approach to how our deep learning model was designed, trained, and evaluated for ulcer detection in endoscopic images. It highlights the strengths of combining a pre-trained VGG-16 backbone with a customized U-Net architecture and thoroughly evaluates the performance using standard medical imaging metrics.

To assess the qualitative performance of our proposed model, a random sample of 50 validation images was selected, where the corresponding predictions are visualized alongside the ground truth. The comparison reveals (Figure 57.3) that our model performs well in identifying ulcer regions, even in complex scenarios with varying shapes and textures. However, minor discrepancies were observed in challenging cases with subtle boundaries or low contrast between ulcers and surrounding tissues.

By adding more localization cues, bounding box annotations improve the model's accuracy by assisting it in more precisely focusing on ulcerous areas as shown in Figure 57.5. By separating ulcer boundaries from non-ulcer regions, this integration helps the model learn spatial constraints and lower false positives. These bounding boxes are added during training, which increases the model's robustness when processing a variety of endoscopic images in addition to its segmentation accuracy.

Conclusion

We provided a thorough pipeline for automated ulcer detection and segmentation in endoscopic images, leveraging a custom architecture based on VGG-16 combined with a U-Net model. The proposed methodology demonstrated significant accuracy in the segmentation of ulcer regions, achieving a remarkable accuracy rate of 95% as shown in Table 57.2. This confirms the robustness of the approach in handling complex medical imaging tasks.

To avoid overfitting, we used a binary cross-entropy loss function to train the model and then adjusted it using a variety of optimization strategies, such as early stopping and model checkpointing Incorporating bounding box annotations in future iterations could refine the segmentation process by first localizing the regions of interest.

References

[1] Aoki, T., Yamada, A., Aoyama, K., Saito, H., Tsuboi, A., Nakada, A., et al. (2019). Automatic detection of erosions and ulcerations in wireless capsule endoscopy images based on a deep convolutional neural network. *Gastrointestinal Endoscopy*, 89(2), 357–363. e2.Epub 2018 Oct 25. PMID: 30670179.

[2] Kumar, A., Upadhyayula, S., Kodamana, H. (2023). A convolutional neural network-based gradient boosting framework for prediction of the band gap of photoactive catalysts. *Digital Chemical Engineering*, 8, 100109. ISSN 2772-5081.

[3] Charisis, V., Tsiligiri, A., Hadjileontiadis, L. J., Liatsos, C. N., Mavrogiannis, C. C., & Sergiadis, G. D. (2010). Ulcer detection in wireless capsule endoscopy images using bidimensional nonlinear analysis. In Bamidis, P. D., & Pallikarakis, N. (Eds.), XII Mediterranean Conference on Medical and Biological Engineering and Computing 2010. IFMBE Proceedings, vol 29. Berlin, Heidelberg: Springer.

[4] Ronneberger, O., Fischer, P., & Brox, T. (2015). U-Net: convolutional networks for biomedical image segmentation. In Medical Image Computing and Computer-Assisted Intervention (MICCAI). Cham: Springer.

[5] Rustam, F., Siddique, M. A., Rehman, H. U., Ullah, S., Mehmood, A., Ashraf, I., et al. (2021). Wireless capsule endoscopy bleeding images classification using CNN-based model. *IEEE Access*, 9, 33675–33688.

[6] Vallée, R., de Maissin, A., Coutrot, A., Normand, N., Bourreille, A., & Mouchère, H. (2019). Accurate small bowel lesions detection in wireless capsule endoscopy images using deep recurrent attention neural network. In 2019 IEEE 21st International Workshop on Multimedia Signal Processing (MMSP), (pp. 1–5). Kuala Lumpur, Malaysia.

58 A literature review of mental health assessment using social media post for depression

Chiranjeeb Das[1,a], Dibakar Pradhan[1,b], Rakesh Patel[1,c], Sohan Kumar Pande[1,d] and Satyabrat Sahoo[2,e]

[1]Department of Computer Science and Engineering, Silicon Institute of Technology, Sambalpur, Odisha, India

[2]Department of Computer Science and Engineering, School of Engineering and Technology, DRIEMS University, Tangi, Cuttack, Odisha, India

Abstract

Social media platforms have become an integral part of daily life, offering a valuable source of information about individuals' mental health. This paper presents a comprehensive review of recent research on leveraging social media data to assess mental health, specifically depression. Various machine learning and natural language processing techniques have been employed to analyze social media posts and identify patterns indicative of depressive symptoms. The review explores the strengths and limitations of different approaches, including traditional machine learning algorithms and advanced deep learning models. Key findings suggest that social media data can be effectively used to detect depression, but challenges like data quality, privacy concerns, and model interpretability remain. Future research directions include incorporating multimodal data, improving contextual understanding, and addressing ethical considerations to enhance the accuracy and reliability of mental health assessments using social media.

Keywords: Data mining, deep learning, depression detection, early intervention, mental health, mental health assessment, natural language processing, public health, sentiment analysis, social media, suicide prevention, text mining

Introduction

Mental health disorders, like depression, are rising. Traditional diagnosis is slow and subjective. Social media data can help early detection [9,13]. This paper reviews machine and deep learning techniques (SVM, Naive Bayes, random forest, CNNs, RNNs, Transformers) [1,3,4] for analyzing social media data to detect depression. We also discuss challenges and ethical issues.

Related Work

Traditional machine learning approaches

Traditional machine learning (ML) like Naive Bayes, SVM, and random forest have been used for sentiment analysis and mental health classification. These methods often use feature engineering techniques such as N-grams, TF-IDF, and Bag-of-Words to represent text numerically.

Refer to Figure 58.1 and Table 58.1 for a concise visual and detailed comparison of traditional machine learning model performance.

Deep learning approaches

Deep learning improves sentiment analysis, enabling accurate text classification. Applied to social media, it can identify mental health patterns like depression. Key techniques include word/character embeddings, sequential models (RNNs, LSTMs), and transformer

Performance Comparison:

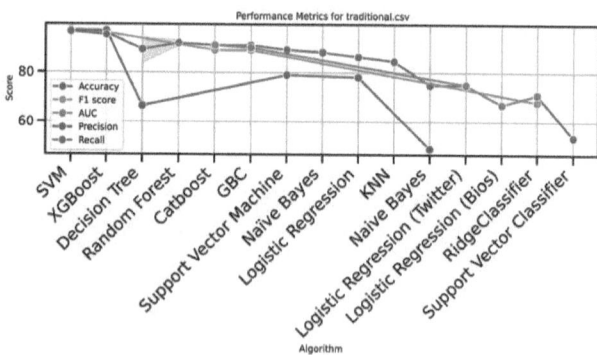

Figure 58.1 Performance metrics for traditional machine learning models
Source: Author

[a]chiranjeebdas2003@gmail.com, [b]deva.pradhan130@gmail.com, [c]rakesh246808@gmail.com, [d]sohan.pande@silicon.ac.in, [e]satyabratsahoo@driems.ac.in

DOI: 10.1201/9781003658221-58

models (BERT, RoBERTa) that capture complex language nuances.

Refer to Figure 58.2 and Table 58.2 for a concise visual and detailed comparison of deep learning model performance.

Hybrid approaches

Hybrid approaches combine the strengths of traditional and deep learning techniques, enhancing accuracy and robustness. Traditional models extract features, while deep learning models learn higher-level

Table 58.1 Traditional model comparison.

Algorithm	References	Accuracy	F1 score	AUC	Precision	Recall
SVM	Ajmal et al. [11]	0.968	0.968	--	0.97	0.965
XGBoost	Ajmal et al. [11]	0.959	0.96	--	0.968	0.952
Decision tree	Bhavya et al. [7]	0.9397	--	--	--	--
Random forest	Bhavya et al. [7]	0.9295	--	--	--	--
Catboost	Pourkeyvan et al. [8]	0.91	0.89	0.91	--	--
GBC	Pourkeyvan et al. [8]	0.91	0.89	0.9	--	--
Random forest	Supriya et al. [5]	0.907	--	--	--	--
Decision tree	Supriya et al. [5]	0.903	--	--	--	--
Support vector machine	Choudhary et al. [12]	0.8919	--	--	0.7907	--
Naïve Bayes	Bhavya et al. [7]	0.882	--	--	--	--
Logistic regression	Choudhary et al. [12]	0.8639	--	--	0.7819	--
KNN	Bhavya et al. [7]	0.8458	--	--	--	--
Decision tree	Choudhary et al. [12]	0.8369	--	--	0.665	--
Naive Bayes	Choudhary et al. [12]	0.75	--	--	0.4908	--
Logistic regression (Twitter, Bios)	Pourkeyvan et al. [8]	0.75, 0.67	--	0.75, 0.67	--	--
RidgeClassifier	Pourkeyvan et al. [8]	0.71	0.68	0.71	--	--
Support Vector Classifier	Supriya et al. [5]	0.5372	--	--	--	--

Source: Author

Table 58.2 Deep learning model comparison.

Algorithm	References	Accuracy	F1 score	AUC	Precision	Recall
BBU (Twitter, Bios)	Pourkeyvan et al. [8]	0.97, 0.95	0.97, 0.94	0.98, 0.96	--	--
DBUFS2E (Twitter, Bios)	Pourkeyvan et al. [8]	0.97, 0.95	0.96, 0.96	0.98, 0.96	--	--
MBBU (Twitter, Bios)	Pourkeyvan et al. [8]	0.96, 96.00	0.96, 96.00	0.98, 0.96	--	--
MLP	Ajmal et al. [11]	0.956	0.956	--	0.955	0.958
DRB (Twitter, Bios)	Pourkeyvan et al. [8]	0.95, 0.95	0.95, 0.95	0.97, 0.96	--	--
DistilBERT (Twitter, Reddit)	Kokane et al. [10]	0.92, 0.85	0.92, 0.85	--	0.92, 0.85	0.92, 0.85
BERT (Twitter, Reddit)	Kokane et al. [10]	0.92, 0.84	0.92, 0.83	--	0.92, 0.84	0.92, 0.84
RoBERTa (Twitter, Reddit)	Kokane et al. [10]	0.91, 0.84	0.9, 0.84	--	0.9, 0.84	0.9, 0.84
XLNet (Twitter, Reddit)	Kokane et al. [10]	0.9, 0.82	0.89, 0.81	--	0.9, 0.81	0.89, 0.81
MLP (Twitter)	Pourkeyvan et al. [8]	0.83	0.82	0.83	--	--
Natural language processing	Choudhary et al. [12]	0.8248	--	--	0.7405	--
CNN	Supriya et al. [5]	0.8154	--	--	--	--
BiLSTM	Philip et al. [6]	0.792	0.793	--	0.762	0.826
BiGRU	Philip et al. [6]	0.787	0.794	--	0.746	0.848
LSTM	Philip et al. [6]	0.782	0.788	--	0.743	0.838
GRU	Philip et al. [6]	0.481	0.65	--	0.481	1

Source: Author

Performance Comparison:

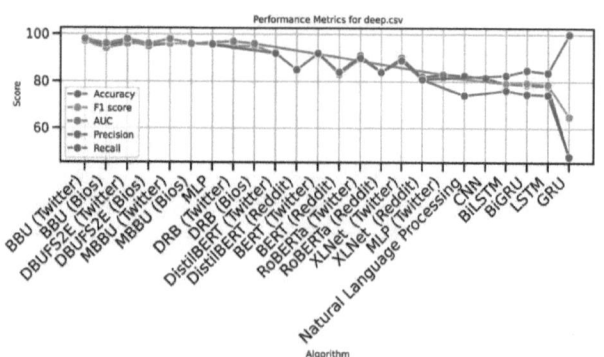

Figure 58.2 Performance metrics for deep learning models
Source: Author

Performance Comparison:

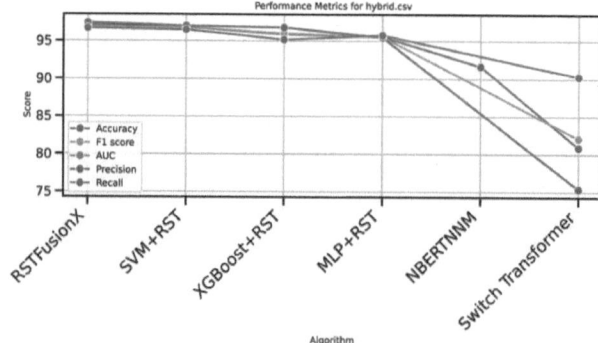

Figure 58.3 Performance metrics for hybrid models
Source: Author

Table 58.3 Hybrid model comparison.

Algorithm	References	Accuracy	F1-score	Precision	Recall
RSTFusionX	Ajmal et al. [11]	0.971	0.969	0.974	0.967
SVM+RST	Ajmal et al.	0.968	0.968	0.97	0.965
XGBoost+RST	Ajmal et al.	0.959	0.96	0.968	0.952
MLP+RST	Ajmal et al.	0.956	0.956	0.955	0.958
N-gram analysis + BERT model Chinese + neural network model	Yao [2]	0.9168	--	--	--
Switch transformer	Philip et al. [6]	0.809	0.821	0.754	0.903

Source: Author

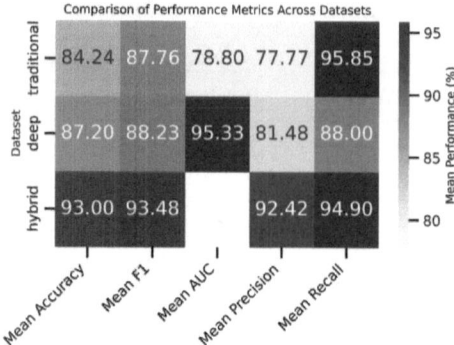

Figure 58.4 Compact heatmap of mean performance metrics across datasets
Source: Author

representations. This fusion often leads to superior performance, especially with complex and noisy social media data.

Refer to Figure 58.3 and Table 58.3 for a concise visual and detailed comparison of hybrid model performance.

Refer to Figure 58.4 for visually compares mean performance metrics across different datasets.

Research Gap

Some key research gaps include:

i. Quality and quantity of data: Limited, noisy, and biased data can hinder model performance.

ii. Cross-Cultural and Linguistic Challenges: Addressing cultural nuances and language variations is crucial.

iii. Ethical considerations: Ensuring privacy, fairness, and responsible AI practices is essential.

iv. Longitudinal studies: Tracking mental health changes over time for effective interventions.

v. Multimodal analysis: Leveraging text, image, and video data for comprehensive understanding.

vi. Explainable AI: Models that can explain their decision-making process for transparency and trust.

Conclusion

Social media data can serve as a rich source for mental health assessment. Traditional machine learning approaches, including SVM and Naive Bayes, as well as deep learning methods, such as RNNs and Transformers, have been used for this purpose. Despite the effectiveness of these approaches, challenges like data quality, privacy, and cross-cultural differences remain. Future research should address these challenges, explore multimodal approaches, and integrate models into clinical settings.

References

[1] Shaik, S., & Bharathi, V. C. (2024). A literature review on the detection of mental illness. In 2024 5th International Conference on Electronics and Sustainable Communication Systems (ICESC), (pp. 1413–1417). IEEE.

[2] Yao, Z. (2024). A multi-model approach to detection of depression in the Chinese social media entries. In 2024 5th International Seminar on Artificial Intelligence, Networking and Information Technology (AINIT), (pp. 2148–2151). IEEE.

[3] Saleem, M., & Afzal, H. (2024). A review of mental health analysis through social media using machine learning and deep learning approaches. In 2024 International Conference on Engineering and Computing Technologies (ICECT), (pp. 1–7). IEEE.

[4] Sundaram, A., Subramaniam, H., Ab Hamid, S. H., & Nor, A. M. (2024). A three-step procedural paradigm for domain-specific social media slang analytics. In 2024 International Conference on Trends in Quantum Computing and Emerging Business Technologies, (pp. 1–7). IEEE.

[5] Supriya, M. S., Aniket, A., Aakanksha, J., & Peter, K. (2024). AI-powered mental health diagnosis: a comprehensive exploration of machine and deep learning techniques. In 2024 International Conference on Distributed Computing and Optimization Techniques (ICDCOT), (pp. 1–6). IEEE.

[6] Philip, A. T., Iyer, R. R., & Nitha, L. (2024). Analysing depression in social media: a study of basic deep learning algorithms and transformer model with a comparative approach. In 2024 Third International Conference on Smart Technologies and Systems for Next Generation Computing (ICSTSN), (pp. 1–6). IEEE.

[7] Bhavya, R., Ramani, R., Royal, G. L. G., Manideep, G., & Sridhar, G. (2024). Exploring depression through social media: a textual analysis. In 2024 International Conference on Knowledge Engineering and Communication Systems (ICKECS), (Vol. 1, pp. 1–7). IEEE.

[8] Pourkeyvan, A., Safa, R., & Sorourkhah, A. (2024). Harnessing the power of hugging face transformers for predicting mental health disorders in social networks. *IEEE Access*, 12, 28025–28035.

[9] Al Meqbaali, M., Ouhbi, S., Jan, R. K., Amiri, L., Al Mugaddam, F., Serhani, M. A., et al. (2024). Mental health professionals' insights on digital health solutions for anxiety in the UAE. In 2024 IEEE 32nd International Requirements Engineering Conference Workshops (REW), (pp. 362–369). IEEE.

[10] Kokane, V., Abhyankar, A., Shrirao, N., & Khadkikar, P. (2024). Predicting mental illness (depression) with the help of NLP transformers. In 2024 Second International Conference on Data Science and Information System (ICDSIS), (pp. 1–5). IEEE.

[11] Ajmal, S., Shoaib, M., & Iqbal, F. (2024). RSTFusionX: leveraging rhetorical structure theory and ensemble models for depression prediction in social media posts. *IEEE Access*, 12, 118389–118404.

[12] Choudhary, J., Awasthi, D., Bhatt, B., Narang, H., Vats, S., & Sharma, V. (2024). SentiSense: pioneering mental health application through advanced technology. In 2024 International Conference on Electronics, Computing, Communication and Control Technology (ICECCC), (pp. 1–5). IEEE.

[13] Mobin, M. I., Akhter, A. S., Mridha, M. F., Mahmud, S. H., & Aung, Z. (2024). Social media as a mirror: reflecting mental health through computational linguistics. *IEEE Access*, 12, 130143–130164.

59 Convolutional neural network approach to emotion recognition in speech

Mousumi Acharya[1,a], Shiba Charan Barik[2,b] and Sudhir Kumar Mohapatra[3,c]

[1]Department of CSE, DRIEMS University, Cuttack, Odisha, India

[2]Department of SOPS, DRIEMS University, Cuttack, Odisha, India

[3]Department of Emerging Technology, Sri Sri University, Cuttack, Odisha, India

Abstract

One of the most natural and efficient ways to communicate emotions is through speech, and identifying emotions is crucial to determining a person's emotional condition. Emotion detection from vocal signals is a complicated and demanding endeavor because of the diversity (pitch, tone, speaking speed, accent) in speech characteristics and the subjectivity (context or environment) related to emotions. Considering the issues related to the various forms of speech traits and models used for emotion detection from vocal signals, this research introduced the identification of emotional states from speech. The paper aims to recognize vocal signals and classify them into four emotional categories: happy, sad, angry, and neutral. This document has employed a three-stage method. In first approach, pre- process the speech signal. In the second approach, extraction of the features from the speech signal using MFCC, Teager Energy Operator, ZCR to improve HCR. Required features are selected and input to the next approach. In third approach, selected features used as input in convolutional neural network to classify the emotion and evaluate the performance.

Keywords: Convolutional neural network, HCR, MFCC, teager energy operator, ZCR, crack detection, internet of thinks, machine learning, SHM, YOLO

Introduction

Emotions communicate substantial information regarding an individual's mental condition and have an important influence on human interactions. During interactions between two individuals, they can readily identify the concealed emotion in the speech of the other person. Years of dedication and observation can assist you in achieving this. Human first examine several aspects of a speech before identifying the speaker's emotion based on prior knowledge or observation. A humanlike system capable of accurately and efficiently detecting emotions must be developed identification of emotion can be accomplished by removing the characteristics or distinct qualities from the speech and need. For detecting these attributes, a portable medical device is required. The proposed paper may be used for recognizing mental state from speech signal. Human meaning through tone, pitch, volume, pace, pauses and many such characteristics (Daniel et al., 2020) [7]. Emotion recognition plays an important measure mental healthcare. The three primary components of the speech emotion system recognition process are the classifier model, feature extraction, and speech preprocessing.

Preprocessing involves identifying speech activity, standardizing the length, and removing unwanted background noise from the speech signal. Prosodic features are extracted from speech, they are represented by pitch, energy, duration and format. In this paper, we proposed the use of i.e. harmonic to noise rate (HNR) with the coefficients ZCR, MFCC, and TEO for feature extraction [6]. We used the convolutional neural network (CNN) as classifier. System is evaluated on the (IEMOCAP) interactive emotional dyadic motion capture database, is an English emotional speech database consisting of audio-visual data. Further preceding relevant research on voice emotion recognition systems is described in the following section. Then, Section 3 offers a perspective on the database that is employed in the system's implementation. Subsequent section, the framework and feature extraction methodology are presented in Section 4. Section 5, the suggested model, CNN described in detail. Conclusion and future work is given in Section 6.

Related Work

Basu (2018) [9] proposed effective classification model like GMM, KNN to identifying relevant speech features like MFCC, LPCC to recognize speech, selecting appropriate speech databases (corpora). The criteria necessary criteria for preparing a database, such as the

[a]mousumi.acharya@driems.ac.in, [b]barikshiba@driems.ac.in, [c]sudhir.mohapatra@srisriuniversity.edu.in

DOI: 10.1201/9781003658221-59

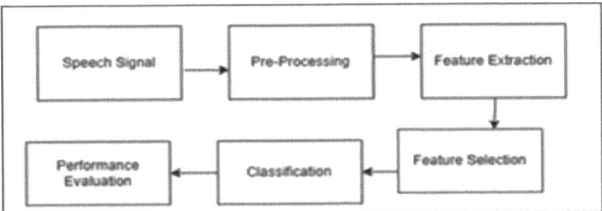

Figure 59.1 Block diagram of speech recognition
Source: Author

scope, physical existence, contents, and language chosen. The paper covers various features excitation from speech like spectral features, and prosodic features, and the classifiers, such as GMM, HMM, and SVM are the work of Swain (2018) [4]. A three-step method includes cleaning the speech signal via framing, windowing, normalization, and noise reduction. The second method focuses on feature extraction using MFCC and TEO. Wani et al. [1] proposed classification methods which include ANN, GMM, SVM, and HMM, as well as deep learning classifiers such as deep neural networks and long short-term memory. Aouani and Ayed [3] proposed an emotion recognition system driven speech recognition method which includes two stages feature extraction which comprising Mel frequency cepstral coefficients, noise ratio, zero crossing rate , harmonic to and Teager Energy Operator and a classification which is represented by SVM.A parallelized CNN-BiGRU model for SER. Maji (2022) [6] proposed a model which integrates CNN (and bidirectional gated recurrent units to seize both spectral and temporal speech features.

Figure 59.1 describes the block diagram of Speech recognotion and Figure 59.3 describes Class-wise (Normal, Anger, Sad, Happy) accuracy of the proposed system using IEMOCAP dataset.

Dataset

For our proposed approach we have considered only the sound part along with English emotion interactive emotional dyadic motion capture (IEMOCAP) Busso et al. [10] includes audio-visual information from ten actors. It has about 12 hours of audio-visual material, including text transcriptions, speech, video, and facial motion capture. It includes dyadic sessions in which actors perform either scripted scenarios or improvisations that are specifically selected to elicit emotional responses. Both dimensional and categorical tags have been assigned to the expressions by various annotators. Among the classification categories are surprise fear, happiness, anger, sadness, neutral.

Methodology

This section elaborates pre-processing and implementation of the proposed work.

Pre-processing
A process of transforming the raw fact into readable and understandable format [5].

1. **Sampling:** Natural sound signals are analog signals i.e. continuous time signals [5].

$$f_s = \frac{1}{T} \qquad (1)$$

Equation (1) signifies the relation between the time period (T) vs sampling frequency (fs).

2. **Pre-emphasis:** These show a slow shift in the direction of signal information extraction over time. H(z) represents the pre-emphasized signal, and equation (2) indicates the pre-emphasis filter that was applied.

$$H(z) = 1 - 0.95z \qquad (2)$$

3. **De-silencing:** These show that the approach to signal information extraction has been gradually evolving all over time. The pre-emphasis filter is indicated by equation (2), where the pre-emphasized signal is indicated by H(z).

4. **Framing:** These instances demonstrate that the approach to signal information extraction has been gradually evolving throughout time. The pre-emphasis filter is indicated by equation (2), where the pre-emphasized signal is indicated by H(z).

5. **Windowing:** Two main goals are usually accomplished with a frame duration of 20 to 30 milliseconds: first, it guarantees that a steady signal value is recorded for a little amount of time, and second, it reduces signal fluctuations throughout that time. In this instance, a frame time of 25 milliseconds has been selected.

$$(n) = 0.54 - 0.46 cos\left(\frac{2\pi n}{M-1}\right) \qquad (3)$$

where w(n) represents windowed signal, M represents the window length-l.

Feature extraction
1 **Cepstral Mel-frequency Coefficient:** The form of the human vocal tract, which dictates the sound qualities, is reflected in this envelope Aouani and Ayed [3] and result is the Mel frequency

cestrum coefficients (MFCC). The following formula is used to determine the ΔMFCC coefficients:

$$\Delta Cep(i) = a \sum_{j=1}^{z} j(Cep(i+j) - Cep(i-j)) \qquad (4)$$

Where α is a constant ≈ 0.2

2   **Zero Crossing Rate:** According to Aouani and Ayed [3] Zero crossing rate ZCR is calculated as the rate of the signal changes within the frame. It represents how many times the signal transitions from a positive value to a negative value and back, divided by the duration of the frame. The formula below is utilized to calculate the ZCR:

$$ZCR = \frac{1}{N-1} \sum_{n=1}^{N-1} \quad sign(s(n)s(n-1)) \qquad (5)$$

3   **Teager Energy Operator:** According to Aouani and Ayed [3] (TEO) Teager Energy Operator is used for examine the speech's qualities when a particular emphasis is present in the utterance. TEO functions analyze the frequency and temporal domain behavior of the statement to calculate the distance between them. The TEO is calculated to extract the total signal as follows.

$$\Psi M\,[xf\,[t]] = (xf\,[t])^2 - (xf\,[t-1]\,\bar{xf}\,[t+1])) \qquad (6)$$

4   **Harmonic to noise ratio:** The proportion of harmonic noise in voice is quantified in decibels by the harmonic to noise ratio (HNR). It explains how the acoustic energy of the emitted voice spectrum is distributed between the harmonic and enharmonic parts [3].

Classification model

Using CNN, we will categorize emotions in this section using features extracted from a sound source. Subsequently, there are three convolutional layers with 128 5 × 5 filters, two of which have ReLU activation layers. Next, we consider batch normalization rate, activation function ReLU and a dropout layer with learning rate 0.2. Using the same set up the final layer of convolution layer follows the flattening layer and dropout layer with a learning rate of 0. 2. A flattering layer with a fully connected model is formed at output of this approach. Softmax activation and batch normalization are subsequently employed. Our model utilizes the RMSProp optimizer, which has a decay rate of 1e-6 and a learning rate of 0. 00001. Figure 59.2 presents a depiction of the proposed architecture.

Simulation Result

This section presents an experimental assessment of the suggested method using the IEMOCAP dataset. For our model, we utilize the IEMOCAP dataset. About 12 hours of audiovisual material from 10 seasoned actors—5 men and 5 women—in both scripted and improvised English dialogues are included [10]. Transcripts, audio, motion capture recordings, and video are all utilized to portray the emotional content of each discussion. We only use audio and transcriptions in this investigation.

There are 5531 utterances in total among the four categories of emotions in this database: angry, sad, neutral, and pleased.

In order to validate the suggested system, the data was divided 80:20 between testing and training. The testing data set was used for accessing the model's overall capacity to identify unknown level of data.

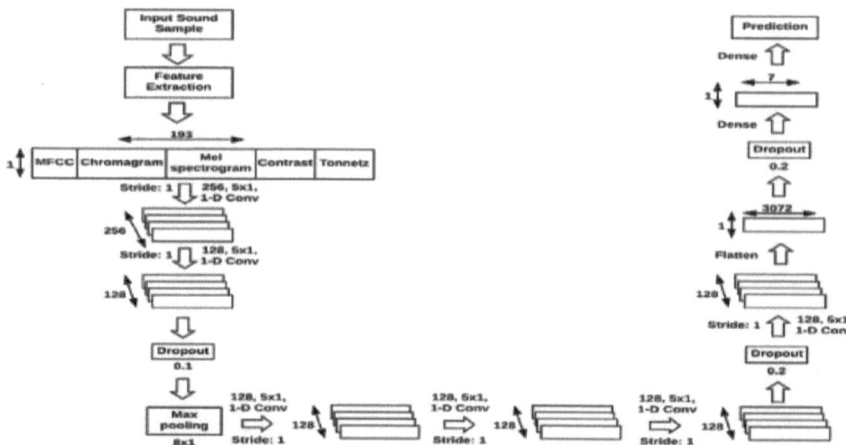

Figure 59.2 A baseline architecture of CNN
Source: Author

Table 59.1 Details of IEMOCAP.

Emotion	Total voices in IEMOCAP
Happy mood	1636
Angry mood	1103
Sad mood	1084
Neutral Mood	1708

Source: Author

Table 59.2 Suggested model's architectures performance using the IEMOCAP dataset.

Dataset	Model suggested	WA(%)	UA(%)
IEMOCAP	CNN Model	68.27	62.75

Source: Author

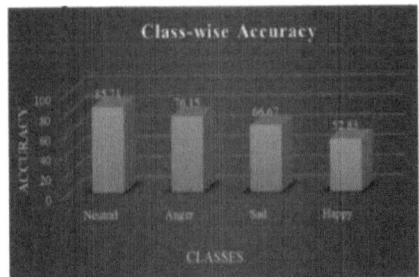

Figure 59.3 Class-wise accuracy of the proposed system using IEMOCAP

Source: Author

This study examined the suggested approach to determining the correctness utilizing a variety of functions employing statistical parameters. By using the confusion matrix TP, FP, FP, FN values are predicted and the model testing was confirmed. In addition to accuracy value, we required precision, memory, and F1-score as evaluation criteria. F1-score depth is defined by the harmonic mean of precision and recall rate.

Table 59.1 describes the block diagram of Speech recognotion.

Table 59.2 shows the comparison result of the suggested recognition method, which is more effective and useful than the baseline using the same voice corpora. The IEMOCAP dataset has a UA accuracy of 62.75%. As seen by the confusion matrix and tables, this model ultimately produced good results for all emotions and better unweighted accuracy than the specified baseline.

Conclusion

In the area of emotion recognition environment, choosing the best cue for emotion recognition presents several difficulties when using voice signals. The restrictions, such as the selection of optimal features, the best model configuration, the best optimizer, and the learning approach, are now being addressed by researchers in order to increase the accuracy of baseline models utilizing a new and simplified high-level technique. The suggested CCN-based SER system learns from a raw audio file and effectively identifies each emotion in accordance with these challenges. By learning additional powerful cues from voice waves, the proposed method shows significant improvements over the typical performance. The technique's unweighted precision was 62.75% when compared to traditional IEMOCAP. The suggested approach showed outstanding generality and recognition rates using unprocessed speech waves. Applications for the proposed system include call centers, safety control systems, internet communication, health care facilities, and many more. In order to solve practical problems in real time, researchers can investigate the CCN model further in the future for speaker recognition, speaker diarization, automatic speaker recognition, and human behavior assessment methodologies.

References

[1] Wani, T. M., Gunawan, T. S., Qadri, S. A. A., Kartiwi, M., & Ambikairajah, E. (2021). A comprehensive review of speech emotion recognition systems. *IEEE Access*, 9, 47795–47814.

[2] An, H., Lu, X., Shi, D., Yuan, J., Li, R., & Pan, T. (2019). Mental health detection from speech signal: a convolution neural networks approach. In 2019 International Joint Conference on Information, Media and Engineering (IJCIME), (pp. 436–439). IEEE.

[3] Aouani, H., & Ayed, Y. B. (2020). Speech emotion recognition with deep learning. *Procedia Computer Science*, 176, 251–260.

[4] Barros, P., Jirak, D., Weber, C., & Wermter, S. (2015). Multimodal emotional state recognition using sequence-dependent deep hierarchical features. *Neural Networks*, 72, 140–151.

[5] Deshmukh, G., Gaonkar, A., Golwalkar, G., & Kulkarni, S. (2019). Speech based emotion recognition using machine learning. In 2019 3rd International Conference on Computing Methodologies and Communication (ICCMC). IEEE.

[6] Maji, B., Swain, M., & Panda, R. (2022). A feature selection based parallelized CNN- BiGRU network for speech emotion recognition in Odia language.

[7] Low, D. M., Bentley, K. H., & Ghosh, S. S. (2020). Automated assessment of psychiatric disorders using speech: a systematic review. *Laryngoscope Investigative Otolaryngology*, 5(1), 96–116.

[8] Nema, B. M., & Abdul- Kareem, A. A. (2018). Pre-processing signal for speech emotion recognition. *Al-Mustansiriyah Journal of Science*, 28(3), 157–165.

[9] Pandey, S. K., Shekhawat, H. S., & Prasanna, S. M. (2019). Deep learning techniques for speech emotion recognition: a review. In 2019 29th International Conference Radioelektronika. IEEE.

[10] Busso, C., Bulut, M., Lee, C. C., Kazemzadeh, A., Mower, E., Kim, S., et al. (2008). IEMOCAP: interactive emotional dyadicmotion capture database. *Language Resources and Evaluation*, 42, 335–359.

60 Review of solar power generation forecasting using deep learning techniques

Somdev Behera[1,a], Gourav Kumar[1,b], Mahaveer Prasad Panigrahi[1,c], Sohan Kumar Pande[1,d] and Satyabrat Sahoo[2,e]

[1]Department of Computer Science and Engineering, Silicon Institute of Technology, Sambalpur, India

[2]Department of Computer Science and Engineering, School of Engineering and Technology, DRIEMS University, Tangi, Cuttack, Odisha, India

Abstract

In this study, we conducted a comprehensive review of various deep learning approaches used to anticipate solar power generation. Because solar electricity generation is intermittent, anticipating it is a difficult challenge on its own itself. Deep learning methods have been shown to be effective and precise in this field because of the increased complexity of data. In our analysis of recent advancements in this field, we mostly concentrated on deep learning methods, which include models like convolutional neural networks, LSTM, ANN, GRU, and others. To give a clear picture of the application of deep learning in this field, we conducted a methodical comparison of the studies and their findings. We also talked about the difficulties and potential applications of this field.

Keywords: ANN, convolutional neural networks, deep learning, feed forward networks, forecasting, GRU, LSTM, machine learning, neural networks, solar power generation

Introduction

The escalating global energy demands, driven by technological advancement and population growth, have necessitated a shift from conventional energy sources to renewable alternatives, particularly solar power. While solar energy offers an abundant and accessible solution, its generation forecasting remains challenging due to various environmental factors such as weather patterns, seasonal changes, and geographical location. Accurate forecasting is crucial for efficient power plant operations, cost reduction, and environmental sustainability. A thorough analysis of deep learning methods for solar power generation predictions is presented in this research. Our goal in doing this thorough evaluation is to give scholars and professionals useful information about the present and potential future paths of deep learning applications in solar power forecasting.

Literature Review

Lin et al. [9] demonstrated the superiority of temporal convolutional neural networks (TCNNs) over RNNs and FFNNs, with lower MAE and RMSE on the UQ and Sanyo datasets. Koprinska et al. [7] reported that CNNs outperformed MLPs and LSTMs in analyzing solar PV data, particularly in error metrics like MAE and RMSE. Rana et al. [12] proposed a novel method combining weather-type clustering with neural network ensembles, achieving enhanced forecasting accuracy. Lim et al. [8] introduced a hybrid CNN-LSTM model capable of delivering high accuracy under both sunny and cloudy conditions. Aslam et al. [1] developed a two-stage attention mechanism with Bayesian optimization over LSTM, achieving superior performance in terms of RMSE, MAE, and R^2. Heo et al. [5] utilized a multi-channel CNN to incorporate meteorological and geographical data, improving forecasting accuracy compared to MLR and ANN models. Wang et al. [14] merged LSTM and Gaussian process regression (GPR), resulting in improved accuracy across two datasets. Du Plessis et al. [2] employed deep learning models such as FFNN, LSTM, and GRU for utility-scale PV forecasting, with the GRU-RNN delivering the best results. Elizabeth et al. [3] used a hybrid multi-step CNN and stacked LSTM model for solar irradiance and POA prediction, showing significant improvements in accuracy. Muhammad et al., 2024 [15] utilized ANN for long-term solar power prediction, demonstrating competitive results across multiple performance metrics. Thipwangmek et al. [13] adopted a 1D CNN-GRU approach for short-term solar forecasting, achieving

[a]somdevbehera3103@gmail.com, [b]gourav2982@gmail.com, [c]pmahaveerp86619@gmail.com, [d]sohan.pande@silicon.ac.in, [e]satyabratsahoo@driems.ac.in

DOI: 10.1201/9781003658221-60

Table 60.1 Comparison of map, R², RMSE and MAE across different methods.

Reference	Method	MAPE (%)	R2	RMSE	MAE	Dataset
[9]	TCNN	-	-	98.118kW	70.015kW	University of Queensland
		-	-	0.721 kW	0.510 kW	Sanyo Datasets
[7]	CNN	-	-	153.91 kW	114.38 kW	University of Queensland
	MLP	-	-	154.16kW	116.64kW	
	LSTM	-	-	161.55kW	127.67kW	
[12]	Weather-Ensemble	6.88	-	-	83.90 kW	St Lucia campus of University of Queensland
[8]	CNN-LSTM	Sunny: 4.58	Both: 0.99	Sunny: 43.87 kW	Sunny: 34.00 kW	Busan, Korea
		Cloudy: 7.21		Cloudy: 9.09 kW	Cloudy: 6.97 kW	
[1]	Attention-LSTM	-	0.8917	0.0638*	0.0324*	Day-ahead (Germany)
[5]	Multichannel CNN	8.639	-	-	-	Combined Data
	MLR	16.187	-	-	-	
	ANN	15.991	-	-	-	
[14]	LSTM-GPR	9.43	-	264.98 kW	201.77 kW	University of Illinois (Dataset 1)
			-	280.89 kW	219.49 kW	University of Illinois (Dataset 2)
[2]	GRU-RNN	-	-	NRMSE: 8.12%	-	Utility-scale
[3]	CNN-LSTM	3.11	0.98	0.36*	0.18*	Sweihan PV Project (Dataset 1)
		6.70	0.96	61.24 kW	29.00 kW	Sweihan PV Project (Dataset 2)
[11]	ANN	-	0.840	10.955%	-	Deployment
[13]	CNN-GRU	-	-	0.025*	-	Winter
		-	-	0.050*	-	Summer
		-	-	0.094*	-	Rainy
[4]	LSTM	2.17	0.9381	374.4 kW	236.35 kW	Winter
		0.275	0.745	883.2 kW	676.34 kW	Summer
[6]	DSE-XGB	-	0.95	1.30 kWh	0.83 kWh	Bunnik, Netherlands
[10]	LSTM	12.45	0.992	0.16 kW	0.05 kW	Standard Dataset

*Normalized value

Source: Author

low RMSE values across different seasons. Elsaraiti and Merabet [4] evaluated LSTM and MLP for short-term solar power prediction, with LSTM consistently outperforming MLP in both summer and winter conditions. Khan et al. [6] proposed a unique stacked ensemble algorithm (DSE-XGB) combining ANN and LSTM, outperforming individual models in terms of R², RMSE, and MAE. Mellit et al. [10] compared a number of deep learning models, such as CNN1D, GRU, BiGRU, LSTM, BiLSTM, and hybrid setups. The best model was found to be LSTM.

Comparative Analysis of Deep Learning Methods

To facilitate a comprehensive comparison, we have organized the performance metrics of different methodologies across studies. Table 60.1 presents a detailed comparative analysis while Figure 60.1 shows an RMSE comparison between the models, and Fig 60.2 shows a performance heatmap between various matrices. Figure 60.3 shows the MAPE comparison of the model performances. Figure 60.4 shows the R^2 score comparison.

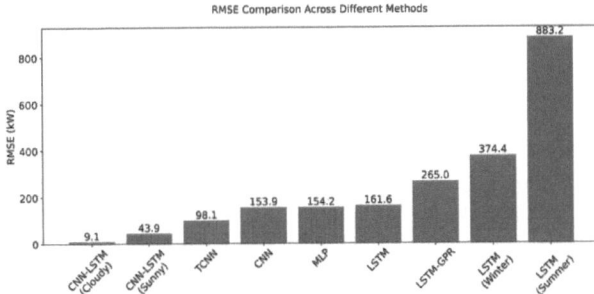

Figure 60.1 RMSE comparison across different methods
Source: Author

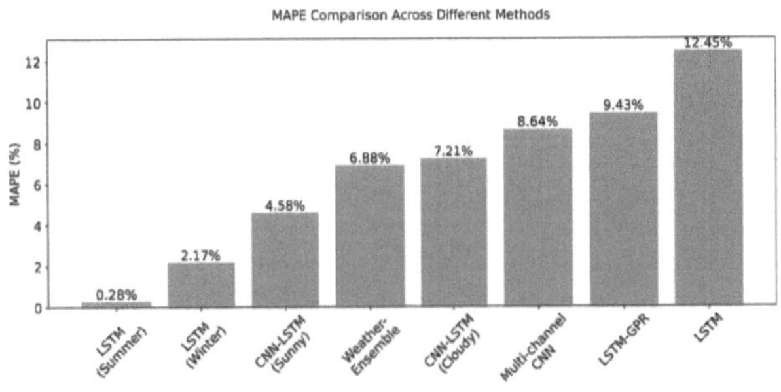

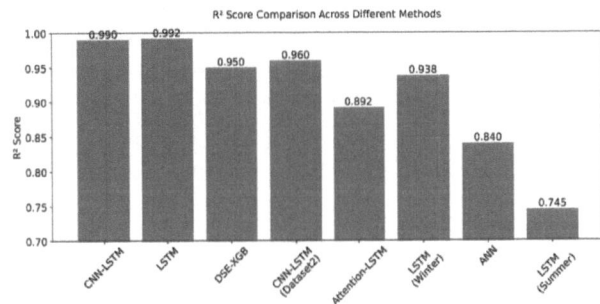

Figure 60.3 MAPE comparison across different methods
Source: Author

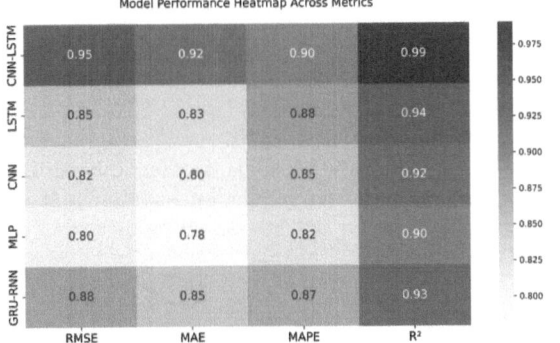

Figure 60.2 Model performance heatmap across metrics
Source: Author

Figure 60.4 Score comparison across different methods
Source: Author

Table 60.2 Statistical summary of performance metrics.

Metric	Best	Worst	Mean	Best Method
RMSE	9.09 kW	883.2 kW	189.47 kW	CNN-LSTM (Cloudy)
MAE	6.97 kW	676.34 kW	147.85 kW	CNN-LSTM (Cloudy)
MAPE	0.275%	16.187%	7.02%	LSTM (Summer)
R^2	0.99	0.745	0.93	CNN-LSTM

Source: Author

Key findings from comparative analysis

Hybrid CNN-LSTM models excelled in solar power forecasting (RMSE: 43.87 kW sunny, 9.09 kW cloudy), outperforming single LSTM models (RMSE: 0.16–883.2 kW). Ensemble methods like DSE-XGB achieved robust performance (R^2: 0.95). Accuracy improved with weather data (MAE: 83.90 kW) and during winter, with utility-scale forecasts maintaining consistent NRMSE (8-9%). Smaller installations showed higher precision. Hybrid models with weather integration and ensemble methods are optimal, but standardized metrics are needed for better comparisons.

Research Gaps

Deep learning for solar power forecasting has made great strides, but there are still a number of important research gaps. Methodologically, there is limited integration of machine learning with statistical methods, insufficient exploration of diverse ensemble approaches, and overreliance on CNNs versus other architectures. Technical limitations include inadequate cross-dataset validation, insufficient analysis of sequence length impacts, and high computational demands restricting real-time applications. Data and environmental challenges manifest through limited geographical and climatic diversity in datasets, insufficient historical data from utility-scale installations, and inconsistent performance across weather conditions. Implementation gaps center on a disproportionate focus on small-scale versus utility-scale systems, limited forecasting horizons, and insufficient optimization for real-world deployment scenarios.

Conclusion

This comprehensive review has examined the application of deep learning techniques in solar power forecasting through analysis of research from prominent journals. Our evaluation reveals that hybrid architectures, particularly CNN-LSTM combinations, demonstrate superior performance across diverse conditions, while RNN-based models excel at capturing temporal dependencies. The analysis highlights varying model performance under different weather conditions. While deep learning shows significant promise in this domain, several challenges persist, including the need for standardized evaluation metrics, improved robustness under varying weather conditions, and enhanced computational efficiency for real-world deployment. The field shows promising advancement with opportunities for further improvements through exploration of advanced ensemble methods and hybrid architectures, suggesting a bright future for deep learning applications in solar forecasting.

References

[1] Aslam, M., Lee, S. J., Khang, S. H., & Hong, S. (2021). Two-stage attention over LSTM with Bayesian optimization for day-ahead solar power forecasting. *IEEE Access*, 9, 107387–107398.

[2] Du Plessis, A. A., Strauss, J. M., & Rix, A. J. (2021). Short-term solar power forecasting: investigating the ability of deep learning models to capture low-level utility-scale Photovoltaic system behaviour. *Applied Energy*, 285, 116395.

[3] Elizabeth Michael, N., Mishra, M., Hasan, S., & Al-Durra, A. (2022). Short-term solar power predicting model based on multi-step CNN stacked LSTM technique. *Energies*, 15(6), 2150.

[4] Elsaraiti, M., & Merabet, A. (2022). Solar power forecasting using deep learning techniques. *IEEE Access*, 10, 31692–31698.

[5] Heo, J., Song, K., Han, S., & Lee, D. E. (2021). Multi-channel convolutional neural network for integration of meteorological and geographical features in solar power forecasting. *Applied Energy*, 295, 117083.

[6] Khan, W., Walker, S., & Zeiler, W. (2022). Improved solar photovoltaic energy generation forecast using deep learning-based ensemble stacking approach. *Energy*, 240, 122812.

[7] Koprinska, I., Wu, D., & Wang, Z. (2018). Convolutional neural networks for energy time series forecasting. In 2018 International Joint Conference on Neural Networks (IJCNN), (pp. 1–8). IEEE.

[8] Lim, S. C., Huh, J. H., Hong, S. H., Park, C. Y., & Kim, J. C. (2022). Solar power forecasting using CNN-LSTM hybrid model. *Energies*, 15(21), 8233.

[9] Lin, Y., Koprinska, I., & Rana, M. (2020). Temporal convolutional neural networks for solar power forecasting. In 2020 International Joint Conference on Neural Networks (IJCNN), (pp. 1–8). IEEE.

[10] Mellit, A., Pavan, A. M., & Lughi, V. (2021). Deep learning neural networks for short-term photovoltaic power forecasting. *Renewable Energy*, 172, 276–288.

[11] Iskandar, M. A., Abd Aziz, M. A. S., Sivaraju, S. S., Borhan, N., Wan, W. A. A. Q. I., & Ahmad, N. (2024). Long-term solar power generation forecasting in the eastern coast region of Malaysia using artificial neural network (ANN) method. *Journal of Advanced Research in Fluid Mechanics and Thermal Sciences*, 117(2), 60–70.

[12] Rana, M., Koprinska, I., & Agelidis, V. G. (2016). Solar power forecasting using weather type clustering and ensembles of neural networks. In 2016 International Joint Conference on Neural Networks (IJCNN), (pp. 4962–4969). IEEE.

[13] Thipwangmek, N., Suetrong, N., Taparugssanagorn, A., Tangparitkul, S., & Promsuk, N. (2024). Enhancing short-term solar photovoltaic power forecasting using a hybrid deep learning approach. *IEEE Access*, 12, 108928–108941.

[14] Wang, Y., Feng, B., Hua, Q. S., & Sun, L. (2021). Short-term solar power forecasting: a combined long short-term memory and gaussian process regression method. *Sustainability*, 13(7), 3665.

[15] Muhammad, A. I., Muhammad Azfar Shamil Abd Aziz, S. S. Sivaraju, Nurdiyana Borhan, Wan Abd Al-Qadr Imad Wan Mohtar, and Nurfadzilah Ahmad. (2024). Long-Term Solar Power Generation Forecasting in the Eastern Coast Region of Malaysia Using Artificial Neural Network (ANN) Method. *Journal of Advanced Research in Fluid Mechanics and Thermal Sciences*. 117 (2): 60–70.

61 A comparative analysis of breast cancer predictive intelligent model using machine learning techniques

Ria Satapathy[1,a], Bhupinder Singh[1,b], Sohan Kumar Pande[1,c] and Satyabrat Sahoo[2,d]

[1]Department of Computer Science and Engineering, Silicon Institute of Technology, Sambalpur, Odisha, India

[2]Department of Computer Science and Engineering, School of Engineering and Technology, DRIEMS University, Cuttack, Odisha, India

Abstract

Breast cancer remains widespread, making early detection vital for survival and treatment. Traditional methods are resource-heavy and lack predictive power. Using data from ACECR and BCRC, this study analyzed 4,004 patients, applying various machine learning techniques. Decision trees achieved 81.62% accuracy, with 89.49% sensitivity and 79.80% specificity, while neural networks matched this accuracy but had better sensitivity (90.80%) and specificity (89.99%). J48 achieved 99% accuracy, and random forest showed 95% sensitivity and 80% accuracy. These results highlight the potential of models like J48 and random forest for more accurate, early cancer detection.

Keywords: Breast cancer, early detection, machine learning (ML), predictive models, treatment efficacy

Introduction

Breast cancer, common in women and less frequently in men, is linked to BRCA1/BRCA2 mutations, genetics, hormonal imbalances, and lifestyle factors Rabiei et al., [10], Venkatesan and Velmurugan, [4]. Research focuses on early detection, treatment, and novel therapies, with the Motamed Cancer Institute using supervised learning, blockchain, NF-net, and CAD systems to improve diagnosis Rabiei et al. [10], Fatima et al. [2].

Machine learning models like U-Net and YOLO enhance image segmentation, while Mask R-CNN ROIAlign and Dense Net CNNs improve mammography analysis [8]. Neural Networks outperformed Decision Trees in predicting breast cancer risk with higher sensitivity (90.8% vs. 89.49%), specificity (89.0% vs. 79.99%), and accuracy (81.62% vs. 80.01%) [3, 8].

Related Work

[1] Machine learning improves breast cancer diagnosis with models like neural networks and decision trees, with CNNs excelling in image detection and combining structured data for better accuracy [8]. Genetic Algorithms optimize Random Forests and SVMs, while CAD systems enhance mammogram analysis [5]. Dataset variability affects performance [9].

[11] Techniques like NF-Net and feature optimization improve precision. SVMs excel in high-dimensional classification, while transfer learning and the J48 algorithm (99% accuracy) show progress [7]. Machine learning outperforms traditional models in distinguishing benign from malignant tumors [6].

Intelligent Prediction Models

Problem statement

Breast cancer is a common, life-threatening disease, and early diagnosis is challenging, especially in women with dense breast tissue. Automated diagnostic systems are essential for accurately analyzing mammograms and MRIs, enabling timely diagnoses to improve outcomes and reduce stress.

Breast cancer diagnosis uses data from sources like Wisconsin Breast Cancer Dataset and MIAS, with noise reduction and dimensionality reduction. Machine learning models (CNNs, SVMs, decision trees) are optimized and evaluated using accuracy, precision, recall, and F1-score.

Feature engineering

Feature engineering is like picking the key puzzle pieces, making the solution faster and more accurate.

- **Multimodal data integration:** Combining clinical, genetic, and imaging data, like mammograms and MRIs, enhances breast cancer risk predictions by providing a complete picture.

[a]satapathyria@gmail.com, [b]bhupindersingh30082001@gmail.com, [c]sohan.pande@silicon.ac.in, [d]satyabratsahoo@driems.ac.in

DOI: 10.1201/9781003658221-61

- **Genetic algorithms:** Genetic algorithms (GAs) optimize models by adjusting parameters, like tuning a radio, to ensure the best performance.
- **Breast density:** Including breast density data in models improves accuracy by adding detail, enhancing early detection of cancer.
- **Getting the most from different image types:** Combining data from different imaging techniques, like mammograms and ultrasounds, improves cancer detection by providing a clearer, multi-angle view.
- **Genetic data:** Researchers are using genetic data to predict cancer risk by extracting features, like mutations linked to higher risk, improving the accuracy of predictions.
- **Combining U-Net and YOLO techniques:**
  (1) **U-Net:** U-Net's encoder-decoder architecture extracts detailed features from images, making it ideal for segmentation tasks in medical images, such as identifying tumors in mammograms.
  (2) **YOLO (You Only Look Once):** YOLO is a real-time object detection algorithm that classifies and localizes objects in an image using a single network, identifying their location with bounding boxes and labels.

Classification algorithms

Traditional machine learning models may struggle to capture complex linguistic nuances and context, especially in social health media data for depression.

- Random forest algorithms: Random forest builds multiple decision trees on random data subsets, improving accuracy and handling noisy or missing data, making it ideal for breast cancer detection.
- Support vector machines (SVM): SVMs are classification algorithms that find the optimal boundary (hyperplane) to separate data into categories, like cancerous and non-cancerous.
- Convolutional neural networks (CNNs): CNNs are deep learning models that analyze images, like mammograms, to detect early breast cancer signs.

They're also being combined with patient data for personalized predictions.

- Artificial neural networks (ANNs): Artificial neural networks (ANNs) mimic the brain to recognize patterns, making them effective for predicting breast cancer by analyzing imaging and patient data.
- Decision trees: Decision trees classify data by splitting it into branches for predictions. Though simple, they can overfit, but pruning and ensemble methods like Random Forests improve performance.
- Early computer-aided detection (CAD) systems: In the early 2010s, CAD systems using machine learning helped radiologists by highlighting suspicious areas in mammograms, aiding in cancer detection.
- Transfer learning: Transfer learning adapts models trained on large datasets to specific tasks like breast cancer diagnosis, leveraging prior knowledge for efficient and accurate predictions.

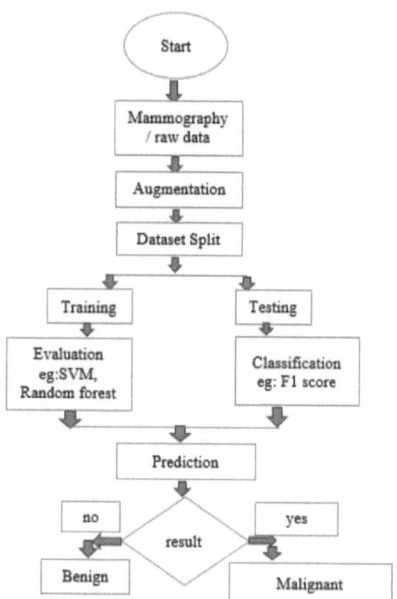

Figure 61.1 Flowchart of breast cancer detection
Source: Author

Table 61.1 Comparative analysis.

Algorithm	Accuracy	Precision	Sensitivity	Specificity
BPBPW with HKF-ABO	99.6%	99.9%	98.68%	95.78%
Yolo + U-Net approach	93.0%	93.9%	94.7%	98.6%
Random forest	86.80%	78.0%	77.7%	-
Multi-layer perception	73%	-	82%	84%

Source: Author

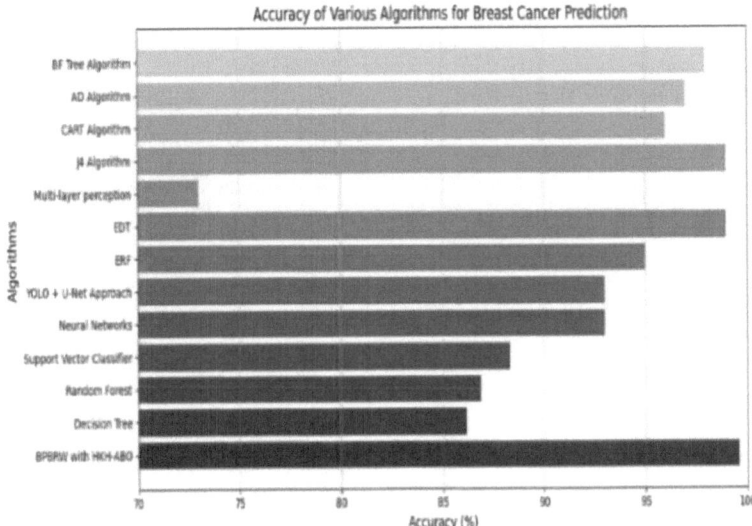

Figure 61.2 Visualization of accuracy
Source: Author

- **Noise filter network (NF-Net):** The noise filter network (NF-Net) reduces data noise to improve prediction accuracy, much like tuning out background chatter to focus on essential information in complex datasets.

Analytical Simulation of ML models

Figure 61.1 shows a machine learning pipeline for breast cancer detection using mammography, with models like SVM and Random Forest. Table 61.1 shows BPBRW with HKH-ABO achieving 99.6% accuracy and 99.9% precision. Neural networks and YOLO + U-Net perform at 93%. Figure 61.2 compares accuracy, with BPBRW leading, followed by BF Tree (98%), AD algorithm (97%), and multi-layer perceptron (73%).

Conclusion

Future advancements in breast cancer prediction include expanding datasets, explainable AI, noise reduction, and multi-modal learning. However, challenges like low accuracy, false negatives in mammography, MRI inefficiencies, and suboptimal optimization still affect diagnostic precision.

References

[1] Rasool, A., Bunterngchit, C., Tiejin, L., & Islam, M. R. (2022). Improved machine learning-based predictive models for breast cancer diagnosis. *International Journal of Environmental Research and Public Health*, 19(6), 3211. https://doi.org/10.3390/ijerph19063211.

[2] Fatima, A., Shabbir, A., Janjua, J. I., Ramay, S. A., Bhatty, R. A., Irfan, M., et al. (2024). Analyzing breast cancer detection using machine learning & deep learning techniques. *Journal of Computing and Biomedical Informatics*, 7(2), 1–11. ISSN: 2710–1606, Research Article https://doi.org/10.56979/702/2024.

[3] Atashi, A., Sohrabi, S., & Dadashi, A. (2018). Applying two computational classification methods to predict the risk of breast cancer: a comparative study. *Multidisciplinary Cancer Investigation*, 2(2), 8–13. DOI: 10.30699/acadpub.mci.2.2.8.

[4] Venkatesan, E., & Velmurugan, T. (2015). Performance analysis of decision tree algorithms for breast cancer classification. *Indian Journal of Science and Technology*, 8(29), 1–18. IPL0625.

[5] Harsha Latha, P., Ravi, S., & Saranya, A. (2024). Breast cancer detection using machine learning in medical imaging – a survey. *Procedia Computer Science*, 239, 2235–2242.

[6] Dewangan, K. K., Janghel, R., Dewangan, D. K., & Sahu, S. P. (2022). Breast cancer diagnosis in an early stage using novel deep learning with hybrid optimization technique. *Multimedia Tools and Applications*, 81, 13935–13960. https://doi.org/10.1007/s11042-022-12385-2.

[7] Al-Imran, M., Akter, S., Mozumder, M. A. S., Bhuiyan, R. J., Rahman, T., Ahmmed, M. J., et al. (2024). Evaluating machine learning algorithms for breast cancer detection: a study on accuracy and predictive performance. *The USA Journals the American Journal of Engineering and Technology*, 6(9), 22. (ISSN – 2689-0984). https://www.theamericanjournals.com/index.php/tajet PUBLISHED DOI: - https://doi.org/10.37547/tajet/Volume06Issue09-04.

[8] Rahman, M. M., Jahangir, M. Z. B., Gupta, K. D., George, R., Rahman, A., Akter, M., et al. (2024). Breast cancer detection and localizing the mass area using

deep learning. *Big Data and Cognitive Computing*, 8, 80. https://doi.org/10.3390/bdcc8070080.

[9] Prerita, Sindhwani, N., Rana, A., & Chaudhary, A. (2024). Breast cancer detection using machine learning algorithms. In 2021 9th International Conference on Reliability, Infocom Technologies and Optimization (Trends and Future Directions) (ICRITO). IEEE. 10.1109/ICRITO51393.2021.9596295.

[10] Rabiei, R., Ayyoubzadeh, S. M., Sohrabei, S., Esmaeili, M., & Atashi, A. (2022). Prediction of breast cancer using machine learning approaches. *Journal of Biomedical Physics and Engineering*, 12(3), 297–308. doi: 10.31661/jbpe.v0i0.2109-1403.

[11] Priyadarshni, V., Sharma, S. K., Rahmani, M. K. I., Kaushik, B., & Almajalid, R. (2024). Machine learning techniques using deep instinctive encoder-based feature extraction for optimized breast cancer detection. *Computers, Materials and Continua*, 78(2), 2441–2468.

62 An efficient CNN based hybrid model for cataract classification on ODIR dataset

Rahul Ray[1,a], Sudarson Jena[1,b], Laxminarayan Dash[2,c] and Sangita Kumari Biswal[3,d]

[1]Department of CSE, SUIIT, Sambalpur University, Burla, Odisha, India

[2]Department of CSE, Odisha University of Technology and Research, Bhubaneswar, Odisha, India

[3]Department of CSE, GITA Autonomous College, Bhubaneswar, Odisha, India

Abstract

Cataract is one of the dangerous eye diseases among more than 40 year age Indian populations which can cause blindness if not treated in its early stage. It is a very tedious and time-consuming process to visit an ophthalmologist in a regular interval for a routine checkup. To build a cost effective and efficient cataract detection and classification method, a hybrid model is proposed in this paper which is constructed using ResNet50 and VGG16 and classification operation is being performed using the Softmax classifier with support of ReLU activation function. This proposed method used a sample of a total of 400 retinal color images which contained 200 normal eye images and 200 cataract eye images. The proposed methodology uses ODIR dataset and achieves to create the classification model of two classes: normal eye class and cataract eye class with the accuracy of 95% for 100 epochs values and 85% for 50 epochs values. This model can be used to build an improved and cost effective cataract detection and classification method with 100 epochs of better accuracy, sensitivity, specificity, and precision of 95.01%, 95.2%, 94.7% and 94.22% respectively.

Keywords: Epoch, F1-score, ODIR, ResNet50, VGG16

Introduction

Cataract is one of the common ocular diseases seen in middle aged Indian population. The main factor of cataract among > 40 years of Indian populations because of biomass fuel usage, excess sunlight exposure and poor diets and in few cases, genetic factor is also another cause among the patients of mid aged Indian. Studies proved that the blindness among cataract affected patients will reach 40 million by 2025 worldwide. In this disease, the lens of the patient's eye is cloudy and fogged up and this happened due to the reason of extra breaking of normal proteins in the lens region of the eye. As a result, the patient cannot able to see properly at night or in a dark environment, is unable to drive vehicles, and has difficulty reading. We are proposing an automated disease diagnosis classification system to detect cataract from the retinal eye images which remained unexplored yet in an efficient manner.

Literature Review

Automatic detection and classification of cataract disease from retinal eye image dataset is very popular in medical imaging domain for disease detection without any help of an ophthalmologist. Different researchers have proposed various methodologies in their research articles in which highlighted processes typically involve mainly three different stages: Pre-processing of input data, feature extraction, and classification of datasets based on extracted features.

There is a deep learning and convolutional neural network (CNN) [1] used for pattern recognition which helps to automate image classification for cataract identification over the traditional methods based on feature extraction involves trained ophthalmologist involvement. In this approach, manipulation process being done by changing epochs values, as a result increase in accuracy value and minimization of data loss happened for cataract identification. A novel approach has been proposed [2] to segment retinal blood vessels which involve pre-processing stage to enhance the quality of image using CLAHE and 2D Gabor wavelet. In segmentation stage, it involves geodesic operators and final resultant segmented image pass through the process of hole filling and removing isolated pixels in the post-processing stage. In the research article [3], the study employed the deep learning image classifier VGG-19 to classify the ODIR-5K dataset which contains 5000 fundus images of eight disease classes like glaucoma, diabetic retinopathy, cataract etc. The study proposed in [4] Utomo et al., 2021) [4] involved convolutional neural network and

[a]rahulray9439@gmail.com, [b]sjena@suiit.ac.in, [c]laxminarayandash20@gmail.com, [d]sangita.biswal95@gmail.com

DOI: 10.1201/9781003658221-62

transfer learning to perform classification between a normal eye and one disease affected eye and the transfer learning if more suitable for multi-class classification with higher accuracy.

Proposed Methodology and Workflow

In this section, the proposed methodology, algorithms and neural network models are employed to establish effectiveness, efficiency and accurate automatic cataract classification system.

The proposed work is being depicted in Figure 62.1 which shows the complete vision of an automatic cataract detection and classification system. This section contains information about the input dataset acquisition, pre-processing of acquired dataset, the CNN based hybrid model for cataract classification and the final output stage for disease classification.

Dataset used

To build automated cataract detection and classification model, images of retinal eye must be available to do the training of different deep learning models. The dataset that has been used in our proposed work is Ocular Disease Intelligent Recognition (ODIR) retinal fundus image collected from Kaggle which is freely available. The ODIR is a database which contains the color fundus retinal eye photographs of 5000 patients from their left and right eyes with their age parameters. This particular dataset is further classified into eight different classes. In our proposed work, two classes named normal (N) and cataract (C) are used for extensive experiments, training and testing our model.

Data pre-processing method

Data pre-processing stage is one of the most crucial phases of any classification operations of images using neural network models. First, normal and cataract classes of ODIR dataset are being considered for the experiment and a single dataset file is being created

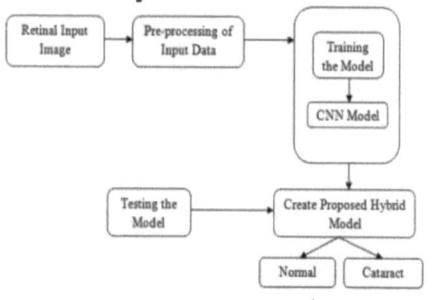

Figure 62.1 General overview of the proposed methodology

Source: Author

by combining 200 number of retinal eye images from each of them. The input image contrasts are enhanced by using contrast limited histogram equalization (CLAHE) which enhances the image quality and discloses the important minor features for an accurate diagnosis system. To detect cataract, detecting eye lens is one of the important parameters and this can be achieved by segmentation process and after that segmented image will be converted to grayscale image format. Now the input image is being resized to height of 224 pixels and width of 224 pixels. Edge enhancement and border detection process carried out and at a later stage, noise reduction filtering methods are being implemented. As a result, the identification of important anatomical features and anomalies can be segregated by the enhanced quality of input retinal image which helps us in building an efficient automated cataract classification model with better accuracy.

Cataract detection and classification using CNN based hybrid model

After getting the results from pre-processing techniques where different morphological operations takes placed of the retinal input fundus image dataset of each image size 224 × 224 × 3, the pre-processed data will go for training process and the training and the testing of the model is being carried out by using a hybrid model made of combined approach of CNN and VGG16.

One of the most popular and commonly used CNN models is ResNet-50 which is used in the proposed model. It is the most powerful model for retinal image dataset classification which can be trained on large size of image dataset. Its architecture is divided into four different major parts: (a)the convolutional layers, (b)the identity block, (c)the convolutional block, and (d)the fully connected layers. The major role of convolutional layers followed by batch normalization and ReLU activation function are to extract the features like edges, shapes, borders and textures etc. from the retinal input image and followed by max pooling layers to reduce the spatial dimensions of the extracted feature maps by keeping important features of input image. The extracted features are processed and transformed by the identity and convolutional block and at last, the fully connected layers are given to a SoftMax activation function to produce the final classifications.

The development of a hybrid model shown in Figure 62.2 that integrates the VGG16 and ResNet50 architectures can effectively harness the unique advantages of both frameworks, thereby enhancing performance in various applications such as image classification and object detection.

To construct this hybrid model, we utilized both the networks as feature extractors. This helps us to achieve either by processing the input image through each network independently and subsequently merging the resulting feature maps or by employing one network to extract features and then passing those features to the other network for further processing. In terms of model composition, there are several strategies to combine the features obtained from VGG16 and ResNet50. Our approach is to concatenate the outputs along the channel dimension, which results in a more comprehensive representation of the input data. Alternatively, one might consider applying pooling techniques, such as average or max pooling, to diminish the dimensionality of the features prior to their integration. This step is crucial as it can help streamline the data while preserving essential information, ultimately leading to improved model performance.

Following the integration of features, it is essential to incorporate one or more fully connected layers to facilitate the processing of the combined feature set. These layers are instrumental in enabling the model to learn and make predictions based on the enriched feature representation. Finally, the architecture should end up in an output layer equipped with a SoftMax activation function for multi-class classification tasks.

Results and Discussion

This section represents the experimental results for detection and classification of cataract from the ODIR input retinal image dataset using pre-processing and pre-trained hybrid model made of ResNet50

and VGG16 with two different Epochs values of 50 and 100. The result section represents different performance metrics such as accuracy, sensitivity, value loss, precision, recall, and f1 Score. The classification results and confusion matrixes for both the epoch values of 50 and 100 are represented.

In the proposed hybrid model, we trained the model with a total of 320 input image datasets of size 224 × 224 × 3. The training operations are being successfully carried out by the combined efficiency of proposed model made of ResNet50 and VGG16 models. The validation and testing dataset size is randomly chosen whose number is 40 of each image size of 224 × 224 × 3. The resultant classifications are being carried out by using ReLu activation function and SoftMax classifier.

Table 62.1 Performance measures for proposed method with epochs 50 and epochs 100.

Performance measures	Class name			
	Normal	Cataract	Normal	Cataract
Accuracy	85%	85%	95.01%	95.01%
Sensitivity	85.71%	85.71%	95.2%	95.2%
Specificity	84.21%	84.21%	94.7%	94.7%
Value loss	15%	15%	4.99%	4.99%
Precision	86.8%	84.7%	96.24%	96.24%
Recall	94.73%	94.73%	95.23%	95.23%
F1-score	85.71%	85.71%	95.24%	95.24%

Source: Author

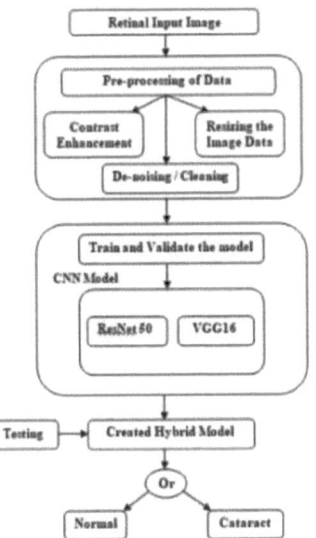

Figure 62.2 Proposed methodology for detection and classification of cataract

Source: Author

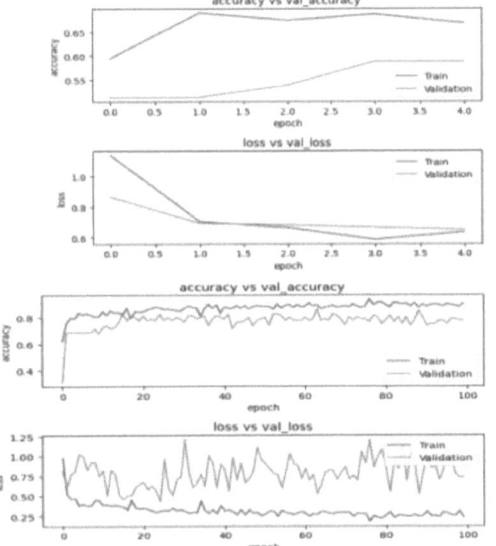

Figure 62.3 The curve of training, value accuracy and value loss for proposed method with epochs 50 and epochs 100

Source: Author

Our proposed method gives accuracy values of 88% with epochs 50 to classify the cataract and normal retinal eye image whereas it gives accuracy values of 90% with epochs 100 to classify the cataract and normal retinal eye image.

The performance measure of our proposed method is represented Table 62.1. The curve for result comparison of value accuracy and value loss of our proposed methodology are represented in Figure 62.3 with epoch 50 and epoch 100 respectively.

Conclusion

The advancement in deep learning based methodologies help in detecting and classifying cataract for an easy and convenient diagnosis process. In the proposed study, the detection and classification performed on ODIR retinal dataset by using the combined operations of ResNet50 and VGG16 CNN models and the classifications are performed efficiently by using SoftMax classifier to create two classes normal eye class and cataract eye class. The proposed method gives 95% of accuracy to detect cataract disease with epochs value 100. In future studies, we will work on the innovative approaches to classify cataract diseases based on their severity levels and will work on real time retinal image datasets to detect the cataract regions in a patient with higher accuracy and efficiency.

References

[1] Babaqi, T., Jaradat, M., Yildirim, A. E., Al-Nimer, S. H., & Won, D. (2023). Eye disease classification using deep learning techniques. *arXiv preprint arXiv*:2307.10501.

[2] Dash, J., & Bhoi, N. (2017). Detection of retinal blood vessels from ophthalmoscope images using morphological approach. *ELCVIA: Electronic Letters on Computer Vision and Image Analysis*, 16(1), 1–14.

[3] Mahmood, S. S., Chaabouni, S., & Fakhfakh, A. (2023). A new technique for cataract eye disease diagnosis in deep learning. Periodicals of Engineering and Natural Sciences, 11(6), 14–26.

[4] Weni, I., Utomo, P. E. P., Hutabarat, B. F., & Alfalah, M. (2021). Detection of cataract based on image features using convolutional neural networks. *Indonesian Journal of Computing and Cybernetics Systems*, 15(1), 75–86.

63 An assessment of vibration in non-prismatic Timoshenko beams with multiple transverse splits

Abinash Bibek Dash[a], Alok Ranjan Biswal[b], Deepak Ranjan Biswal[c], Rasmi Ranjan Senapati[d] and Satyakam Acharya[e]

Department of Mechanical Engineering, DRIEMS University, Tangi, Cuttack, Odisha, India

Abstract

This review article presents a detailed synthesis of research on the vibration analysis of non-prismatic Timoshenko beams with multiple transverse fractures. It integrates findings from analytical, numerical, and experimental studies, examining essential aspects such as crack modeling, vibration-based damage detection, the influence of crack parameters on beam dynamics, and the effects of crack interactions, beam geometry, and material properties. The article also highlights knowledge gaps and proposes future research directions, including the development of nonlinear crack models, refinement of numerical methods, and further experimental validation. By consolidating existing research and identifying areas for advancement, this review aims to support and shape future studies in this field.

Keywords: Geometric modeling, non-prismatic beam, Timoshenko beams, transverse cracks

Introduction

Non-prismatic Timoshenko beams, characterized by varying cross-sectional geometry along their length, exhibit complex structural behavior due to non-uniform properties such as moment of inertia and area [3, 6]. The Timoshenko theory accurately models real-world structures by accounting for shear deformation and rotary inertia. These beams, including tapered, stepped, and curved types, are crucial in aerospace, civil, mechanical, and biomechanical engineering applications. Researchers focus on developing advanced numerical methods, analytical solutions, and experimental investigations to tackle mathematical complexity and simulation challenges, ultimately enhancing structural integrity, vibration control, safety, and material distribution optimization. Multiple transverse cracks in structural components, such as beams and bridges, significantly compromise their integrity and performance [5, 13]. These cracks, oriented perpendicular to the longitudinal axis, can arise from various factors including fatigue, corrosion, and external loading. The presence of multiple transverse cracks significantly compromises structural integrity, exacerbating stress concentrations to accelerate crack propagation, weakening rigidity to induce excessive vibrations, increasing the risk of catastrophic failure, and compromising load-carrying capacity [2]. Furthermore, these cracks necessitate frequent inspections and maintenance to mitigate potential disasters, underscore the need for advanced analysis techniques and monitoring strategies to ensure structural reliability and safety. The integrity of mechanical and civil engineering systems relies heavily on vibration analysis, particularly for non- prismatic Timoshenko beams used in high-stress industries. Cracks from fatigue, corrosion, or accidental damage significantly alter beam dynamics, affecting frequency, shape, and damping, and require precise evaluation [7]. Accurate vibration study of non-prismatic Timoshenko beams possessing several transverse splits is essential for structural health monitoring (SHM), damage detection and localization, condition-based maintenance and performance optimization etc. However, the complex interactions between crack parameters, beam geometry, and material properties pose significant challenges. Existing research has addressed various aspects of this problem, including analytical and numerical modeling, experimental validation, and vibration-based damage detection. This observation purposes to deliver a widespread synopsis of the recent state-of-the-art in the domains like (i) Crack interactions significantly affect beam dynamics, leading to changes in natural frequencies and mode shapes, (ii) beam geometry and material properties influence crack behavior and vibration response, (iii) nonlinear crack models outperform linear models in predicting vibration behavior, (iv) multiple cracks can lead to complex mode shapes and frequency veering phenomena, (v) vibration- based damage detection methods can accurately identify crack locations and depths.

[a]abinashdash@driems.ac.in, [b]dr.alokranjanbiswal@driems.ac.in, [c]deepak.biswal@driems.ac.in, [d]rashmisenapati@driems.ac.in, [e]free2mailsatya@gmail.com

DOI: 10.1201/9781003658221-63

Methodology and Research Gap

Research on vibration analysis of non-prismatic Timoshenko beams holds several transverse splits reveals significant gaps. Firstly, there is a lack of closed-form solutions, with current solutions relying on numerical methods or approximate analytical techniques [4, 12]. Developing exact, closed-form solutions is essential to accurately predict vibration behavior. Understanding crack interaction effects on vibration behavior remains limited, with factors like crack orientation, depth, and location needing study. Challenges include modeling complexity, limited experimental validation, and oversimplified models, highlighting the need for advanced nonlinear, 3D, and dynamic approaches [14]. Additional research gaps include inadequate consideration of boundary conditions, lack of experimental validation, insufficient investigation of nonlinear dynamics, and limited development of advanced numerical methods. Addressing these gaps will significantly advance vibration study of non-prismatic Timoshenko beams with several transverse ruptures. The interdisciplinary research opportunities is presented in the Figure 63.1

The challenges and opportunities are presented in detail in Figure 63.2 and 63.3.

The future direction of crack modelling and the allied aspects are presented in the Table 63.1.

The emerging trends in this regard are presented in Table 63.2.

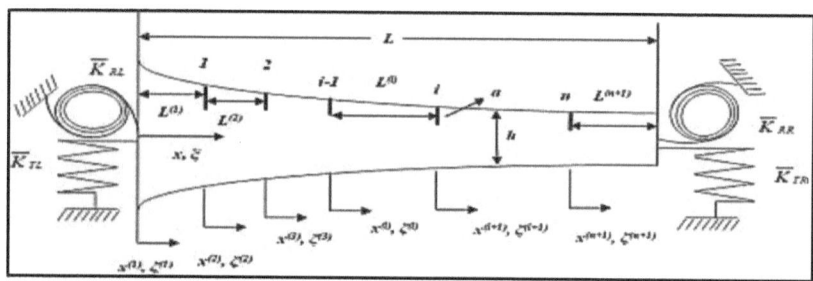

Figure 63.1 Non-uniform Simply supported beam with multiple open splits
Source: Author

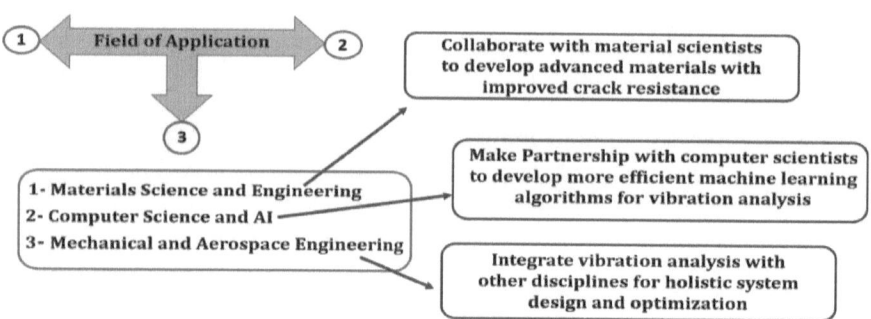

Figure 63.2 Interdisciplinary research opportunities
Source: Author

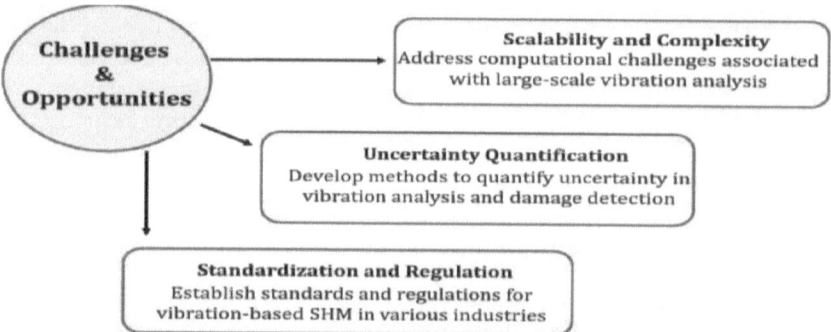

Figure 63.3 Challenges and opportunities
Source: Author

Table 63.1 Future direction.

Sl. No.	Direction	Description
1	Advanced crack modeling	Develop more accurate crack models incorporating nonlinear material behavior, 3D crack propagation, and dynamic effects.
2	Nonlinear dynamics and chaos	Investigate complex phenomena such as bifurcations, instability, and chaos in cracked non-prismatic beams.
3	Machine learning and AI applications	Integrate machine learning algorithms for vibration-based damage detection, crack localization, and predictive maintenance.
4	Multi-scale and multi-physics analysis	Develop coupled models incorporating structural, thermal, and fluid dynamics effects.
5	Experimental validation and testing	Conduct comprehensive experimental studies to validate numerical models and investigate novel testing techniques.

Source: Author

Table 63.2 Emerging trends.

Sl. No.	Trend	Description
1	Smart materials and structures	Integrate piezoelectric, shape memory alloy, or other smart materials for self-sensing and adaptive vibration control.
2	Internet of Things (IoT) and cyber-physical systems	Develop networked vibration monitoring systems for real-time SHM and condition-based maintenance.
3	Additive manufacturing and 3D printing	Investigate vibration behavior of 3D-printed non-prismatic beams with complex geometries.
4	Nano- and micro-scale vibration analysis	Explore vibration behavior of Nano- and micro-scale beams with applications in MEMS and NEMS.
5	Biologically inspired designs	Develop biomimetic beam designs incorporating nature-inspired crack-resistant mechanisms.

Source: Author

Geometric Modelling

Geometric modeling in vibration analysis involves creating precise representations of structures to study their dynamic behavior. It enables accurate simulation of natural frequencies, mode shapes, and response to external forces, enhancing design optimization and ensuring structural stability under dynamic conditions. The detailed conditions of the geometric models of non-prismatic beams are presented in Table 63.3. This module involves developing accurate geometric models of non- prismatic beams and characterizing material properties, including nonlinear behavior and damage evolution [11].

Where parameters being represented as:
A(x): Cross-sectional area at distance x
A0: Initial cross-sectional area
A: Taper ratio or shape parameter
X: Distance along beam axis
L: Beam length
ΔA: Change in cross-sectional area
θ(x − x0): Heaviside step function

e: Exponential function
π: Mathematical constant (pi)

Crack Modeling and Simulation

This module emphasizes advanced crack models integrating nonlinear mechanics, 3D propagation, and dynamic effects, essential for accurately predicting crack behavior in complex structures. [1, 10] advanced models overcome linear elastic fracture mechanics (LEFM) limitations by integrating nonlinear behaviors, 3D crack propagation, and dynamic effects. They analyze plasticity, creep, mixed-mode fractures, out-of- plane growth, crack interactions, high-velocity propagation, and dynamic arrest, offering comprehensive fracture insights. Advanced crack models have significantly enhanced the accuracy of crack propagation simulations [9]. Notably, five prominent models have emerged and shown underneath: (i) The extended finite element method (XFEM) seamlessly integrates crack discontinuities within elements for precise simulations, (ii) Phase field modeling leverages

Table 63.3 Geometric models of non-prismatic beams.

Model	Description	Mathematical representation	Advantages	Limitations
Tapered beam	Linearly varying cross-sectional area	$A(x) = A_0(1 - \alpha x)$	Simple, accurate for small taper	Inaccurate for large taper
Step beam	Abrupt changes in cross-sectional area	$A(x) = A_0 + \Delta A\, \theta(x - x_0)$	Accurate for sudden changes	Discontinuous, difficult to analyze
Exponential beam	Exponentially varying cross-sectional area	$A(x) = A_0 e^{(-\alpha x)}$	Accurate for gradual changes	Difficult to solve analytically
Parabolic beam	Parabolically varying cross-sectional area	$A(x) = A_0(1 - \alpha x^2)$	Accurate for curved beams	Difficult to solve analytically
Hyperbolic beam	Hyperbolically varying cross-sectional area	$A(x) = A_0(1 - \alpha x^{-1})$	Accurate for rapidly changing beams	Difficult to solve analytically
Polynomial beam	Polynomially varying cross-sectional area	$A(x) = A_0(1 - \alpha x^n)$	Flexible, accurate for various shapes	Difficult to solve analytically
Sinusoidal beam	Sinusoidally varying cross-sectional area	$A(x) = A_0(1 - \alpha \sin(\pi x/L))$	Accurate for wavy beams	Difficult to solve analytically

Source: Author

continuous damage variables to effectively simulate crack propagation, (iii) Peridynamics pioneered non-local interactions to meticulously model crack growth, (iv) Cohesive zone modeling employs nonlinear traction-separation laws to capture nuanced crack behavior, (v) Damage mechanics comprehensively simulates material degradation and crack initiation. Crack modeling advancements transform industries by enhancing structural safety and enabling stronger materials. They offer improved accuracy, predictive capabilities, optimized designs, reduced conservatism, and increased safety, driving innovation in engineering and scientific applications [15]. These advancements enable informed decision-making, efficient design and reduced risk.

Vibration Analysis and Dynamics

Vibration analysis and dynamics involves applying analytical and numerical approaches (e.g., finite element, boundary element) to simulate vibration response, incorporating effects of crack interactions, beam geometry, and material nonlinearity.one of the important aspects in this regard is damage detection and structural health monitoring (SHM)This module integrates vibration analysis with machine learning, signal processing, and SHM techniques to detect and localize cracks, predict remaining lifespan, and optimize maintenance [8]. By integrating these modules, researchers can: Develop accurate predictive models for cracked non-prismatic beams, investigate complex phenomena (e.g., nonlinear dynamics, chaos), design optimized beam geometries and materials and can implement effective SHM strategies. This unified framework facilitates a holistic understanding of cracked non-prismatic beam dynamics, enabling advancements in vibration-based SHM, condition-based maintenance, and structural integrity. By consolidating existing knowledge and identifying research needs, this review will facilitate advancements in vibration-based SHM and condition-based maintenance of non-prismatic Timoshenko beams.

Conclusion

Research on vibration analysis of non-prismatic Timoshenko beams with multiple transverse cracks explores advanced models, nonlinear dynamics, and machine learning to improve understanding of cracked beam behavior. Key considerations include crack interactions, geometry, and material nonlinearity, with experimental validation crucial for model reliability. These advancements enhance structural integrity, enable condition-based maintenance, reduce costs, extend component lifespan, and ensure safety. Despite progress, challenges remain, necessitating further research to refine structural health monitoring and vibration-based damage detection techniques. This study aims to inspire future work in this evolving field.

References

[1] Aabid, A., Parveez, B., Raheman, M. A., Ibrahim, Y. E., Anjum, A., Hrairi, M., & Mohammed Zayan, J. (2021, May). A review of piezoelectric material-based structural control and health monitoring techniques for en-

gineering structures: Challenges and opportunities. *In Actuators,* 10(5), 101.

[2] Abdullah, A. (2024). Review of vibration-based damage assessment in wind turbines. Diss. Politecnico di Torino.

[3] Balduzzi, G., Aminbaghai, M., Sacco, E., Füssl, J., Eberhardsteiner, J., & Auricchio, F. (2016). Non-prismatic beams: a simple and effective Timoshenko-like model. *International Journal of Solids and Structures,* 90, 236–250.

[4] Ebrahimi, A., Heydari, M., & Behzad, M. (2017). Forced vibration analysis of rotors with an open edge crack based on a continuous vibration theory. *Archive of Applied Mechanics,* 87, 1871–1889.

[5] French, C., Eppers, L., Le, Q., & Hajjar, J. F. (1999). Transverse cracking in concrete bridge decks. *Transportation Research Record,* 1688(1), 21–29.

[6] Ke, W. U., Qibo, P. E. N. G., Xinfeng, W. U., Pengbo, L. I. U., & Yajun, K. O. U. (2024). Analytical method for vibration analysis of multi-cracked Timoshenko beam structures with elastic foundations. *Acta Aeronautica et Astronautica Sinica,* 45(22).

[7] Kim, H., & Melhem, H. (2004). Damage detection of structures by wavelet analysis. *Engineering Structures,* 26(3), 347–362.

[8] Manoach, E., Warminski, J., Kloda, L., Teter, A. (2017). Numerical and experimental studies on vibration based methods for detection of damage in composite beams. *Composite Structures,* 170, 26.

[9] Muttillo, M., Stornelli, V., Alaggio, R., Paolucci, R., Di Battista, L., de Rubeis, T., & Ferri, G. (2020). Structural health monitoring: An IoT sensor system for structural damage indicator evaluation. *Sensors,* 20(17), 4908.

[10] Navadeh, N., Sareh, P., Basovsky, V., Gorban, I., & Fallah, A. S. (2021). Nonlinear vibrations in homogeneous nonprismatic timoshenko cantilevers. *Journal of Computational and Nonlinear Dynamics,* 16(10), 101002.

[11] Rezaiee-Pajand, M., & Gharaei-Moghaddam, N. (2019). Vibration and static analysis of cracked and non-cracked non-prismatic frames by force formulation. *Engineering Structures,* 185, 106–121.

[12] Saavedra, P. N., & Cuitino, L. A. (2001). Crack detection and vibration behavior of cracked beams. *Computers & Structures,* 79(16), 1451–1459.

[13] Sekhar, A. S. (2008). Multiple cracks effects and identification. *Mechanical Systems and Signal Processing,* 22(4), 845–878.

[14] Singh, R., & Sharma, P. (2022). Vibration analysis of an axially functionally graded material non-prismatic beam under axial thermal variation in humid environment. *Journal of Vibration and Control,* 28(23–24), 3608–3621.

[15] Zhang, Y., Ruiz, C., Rohal, S., Pan, S. (2023). Cyberphysical augmentation for vibration sensing in autonomous retails. In Proceedings of the 24th International Workshop on Mobile Computing Systems and Applications (8–14).

64 Predictive analytics in healthcare systems

Subasish Mohapatra[1,a], Subhadarshini Mohanty[1,b], Jyoti Ranjan Nayak[1,c] and Sanjeeb Kumar Nayak[2,d]

[1]School of Computer Sciences, OUTR, Bhubaneswar, Odisha, India

[2]Department of IT, VSSUT, Burla, Odisha, India

Abstract

A robust healthcare system is crucial for a thriving society. Using computers in healthcare has transformed administrative procedures, medical research, and patient care. Artificial intelligence (AI) and machine learning (ML) models and algorithms are increasingly used in disease prediction, prevention, and population health management. The coronavirus disease (CO-VID-19) outbreak has prompted an urgent need for accurate and reliable forecasting models to predict the spread of the virus. This paper uses a comparative analysis between long short-term memory (LSTM) and autoregressive integrated moving average (ARIMA) models to forecast the daily time series of COVID-19 cases. These two distinct forecasting techniques are implemented and evaluated against the COVID-19-time series data. LSTM is a type of recurrent neural network (RNN) selected for its ability to capture nonlinear dependencies and long-term patterns within sequential data. At the same time, ARIMA is a classical time series analysis method chosen for its simplicity and interpretability. The evaluation metrics employed include mean absolute error (MAE), root mean squared error (RMSE), and mean absolute percentage error (MAPE), providing comprehensive insights into the predictive performance of each model.

Keywords: Autoregressive integrated moving average, long short-term memory, mean absolute percentage error, recurrent neural network, root mean squared error

Introduction

A robust healthcare system is crucial for a thriving society, promoting well-being, economic stability, social cohesion, and individual prosperity. It ensures timely access to medical services, prevents disease spread, safeguards public health, and promotes preventive care. Using computers in healthcare has transformed administrative procedures, medical research, and patient care, changing from manual systems to sophisticated digital networks, and enhancing patient outcomes, diagnostic accuracy, and treatment efficacy. In the 1960s, paper-based records were the primary method for storing healthcare data, making time-consuming and cumbersome retrieval. Early computers were used for administrative tasks and statistical analyses, but their size, cost, and limited processing power confined their complexity. From the 1980s, computers were used for medical imaging, improving diagnostic accuracy and faster processing. From the 1990s to the 2000s, hospitals integrated computers into networked information systems, enhancing collaboration and efficiency.

Literature Review

Aci et al. [1] proposed a hybrid technique that combines k-nearest neighbor (kNN), Bayesian methods, and genetic algorithms. Their technique improves classification performance by removing data, which makes learning harder and allows for fresh data collections based on available data. In 2017, Liu et al. [9] created a hybrid classification system combining Relief and rough set (RFRS) techniques to detect heart disease. The system performed better than other common classifiers regarding accuracy, sensitivity, and specificity. Again, in 2017, Lal et al. [8] developed a hybrid classification method that combines decision tree (DT) and Naive Bayes (NB) techniques. The method divides fitness data into two groups based on unique properties and then submits the categorized data to the Naive Bayes method for further classification. The hybrid classifier improves accuracy by 15.79% compared to DT and 3.6% compared to NB. Heart disease is a major global health concern that affects millions of lives. Early detection of cardiovascular problems is crucial. Regular clinical data analysis can help. Kavita et al. used ML algorithms, namely random forest (RF), DT, and a hybrid model, to make educated judgments and accurate forecasts. Their hybrid prediction model achieved an accuracy of 88.7%.

Mohan et al. proposed a model combining RF and linear method properties to enhance the accuracy of prediction for cardiovascular disease detection. They achieved an accuracy of 88.7%. In 2021, Zheng et

[a]smohapatra@outr.ac.in, [b]sdmohantycse@outr.ac.in, [c]jrn26497@gmail.com, [d]sknayak_ca@vssut.ac.in

DOI: 10.1201/9781003658221-64

al. created a hybrid AI model for COVID-19 prediction by enhancing the susceptible-infected model with LSTM networks and natural language processing (NLP). The model aims to minimize prediction errors and improve accuracy, with predictions compatible with actual epidemic instances compared to classic epidemic models. Pham et al. developed DeepCare, an LSTM-based deep dynamic neural network (NN), to address the issue of episodic and time-inconsistent healthcare observations in electronic medical records. The network evaluates medical data, predicts future medical outcomes, and determines current sickness states. Nilashi et al. worked on a new approach for predicting Parkinson's disease using an incremental support vector machine, non-linear iterative partial least squares, and a self-organizing map for clustering. Healthcare executives must anticipate time series data like fertility, illness, and death rates. Using linear or NN models may be challenging due to complex patterns. Purwanto et al. suggested a dual hybrid forecasting model, including linear regression, fuzzy models, and NNs, for decision-making in the healthcare business. The fuzzy model provides qualitative output. Categorization techniques in model development are built on DT, ANN, NB, and SVM algorithms. Kumar et al. [7] studied four countries' numerical models for future trends and COVID-19 data. They found that ARIMA exceeded linear regression and RF in forecasting. They also evaluated mortality-predicting tasks, finding that LSTMs significantly outperformed ARIMA.

Comparative Analysis of Different Models

The landscape of predictive analytics in healthcare has witnessed a surge in interest, with numerous studies exploring diverse methodologies to enhance forecasting accuracy. Having done all the literature review, it became apparent that a singular approach may need to be revised in capturing the intricate dynamics inherent in healthcare data. Commonly, LSTM models exhibit prowess in capturing temporal dependencies, while ARIMA models excel in modelling linear trends. This motivated us to see how the ARIMA and LSTM models would perform on the COVID-19 dataset. The comparison of hybrid, non-hybrid and statistical models is shown in Figure 64.1. As seen from this Figure, the performance of a hybrid model reaches 100% accuracy.

Implementation and Result Analysis

We evaluated the performance of both ARIMA and LSTM models in worldwide COVID data and analyzed their output. The details about the parameters used for the implementation are shown in Table 64.1. The time series graph of the COVID-19 data is shown in Figure 64.2 by considering the total number of global:

$$MSE = \sum_{i=1}^{n}(error * error/n), \tag{1}$$

$$RSME = \sqrt{\left\{\sum_{i=1}^{n}\left(\frac{error*error}{n}\right)\right\}}, \tag{2}$$

$$Absolute = \sum_{i=1}^{n}(|error|/n), \tag{3}$$

Figure 64.2 compares the actual versus predicted number of cases using the ARIMA model. This figure shows that the error is slightly more than the LSTM model. When dealing with 10 million or more numbers, the MAPE is often more meaningful than the MAE. This is because MAPE is a percentage-based metric that provides a relative measure of the error, making it more interpretable and comparable across different data scales. As an absolute error measure,

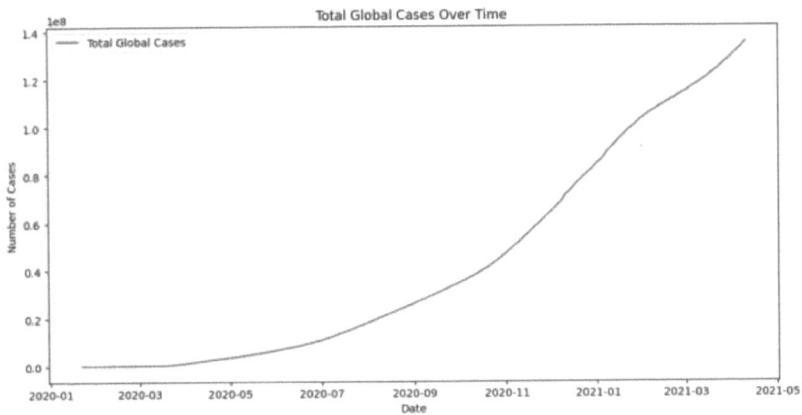

Figure 64.1 Time series graph of the COVID-19 data
Source: Author

Table 64.1 Parameters and their descriptions of cases.

Sl. No.	Dataset parameters	Description
1	Document(s) size	555 Kb
2	Dataset dimension	275 × 448
3	Data source	World meter, WHO

Source: Author

Table 64.2 Comparison of MAPE, MAE, and RMSE for LSTM and ARIMA models.

Sl. No.	Parameters	LSTM	ARIMA
1	MAPE	0.0015	0.0276
2	MAE	1,48,168.3609	31,25,344.0535
3	RMSE	1,79,141.8826	41,37,693.4809

Source: Author

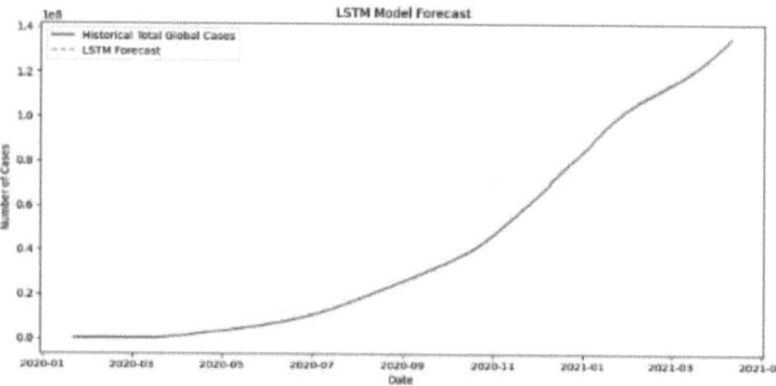

Figure 64.2 LSTM actual vs. LSTM predicted graph
Source: Author

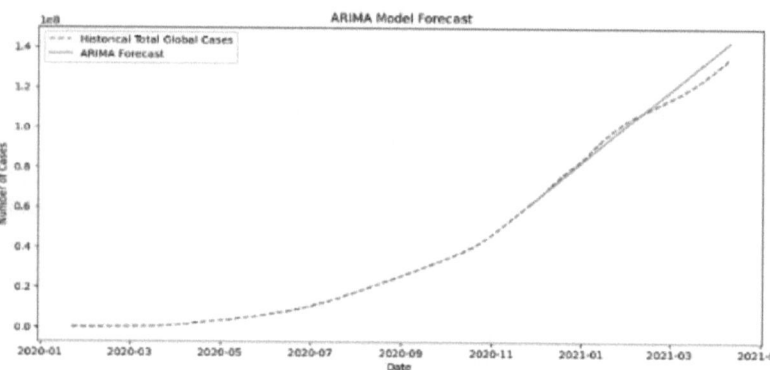

Figure 64.3 ARIMA actual vs. ARIMA predicted graph
Source: Author

MAE may not give you a clear sense of relative accuracy when dealing with significantly large values. In contrast, MAPE expresses the error as a percentage of the actual values, which can be more informative in practical terms, especially when dealing with large magnitudes. When dealing with large numbers, especially when the magnitude of the data is on the order of millions or more, RMSE may not always be the most intuitive metric for assessing the performance of a model. The reason is that RMSE penalizes more significant errors more heavily, and in the context of huge numbers, even a small absolute error can result in a substantial RMSE. This might not always align with the practical significance of the error. Table 64.2 compares MAPE, MAE and RMSE for LSTM and ARIMA models. It is clear from this table that LSTM outperforms ARIMA in terms of all the performance metrics.

Conclusion

This paper explores the time series forecasting methods for the COVID-19 dataset sourced from World Meter and WHO. We undertook a comparative analysis of two prominent models: LSTM and ARIMA. The primary objective is to determine the efficacy of each model in capturing the dynamic patterns inherent in the pandemic data. Our findings demonstrate that LSTM outperformed ARIMA in providing more accurate and robust predictions. Our exploration supports the efficacy of LSTM as a potent tool for forecasting COVID-19 trends, offering enhanced accuracy and adaptability compared to the ARIMA model.

References

[1] Aci, M., Inan, C., & Avci, M. (2010). A hybrid classification method of k nearest neighbor, Bayesian methods, and genetic algorithm. *Expert Systems with Ap plications*, 37(7), 5061–5067.

[2] Kavitha, M., Gnaneswar, G., Dinesh, R., Sai, Y. R., & Suraj, R. S. (2021). Heart disease prediction using hybrid machine learning model. *In 2021 6th international conference on inventive computation technologies (ICICT)* (pp. 1329–1333). IEEE.

[3] Kumar, S. L. (2021). Predictive analytics of Covid-19 pandemic: statistical modelling perspective. *Walailak Journal of Science and Technology (WJST)*, 18(16), 15583–14.

[4] Lal, A., & Kumar, C. R. S. (2017, April). Hybrid classifier for increasing accuracy of fitness data set. *In 2017 2nd International Conference for Convergence in Technology (I2CT)* (pp. 1246–1249). IEEE.

[5] Liu, X., Wang, X., Su, Q., Zhang, M., Zhu, Y., Wang, Q., & Wang, Q. (2017). A hybrid classification system for heart disease diagnosis based on the RFRS method. *Computational and Mathematical Methods in Medicine*, 8272091

[6] Mohan, Senthilkumar, Chandrasegar Thirumalai, and Gautam Srivastava. Effective heart disease prediction using hybrid machine learning techniques. *IEEE access*. 7 (2019): 81542-81554.

[7] Nilashi, M., Ibrahim, O., Ahmadi, H., Shahmoradi, L. and Farahmand, M., 2018. A hybrid intelligent system for the prediction of Parkinson's Disease progression using machine learning techniques. *Biocybernetics and Biomedical Engineering*, 38(1), pp.1–15.

[8] Gupta, M., Phan, T.L.T., Bunnell, H.T. and Beheshti, R., 2022. Obesity Prediction with EHR Data: A deep learning approach with interpretable elements. *ACM Transactions on Computing for Healthcare (HEALTH)*, 3(3), pp.1–19.

[9] Ma, R., Zheng, X., Wang, P., Liu, H. and Zhang, C., 2021. The prediction and analysis of COVID-19 epidemic trend by combining LSTM and Markov method. Scientific Reports, 11(1), p.17421.

65 Electromagnetic interference shielding of Zn-50%-Al alloy-coated polypropylene flexible conducting film

Rajendrakumar Sharma[a], Dibyaranjan Das[b], Prafulla K. Dash[c], Asutosh Acharya[d], Kajal Parashar[e] and SKS Parashar[f]

Nanosensor Lab, School of Applied Sciences, Kalinga Institute of Industrial Technology (KIIT) Deemed to be University, Bhubaneswar, Odisha, India

Abstract

The increasing usage of electronic gadgets in modern society may produce harmful electromagnetic pollution. Flexible EMI shielding materials are gaining popularity due to advances in flexible electronics. PAN Electronics (India) Ltd acquired a 6 μm polypropylene (PP) film and coated it with 0.05 μm of zinc and aluminum alloy (50-–50 wt%) using chemical vapor deposition. Zinc-aluminum coatings with a thickness of 0.05 μm were applied to a 6 μm polypropylene (PP) sheet for effective EMI shielding. Zn-Al alloy coatings considerably increased the conductivity and shielding characteristics of the PP film. The Zn-Al coated PP film had a high total EMI shielding efficiency (SE_T) of 24.13 dB, whereas the PP film had a SE_T value of 0.40 dB. As a result, the Zn-Al coated PP film might be a promising option for researching high-performance EMI shielding.

Keywords: Conducting film EMI shielding, flexible, , Zn-Al alloy coated polypropylene (PP) film

Introduction

Modernity is brought about by advancements in electronic and communication technology, but they also create EMI, which can lead to the breakdown of sensitive devices, hospitals, military bases, buildings, and damage human health [5]. It is critical to produce EMI shielding materials with outstanding flexibility, lightweight design, great stability, and effective EMI shielding effectiveness (SE) so they may be used in the future generation of electronic devices in order to tackle these problems [4]. Conductive polymer composites (CPCs) are often used as shielding materials owing to their affordability, lightweight nature, ease of fabrication, superior tensile strength, and remarkable dimensional stability [12] suited for applications in the Packaging and Food and Beverage industries, supercapacitor and battery industry. Conductive polymer film composites (CPFCs) may be fabricated using a straightforward method involving the application of conductive fillers onto the surface of a flexible polymer film [10]. Polypropylene (PP) fabric, a recyclable eco-friendly industrial material, meets the standards of flexible polymer film, characterized by economic advantages, lightweight, high tensile strength, and chemical resistance. The establishment of a robust conductive network significantly affects EMI shielding effectiveness [2]. Selecting metal coating materials with good electrical conductivity, such as metal nanoparticles and carbon, might enhance the EMI shielding efficiency of CPFCs [7]. Zinc, because of its exceptional electrical, mechanical, and chemical resilience, has emerged as an appropriate coating material for CPFCs. Kaushal et al. [8] examined the shielding of -10.30 dB at 20 wt% graphite powder (GP) in polypropylene polymer inside the X-band (8.2 GHz–12.4 GHz) [8]. In the X-band, Sun-Kuk Kim et al. [9] investigated a shielding efficiency of 10.2 dB at a 5% weight percentage of reduced graphene oxide in a polypropylene polymer matrix [9]. ZnO has shown promising microwave absorption properties in several composites such as ZnO nano-wire polyester composites [11], tetra needle-like ZnO whiskers (Z. Zhou et al., 2006) [13], and cage-like ZnO/SiO_2 nanocomposites [3].

In this work, zinc-aluminum alloy coated with polypropylene was studied for high microwave absorption. Various characterization technologies such as a XRD were employed to study the phase. The conductivity and EMI shielding were examined using a vector network analyzer (VNA) in the X-band. This study convinced that high Absorption due to shielding values can significantly demonstrate the effectiveness of materials in absorbing electromagnetic (EM) waves.

Experimental Procedure

The polypropylene (PP) film of 6 μm a thickness, that had been coated with 0.05 μm of zinc and aluminum

[a]spel.capacitor@gmail.com, [b]dibyaranjandas74@gmail.com, [c]dashprafulla@gmail.com, [d]ashu1492@gmail.com, [e]kparasharfch@kiit.ac.in, [f]sksparasharfpy@kiit.ac.in

DOI: 10.1201/9781003658221-65

alloy (50–50 wt%) using the chemical vapor deposition process, which was by supplied by PAN Electronics (India) Ltd. Metallization of Zn-Al (50–50 wt%) alloy was done in high vacuum under controlled vapor deposition technique in metallurgy plant Leybold made Germany. The metallization Zn-Al (50–50 wt%) alloy composition of the film varies with 10% and the coating thickness varies with 5%. X-ray diffraction (XRD) measurements of polypropylene film and Zn-Al coated polypropylene film was performed using an XRD (Bruker AXS D8 advanced) with Cu-K$_\alpha$ radiation. The scattering parameters were measured using a VNA of the Rhode and Schwarz model (ZNB20).

Results and Discussion

Phase and conductivity analysis
Figure 65.1(a) displays the XRD patterns of PP and PP@Zn-Al film, measured at 2θ angle of 20–80° with step size 0.01. The diffraction pattern of α-PP reveals distinct peaks at certain angles 2θ of 14.1°, 16.4°, 18.1°, and 25.2° and these peaks correspond to the

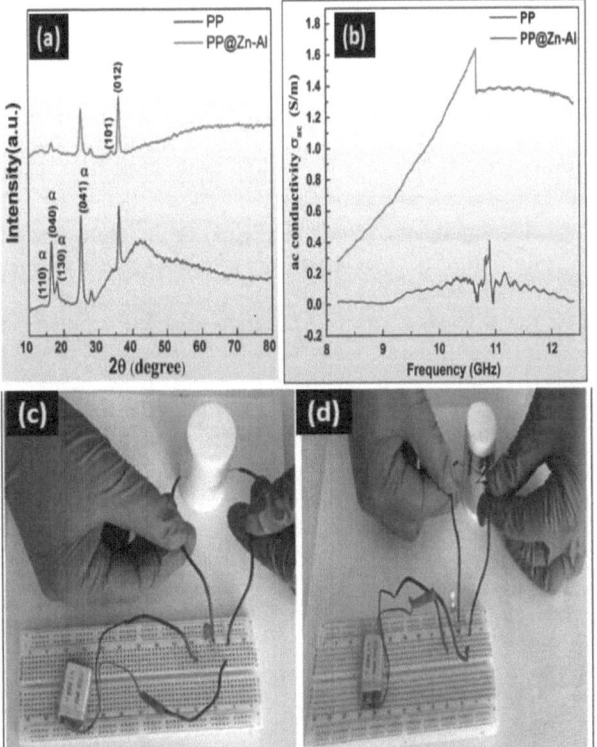

Figure 65.1 (a) XRD of polymer film PP and PP@Zn-Al film of, (b) Variation of ac conductivity σ$_{ac}$ (S/m) with frequency (c) PP film showing non-conducting nature, (d) PP@Zn-Al) glow of red LED showing conducting nature
Source: Author

crystallographic planes (110), (040), (130), and (041), respectively. Based on the pure ZnO reference (JCPDS 00-048-1026) [1], all observed peaks may be attributed to the hexagonal wurzite structure of ZnO. The phase achieved is entirely pure, with no visible peak corresponding to aluminum. This suggests that aluminum atoms have been properly integrated into the ZnO lattice. The diffraction peaks observed at 35° in Figure 65.1 can be related to the presence of ZnO as a minor phase. Figure 65.1(b) displays the photograph of propylene film (PP) with circuit showing non-conducting nature and 65.1(c) displays the image of Zn-Al alloy coating propylene film (PP@Zn-Al) glow of red LED showing conducting nature. Figure 65.1(d) shows the variation ac conductivity (σ$_{ac}$) with frequency (GHz) for propylene film and Zn-Al alloy coating propylene film. The ac electrical conductivity is determined using the formula, σ$_{ac}$=2πfε$_0$ε″, where ε$_0$ is the permittivity of empty space. The fluctuation of σ$_{ac}$ was comparable to that of ε″ as a function of frequency. The extraordinary electrical conductivity is due to the development of a three-dimensional segregated architecture of Zn-Al alloy in the polypropylene matrix.

The conductivity increases with frequency due to the electrical conductivity or hopping process. Without the presence of Zn-Al alloy, conducting networks are uncommon, and conductivities are mostly determined by electron hopping.

EMI shielding
Shielding refers to the technique by which a substance prevents electromagnetic radiation from passing through it. The EMI shielding effectiveness can be expressed by the following equations is the logarithm of the ratio of power incident (P$_i$) to power transmitted (P$_t$) in decibel SE$_T$ = -10log (P$_i$/P$_t$).

$$SE_T = SE_A + SE_R + SE_M \approx SE_A + SE_R \quad (1)$$

Here, SE$_A$ is the absorption shielding effectiveness, SE$_R$ is the reflection shielding effectiveness and SE$_M$ is the multiple internal reflections shielding effectiveness (with SE$_A$> 10 dB, SE$_M$ can be ignored because of the internal inhomogeneity of the shielding materials). SE$_R$ and SE$_A$ are given by the equations (2) and (3) respectively [6].

$$SE_R = -10 \log(1-R) \quad (2)$$

$$SE_A = -10\log (T/(1-R)) \quad (3)$$

Where R= |S$_{11}$|^2= |S$_{22}$|^2 is the reflection coefficient, T= |S$_{12}$|^2= |S$_{21}$|^2 is the transmission coefficient and S$_{11}$,

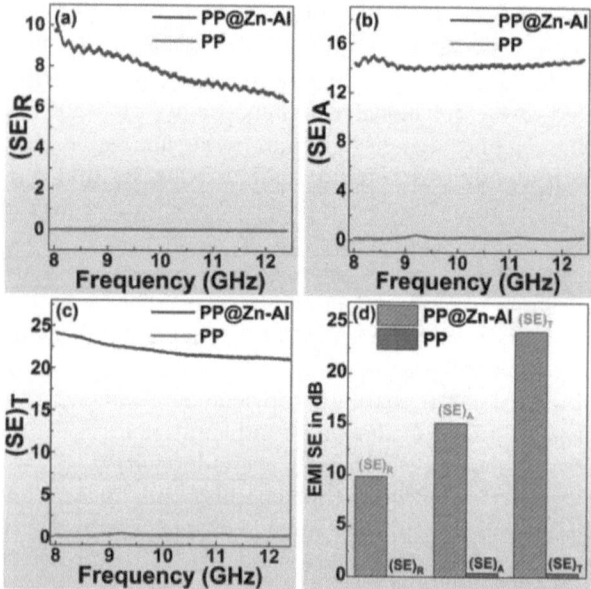

Figure 65.2 Frequency dependence (a) SE_R (b) SE_A and (c) SE_T, (d) comparative bar diagram of SE for propylene film (PP) and Zn-Al alloy coating propylene film (PP@Zn-Al)

Source: Author

S_{22}, S_{12}, and S_{21} are the measured scattering parameter of the material. propylene film shielding mechanism in the film, rather than reflection. The conducting permanent network pathways were developed by coating with Zn-Al throughout the PP matrix, leading to higher values of absorption parameters in SE. The magnetic and dielectric loss may result in improved shielding properties. Figure 65.2 displays the frequency-dependent behavior of SE_R, SE_A, SE_T, and comparative bar diagram of shielding effectiveness of propylene film (PP) and Zn-Al coating (PP@Zn-Al). The observed maximum values of SE_A, SE_R, and SE_T are found to be 15.1 dB, 9.83 dB, and 24.13 dB for Zn-Al coating propylene film (PP@Zn-Al) and 0.39 dB, 0.01 dB, and 0.40 dB for propylene film (PP). The results suggest that absorption dominant shielding.

Conclusion

This study examined the electrical conductivities, complex electromagnetic parameters, and EMI shielding effectiveness in PP and PP@Zn-Al film. The observed maximum values of total shielding effectiveness SE_T were found to be 24.13 dB for Zn-Al alloy coating propylene film (PP@Zn-Al) and 0.40 dB for propylene film (PP). Due to the superior electrical conductivity of zinc oxide, the Zn-Al coating propylene film (PP@Zn-Al) shows much improved total

EMI shielding effectiveness (SE_T) (~ 25-fold) as compared to propylene film (PP) throughout the whole frequency range.

References

[1] Alibakhshi, E., E. Ghasemi, M. Mahdavian, B. Ramezanzadeh, & Farashi, S. (2016). Fabrication and characterization of PO43– intercalated Zn-Al-layered double hydroxide nanocontainer. *Journal of The Electrochemical Society*, 163(8), C495.

[2] Awad, S., Chen, H., Chen, G., Gu, X., Lee, J. L., Abdel-Hady, E. E., & Jean, Y. C. (2011). Free volumes, glass transitions, and cross-links in zinc oxide/waterbornepolyurethanenanocomposites. *Macromolecules*, 44(1), 29–38.

[3] Cao, M.-S., Shi, X.-L, Fang, X.-Y., Jin, H. B., Hou, Z. L., Zhou, W., & Chen, Y. J. (2007). Microwave absorption properties and mechanism of cagelike ZnO/SiO$_2$ nanocomposites. *Applied Physics Letters*, 91(20), 203110.

[4] Das, D., Samal, R. R., Dikshit, A. P., Parashar, K., & Parashar, S. K. S. (2024). Design and electromagnetic shielding of Bi$_{0.5}$Na$_{0.5}$TiO$_3$ ceramics as meta-surface absorber for high-frequency microwave application. *Ceramics International*, 50(22), 46621–46631.

[5] Dikshit, A. P., Das, D., Samal, R. R., Parashar, K., Mishra, C., & Parashar, S. K. S. (2024). Optimization of (Ba$_{1-x}$Ca$_x$)(Ti$_{0.9}$Sn$_{0.1}$)O$_3$ ceramics in X-band using machine learning. *Journal of Alloys and Compounds*, 982, 173797.

[6] Dikshit, A. P., Das, D., Samal, R. R., Tyagi, A., Parashar, K., &Parashar, S. K. S. (2024). Structural and electromagnetic shielding of ZnO ceramics in X-band. *Journal of Materials Science: Materials in Electronics*, 35(13), 933.

[7] Jiang, C., Tan, D., Li, Q., Huang, J., Bu, J., Zang, L., Ji, R., Bi, S., & Guo, Q. (2021). High-performance and reliable silver nanotube networks for efficient and large-scale transparent electromagnetic interference shielding. *ACS Applied Materials & Interfaces*, 13(13), 15525–15535.

[8] Kaushal, A., &Singh, V. (2020). Melt-processed graphite-polypropylene composites for EMI shielding applications. *Journal of Electronic Materials*, 49(9), 5293–5301.

[9] Kim, S.-K., Kim, J. T., Kim, H. C., Rhee, K. Y., & Kathi, J. (2012). Thermal and mechanical properties of epoxy/carbon fiber composites reinforced with multiwalled carbon nanotubes. *Journal of Macromolecular Science, Part B*, 51(2), 358–367.

[10] Luo, J., Huo, L., Wang, L., Huang, X., Li, J., Guo, Z., Gao, Q., Hu, M., Xue, H., & Gao, J. (2020). Superhydrophobic and multi-responsive fabric composite with excellent electro-photo-thermal effect and electromagnetic interference shielding performance. *Chemical Engineering Journal*, 391, 123537.

[11] Shi, F., Chen, H., & MacLaren, S. (2004). Wafer-bonded semiconductors using In/Sn and Cu/Ti metallic interlayers. *Applied physics letters*, 84(18), 3504–3506.

[12] Huang, W., Dai, K., Zhai, Y., Liu, H., Zhan, P., Gao, J., Zheng, G., Liu, C., & Shen, C. (2017). Flexible and lightweight pressure sensor based on carbon nanotube/ thermoplastic polyurethane-aligned conductive foam with superior compressibility and stability. *ACS Applied Materials Interfaces*, 9(48), 42266–42277.

[13] Zhou, Z., Chu, L., & Hu, S. (2006). Microwave absorption behaviors of tetra-needle-like ZnO whiskers. *Materials Science and Engineering*, 126(1), 93–96.

66 Robust digital image watermarking using hybrid optimization technique

Alina Dash[a], Kshiramani Naik[b] and Priyanka Priyadarshini[c]

Department of Computer Science and Engineering, VSSUT, Burla, Sambalpur, Odisha, India

Abstract

Multimedia data, accessible online through images and videos, is highly susceptible to duplication, modification, and easy dissemination due to its digital format. The expansive reach of the internet enables swift transfer of data with a simple click, prompting the need for safeguarding multimedia content. Digital watermarking has emerged as a crucial solution to address this challenge. Its applications span copy control, broadcast monitoring, video authentication, fingerprinting, and copyright protection, witnessing a substantial rise. Watermarking involves embedding secret data, known as watermarks, within multimedia files without altering the perceptibly of the original content. This technique acts as a shield against illegal copying and alterations, ensuring the integrity of multimedia documents. Specifically, the proposed method, termed robust digital image watermarking using hybrid optimization, amalgamates discrete wavelet transform (DWT), discrete cosine transforms (DCT), bacterial foraging optimization (BFO), and grasshopper optimization algorithm (GOA), which are conventionally utilized individually. Through a comparative analysis highlighting various attributes, the combined approach demonstrates superior performance over preceding methods. By integrating these techniques, the shortcomings of each approach are mitigated and optimized scaling factors are found for embedding data.

Keywords: Digital image watermarking, discrete wavelet transform, discrete cosine transform, bacterial foraging optimization, grasshopper optimization algorithm, image fidelity

Introduction

In the digital age, the vast proliferation of multimedia data, such as images and videos, over the internet has brought unprecedented convenience and accessibility to information sharing. However, this ease of access has also given rise to concerns regarding the unauthorized duplication, modification, and distribution of multimedia content. With the internet's expansive reach, virtually any piece of data can be effortlessly transmitted from one person to another with just a single click. As a result, ensuring the security and protection of multimedia data has become an increasingly prominent challenge in the contemporary digital landscape. To address these concerns, digital watermarking has emerged as a powerful solution.

Watermarking involves the imperceptible embedding of additional data, known as watermarks, within existing multimedia content, such as images, videos, or music. This process enables the secure attachment of identifying information or copyright protection to the media without altering its original format. Digital watermarking is a crucial technique aimed at safeguarding multimedia documents and files from illegal copying and manipulation.

The proposed technique combines two distinct watermarking and two optimization methods that are typically employed individually. By amalgamating these techniques, the approach seeks to mitigate the limitations of each method, ultimately resulting in enhanced performance and robust protection for multimedia data. This research represents a significant step forward in the ongoing efforts to fortify the security of multimedia data in an ever-evolving digital landscape.

Literature Survey

Work has been done in recent years in digital watermarking to increase the watermark's sensitivity. This region provides a review of the literature on this work. The general process is almost the same for all of the papers, some of the work on watermarking proposed recently are mentioned below:

Several hybrid watermarking techniques have been proposed to enhance robustness and imperceptibility. Barnouti et al. [1] combined DWT and DCT for embedding and extracting watermarks with high durability and performance. Deeba et al. [2] added a security layer using the Arnold transform and embedded the encrypted watermark in the host image's LL band using DWT and SVD. Devi et al. [3]. Jain et al.[4] proposed a watermarking algorithm utilizing bacterial foraging optimization, demonstrating its effectiveness in embedding and extracting watermarks while maintaining image quality. Employed RDWT-SVD with JAYA-Firefly optimization to ensure robustness and

[a]alinadash_cse@vssut.ac.in, [b]kshiramaninaik_it@vssut.ac.in, [c]ppriyadarshini398@gmail.com

DOI: 10.1201/9781003658221-66

imperceptibility by embedding in low-information blocks. Kaushik et al. [5] integrated DCT-DWT with Bacterial Foraging Optimization (BFO) to fine-tune gain, enhancing security and resistance to attacks. Kumar et al. [6] used IWT-SVD to embed watermarks in the LL sub-band and demonstrated resilience against noise and resizing. Sharma et al. [7] applied DCT-DWT and optimized scaling factors using the grasshopper optimization algorithm (GOA) for improved invisibility. Verma et al. [8] utilized DWT-SVD with particle swarm optimization (PSO) for embedding watermarks in medium frequency bands with optimal scaling factors. Yadav et al. [9] and Yildiz et al. [10] leveraged DWT-SVD with Arnold transform and hybrid optimization techniques like BFO-PSO to enhance robustness against image processing attacks.

Proposed Work

There are various techniques to hide a key in an image to protect it from duplicity. Considering the techniques used, the main objective of this project is to embed an invisible watermark. The proposed work contains a combination of four methods which are generally used independently. At last, there is a comparison based on various parameters to visualize that the combination is the advanced outcome to the previous approaches. Every technique has its own cons which are reduced by combining them. The methods used here are discrete wavelet transform (DWT), discrete cosine transform (DCT), bacterial foraging optimization (BFO) and grasshopper optimization algorithm (GOA). The performance is calculated by comparing the (PSNR) value, the Image Fidelity (IF) value, and normalized cross-correlation (NCC) value. The proposed watermark algorithm goes through two processes i.e. Digital watermark embedding and digital watermark extraction. The embedding of a message through any technique comes with a constraint that the message has to be recovered at the receiver end correctly and this depends on validation with NCC. So, a gain factor is introduced which measures the intensity of message hiding and retrieval.

Figure 66.1 represents the proposed embedding and extraction process.

Both DCT and DWT techniques simultaneously to get more robustness and security. DCT is implanted in the host image which divides image to different frequencies. The message needs to be embedded in an image having lower frequency which will not interfere with the host image. At first, the DCT principle is applied which uses Fourier transform to get images with two band frequency- high and low. Various parameters are calculated to get the strength of encryption. After that DWT is enforced which splits the host image into the frequency coefficients such as approximation, horizontal, vertical and detailed? The watermark picture is added using a horizontal coefficient, and the cH1 component of the DWT altered image is then further changed using DCT. The resilience of the message is set by the ideal gain factor required for the embedding of the watermark image. BFO and GOA are implemented to trade-off the gain factor when embedding the watermark.

Proposed watermark embedding and extraction algorithm

1. Select and reinitialize the key into an order of 1 and 0.
2. Convolution is used to encrypt the message, and the processed message is then used.
3. A PN sequence is generated using the key. Every step of the embedding procedure uses the same PN order.
4. Two highly uncorrelated PN sequences should be generated.
5. DWT is used to divide the host image into four well separated coefficients.
6. For DCT, a bar sized and central belt coefficient matrix is decided. The work uses an 8-block format, and the literature research informs the selection of an 8*8 mid band matrix.
7. DCT is now applied to cH1, and both PN sequences are embedded by advantage factors of y based on the message's unit value. A method for implanting PN sequences is given in equation (1) and (2).

$$I = i + y^* \ pn \ 0 \ if \ bit = 0 \tag{1}$$

$$I = i + y^* \ pn \ 1 \ if \ bit = 1 \tag{2}$$

Here, i is host image, y is watermarked bit, i = embedded image.

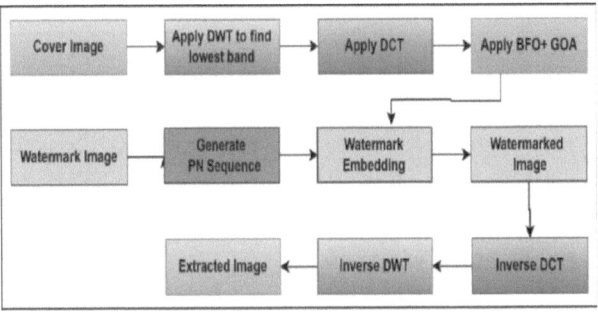

Figure 66.1 Proposed embedding and extraction method
Source: Author

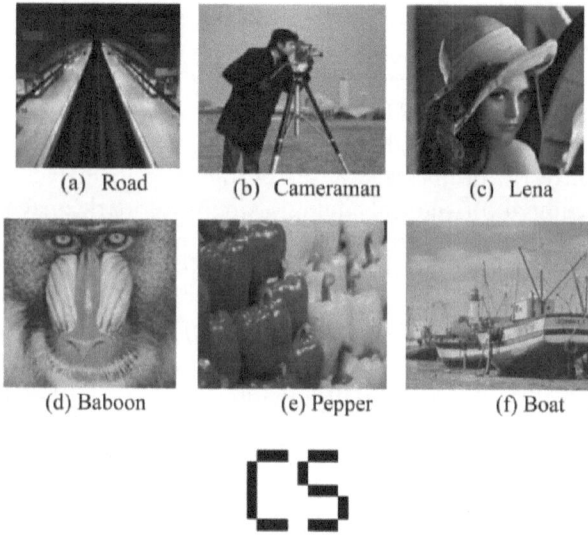

(a) Road   (b) Cameraman   (c) Lena

(d) Baboon   (e) Pepper   (f) Boat

(g) Watermark message

Figure 66.2 Host images : (a) Road, (b) Cameraman, (c) Lena, (d) Baboon, (e) Pepper, (f) Boat, watermark image: (g) CS logo
Source: Author

8.   IDCT is executed now on ch1.
9.   The same process is done for cv1.
10. IDWT is applied using the most recent values for atomization coefficient, horizontal and vertical coefficient and detailed coefficient and an encrypted image is constructed.
11. Extraction process of watermark image from the watermarked image is the reverse of the embedding algorithm
12. The picture quality of an encrypted image is discovered by PSNR and normalized cross correlation and image fidelity.

Experiment Results

The following table (Table 66.1) depicts various image quality metrics which are calculated by considering the input image and the generated watermarked image and the extracted watermark.

Mean square error (MSE) measures the collective squared error among the disturbing noise and the highest power of a particular signal. Lower the MSE value indicates high quality. The mean square error value can be calculated using Equation 3 as mentioned below. The MSE quantifies the total squared discrepancy between unwanted noise and the most dominant signal power. A lower MSE value signifies higher quality in the assessment of the signal. Equation 3, detailed below, allows for the calculation of the mean square error value.

Table 66.1 Results obtained using DWT method.

Original image	Watermark	Watermarked image	Extracted watermark

Source: Author

Table 66.2 Values for PSNR, NCC, IF for different images.

Image	PSNR	NCC	IF
(a)	26.2259	0.0018	-0.0364
(b)	20.7638	0.0020	-0.0304
(c)	25.6236	0.0019	-0.0461
(d)	25.0722	0.0020	-0.0252
(e)	25.0150	0.0019	-0.0331
(f)	25.2095	0.0019	-0.2626

Source: Author

$$MSE = \frac{1}{M \times N} \sum_{p=0}^{N-1} \sum_{q=0}^{M-1} (I(a,b) - I_w(a,b))^2 \qquad (3)$$

where I(a,b) represents the host image, Iw(a, b) signifies the watermark image, and M×N denotes the size of the image.

Table 66.3 Results obtained using DWT+DCT method.

Original image	Watermark	Watermarked image	Extracted watermark

Source: Author

Table 66.5 Results obtained using DWT+DCT+BFO method.

Original image	Watermark	Watermarked image	Extracted watermark

Source: Author

Table 66.4 Values for PSNR, NCC, IF for different images.

Image	PSNR	NCC	IF
(a)	44.9088	0.0039	-1.8056e-04
(b)	40.1234	0.0039	-3.5222e-04
(c)	44.9088	0.0039	-1.8056e-04
(d)	44.7945	0.0036	-9.0563e-05
(e)	44.6799	0.0039	-1.1887e-04
(f)	44.6799	0.0038	-1.0577e-04

Source: Author

Table 66.6 Values for PSNR, NCC, IF for different images.

Image	PSNR	NCC	IF
(a)	48.9510	0.0039	-6.4857e-05
(b)	43.8978	0.0039	-3.5222e-04
(c)	48.8153	0.0039	-7.3445e-05
(d)	48.628	0.0033	-3.7074e-05
(e)	48.558	0.0039	-4.8861e-05
(f)	48.6645	0.0039	-3.9354e-05

Source: Author

PSNR gauges the fidelity of a watermarked image by comparing original data to the watermarked version, using MSE-based calculations, as depicted in the Equation 4.

$$PSNR\ (I, I_w) = 10 \times log_{10} \frac{MAX_I^2}{MSE} \qquad (4)$$

Table 66.3, 66.5, 66.7 shows the cover/original image, watermark image and watermarked image and extracted image for DWT, DWT + DCT, DWT + DCT + BFO, DWT + DCT + BFO + GOA respectively. Table 66.2, 66.4.66.6 and 66.8 shows the results of PSNR,

Table 66.7 Results obtained using DWT + DCT + BFO + GOA.

Original image	Watermark	Watermarked image	Extracted watermark

Source: Author

Table 66.8 Values for PSNR, NCC, IF for different images.

Image	PSNR	NCC	IF
(a)	54.844	0.0039	-1.6696e-05
(b)	49.8417	0.0039	-3.5222e-04
(c)	48.8153	0.0039	-7.3445e-05
(d)	54.8175	0.0026	-8.9149 e-05
(e)	54.7089	0.0036	-1.1808e-05
(f)	54.5859	0.0039	-1.0066e-05

Source: Author

NCC and IF for DWT, DWT + DCT, DWT + DCT + BFO, DWT + DCT + BFO + GOA respectively.

Table 66.9 compares the proposed technique with other state of arts.

Here, 'I' stands for the original host image, 'Iw' signifies the watermarked image, and 'MAX' represents the maximum potential value of the signal.

Table 66.9 Comparison with existing method.

	DWT+DCT+ GOA [1]	DWT+DCT+ BFO [4]	DWT+DCT+ BFO +GOA
PSNR	42.2250	48.62	55.8178
NCC	0.0028	0.0033	0.0026
IF	-5.9966e-04	-3.7074e-05	-8.9149e-06

Source: Author

Comparative analysis of proposed method with existing method

Table 66.9 provides a summary of the comparative study of the suggested approach. It is evident that the different images, the suggested procedure, produce the best results. In comparison to other methods, the PSNR value for the proposed method is high.

Percentage increase in PSNR with respect to DWT+DCT+BFO method

Increase in PSNR: 12.592 dB

Percentage increase in PSNR: $\frac{Increase}{Original\ value} \times 100\ \% =$ 29.82%

Percentage increase in PSNR with respect to DWT + DCT + GOA method

Increase in PSNR: 6.1978 dB

Percentage increase in PSNR: $\frac{Increase}{Original\ value} \times 100\%$ =12.75%

Conclusion

The technique proposed a digital watermark to images is based on the DCT-DWT optimized with bacterial foraging optimization (BFO) and grasshopper optimization algorithm (O). The combination of two optimization techniques helped to find out proper optimized scaling factor, used for embedding the watermark into cover image. The results show that the proposed method gives good quality image with higher PSNR values like 12.75% and 29.82% increased value than existing methods, which helps in providing better security to the image or data. Also, through various quality assessment metrics we check the amount of similarity and dissimilarity between the implanted watermark and recovered watermark. This helps in claiming ownership, quality assessment, and temper proofing. Hence, the proposed watermarking procedure can be implemented in various fields with the combination with other techniques.

References

[1] Barnouti, N. H., Zaid, S. S., & Khaldoun, L. H. (2018). Digital watermarking based on DWT (discrete wavelet

transform) and DCT (discrete cosine transform). *International Journal of Engineering & Technology*, 7(4), 4825–4829.

[2] Deeba, F., She, K., Dharejo, F. A., & Zhou, Y. (2020). Lossless digital image watermarking in sparse domain by using K-singular value decomposition algorithm. *IET Image Processing* 14(6), 1005–1014.

[3] Devi, K. J., Singh, P., Thakkar, H. K., & Kumar, N. (2022). Robust and secured watermarking using Ja-Fi optimization for digital image transmission in social media. *Applied Soft Computing*, 131, 109781.

[4] Jain, C. (2014). Watermarking algorithm using bacterial foraging optimization. *Image*, 256, 256.

[5] Kaushik, P., & Dua, S. (2014). Digital image watermarking using BFO optimized DWT and DCT & comparison between DWT, DWT+ DCT, DWT+ DCT+ BFO. *International Journal of Recent Research Aspects*, 1(3), 13–16.

[6] Kumar, A., Parmar, G., & Bhatt, R. (2018). IWT-SVD based image watermarking under various attacks. *International Journal of Computer Sciences and Engineering*, 6(6), 437–441.

[7] Sharma, S., Chauhan, U., Khanam, R., & Singh, K. (2021). Digital watermarking using grasshopper optimization algorithm. *Open Computer Science*, 11(1), 330–336.

[8] Verma, V., Srivastava, V. K., & Thakkar, K. (2016). DWT-SVD based digital image watermarking using swarm intelligence. *2016 International Conference on Electrical, Electronics, and Optimization Techniques (ICEEOT)*. IEEE, 2016.

[9] Yadav, M., & Jain, N. (2017). An invisible, robust and secure DWT-SVD based digital image watermarking technique with improved noise immunity. *IOSR Journal of Electronics and Communication Engineering*, 12, 7–11.

[10] Yildiz, S., Üstünsoy, F., and Sayan, H. H. (2023). Digital image watermarking with hybrid structure of DWT, DCT, SVD techniques and the optimization with BFO algorithm. *Politeknik Dergisi*, 1–1.